NFPA®

THIRD EDITION

P9-BHU-190

STRUCTURAL FIREFIGHTING

STRATEGY AND TACTICS

BERNARD "BEN" J. KLAENE

Chief of Training/Safety, Cincinnati Fire Department, Retired
Cincinnati, Ohio

JONES & BARTLETT
LEARNING

World Headquarters
Jones & Bartlett Learning
5 Wall Street
Burlington, MA 01803
978-443-5000
info@jblearning.com
www.jblearning.com

Jones & Bartlett Learning books and products are available through most bookstores and online booksellers. To contact Jones & Bartlett Learning directly, call 800-832-0034, fax 978-443-8000, or visit our website, www.jblearning.com.

Substantial discounts on bulk quantities of Jones & Bartlett Learning publications are available to corporations, professional associations, and other qualified organizations. For details and specific discount information, contact the special sales department at Jones & Bartlett Learning via the above contact information or send an email to specialsales@jblearning.com.

Copyright © 2016 by Jones & Bartlett Learning, LLC, an Ascend Learning Company

All rights reserved. No part of the material protected by this copyright may be reproduced or utilized in any form, electronic or mechanical, including photocopying, recording, or by any information storage and retrieval system, without written permission from the copyright owner.

The content, statements, views, and opinions herein are the sole expression of the respective authors and not that of Jones & Bartlett Learning, LLC. Reference herein to any specific commercial product, process, or service by trade name, trademark, manufacturer, or otherwise does not constitute or imply its endorsement or recommendation by Jones & Bartlett Learning, LLC and such reference shall not be used for advertising or product endorsement purposes. All trademarks displayed are the trademarks of the parties noted herein. *Structural Firefighting: Strategy and Tactics, Third Edition* is an independent publication and has not been authorized, sponsored, or otherwise approved by the owners of the trademarks or service marks referenced in this product.

There may be images in this book that feature models; these models do not necessarily endorse, represent, or participate in the activities represented in the images. Any screenshots in this product are for educational and instructive purposes only. Any individuals and scenarios featured in the case studies throughout this product may be real or fictitious, but are used for instructional purposes only.

The publisher has made every effort to ensure that contributors to *Structural Firefighting: Strategy and Tactics, Third Edition* materials are knowledgeable authorities in their fields. Readers are nevertheless advised that the statements and opinions are provided as guidelines and should not be construed as official policy. The recommendations in this publication or the accompanying resource manual do not indicate an exclusive course of action. Variations taking into account the individual circumstances and local protocols may be appropriate. The publisher disclaims any liability or responsibility for the consequences of any action taken in reliance on these statements or opinions.

05777-5

Production Credits

Chief Executive Officer: Ty Field
President: James Homer
Chief Product Officer: Eduardo Moura
Vice President, Publisher: Kimberly Brophy
Vice President of Sales, Public Safety Group: Matthew Maniscalco
Executive Acquisitions Editor—Fire: William Larkin
Senior Acquisitions Editor—Fire: Janet Maker
Senior Development Editor: Janet Morris

Production Editor: Jessica deMartin
Senior Marketing Manager: Brian Rooney
VP, Manufacturing and Inventory Control: Therese Connell
Composition: Cenveo® Publisher Services
Cover Design: Kristin E. Parker
Rights and Photo Research Coordinator: Ashley Dos Santos
Cover Image: © Glen E. Ellman
Printing and Binding: Courier Companies

Library of Congress Cataloging-in-Publication Data

Klaene, Bernard J.
  Structural firefighting : strategy and tactics / Bernard Klaene.—Third edition.
    pages cm
  "National Fire Protection Association."
  ISBN 978-1-4496-4239-6 (pbk.)
1. Fire extinction. 2. Fire prevention. 3. Fire departments—Equipment and supplies. I. Title.
  TH9310.5.K535 2015
  628.9'25—dc23

2014036064

6048

Printed in the United States of America
18 17 16 15 14   10 9 8 7 6 5 4 3 2 1

BRIEF CONTENTS

TABLE OF CONTENTS

Flames: Drx/Dreamstime.com; Steel texture: © Sharpshot/Dreamstime.com

CHAPTER 3

CHAPTER 4

CHAPTER 5

CHAPTER 6

Flames: Drx/Dreamstime.com; Steel texture: © Sharpshot/Dreamstime.com

CHAPTER 7

CHAPTER 8

Flames: Dn/Dreamstime.com; Steel texture: © Sharpshot/Dreamstime.com

CHAPTER 9

CHAPTER 10

CHAPTER 11

The Role of Occupancy

CHAPTER 12

High-Rise Buildings

Flames: Drx/Dreamstime.com; Steel texture: © Sharpshot/Dreamstime.com

Flames: Dri/Dreamstime.com; Steel texture; © Sharpshot Dreamstime.com

ACKNOWLEDGMENTS

Advisory Committee

Kelly Joe Aylor
Fire Chief
Florence, Kentucky, Fire Department

David Beste
Captain
Bellevue, Washington, Fire Department

John Bostwick
Battalion Chief
Cedar Falls, Iowa, Fire Department

Terence Delaney
Safety Battalion Chief
Louisville, Kentucky, Fire Department

Fernando M. Gray, Sr.
Executive Deputy Chief
Dallas, Texas, Fire-Rescue Department

Ron L. Jamison
Deputy Chief
Phoenix, Arizona, Fire Department

Jesse C. Kruckeberg
Captain—Training Division
Colorado Springs, Colorado, Fire Department

Thomas Lakamp
District Chief
Cincinnati, Ohio, Fire Department

Nick Morgan
Captain
St. Louis, Missouri, Fire Department

John J. O'Donoghue
Operations Assistant Chief (Retired)
Cambridge, Massachusetts, Fire Department

Mario Rueda
Deputy Chief
Los Angeles, California, Fire Department

Matt Smith
Assistant Chief
Brighton, Michigan, Area Fire Authority

Reviewers

Robert Jay Alley
Blue Ridge Community College
Flat Rock, North Carolina

Brian Bagwell
Metropolitan State University of Denver
Denver, Colorado

Timothy Baker
Lansing Community College
Lansing, Michigan

Greg Barton
Deputy Fire Chief
Beverly Hills, California, Fire Department

Paul M. Berens
Fire Prevention Technician, Cunningham Fire
    Protection District
Denver, Colorado

David L. Bullins
Guilford Technical Community College
Jamestown, North Carolina

Brian R. Carlson
Anderson Township Fire and Rescue
University of Cincinnati
Cincinnati, Ohio

Richard Carroll
Fire Protection Instructor, Cleveland Community College
Shelby, North Carolina

Earl R. Cumley
Chief, Bennett Fire Protection District
Bennett, Colorado

Sean DeCrane
Cleveland Fire Training Academy
Cleveland, Ohio

Peter Van Dorpe
Assistant Chief, Algonquin–Lake in the Hills FPD
Lake in the Hills, Illinois

Timothy Dorsey
Deputy Chief, West County EMS and Fire Protection
    District
St. Louis, Missouri

Kevin A. Gallagher
Chief, Acushnet Fire and EMS Department
Acushnet, Massachusetts

Todd Gilgren
Captain, Arvada Fire Protection District
Arvada, Colorado

Ian K. Harrell
Surry County Hazardous Materials Team
Mount Airy, North Carolina

Bradley Iverson
Las Vegas Fire and Rescue, College of Southern
    Nevada
Las Vegas, Nevada

Gary R. Johnson
Fire Science Instructor, Central Ohio Technical College
Newark, Ohio

Shad Johnson
Cleveland Community College
Shelby, North Carolina

Stephen Kerber, PE
Director, UL Firefighter Safety Research Institute

Shawn C. Koser, MBA, NREMT-P, FSI
Battalion Chief, Columbus Division of Fire
Columbus, Ohio

James D. Lea
Captain/Training Officer
Spartanburg, South Carolina, Fire Department

Robert Lindstedt
Southern Maine Community College
South Portland, Maine

Paul R. Martin
Deputy District Chief (Retired)
Chicago, Illinois, Fire Department

Tony Mecham
Fire Chief, CAL FIRE/San Diego County Fire
Fire Science Instructor, Riverside Community College
Riverside, California

William Kenneth Nelms
Chief
Nashville, Tennessee, Fire Department, EST

Paul Nelson
Fire Chief
DeForest, Wisconsin, Fire and EMS

Derek Parker
Fire Captain
Sacramento, California, Fire Department

Richard M. Siebel
Captain
Raleigh, North Carolina, Fire Department/Keeter
Training Center

Landon Stallings
Fire Marshal
Fort Worth, Texas, Fire Deparment

Robert Swiger
NCCRS/ISO Consultation Management Solutions
for Emergency Services

Jimmy Talton
Fire Chief
Reece City, Alabama, VFD

Gary "Skip" Tinagero
Battalion Commander
Albuquerque, New Mexico, Fire Department

William Wren
State Fire Instructor, New York
Member, New Hartford Fire Department
New Hartford, New York
Instructor, NYS Academy of Fire Science
Montour Falls, New York

DEDICATION

Flames: Drx/Dreamstime.com; Steel texture: © Sharpshot/Dreamstime.com

*This book is dedicated to colleagues
who gave their lives serving others.
Hopefully, by sharing many years
of fire service training and experience,
future tragedies can be prevented.*

The goal of this text is to explain proven tactics and strategies used at structure fires. The guiding objective throughout this text is to prepare the fire officer to apply appropriate tactics and strategies at structure fires, fully using available resources in a safe and effective manner.

It is assumed that the reader has a basic understanding of firefighting and associated tasks as outlined in *Fundamentals of Fire Fighter Skills* and such National Fire Protection Association (NFPA) documents as NFPA 1001, *Standard for Fire Fighter Professional Qualifications*; NFPA 1002, *Standard for Fire Apparatus Drive/Operator Professional Qualifications*; NFPA 1021, *Standard for Fire Officer Professional Qualifications*; or other similar firefighting manuals and standards. Qualified fire fighters with basic training and some level of experience who are seeking positions within a fire department as company and/or chief officers are the intended audience for this text.

Many fire departments adopted the first and second editions of *Structural Firefighting* as part of their reading list for promotional examinations. To ensure consistency throughout a fire department, it is best when all department members use the same strategy and tactics text. The lessons learned in *Structural Firefighting* apply to all members of the department, and it is an excellent text for providing consistent, department-wide strategy and tactics training.

The basis for an incident action plan that leads to a safe and effective fire-ground operation is an operational priority list. This text is designed around a three-point list:

1. Life safety
2. Extinguishment
3. Property conservation

Several other priority lists exist, but each list basically prioritizes life safety, followed by a logical sequence of activities aimed at reducing property loss. It must be stated that crossover exists among the listed priorities, and quite often, a lower-priority item provides the means to achieve a higher priority. For example, when the fire is extinguished, occupants and fire fighters are in less danger. Thus, extinguishment becomes an essential part of the life safety priority. The importance of extinguishment cannot be overemphasized and is a recurring theme throughout this text.

An important goal of any fire strategy and tactics text, including this one, is to give the reader the necessary tools to achieve maximum productivity under adverse fire-ground conditions. While fire-ground experience is a prerequisite to using the tools presented in this text, this text will enable readers to learn fire-ground procedures at an accelerated pace, thus reducing the cost in lives and property associated with learning by experience only. In addition to updating materials, answers to frequently asked questions are imbedded in this third edition. Further, explanations of many of the topics covered in the first and second editions have been expanded to include information from several recent studies by the National Institute of Standards and Technology (NIST), Underwriters Laboratories (UL), and others.

This is a dynamic and exciting time in the fire service. Many research studies with direct application to structural firefighting have been conducted since the publication of the second edition of this text. Staffing studies by the NIST prove the need for adequate staffing as a result of quantifying the effects of understaffing. Positive-pressure ventilation studies increase our understanding of this tactical method. The UL ventilation and fuel-load studies indicate a need to change fire-ground tactics. Future studies are planned to further promote scientific knowledge in ways that directly affect the way fire fighters "go to work" at structure fires. Of particular interest to this author is a proposed study investigating rate-of-flow requirements. Some old myths, including some held by this author, have been discredited.

The information gained through these various studies should improve safety and effectiveness at the fire scene. However, it is essential to closely study the parameters of research studies. As an example, most of the recent research involves one- and two-family residential occupancies. To take maximum advantage of the knowledge gained through research, we must share experiences, especially experiences in structures that do not exactly correspond to the research facilities and conditions.

The development of an effective fire officer or incident commander involves several steps over time. Training, education, and experience all play a role in the developmental process. A successful fire officer's background should include the following:

- Basic fire fighter training (as outlined in NFPA 1001, *Standard for Fire Fighter Professional Qualifications*)
- Experience as a fire fighter
- Advanced fire fighter training
- Company officer training (NFPA 1021, *Standard for Fire Officer Professional Qualifications*)
- Education in tactics, building construction, command, and organization

This list is not all inclusive, and the mix of experience, education, and training will vary in the developmental process. Experience is especially difficult to categorize. The needed experience may not be provided by serving several years with a department or company that has few structure fires. In turn, the assignment of a fire fighter or officer to a very busy unit does not guarantee diverse learning. For example, a fire fighter on a busy company may always be in the attack position with a 1¾-inch (45-mm) hose line. This fire fighter will be very proficient at advancing hose lines but may not be knowledgeable about other aspects of firefighting such as truck company operations.

While it is not possible (nor desirable!) to totally eliminate experience from the process of becoming an effective company officer or incident commander, training and education do reduce the experience necessary to become proficient. Of particular value is live fire training and realistic simulations. Some level of experience is necessary to "read the fire ground" and to understand the capabilities and limitations of resources and the impact of lead time. Ongoing training and education play a key role in the development of even the most experienced fire officer.

By applying the principles described in this text, a company officer or incident commander can use his or her experiences more effectively at the scene of structure fires. The fire officer who relies entirely on experiential learning essentially places the lives and property of others at a greater risk while he or she learns the trade. Sadly, many departments do not provide adequate structural firefighting training. After the attack on the World Trade Center on September 11, 2001, the NFPA conducted a fire department needs assessment for the U.S. Fire Administration. The study found, in part, that more than half of the fire departments (55%) surveyed did not provide structural firefighting training to all members. A follow-up report, *Four Years Later—A Second Needs Assessment of the U.S. Fire Service*, found that 53% of departments still failed to provide structural firefighting training to all members. A third needs assessment was conducted by the NFPA in 2010 showing a slight improvement, indicating that 46% of the departments responsible for structural firefighting have not formally trained all their personnel. With nearly half of all departments lacking adequate training, there is a definite need to better train and educate members of the fire service in strategy and tactics. This text provides a readily available tool to improve this unacceptable statistic.

Fire is an integral part of our society. We depend on it for warmth and convenience. Fire allowed humans to progress from cavemen to the computer age. Yet uncontrolled fire is feared, and rightfully so, as it continues to destroy life and property. The

Flames: Orv/Dreamstime.com; Steel texture: © Sharpshot/Dreamstime.com

modern fire department uses public safety education and code enforcement to help prevent and reduce the impact of fire. When these proactive efforts fail, fire fighters are forced into a reactive suppression mode. Even where considerable effort is expended to prevent fires, they do occur. In this text, fire is viewed as a formidable adversary, and understanding "the enemy" is crucial if we are to be successful. Firefighting is as much an art as a science, but applying the available science can aid us in our war against fire.

Many have compared fire-ground management to commanding a military battle or coaching a football game. These analogies are correct and useful, but with one significant difference: The military commander and football coach face an unpredictable enemy; in contrast, the fire officer, if he or she has done effective preincident planning, has a very predictable enemy. Fire is bound by the laws of chemistry and physics. The surprises that may occur are a result of incomplete information or failing to understand basic natural laws governing the fire, building construction, and the impact of the fire on the structure.

This text is by no means the only one a fire fighter endeavoring to be a proficient fire officer should read. The modern fire department may respond to hazardous materials releases, medical emergencies, technical rescues, wildland fires, and a wide variety of other emergencies. The scope of this text is limited to structural firefighting. The incident commander should be a qualified safety officer and be well versed on fire fighter safety laws and standards. The NFPA, the Occupational Safety and Health Administration (OSHA), and the National Institute of Occupational Safety and Health (NIOSH) have developed standards related to fire fighters' health and safety that should be familiar to the fire officer.

This text includes numerous case studies, incident summaries, and references to other cases discussed in fire service periodicals, NFPA Fire Investigations,

the NIOSH Firefighter Fatality Investigation and Prevention Program, and the Technical Report Series from the U.S. Fire Administration. Since some types of fires may occur only once in a lifetime (conflagration, large area fire, major high-rise fire), fire officers relying strictly on experience may be at a loss when confronted with these rare incidents. Collectively, the fire service has learned many lessons about strategies and tactics that were applied at these major events. Only by reviewing case studies and ongoing education can most fire officers learn about these most challenging incidents.

Preincident planning and standard operating procedures (SOPs) are essential elements of successful operations. Preincident planning is particularly important in large and complex structures. This text reiterates the importance of preincident planning to successful operations. Advanced knowledge of some of the factors leading to strategic decisions before an incident occurs will reduce the information processing needs of the fire officer in the crucial early moments of an operation. Predesignating initial operations using SOPs allows the incident commander to process information while a predictable course of action takes place.

Although a great deal of science is associated with fighting fires in structures, the application of this science requires a less-than-scientific approach. As a matter of fact, the rigid application of scientific principles sometimes leads to theories that are technically correct but empirically fallacious. As you progress through this text, you will find several places where a fallacy is introduced but then disputed. Some commonly held fallacies are described for the purpose of disproving misconceptions. "Fallacy/ Fact" boxes alert you to commonly held beliefs that are fallacious.

Added to this edition are the "IC Rules of Engagement for Fire Fighter Safety" and "Rules of Engagement for Fire Fighter Survival," developed by the International Association of Fire Chiefs. The

Firefighter Safety chapter contains a complete list of the rules of engagement as well as the objective for each rule. The rules are also placed throughout the book with topics related to a specific rule of engagement.

This text is designed to be used in a classroom environment or in a distance-learning format, with an instructor providing additional insight and practice materials. _Challenging Questions and Activities_ assist the student and instructor in applying the information contained in each chapter. An Instructor's ToolKit provides slide presentations, lecture outlines, and additional activities. Given the distance-learning design, the text is ideal for self-study.

The author has many people to thank. First and foremost is Russ Sanders, my coauthor for the first and second editions of _Structural Firefighting_. Russ not only contributed mightily to the text, he remains a mentor and information source. Although he was not involved in the actual writing of this third edition,

his experience as Chief of Department in Louisville, Kentucky, and as an NFPA staff member are evident on every page. Cincinnati District Chief Thomas Lakamp served as a special advisor and reviewer. In addition to his field experience as a company and chief officer, Adjunct Professor Lakamp uses _Structural Firefighting_ as a textbook at the University of Cincinnati. He provided invaluable technical advice and guidance in every chapter of this text. Photographers Denny Baker, Mike Carey, Bernard Erwin, Nick Morgan, Jeff Neal, David J. Jones, and Bill Strite provided live-action fire photographs.

Dr. Rita Fahy, manager of NFPA Fire Data Bases and Systems, researched and provided much of the statistical information contained in this text. Bob (Robert) Duval, Senior Fire Investigator for the NFPA, provided background information related to fire investigations, as well as pointing me in the right direction when searching for "real-life" fire experiences related to specific topics.

Flames: © Drx/Dreamstime.com; Steel texture © Sharpshot/Dreamstime.com; Chapter opener photo: © Jones & Bartlett Learning, Photographed by Glen E. Ellman

Organizing, Coordinating, and Commanding Emergency Incidents

LEARNING OBJECTIVES

- Identify and define the main functions within the National Incident Management System (NIMS) and how they interrelate during an incident.
- Given different scenarios, organize an operation using NIMS.
- Discuss and contrast fire-ground management compared to administrative management.

LEARNING OBJECTIVES (CONTINUED)

- Discuss the history and evolution of incident management systems, including the development of the NIMS.
- Define unified and single command, listing the advantages and disadvantages of each.
- Compare command modes available to the first-arriving officer, determining situations where each mode would be appropriate.
- Develop an initial report.
- Explain the importance of and develop a status report.
- Analyze the command transfer process, discussing when and how command should be transferred.
- Define and list the problems associated with freelancing.
- List the attributes of a good command post.
- Define and explain the importance of maintaining a reasonable span of control.
- Describe and enumerate the importance of staging.
- Compare a staged company to a parked apparatus.
- Define and explain the role of the incident commander.
- Identify, define, and place command staff positions on a NIMS organizational chart.

- Identify, define, and place the five sections on a NIMS organizational chart.
- Describe the position of and function of a chief's aide.
- Define and describe the functions of branches, divisions, groups, task forces, and strike teams.
- Explain the two-in/two-out rule.
- Organize an operation using geographic and functional assignments, and describe when each should be used.
- Given a fire situation, apply an intuitive naming system for various tactical-level management units.
- Recognize and articulate the importance of fire-ground communications.
- List general rules for incident scene communications.
- Define and explain unity of command.
- List and compare various means of communications that could be used at the incident scene.
- Develop a communications network that supports a NIMS organization.
- Explain methods that can be used to reduce radio communications to and from the incident commander.

Case Study

Incident scene when fire was first discovered.

Courtesy of Julie Miklaszewicz

On March 4, 2012, a fire lieutenant from a fire department providing mutual aid was killed at a theater fire when a bowstring roof collapsed. The first-arriving fire department from the department having jurisdiction attacked the fire on side A, directing the responding mutual aid department to attack the fire on side C. The two departments attacked the theater fire from opposite sides, establishing their own incident command and tactical operations plan with separate accountability systems. The first-due fire department initially fought the fire defensively from the side A, while the mutual aid lieutenant and two additional fire fighters entered the building to conduct an offensive operation on side C. The first-due fire department eventually directed an elevated master stream into the lobby and then onto the roof while mutual aid fire fighters were operating inside the building. Roof conditions deteriorated until the roof collapsed into the structure, trapping the mutual aid lieutenant.

1. How important is to have a single incident commander in control of an incident scene?
2. What are the dangers of a combination offensive–defensive attack strategy?

Data from: NIOSH FIRE FIGHTER FATALITY INVESTIGATION AND PREVENTION PROGRAM, 2012-08. Volunteer Lieutenant Killed and Two Fire Fighters Injured Following Bowstring Roof Collapse at Theatre Fire – Wisconsin.

Flames: © Jag_cz/ShutterStock, Inc.; Steel texture: © Sharpshot/Dreamstime.com; Stacked photo background: © Vitaly Korovin/ShutterStock, Inc.; Paper: © silver-john/ShutterStock, Inc.; Photo: Courtesy of Julie Miklaszewicz

Introduction

This chapter discusses the importance of adopting and implementing the **National Incident Management System (NIMS)** for use at all structure fires. Use of NIMS is essential to enable the **incident commander (IC)** to safely and effectively manage incident resources. NIMS is instrumental in ensuring that fire fighter safety, as well as the three operational priorities of *life safety*, *extinguishment*, and *property conservation*, are addressed at the fire scene. Because even the best organization is at a distinct disadvantage without good communications, which is probably the most cited deficiency at actual emergencies and exercise scenarios, communications is also addressed in this chapter.

Decision making at the incident scene is different than day-to-day administrative decision making. The IC must decide on the proper course of action with limited information available in a relatively short period of time. This process is referred to as **recognition-primed decision making (RPD)**. Administrative decision making is less time sensitive; therefore, input should be included from diverse sources, and all options should be carefully considered. This process is referred to as **rational decision making (RDM)**. The time constraints of the fire ground require RPD. Proficient fire officers make sound decisions within the time and information constraints of the fire ground by taking advantage of previous training, education, and preincident planning and by matching the situation at hand to similar experiences at previous fires. Variations in building types, occupancies, and potential fire scenarios are nearly endless, making it almost impossible to gain experience in handling every possible fire scenario. Well-constructed fire simulation training is highly recommended to increase experiential learning.

The adoption of an incident management system meeting the criteria outlined in NFPA 1561, *Standard on Emergency Services Incident Management System and Command Safety*,[1] is a must. Such a system will allow for the expansion of forces while maintaining a reasonable span of control. The National Incident Management System (NIMS)[2] is highly recommended and should be used at every incident regardless of size. Fire-ground operations have often been compared to military operations. A military commander uses a battle plan much as an IC uses a preincident plan. Standard military maneuvers (the equivalent of fire department standard operating procedures, or SOPs) are developed in advance. The military has a very rigid organizational structure that limits the span of control to a relatively small number. A general trying to give orders to each of the thousands of men and women under his or her command would soon be overwhelmed and lose control. This general may have success with skirmishes in which few soldiers are actually involved in battle, but he or she could not hope to win a major campaign.

On the fire ground, it is a mistake for a chief officer to attempt to control all aspects of an incident. Like the general, a chief may be successful in handling a skirmish (e.g., a one-alarm fire involving only four or five companies), but he or she will quickly become overwhelmed when trying to apply a micromanaging style to a major incident.

NFPA 1500, *Standard for Fire Department Occupational Safety and Health Program*,[3] requires the use of an incident management system, as do current Occupational Safety and Health Administration (OSHA) regulations dealing with hazardous materials response.[4]

Evolution of the Incident Management System

A significant amount of work has gone into developing systems to manage emergency incidents of all types. Forest fires in California in 1970 prompted the development of the FIRESCOPE Incident Command System (ICS).[5] FIRESCOPE (which stands for FIrefighting REsources of Southern California Organized for Potential Emergencies) greatly improved operations at large wildland fires, and its use by fire departments of all sizes in California provided the testing ground that led to the urbanization of the FIRESCOPE ICS. As use of the system expanded, improvements were made based on the experience gained at actual incidents. The result was an extremely useful tool that is available to the fire service and others involved in emergency response. This is the system used to coordinate resources within NIMS.

Command
National Incident Management System (NIMS)

It is essential that the entire response community, not just the fire service, be familiar with and trained in using NIMS. Agencies such as police, health department, and local disaster agencies, as well as mutual aid departments, should all work under the same system. The use of NIMS provides common terminology and operational assignments for all agencies at the incident scene. This uniformity can greatly facilitate the successful conclusion of an incident. Homeland Security Presidential Directive (HSPD)-5 establishes NIMS as the national system, requiring its use by federal agencies and making its adoption mandatory when the federal government is providing preparedness assistance.

NIMS should be used for all incidents, regardless of their size. If personnel are using NIMS for smaller, routine incidents, they will be better prepared to use it for large-scale incidents. It is a mistake to wait until

Incident Summary

When the Alfred P. Murrah Federal Building in Okla-homa City was blown up by a terrorist's bomb on April 19, 1995, myriad agencies became involved in many different stages of the incident. Overall com-mand of the incident was taken and held by the Oklahoma City fire chief, Gary Marrs. However, as the incident evolved, other jurisdictions became involved in the cause and origin determination, search for evidence, and other law enforcement functions. Despite the size and complexity of the incident, the fire department retained command of the incident, coordinating all operations with the different juris-dictions involved.

Managing the multitude of agencies involved in the rescue and recovery operations following the Oklahoma City bombing was successfully accomplished using an incident management system. The Oklahoma City Fire Department retained command of the entire incident until its completion.

Courtesy of FEMA. Edward R. Comeau and Stephen Foley, "Oklahoma City, April 19, 1995," NFPA Journal, July/August, 1995.

an incident grows to a certain size before implement-ing NIMS; by then, it is too late to start trying to put together the organizational segments required to support the incident. It is much easier to grow NIMS as the incident expands, integrating other agencies and jurisdictions as needed.

The command system should be regional in scope and capable of handling large numbers of resources beyond the local jurisdiction. NIMS is national in scope, and because of its flexibility and adaptabil-ity, it is capable of handling the largest imaginable structure fire, including fires that involve multiple buildings. The focus of this text is on structure fires, but NIMS is an all-hazards system capable of han-dling any emergency situation that the fire officer may encounter.

Unified Command

NIMS addresses situations in which more than one jurisdiction or agency has responsibility by estab-lishing a **<u>unified command</u>**. Unified command pro-vides an invaluable method for controlling very large incidents in which multiple agencies or jurisdictions are involved. In a unified command scenario, differ-ent jurisdictions and/or agencies share responsibility for developing the incident action plan.

A word of caution about unified command is in order. Whenever possible, a **<u>single command</u>** (i.e., one person is designated as the overall IC) is pre-ferred to a shared or unified command. However, some situations dictate the use of a unified command system, such as when the incident objectives are not clearly defined as relating to fire or law enforcement.

Flames: © Jag_cz/ShutterStock, Inc.; Paper: © silver-john/ShutterStock, Inc.

Structure fires are clearly under the jurisdiction of the local fire department unless the structure or structures span more than one jurisdiction. Much has been written about the difficulty in communications and coordination between fire and police at the World Trade Center on September 11, 2001. Some feel this event called for a unified command. The police department played a major role, but the emergency phase of the operation clearly should have been under the jurisdiction of the fire department. It is not unusual for police to assist in evacuating or isolating the area during a structure fire; they also control vehicular and pedestrian traffic around the incident scene. There could be a large contingent of police officers at the fire scene, but the fire department should retain control, with a police department supervisor coordinating police activities from the fire department incident command post. The police department supervisor (law branch director or law group supervisor) should be equipped with a radio capable of communicating with all police personnel, enabling the IC to communicate through the police supervisor. This police supervisor could also be assigned to the operations section. Though the World Trade Center events on September 11, 2001, may have been a case for unified command, seldom would a unified command be warranted for a structure fire.

NOTE

Names of tactical-level management units (i.e., groups, divisions, and branches) should indicate the area or function being managed. Tactical-level management units that manage police personnel are referred to as "law branch" or "law group" in this text. Calling them a "police branch" or "police group" would also be correct.

Texture: © Eky Studio /Shutterstock, Inc; Steel: © Sharpshot/Dreamstime.com

A unified command can also evolve after the emergency has been mitigated and the incident has moved into the cleanup phase. For example, once the life safety issues have been resolved, several agencies might become engaged in a fire involving hazardous materials.

Problems can arise when various agencies with different priorities are charged with developing and implementing an incident action plan. When using the unified command structure, plan development is shared, but NIMS requires that one operations chief direct all field units.

Fallacy	Fact
Whenever more than one agency responds or the fire spreads over more than one jurisdiction, always establish a unified command.	Whenever possible, establish a single command with one person assuming the role of incident commander.

Steel: © Sharpshot/Dreamstime.com

Initial Command

Major incidents typically begin with response by one or two fire companies. It is critical that the person serving as the initial IC get the operation off to a good start by establishing command and following department procedures. It is very difficult to recover once an operation becomes chaotic.

NOTE

The ultimate success or failure of a fire operation is quite often dependent on the actions or inactions of the first-arriving company.

Texture: © Eky Studio /Shutterstock, Inc; Steel: © Sharpshot/Dreamstime.com

It is certainly desirable for the first-arriving company officer to establish a formal command post outside of the structure and concentrate on directing operations. However, few fire departments have sufficient staff to allow the company officer to establish a stationary command post on arrival. It is important to recognize that the officer of a first-in unit, who is performing task-level functions and commanding the operation, is at a distinct disadvantage compared to the chief officer, who is at a stationary command post with a single function. If the first-in officer delays his or her task-level actions to establish a stationary command post, the fire may make substantial progress during this delay.

It is essential that someone always be in command. Most departments respond to structure fires as fire companies, with a company officer or member in charge of the company. Most departments permit three command options for this first-arriving company officer:

- *Investigation:* Generally, nothing showing on arrival, no need to take immediate action to save lives or protect property (command post not established)
- *Fast attack:* Obvious signs of fire or some other dangerous conditions that require immediate action where the company officer's direct involvement will be beneficial to the outcome (command post not established)
- *Command:* A situation of such a magnitude or complexity that the company officer must assume a command position and not become directly involved in the operation so that he or she can coordinate the actions of personnel and other incoming units (command post established)

When fire companies are staffed by four or more personnel, the investigative or fast attack options allow the operation to continue while the company officer is busy performing initial tasks as part of the entry team. Some departments assign the first-arriving officer to establish an exterior command post with two or more fire fighters making entry to conduct an offensive fire attack. This procedure has the advantage of providing a better command function but often results in less-experienced personnel conducting the initial interior attack with no apparent leader. For these reasons, the preferred fast attack mode places the company officer as part of the interior attack crew. There are some situations in which an individual fire fighter or officer arrives at the scene prior to the arrival of a fire company. In such cases, that person is the initial IC.

Company officers must be reminded of the importance of establishing command at fires and other emergencies. To ensure fire fighter safety, there should always be a strong command presence. Establishing command during training does much to remind the company officer of this important concept. Postincident critiques and tactics training are also necessary if company officers are expected to establish a stationary command post when circumstances permit or require it.

Most departments adhere to the strict interpretation of establishing command by announcing command. Others are satisfied with carrying out the duties, without an announcement of command, when the officer is in the investigation or fast attack mode. Announcing command reinforces the fact that this first-arriving officer is in command and is therefore recommended. Others believe that the command announcement is unnecessary, as command is assigned to the first-in officer per SOPs, but confusion could result when two or more companies arrive simultaneously. Regardless, there must always be someone in command of every single incident, no matter how large or small. The first-arriving officer must report conditions upon arrival at the scene. The expected initial report must be included in the department SOPs. Some of the common items included in an initial report are:

- Confirmation of command
- Confirmation of address
- Announcement of command mode (investigation, fast attack, or command)
- Brief description of building or specific name of well-known structures
- Occupancy type
- Conditions—e.g., heavy smoke with occupants at windows
- Actions being taken
- Resources needed

The following is an example of an initial report:

Engine 1 is on the scene at 1234 Main Street. Heavy fire and smoke are visible from the second floor of a three-story, wood-frame apartment building; one victim can be seen at a second floor window at the front of the building. Engine 1 has a water supply and we are advancing a 1¾-in. (45-mm) hose line to the second floor. Request a second alarm assignment at this location. Engine 1 is fast attack Main Street Command.

When department SOPs outline initial actions to be taken, the initial report could be shortened. As an example, SOPs could direct the first-arriving engine company to obtain a water supply and assume command in the fast attack mode. Therefore, water supply and command mode could be eliminated from the initial report, provided the company officer is following the outlined procedures.

Whenever it is feasible, the first-in company officer should attempt to view all sides of the building involved. The officer can often view three sides of a building by having the driver pull just past the building involved. The officer then needs to get only a view of one remaining side. In some cases, due to the size or complexity of the building or area, having first-arriving units view all four sides of a building would result in a considerable delay in initiating operations. In such cases, it may be best to begin operations immediately. The IC should then assign later-arriving personnel to check the rear and other sides of the building. Sometimes the best route to the rear of the building is through the building.

Figure 1-1 Closely spaced multistory residential buildings.
Courtesy of Bernard J. Klaene.

be considerably delayed in waiting for the officer to reach the rear of the building. While walking through the fire building or adjacent building, an officer or fire officer could check the rear and connecting walls of these buildings. In some departments, the roof crew is assigned to check the rear of the building, which is a good procedure if a roof crew is actually needed.

The information required in the initial report should match department needs and resources. As an example, it may be necessary to include additional information regarding water supply when parts of the jurisdiction do not have nearby fire hydrants.

Use good communications techniques when using the radio:

- Take a deep breath.
- Think about what you are going to say before transmitting the message.
- Use clear text—no **10-codes**.
- Key the radio, followed by a very short delay. *This is very important; with some radios, part of your transmission will be lost if you start speaking immediately.*
- Speak slowly and distinctly.

It is imperative that all members who could possibly be in charge of the first-arriving unit practice transmitting an initial report. The *Challenging*

IC Rule of Engagement for Fire Fighter Safety

Rapidly conduct, or obtain, a 360-degree situational size-up of the incident.

Reproduced from: © 2011, IAFC Safety, Health and Survival Section. All rights reserved.

Texture: © Eky Studio./Shutterstock, Inc; Steel: © Sharpshot/Dreamstime.com

As an example, refer to the buildings in Figure 1-1, which are closely grouped, multistory residential buildings. If a complete walk-around of a building located mid-block were carried out prior to commencing operations, the initial attack would

Questions and Activities at the end of this chapter provide an opportunity to do just that.

Command by a Chief Officer

The three command mode options (investigation, fast attack, and command) apply only to company-level operations. When a chief officer assumes command, it is critical to establish a stationary command post. As higher-ranking chief officers arrive, they have the option of accepting the command role and reassigning the previous command officer to a position that may be mobile (e.g., safety officer), to an interior tactical-level management unit, or to another general and/or command staff position. This change should occur only after a formal transfer of command.

The current IC must provide a status report to the chief officer assuming command. As soon as the chief officer taking command has an opportunity, he or she should transmit a progress report to dispatch. Part of the size-up and status report process involves reconnaissance from units at the scene. This is most often done using the radio system to request a progress report from operating units. For example:

"Command to Engine 1, what is your status?"

"Engine 1 to command, we have the fire knocked down on the second floor.

"Command to Truck 1, what is your status?"

"Truck 1 to command, one victim has been rescued and a primary search completed for the entire structure."

The IC would then communicate a status report:

"Main Street Command to dispatch, we have the fire under control, one victim rescued, and the primary search is complete."

The IC will also need to use the communications network to relay orders to tactical-level management units or individual companies, depending on the size and complexity of the incident. It is important to first identify the unit you are calling. For instance, you may say, "Command to Engine 1." When Engine 1 replies, issue brief, specific, clear orders. The unit being assigned should repeat the order. ICs must remember that completing a task will take time, and on occasion the unit assigned to complete the task will

be unsuccessful. When an order is issued, the IC will assume that the objective is being successfully completed unless the company given the directive advises otherwise. However, the IC should request a status report from the company assigned the task after reasonable time has elapsed to complete the assignment.

When available, a chief's aide or planning section chief can estimate the time needed for task completion and track elapsed time. Once the IC gives an order, the companies assigned the task(s) must complete the objective(s) and notify the IC that the assignment is complete or advise the IC that they are unable to complete the assignment and briefly explain the reasons why.

On occasion, a crew could be redirected by another officer who is not within their **chain of command**. NFPA 1561, *Standard on Emergency Services Incident Management System and Command Safety*, specifically addresses this issue:

- **NFPA 1561.5.8.8.1**—Where conflicting orders are received at any level of the incident management system, the individual receiving the conflicting order shall inform the individual giving the order that a conflict exists.
- **NFPA 1561.A.5.8.8.1**—The guideline for clarifying conflicting orders shall not apply to imminent hazard situations where immediate action is needed to avoid a dangerous situation.
- **NFPA 1561.5.8.8.2**—If the conflicting order is to be carried out, the individual giving the conflicting order shall so inform the individual who provided the initial order.

This practice of working outside the chain of command can have severe negative consequences and should be avoided.

Transfer of Command

Command transfer must be addressed in department SOPs. In larger departments where there is a multi-level rank structure, it is essential that later-arriving company officers know whether they are required to assume command or whether they have other options. The first-arriving company officer may not be at a command post. When command is transferred to another officer, it should be formalized at a stationary command post.

Unless there is a compelling reason, command should not be transferred between company officers. The situations where command transfer between company officers is permitted should be clearly outlined in department SOPs. Multiple command transfers in the initial stages of the operation generally result in confusion and could reduce the number of people engaged in hands-on life safety and extinguishment activities. Later-arriving company officers should reevaluate conditions upon arrival. An officer who finds an unsafe operation in progress upon arrival must immediately take command. In this case, a formal command post must be established and immediate corrective actions taken.

When command is transferred, there is the possibility of information being lost in the process. It is necessary for the person assuming command to communicate with the previous IC to determine both situation and resource status of the operation in progress. If a tactical worksheet is in use, it should be passed to the new IC. While this communication is taking place, arriving units should be awaiting orders. The longer it takes to transfer command, the greater the chance of **freelancing**. A strong command presence and efficient transfer process will ensure a smooth transition and eliminate independent actions. Keeping everyone focused on achieving the tactical objectives outlined in the incident action plan is imperative.

The IC assuming command must evaluate the safety and effectiveness of the operation in progress. If the operation is unsafe or is not focused on accomplishing the incident priorities of life safety, extinguishment, and property conservation, the IC assuming command must reorganize the operation and reassign operating units as needed. It is very difficult to reorganize and redirect operations in the heat of battle, but it is sometimes necessary. If the current operation is being conducted safely and effectively, the task of assuming command is much easier. The person assuming command should improve the operation by updating the size-up and augmenting

operations in need of additional resources. Information available to the IC improves as companies relay critical information from the interior or remote positions and more time is available to review preincident plans and other information sources.

If department procedures give a more senior or higher-ranking officer the choice between taking command and allowing the present IC to continue serving in the position, allowing the present IC to continue could avoid numerous and unnecessary command changes. However, the higher-ranking officer, especially chief officers, face a dilemma in allowing a lower-ranking officer to retain command: If there is a major problem in resolving the incident, the highest ranking chief officer present onsite may be held accountable, even if he or she did not assume command. Thus, the department and its officers should balance the potential efficiency gained in retaining command with the possibility that higher-ranking officers may be held responsible and accountable. A possible solution to this dilemma is assigning the lower-ranking officer as the operations section chief. The operations section delegation allows the higher-ranking officer to establish to overall command while providing tactical-level experience to the lower-ranking officer.

Command transfers tend to disrupt the continuity of operations, but department culture or rules may require command changes, particularly among higher-ranking chief officers.

On rare occasions, command of a structure fire could be transferred to another agency. Fires and explosions are used as weapons of terror. This could be a case where law enforcement would assume command from the fire department after the fire situation is controlled. The February 26, 1993, explosion and fire at the World Trade Center is an example of a structure fire where command might eventually be transferred to police authorities.[6] Fires or other fire department responses where an "active shooter" is on scene could result in command by a law enforcement agency or a unified command.

Fires involving hazardous materials can also result in transferring command to another agency. The accidental fire involving a large quantity of solvents at the BASF plant in Cincinnati is an example. Once the fire was brought under control and all victims removed, the Ohio Environmental Protection Agency (EPA) assumed command and maintained a presence at the scene for 5 months during cleanup operations. The Ohio EPA retained command during this 5-month period even though fire units were at the scene whenever cleanup operations were in progress.

Information passed on during a transfer of command should include, at a minimum, the following:

- Progress report
- Company location and assignments
- Current ICS organization structure/assignments
- Known hazards, such as hazardous materials, downed power lines, building hazards, and terrain issues
- Additional resources requested, but not assigned

Delegation

Establishing control over available resources entails the delegation of authority. The IC develops the strategy while division/group supervisors or branch directors develop tactics within the overall strategy by assigning and coordinating tasks for units that are working under their supervision. Information exchange through good communications is the key to managing the fire ground.

Command Post

The first-arriving company officer will be in command but generally will be working inside the building during an offensive fire attack. However, fires that are clearly defensive operations from the beginning are normally handled by the first-arriving officer establishing a stationary command post and assigning other first-alarm companies.

A good command post will:

- Be established in an identified location that is easily found

- Be outside the hot zone, in an area where personal protective clothing is not required (cold or green zone)
- Provide a view of the two most important sides of the building
- Never hinder apparatus movement

In most cases, companies and agencies will report to the command post for instructions and information. Therefore, the command post should be easily found. When establishing a command post, the IC should communicate the location using the street name for an exterior command post (e.g., First Street Command) or the building name when using a command center within a building (e.g., Scripts Command) **Figure 1-2** and **Figure 1-3**.

For outside command post locations, the IC should identify the side of the building where the

Figure 1-2 Exterior view of a high-rise building.

Courtesy of Bernard J. Klaene.

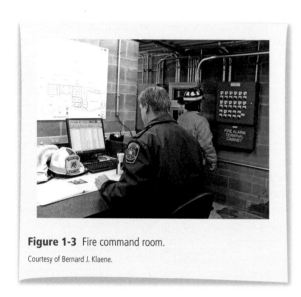

Figure 1-3 Fire command room.

Courtesy of Bernard J. Klaene.

command post is located (e.g., Third Street Command). There are times when some degree of isolation is preferred. Companies may be assigned to report directly to group and division supervisors or branch directors. Outside agencies should report to a liaison officer, and the public information officer should manage the media.

It is usually good practice to position the command post so that two sides of the building can be viewed. There is a distinct advantage in being able to see the effects of tactical decisions. However, there are times when seeing the scene can be a distraction, causing the IC to focus on the visible while failing to deal with higher priorities that cannot be seen. Generally, the larger and the more complex the situation, the farther away and more isolated the command post should be. Isolation from distractions at the scene can be an important factor in developing good incident action plans. Most often, the command post will be located at or near the scene of a structure fire. When confronted with a large-scale operation involving very large buildings or multiple buildings, it is sometimes a good idea to place the command post in a nearby building with good lighting and communications. For a catastrophic incident, such as the September 11, 2001, attack on the World Trade Center, establish a

command post at a location well away from the incident. Most city and county governments have pre-planned emergency operations centers (EOCs), which could be used for this purpose. The command post is at the location of the IC; even if an EOC is functioning for a structure fire, when the IC is at the incident scene, then so is the command post.

NOTE

Wherever the incident commander is located—that is the command post.

Texture: © Eky Studio /Shutterstock, Inc; Steel: © Sharpshot/Dreamstime.com

Managing an incident requires the IC's undivided attention. The command post should be in a location that supports command, control, and coordination of all incident activities. Good communication is critical to the command function.

An IC manages most fires with a few command staff, sections, or tactical-level management units. ICs at these everyday incidents generally use an automobile or sports utility vehicle as a command post. Some ICs prefer to establish command by establishing an outside position away from their vehicle or on the outside at the rear of their vehicle. Many command vehicles have status boards and other command-post equipment in the rear vehicle compartment or trunk. Other departments require or strongly suggest that the IC stay inside the vehicle and place command equipment in the front seat area to support this policy. Sports utility vehicles are particularly well suited for placing command post materials and equipment in the front seat or rear compartment. There is also an advantage in using four-wheel drive vehicles with a higher road clearance in areas where snow is to be expected or where some responses require off-road travel. Working from the vehicle generally provides better communications, as vehicle radios are usually more powerful and the vehicle engine maintains a constant charge. Working inside the vehicle also affords a measure of security and safety for the IC, in addition to climate control. When commanding from an exterior position, conditions such as high winds, heavy snow, torrential rains, or other types of inclement weather can be distracting or even dangerous. The larger and more complex the incident, the greater the need for command post facilities and space Figure 1-4 .

Figure 1-4 Exterior command post.
Courtesy of David J. Jones, Cincinnati, Ohio.

Span of Control

Both safety factors and sound management planning influence and dictate span-of-control considerations. The **span of control** is the number of people reporting to a supervisor. For any individual with emergency management responsibility, the span of control should range from three to no more than seven, with five being the rule-of-thumb average.

The type of incident, the nature of the task, hazard and safety factors, and the distance separating tactical units will influence span-of-control decisions. Maintaining an effective span of control involves anticipating change rather than reacting to it. This foresight is especially important during rapid buildup of the command structure, when good management is made difficult because of numerous reporting units.

Large-scale operations where the span of control exceeds manageable limits become chaotic and unsafe. Span-of-control limitations are generally in

the area of five subordinates reporting to a single supervisor under emergency fire conditions. NIMS recommends a span of control ranging from three to seven people reporting to a supervisor. A one-to-one span of control serves no useful purpose, making the NIMS organizational structure more complex than necessary. The larger and more complex the NIMS organization, the more difficult it is to control. An important rule to remember is "keep it simple." Form tactical-level management components as necessary to maintain a reasonable span of control and to coordinate specific geographic and functional operations. However, remember that each organizational layer places the IC farther away from having direct contact with operating units. In some large, complex incidents, this is necessary and desirable; however, at most incidents it is not.

A question sometimes arises regarding a two-to-one span of control. If there is a need to sector an operation to maintain a reasonable span of control, and only two units are needed in a geographic area or to perform a specific function, a two-to-one span of control may be acceptable Figure 1-5 .

Calling for Additional Resources

Some in the fire service would recommend a piecemeal approach when requesting help to avoid being overwhelmed by units reporting for assignments. A much better approach is to stage arriving resources until the IC identifies tactical needs. Companies or crews can be sent directly from staging to a group, division, or other assignment. Remember, it is best to call for additional help before it is needed. If calls for additional resources are delayed until after the need is obvious, the resources will arrive too late. The need for additional units must be anticipated and requested well in advance of when the units should actually be in position. Consider how traffic, weather conditions, distance, and other incidents affect response times.

Smaller departments do not have the benefit of the large number and variety of resources that are typically part of a larger department. In these situations, smaller departments enter into a mutual

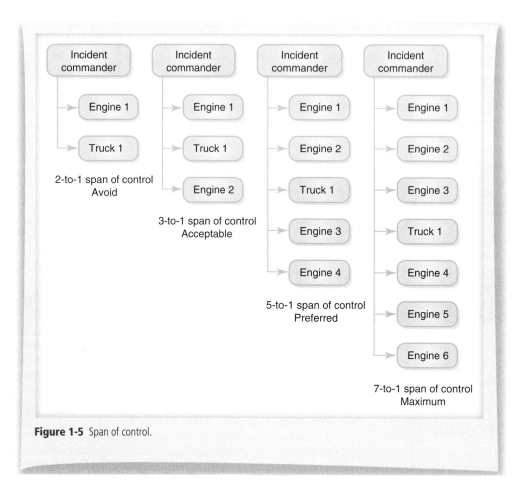

Figure 1-5 Span of control.

aid agreement, which will ensure that resources needed for larger events, including a tactical reserve, are available when requested. Mutual aid resources can be dispatched at the time the alarm is received from the public (automatic mutual aid) or dispatched only when requested by the responding department. Automatic mutual aid is gaining popularity, as many departments do not have the resources needed to combat a working structure fire. Mutual aid agreements should be written and signed by community leaders to alleviate any misunderstanding that may arise. Agreements should spell out what type of aid will be available, staffing levels, financial obligations, liability related to injuries and equipment loss, and who is in charge,

to name a few details. In addition, these departments should periodically train together. Training together can reduce costs, uncover incompatible equipment (e.g., hose threads and hydrant connections), increase comfort levels when working with fire fighters in other departments, and provide an opportunity to become familiar with or reconcile differences in SOPs.

Demobilization

As the incident is brought under control and essential tasks are completed, it is time to start returning units to stations and "available" status. There is a natural tendency to return later-arriving units first. Company apparatus that arrived first are pumping water, are providing access via ladders, or are otherwise committed to needed operations. The first-arriving units are more likely to be fatigued or exhausted and are more prone to injury, whereas later-arriving units may have been held in reserve and may be fresh.

A word of caution regarding the release of resources: Maintaining an adequate reserve and rapid intervention crew should continue as long as companies remain on the scene. Many injuries occur after emergency phase operations appear to be complete. Always expect the unexpected! As an example, on June 17, 1972, Boston Fire Department lost nine members when the Hotel Vendome collapsed during overhaul operations.

Staging

Staging areas should be established to locate resources that are immediately available for an assigned task. There are two types of staging. The most common type of staging places unassigned, fully staffed units at an exterior location near the fire scene ready for immediate assignment. Some departments also stage later-arriving initial response companies while the first-in company completes their size-up or investigation. This is sometimes referred to as Level 1 staging. The second-due engine is often assigned to locate at a fire hydrant. When staged at a hydrant, it is good practice to check the hydrant. In areas where hydrants are not available, the water source may be a static source such as a cistern, lake, pond, or swimming pool. When assigned to stage at a static water supply, ensure that water is available. Static water sources are known to dry up during droughts, and a pool or cistern could be drained for periodic maintenance. In any case, make sure the water supply is viable when staged at a source of water.

SOPs must address staging, including interior staging at high-rise buildings, which is addressed in the *High-Rise Buildings* chapter. Staging allows the IC to control access to an incident scene, while deploying resources in a safe and effective manner. A staging area can be located anywhere mobile equipment

Incident Summary

On June 17, 1972, at 2:35 PM, the Boston Fire Department responded to a fire alarm at the 101-year-old Hotel Vendome. A few minutes later, a working fire was called in with subsequent alarms sounded. A total of 24 fire companies responded to the fire and were able to bring it under control by 4:30 PM. Three companies remained on the scene to complete the overhaul process. At 5:28 PM, suddenly and without warning, a large section of the five-story building collapsed on top of a ladder truck and 17 fire fighters. Nine fire fighters were killed.

The Hotel Vendome fire and collapse.
© The Boston Globe/Getty Images

can be temporarily parked while awaiting assignment. It is best to stage units two or more blocks from the actual incident to avoid the temptation to freelance into action and to ensure that staged units are in a safe area. When establishing a staging area, locate it far enough away to avoid obstructing or slowing access to the incident scene, but close enough to allow companies to arrive quickly once summoned. When weather conditions are extreme, such as freezing temperatures, shelter should be provided for companies that could be staged for a long period of time.

The best time to identify good staging locations is during preincident planning—not in the middle of the night during the heat of battle. Predetermining the location of the staging area is especially important for large, complex properties. Depending on factors such as property size, access issues, and weather conditions, the staging area could be an open-area parking lot, a nearby firehouse, or a predetermined location that would provide some measure of security and protection from the elements.

When a chief officer arrives on the scene, there is an immediate need to reevaluate the situation and modify the incident action plan if necessary. During these early stages of an operation, the IC can be easily distracted by information overload from several sources, including radio communications from both on-scene and incoming units. If the IC has not completed the incident action planning process, including the determination of how responding fire forces are to be deployed to carry out the plan, incoming units will have a tendency to freelance into action. Further, an IC who is already overwhelmed with radio traffic is often reluctant to call for additional assistance, even when the need is obvious. Establishing a staging area early on, and directing all responding companies to that location unless otherwise ordered, allows the IC to better manage on-scene units, establish a ready tactical reserve, and eliminate potentially dangerous freelancing. Establishing a tactical reserve is an important command consideration. If all units at the scene are committed to the operation, a tactical reserve is needed to cover unanticipated problems and provide relief to operating crews. Staging is an important tactical tool that is underutilized by many departments.

The staging area can also be used as a parking area for out-of-service apparatus. During a large-scale offensive operation, later-arriving units often are needed only to deliver additional fire fighters. The apparatus pumps, ladders, and tools are not needed; therefore, the unneeded apparatus are parked out of the way in the staging area, or at base at a high-rise fire when base is established. It is important to distinguish between an out-of-service apparatus parked in staging and a staged unit that is part of the tactical reserve. A staged unit is a fully staffed unit, such as an engine company, truck company, or medic unit that is ready to respond immediately to the incident scene. Apparatus without adequate staffing are classified as out of service, rather than staged.

> **NOTE**
>
> Staging is an important management tool that increases incident scene safety and efficiency by providing a means to immediately deploy forces to achieve identified tactical objectives while preventing dangerous freelancing.

Texture: © Eky Studio /Shutterstock, Inc; Steel: © Sharpshot/Dreamstime.com

SOPs should outline how the staging area is managed, the duties of the staging manager and of the company officers whose units are assigned to staging, as well as the physical characteristics of an effective staging location. During large-scale incidents, a staging manager is assigned to manage staging and a more formal staging area is established in an area that provides adequate space and access to the incident scene. The staging manager is responsible for managing all incoming resources and dispatching resources at the request of the IC or operations section chief. Crews in staging must remain together with their apparatus and equipment, ready for immediate deployment. Span of control is not a problem in staging. The staging manager can manage numerous companies because of their inactive status.

Interior staging is typically used for a fire on an upper floor of a high-rise building. The *High-Rise Buildings* chapter addresses high-rise staging procedures.

NIMS Organization and Positions

As the need for resources increases, there is a need for a larger and more complex organization. NIMS is a tool used to organize an operation. NIMS is not a tactical objective, but a means to command and control an incident to achieve incident action plan objectives. With this in mind, the NIMS organization should be as simple as possible.

Modular Organization

The NIMS organizational structure develops in a modular fashion based on the type and size of the incident. First and foremost, there must always be an IC. Other line and staff positions are assigned by the IC according to incident priorities.

The specific organizational structure that is established for any given incident will be based on the management needs of the incident. If one individual (the IC) can simultaneously manage all of the major geographic and functional areas, no further organization is required. However, if one or more of the areas require independent management, then a qualified individual should be assigned to manage each of the necessary areas. If the IC does not specifically delegate the geographic areas or command functions, then he or she retains responsibility for all that are not delegated. NIMS positions should be thought of as a job description for the IC. The IC decides whether to do the task or to have someone else perform that function. NIMS positions are divided into command staff and general staff (sometimes referred to as line officers).

> **NOTE**
>
> The only NIMS position that must be established at every incident is IC.

Texture: © Eky Studio /Shutterstock, Inc. Steel: © Sharpshot/Dreamstime.com

Command Staff

All command staff positions report directly to the IC. Command staff positions are established to assume responsibility for key activities that are not included in the line organization. NIMS identifies three command staff positions:

1. Safety officer
2. Liaison officer
3. Public information officer

These positions are generally placed in "staff" position on the NIMS organizational chart **Figure 1-6**.

Figure 1-6 NIMS showing command staff positions.

Incident Safety Officer

The **incident safety officer** is a key position on the fire ground and is important to fire fighter safety particularly at large-scale or complex incidents. This position should be staffed by an experienced officer who meets the requirements outlined in NFPA 1521, *Standard for Fire Department Safety Officer.*[7]

The function of incident scene safety must be carried out at all incidents. It is the responsibility of the incident commander (IC) who cannot perform this function due to the size or complexity of the incident to assign or request response of an incident safety officer. There are, however, incidents that require immediate response or appointment of an incident safety officer.

These types of incidents should be defined in the fire department's response policy or procedure to ensure that the incident safety officer responds. Likewise, some situations require an incident safety officer to respond after members are on the scene, such as a working fire or at the request of the incident commander.

Fire departments should develop response procedures for an incident safety officer that is on call or designated to respond. Examples could be as follows:

a. Commercial fires
b. Multiple alarm fires
c. Fire fighter injury or fire fighter transported for treatment
d. Hazardous materials incident
e. Active shooter incident
f. Technical rescue incident
g. At the request of the IC

Safety is the command staff position that should be staffed most often. The IC should be at the command post, but the incident safety officer must monitor all areas where fire fighters are conducting operations. Therefore, it is virtually impossible for one person to effectively handle both functions (IC and safety officer) at a large or long-term incident. Some incidents may require assistant safety officers. When more than one safety officer is assigned, assistant safety officers are generally appointed to monitor specific areas, with all assistant safety officers reporting to the incident safety officer responsible for monitoring and advising the IC regarding overall incident safety. Seldom is more than one safety officer assigned at a structure fire. Appointing an incident safety officer does not relieve the IC of responsibility for safe operations. The IC has overall responsibility for scene safety, tactical-level management supervisors are responsible for the safety of personnel assigned to them, company officers are responsible for the safety of their crew, and each member has a safety responsibility to look after themselves and other members of their crew.

It is imperative that the incident safety officer focus on the overall operation and major risks to fire fighters. There is a tendency for safety officers to spend too much time concentrating on minor safety infractions while failing to recognize potentially deadly hazards. As an example, a safety officer

checking the proper donning of protective clothing might lack the time necessary to properly evaluate structural stability.

The safety officer is charged with monitoring all areas where fire fighters are working at the incident. This monitoring provides the incident safety officer with a unique overview of the entire operation that should be shared with the IC as the safety officer completes a tour of the incident scene. It also helps the safety officer review the incident action plan in regard to safety issues, including checking the NIMS organizational chart at the command post.

Liaison Officer

Liaison is the point of contact for non-fire agencies that are assigned to non-operations functions. If more than one person or unit from an agency responds to the incident, one person from that agency should be appointed to communicate with the **liaison officer** (or the IC if liaison is not staffed).

This person would represent the agency and be responsible for coordinating other people and units from the agency that are working to resolve the incident. The agency representative must have the authority to speak on behalf of his or her agency, although sometimes the agency representative will find it necessary to consult with the agency prior to making a commitment. In most cases, the police department reports to the liaison officer when this command staff position is staffed. However, if law enforcement personnel are playing a major role in meeting incident action plan objectives, they would then report to a general staff position within the operations section or directly to the IC.

If the IC does not staff liaison, representatives from all responding agencies will communicate directly with the IC. Most structure fires do not require the separate staffing of the liaison officer position. However, when many agencies are assisting and all of them are reporting to the IC, it will become extremely difficult for the IC to effectively communicate with emergency resources operating at the scene. When this is the case, the IC should staff the liaison position.

Public Information Officer

The **public information officer (PIO)** disseminates information to the public, usually via the media. This officer provides critical information to the community regarding protective actions that need to be taken as well as information of general interest. There should be only one public information officer per incident to ensure continuity of information being released to the public. An efficient public information officer will monitor communications at the command post and keep abreast of the current status of the operation. It is essential that all communications to the public concerning emergency operations related to the incident be from the IC or from the public information officer in consultation with the IC. The IC must clear all information relayed to the public. This is particularly critical when there are injuries or fatalities, or when there is suspicion that a crime has been committed.

If there is a high level of public interest in the incident, the media will require a substantial amount of time, thereby making it difficult for the IC to maintain communications with operating units and the media. In this case, a public information officer should be appointed or other general NIMS positions handed off to allow the IC time to communicate with the media. Designating a media area **Figure 1-7** that provides a view of the fire or operating equipment assists in developing good relations with the media and discourages them from seeking out prime locations on their own. The IC or public information officer must prepare to provide factual information during news briefings while not disclosing information that should not be made public. The media may turn to unreliable sources of information if the fire department fails to provide the information they need for their report. Many departments now take advantage of social media sites to better address the public's need to know. The media may also have

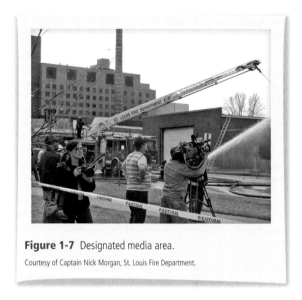

Figure 1-7 Designated media area.
Courtesy of Captain Nick Morgan, St. Louis Fire Department.

information useful to the IC such as a fire reported in a nearby community.

General Command Staff Considerations

There is debate as to whether the command staff positions should be counted in the IC's span of control. A valid argument can be made for not counting these positions against the recommended three to seven reporting positions in the IC's span of control, as they serve roles similar to those of command aides or adjuncts.

Some departments preassign staff officers to command staff positions. For example, the training chief may function as the safety officer at the scene. These individuals receive special training and often develop useful contacts prior to the incident. As an example, the liaison officer might attend local emergency planning committee (LEPC) meetings to meet individuals representing nongovernment agencies. The department public information officer could develop a rapport with media personnel.

There is value in preassignment. However, preassignments may delay the staffing of important positions. To ensure that positions such as safety officer and liaison officer are filled, it is often necessary to make temporary assignments until the preassigned personnel arrive at the scene. Whoever is assigned to fill a command staff position must be thoroughly trained and qualified to carry out the responsibilities of the appointment.

Large fire events often require the staffing of all three command staff positions. In most cases, a person can be assigned to each position. Only one person should be assigned to each command staff position. If the safety, liaison, or public safety officer position involves more work than can be effectively completed by a single person, assistant safety, liaison, and public information officers can be assigned, but they are subordinate to the incident safety, liaison, and public information officer. At times it is possible to combine assignments. If the workload is not too great, one person may be able to manage both liaison officer and public information officer positions.

However, when delegated, the safety officer should always be a separate assignment. Command staff officers and section chiefs are frequently located at the command post, whereas personnel who are assigned as safety officers should be mobile and able to monitor the entire operation.

General Staff

General staff positions provide a pyramid-structured hierarchy capable of coordinating and controlling the incident. The IC is at the top of the pyramid with five possible organizational layers between the IC and individual responders **Figure 1-8**. Rarely would it be necessary to use all NIMS hierarchical levels and all subordinate units at a structure fire. Positions are staffed when incident conditions require separate management of geographic areas or functions.

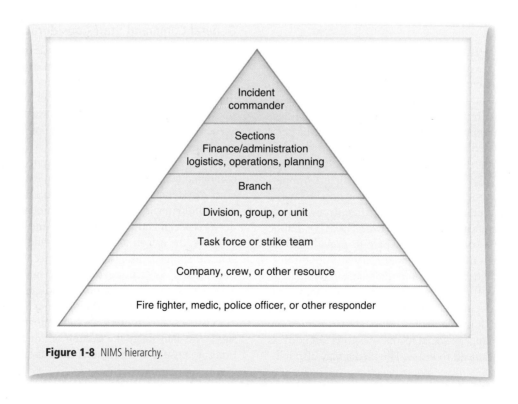

Figure 1-8 NIMS hierarchy.

Sections

Section chiefs report directly to the IC **Figure 1-9**. As priorities are established, five separate sections can be assigned (finance, logistics, operations, planning, and information/intelligence). Each section can have several subordinate units. Only in the most extreme circumstances would all section positions be staffed at a structure fire. When established, section chiefs can further delegate management authority for their area of responsibility as required. If necessary, functional units may be established within the section by the section chief. Similarly, each functional unit leader can assign individual tasks within the unit as needed. Units subordinate to each section are staffed depending on incident requirements. When the section leader requires assistance in handling section duties, a deputy section chief is assigned.

The five main sections of the NIMS organization under the direction of the IC are as follows:

- Finance/administration section
- Logistics section
- Operations section
- Planning section
- Information/intelligence section

It is important to note that these positions are *sections*, not sectors.

Finance/Administration Section

The **finance/administration section** is unlikely to be separately staffed at the scene of a structure fire. Finance/administration, as the name indicates, manages financial matters and/or provides administrative services; the functions provided by this section are shown in **Figure 1-10**. There are some cases in which the fire department is able to recover expenses, such as incidents involving hazardous materials. For instance, the Cincinnati Fire Department was able to recover expenses after the previously referenced BASF chemical company fire. However, it is important to remember that, even at smaller incidents where the finance/administration section is not staffed, these duties are important and remain the IC's responsibility. In these instances, finance/administration duties might be limited to completing the National Fire Incident Reporting System form that could be later used by the property owner and/or insurance company. However, at larger, more complex incidents, detailed

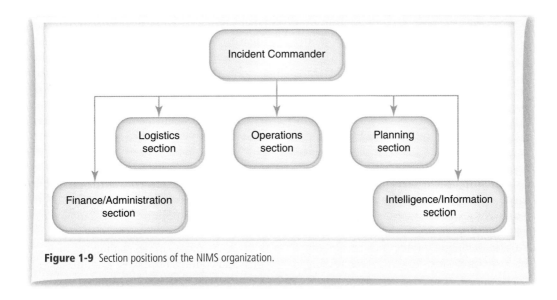

Figure 1-9 Section positions of the NIMS organization.

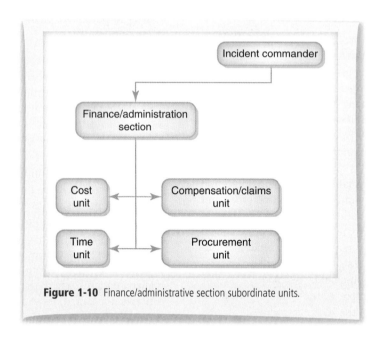

Figure 1-10 Finance/administrative section subordinate units.

documentation might be necessary to secure reimbursement for damaged equipment, payroll for responding or back-fill personnel, file compensation for personal injuries, or many other reasons. At these incidents, the finance/administration position should be staffed.

Logistics Section

The **logistics section** locates and provides the materials, equipment, supplies, and facilities required to support incident operations. The logistics section chief can be thought of as the supply sergeant or quartermaster. Most of the resources needed to support incident operations are already on the scene or provided by additional fire companies; therefore, the logistics section is seldom staffed at small structure fires. However, there are circumstances where supplies or facilities are needed. For example, a fire situation where companies are on the scene for an extended time would require rehabilitation with food, water, sheltered areas, and possibly toilet facilities. The logistics section chief would be responsible for bringing these resources to the scene.

One of the most important logistics section units is the communications unit. This unit assists in setting up the communications network, as well as providing and maintaining communications equipment. Some departments have a communications specialist respond to the scene of major or long-term emergencies with spare radios, batteries, and chargers. The need to staff the communications unit increases as the communications network becomes more complex.

The ground support unit would provide items such as fuel at the scene of a structure fire. **Figure 1-11** shows the logistics section subordinate units. Some charts further subdivide the logistics section into service and support branches.

Planning Section

The **planning section** chief is the information manager. Planning should be the most frequently implemented section position. The planning chief gathers information about the incident, tracks resources, and assists the IC in developing the incident action plan and contingency plans. Planning should be one of the first sections staffed during a major incident.

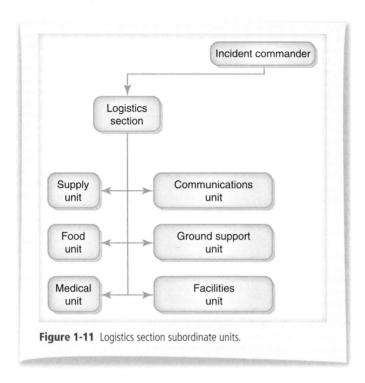

Figure 1-11 Logistics section subordinate units.

When transferring command, consider assigning the previous IC as the planning section chief, as this person should know the status of the operation as well as the number and location of incident resources. A major part of the planning section chief's responsibilities is tracking and documenting incident status and on-scene resources, oftentimes referred to as "sit-stat" and "re-stat," or situation status and resource status, as shown in **Figure 1-12**.

The demobilization unit prepares and implements a plan to return personnel and resources to service as an incident is being brought under control. This is an important planning section function when large geographic areas have been stripped of resources.

The documentation unit would collect incident information, which is important to a formal incident evaluation. The documentation unit becomes more important when there will be a formal investigation or study following the incident and the finance/administration section is not staffed.

Most technical specialists report to the planning section, but they could report to other sections, command staff, or the IC as needed.

The Chief's Aide Position

Closely related to planning is the position of chief's aide. At one time, most large city fire departments provided chief officers with a chief's aide, but due to budget constraints, many cities eliminated this position. Some departments called the chief's aide a driver or chauffeur, which made it more difficult to defend retaining the position. As administrators examined fire departments in search of ways to cut the budget, the "chauffeur" position was a likely candidate for elimination.

The chief's aide drives the chief's vehicle. Driving is a small but important part of the duties assigned to an aide. Having an aide drive the chief officer to the fire scene allows the chief officer to concentrate on communications as well as accessing pre-incident plans and other sources of information to

Figure 1-12 Planning section subordinate units.

initiate the size-up process. At the scene of an incident, the chief's aide can manage command tasks so the chief officer can concentrate on developing an incident action plan and deployment of forces. The chief's aide can also be assigned to do a complete walk-around of the fire building, handle communications, manage the accountability system, maintain the NIMS organizational chart, and research information sources.

The aide's duties could be described as the planning section or as a planning section subordinate unit, such as re-stat or sit-stat. Departments who have chief's aides strongly defend keeping the position and usually refer to the chief's aide by a title more suitable to the duties they perform, such as field incident technician, staff aide, or command adjunct.

The Phoenix Fire Department assigns each battalion chief a fire captain to perform the duties of a field incident technician. This program not only provides the battalion chief assistance at the command post, but also provides invaluable command post experience for future battalion chiefs. Until recently, the Los Angeles City Fire Department assigned a staff aide to each division and battalion chief. Budget constraints resulted in the elimination of 7 of the 16 on-duty staff aides. During times defined as "high-hazard days," staff aides are hired back for battalion chiefs not assigned a staff aide. In addition, Los Angeles City dispatches a command post company to all multiple-unit responses.

A chief's aide can assist the IC in organizing and coordinating a safe and effective operation. Departments that do not have members assigned as chief's aides often assign responding personnel to these duties at the incident scene. This temporary assignment is highly recommended as an alternative. However, this temporary assignment is not equal, as command assistance will be delayed and will lack the team concept developed between a chief officer and a regularly assigned aide. To overcome some of the shortcomings of using a member from a responding company, it is recommended that these selected members attend additional training as a chief's assistant.

It does not make much sense to assign a fire fighter to assist the IC if the IC has to stop and train the fire fighter while the fire is in progress. This is definitely not a position for the newest member of the crew. Additional training might include maintaining an ICS chart. Attendance at a formal Planning Section Chief class may fulfill this requirement. Knowledge of communications procedures and the ability to communicate effectively on the radio are important attributes for an aide. The member should have a good working knowledge of firefighting tactics to enable him or her to give the IC accurate information that can be applied to the strategy selected. The aide must know the procedures and methods used to communicate with other agencies and/or the media to allow the aide to serve as a liaison or public information officer at emergencies. The department should maintain a list of required qualifications for the chief's aide position.

Information/Intelligence Section

The intelligence/information section provides information related to the prevention of criminal activity, including terrorism, or the apprehension of perpetrators of criminal activity. In a fire situation, this section could be used to determine the fire cause or lead to the apprehension of an arsonist. Depending on the situation, the information/intelligence function could be located in one of four places on a NIMS organizational chart. It could be part of the general staff. When information gathered is vital to the safety and welfare of the community and when responders will play a significant role in the investigation of a terrorist or criminal event or the event involves specialized and technical information, the information/intelligence section chief would be assigned. When the information gathered will affect firefighting tactics, this function could be located within the operations section. If the information is neither critical nor sensitive, or if the information has little tactical benefit but does provide some real-time information to the IC, this function could be part of the command staff. However, in most situations the appropriate location of the information/intelligence function would be within the planning section.

Operations Section

The **operations section** manages all tactical operations such as search and rescue, extinguishment, and provision of medical care to victims. When the IC hands off the operations section, the operations chief will make all tactical assignments and control all resources working to resolve the incident. Delegating the operations section allows the IC to focus on the overall strategy and other command functions. However, it is important to note that staffing the operations section means the IC no longer has direct control over operating companies, so there is another layer of organization between the IC and operating units. Typically, the IC delegates authority for operations only at large-scale incidents. Senior-ranking officers sometimes assign the person they are relieving of command to the operations section to provide the lower-ranking officer with experience in managing larger operations, while the senior officer maintains overall control as the IC.

The operations section generally has the most resources, and the most complex organization, possibly requiring several hierarchical levels to effectively maintain a reasonable span of control. The hierarchical levels within the operations section are shown in **Figure 1-13**.

The operations section is capable of controlling as many as 625 companies or crews with a five-to-one span of control **Figure 1-14**. At the maximum recommended span of control (seven to one), a total of 2,401 companies or crews can be effectively controlled. Numerous large-scale disasters, including hurricanes, wildland fires, and earthquakes have demonstrated that, when properly implemented, NIMS includes all of the tools necessary to effectively manage the largest and most complex operations. NIMS has the capacity to manage the largest imaginable structure fire.

Which sections the IC should hand off depends on the incident. Some officers may be reluctant to hand off the operations section, as they lose direct contact with tactical operations. Some departments require the highest-ranking officer at the scene to assume command. Other departments give a higher-ranking

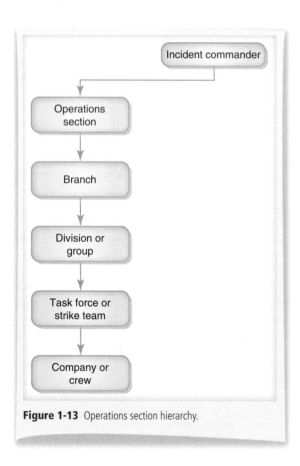

Figure 1-13 Operations section hierarchy.

officer the option to assume command of the incident, to fall into the command structure under the direction of the present IC, or to serve as a senior advisor.

Allowing a lower-ranking officer to retain command shows confidence in the officer and provides experience in commanding a large-scale situation. However, a better option might be assigning the lower-ranking officer as the operations section chief. By handing off operations to the lower-ranking officer, the senior officer can assume command and direct the general strategy while the lower-ranking officer gains experience in handling dynamic, tactical assignments.

Command staff and section chiefs make up the **incident management team (IMT)**, which provides the necessary staff and line functions for an incident.

Incident management teams for most structure fires are made up of members from the responding department or from departments that normally provide mutual aid assistance. However, FEMA is encouraging the development of all-hazards incident management teams on a regional and state level. Regional, state, or national incident management teams would include members certified to function in the various command staff and section positions, much like the "Red-Card" system used by the forest service when combating large-scale wildland fires. The U.S. Fire Administration provides NIMS organization and other incident management team training to facilitate certification as an all-hazards member of an incident management team. Some regional teams establish training and certification standards identified within the region. Regional teams are particularly valuable when the responding department does not have personnel specifically qualified to staff a command staff or section position. As an example, members of regional departments could attend training and educational sections to qualify as safety officers. A qualified safety officer could then be dispatched to any report of a working structure fire within the region.

A more detailed description of each section's responsibilities and subordinate units is provided in the U.S. Homeland Security publication *National Incident Management System*.[8]

Throughout an incident, from inception to conclusion, the duties of each section are the ultimate responsibility of the IC. The IC chooses to retain the duties or delegates sections that will assist in managing the incident, depending on the incident and resource requirements.

Branches, Divisions, and Groups

In using NIMS, the first management assignments by the IC will normally be to hand off geographic and functional areas of responsibility at the **division** or **group** level. A division is in charge of units working in a geographic area, whereas a group is in charge of units assigned to complete a function (e.g., search and rescue).

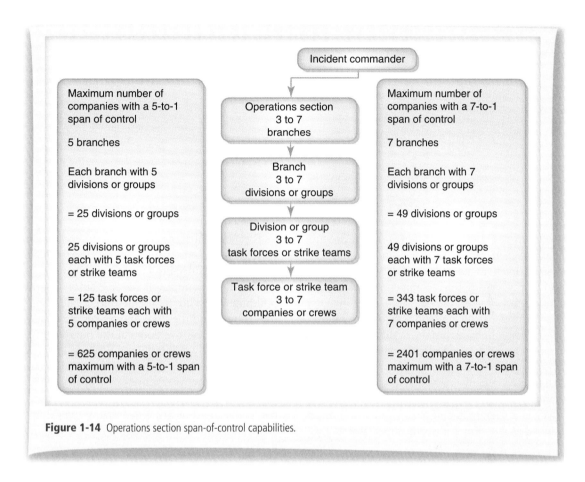

Figure 1-14 Operations section span-of-control capabilities.

As the fire progresses from a first-alarm response to extra alarms, there may be a need to sector the operation to maintain a reasonable span of control or to better manage specific functions of the operation. For example, a fire department responds to an apartment fire with three engine companies, two truck companies, and a heavy rescue company. The command structure would be as shown in **Figure 1-15**.

Engine 1 represents the first-arriving engine company, which arrives nearly simultaneously with Truck 1 from the same fire station. Given the fact that an interior operation is indicated, the **two-in/ two-out rule** must be followed with at least two members assigned as the rapid intervention crew.

NOTE

Two-in/two-out rule: Fire fighters working inside the hazard area must work in crews of at least two people (two-in). During the initial stages of an operation, the two people working in the hazard area must be backed up by at least two people outside the hazard area (two-out) who are properly equipped and immediately available to come to the aid of the inside crew. The two-out rule does not apply to subsequent arriving units. The two-out team is often referred to as the initial rapid intervention crew, or IRIC, which is replaced by the rapid intervention crew, or RIC, as additional units arrive.

Texture: © Eky Studio /Shutterstock, Inc; Steel: © Sharpshot/Dreamstime.com

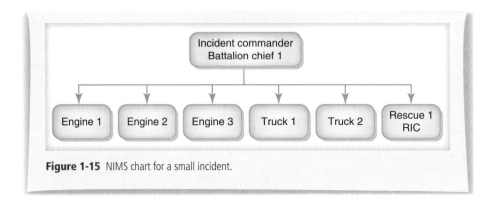

Figure 1-15 NIMS chart for a small incident.

Notice that a rapid intervention crew and five fire companies are reporting to the IC in Figure 1-15. This small incident exceeds a five-to-one span of control and is nearing the recommended maximum span of control of seven to one.

Many fire departments send a second chief officer to every fire. In many instances, this second chief officer is assigned to supervise the fire floor. Using Figure 1-16 as an example, for a fire on the second floor, a truck company is assigned to search and rescue, and two engine companies are assigned to lay an attack and backup line. If the second chief officer is assigned as Division 2, division supervisor is now in direct control of the critical second floor operation with a three-to-one span of control, and the IC's span of control is reduced to four to one (Figure 1-16).

Another example is a five-story apartment building with a fire on the first floor extending into the hallway and the floor above. Rescue operations are needed on all five floors of the building. The IC would normally assign forces working in the same geographic area to a division to reduce the span of control. For example, suppose the apartment building has a fire on the first floor with three engine companies and a truck company working that area. An engine company and a truck company are

Figure 1-16 NIMS chart for a small incident with division supervisor assigned to fire floor.

working on the second floor, and an engine company is assigned to each of the remaining three floors. An additional truck company is assigned to ventilate the building. The incident could be managed in several different ways, but with 11 companies working at the scene, there is a definite need to assign tactical-level management units. There are many correct ways to organize the incident; one of many correct methods is depicted in the organizational chart in **Figure 1-17** .

All companies working on the first floor should be under the supervision of a single branch or division. Companies performing search and rescue tasks could be assigned to the search and rescue group. However, having units in the same general area reporting to different supervisors is not a good practice. In most cases, geographic rather than functional

assignments are preferred. Most fire officers share this preference because communications are improved and the division supervisor can personally monitor the status of operations, thus eliminating the need for status reports within the division.

The three companies working on the upper floors in the apartment building example could report directly to command, but functional assignments also assist the IC in managing this important component of the incident action plan. The supervisor in charge of the search and rescue group will manage search and rescue assignments on all three floors, reassigning or requesting assistance as needed to complete the search and rescue on upper floors. The IC's incident action plan calls for removing the occupants. Therefore, the search and rescue group supervisor applies the tactics necessary to achieve command's objectives

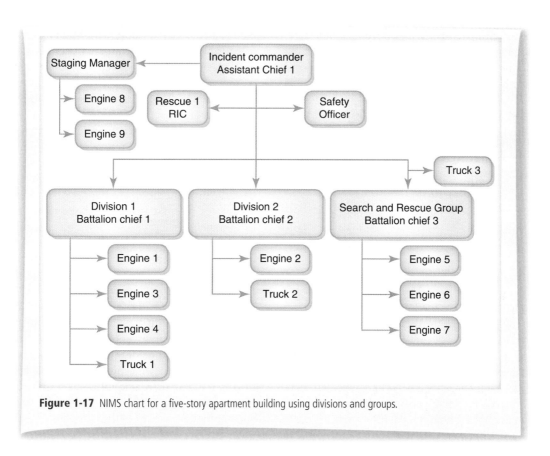

Figure 1-17 NIMS chart for a five-story apartment building using divisions and groups.

and issues the necessary task-level assignments to companies within the search and rescue group. As an example, the company assigned to the third floor finds very little smoke and can quickly complete the primary search on that floor. In this example there is the possibility that many occupants who need assistance remain on the top floor, and the primary search is difficult due to heavy smoke conditions. The search and rescue group supervisor could then reassign the company from the third floor to assist personnel assigned to the top floor. After the primary search is complete, the group supervisor could assign companies to the secondary search, avoiding repeat searches by the same companies.

All tactical operations within this group are managed by a single supervisor, who organizes and coordinates all activities related to the search and rescue on the designated floors. The IC is regularly informed of progress being made but does not assign or reassign companies, as these responsibilities are assigned to the group or division supervisors. Truck 3 is assigned to ventilate as a single resource reporting directly to the IC.

Fallacy	Fact
A floor level is always named as a division.	A floor level is named as a division only if a tactical-level management unit supervisor is assigned to direct multiple units on that level.

Steel: © Sharpshot/Dreamstime.com

Branches can be used in place of divisions or groups, but this is not the recommended way to initially reduce the span of control. Branches are generally reserved for operations beyond the span of control of a single division or group, or when a contingent of units from another agency are working together, such as a law branch for police personnel. In the apartment building example, if nine companies were being used

on the first floor, Division 1 would become Branch 1, with a search and rescue group or groups and a suppression group reporting to the branch director.

A large medical or police contingent could also dictate the establishment of branches. If a separate agency provides medical services, a branch assignment is even more likely for the medical component of the operation.

Nine companies working on the first floor under a single branch, as well as the entire medical operation under a medical branch, are shown in the organizational chart in **Figure 1-18**.

In 1990 when the BASF fire occurred, ICS terminology included the term "finance section" instead of the present "finance/administration section." The IC began by dividing the BASF operation geographically, but there were 15 companies working on the Dana Avenue side of the fire. The IC recognized that the recommended span of control was being exceeded; therefore, the Dana Avenue side was further subdivided into the Dana Division East and Dana Division West. Even with this subdivision, there were eight companies reporting to Dana Division East, which exceeds the recommended seven-to-one maximum span of control. To reduce the span of control, the Dana Avenue side could have been assigned to a branch with three or more subordinate tactical-level management units. Also worth noting is that not everyone reporting to a branch needs to be at the division/group level. Individual companies, task forces, or strike teams could also report directly to a branch. **Figure 1-19** shows a revised version of the organizational chart for the Dana Avenue side of this fire. There are many correct ways to organize this operation, but the geographic sectoring allowed division supervisors, many of whom were company officers, to manage the operation in their area of responsibility using face-to-face communications. Note that a rapid intervention crew is not shown on the chart. At the time of this fire, rapid intervention crews were not in widespread use. If this fire were to occur today, there should be at least two rapid intervention crews, with one stationed at the front and another at the rear of the complex.

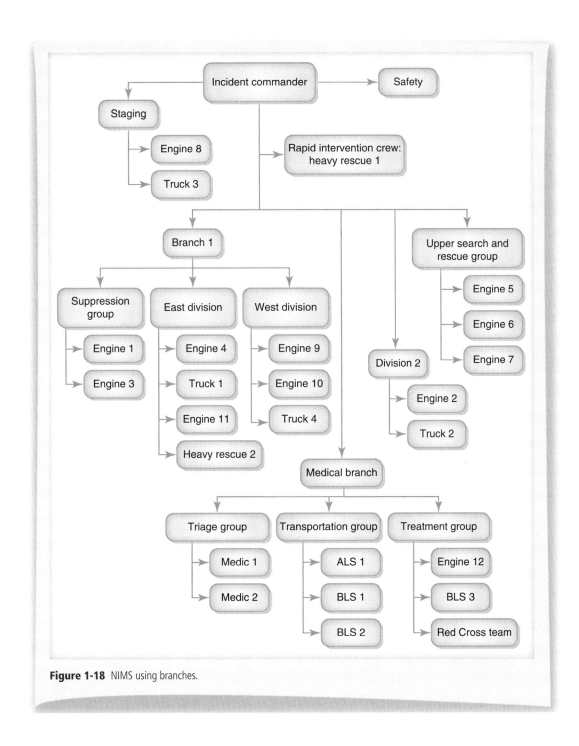

Figure 1-18 NIMS using branches.

Incident Summary

On the afternoon of July 19, 1990, an explosion and fire at the BASF Chemical Plant in Cincinnati, Ohio, required the services of 25 fire companies, emergency medical services (EMS) units, police, and other agencies. The ICS was used to coordinate and control activities at the scene. The rear of the plant bordered Norwood, Ohio. As companies arrived, they found severely injured employees, a massive fire, and continuing explosions. Two plant employees were killed, and 88 people were injured. The 16 buildings on the BASF complex were heavily damaged or destroyed; 161 non-BASF buildings were damaged. Command was transferred to the Ohio EPA on the evening of July 20, 1990. Cleanup efforts with fire companies in a stand-by mode continued for nearly 5 months.

BASF chemical plant operations after explosion.

Flames: © Jag_cz/ShutterStock, Inc.; Paper: © silver-john/ShutterStock, Inc.

Figure 1-19 Dana Avenue NIMS organization.

A Word About Sectors

Prior to the formal implementation of NIMS, the term "sector" was commonly used to identify geographic or functional tactical-level management units at the division or group level. NFPA 1561, 2014 A.3.58.2 and 1561A.3.58.3 describe groups and divisions and add: "Based upon current federal guidelines, agencies currently using the term *sector* are encouraged to change terminology to become NIMS compliant for their incident and daily operations by using the term *division* for reference to organizational components based on geographic area and *group* for organizational components base on function."

Task Force and Strike Team

The use of **task forces** and **strike teams** is an additional way to reduce the span of control, thus reducing the communications load at an incident **Figure 1-20** . Task forces can be any combination of resources, whereas strike teams must be resources of the same type. Strike teams are a common way to organize large numbers of mutual aid companies or private sector resources. Some states define strike teams as a specific number of companies to be used as a statewide resource. A strike team could be any number of units of the same type, but the jurisdiction defining the strike team must predetermine the number of units. As an example, some state

and local governments assign five engine companies and a strike team leader to an engine company strike team. They also address minimum staffing, communications, and other factors. Task forces are much more flexible and are more likely to be formed at the time of the incident. A common application of the task force concept would be two engine companies and a truck company working as a team.

Naming Geographic Areas of a Building

Notice that in Figure 1-18, Branch 1 is in charge of operations on the first floor. The upper search and rescue group is distinguished from the east and west groups assigned to search and rescue on the first floor. Both of these assignments are inside the building.

When naming areas outside the building, the normal designation is to name the sides of the building using an alphabetical designation, where side A is designated as the front of the building. Continuing clockwise around the building, side B is the side left of side A, side C is the rear of the building, and side D is the right side of the building, as shown in **Figure 1-21** . This commonly used system is useful when assigning locations, providing progress reports, and undertaking other fire-ground communications. However, there is sometimes confusion

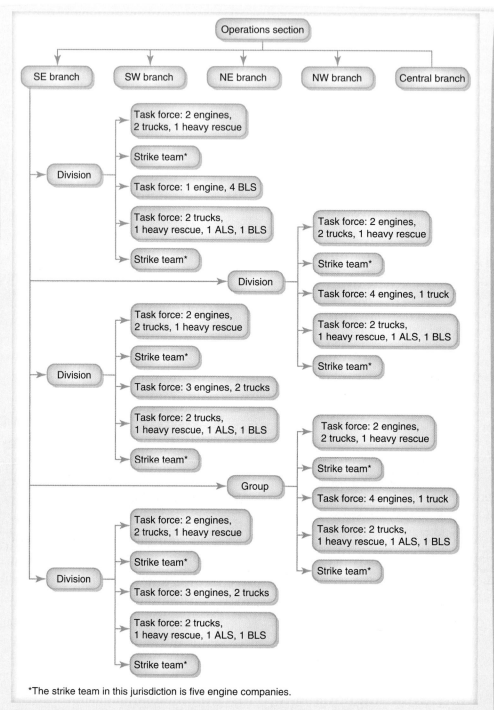

*The strike team in this jurisdiction is five engine companies.

Figure 1-20 Use of task forces and strike teams to increase organizational capabilities.

Figure 1-21 Naming the sides of a building.

regarding which side to name as side A. As shown in Figure 1-21, there are two entrances to the hotel part of the building, and the command post is at a good location for a fire in the hotel viewing both of these sides. Some departments name the side of the building where the command post is located as side A. Using this approach, either the Walnut Street or the 13th Street side could be named side A. Confusion when establishing the location of fire companies or when providing progress reports, maydays, and so on can negatively affect operations, resulting in unsafe and ineffective tactics. Naming the address side of the building as side A eliminates much of the confusion. In Figure 1-21, the office building address is 1210 13th Street and the hotel's address is 1220 13th Street. Therefore, side A of either building is the 13th Street side, regardless of the location of the command post. The two attached buildings in Figure 1-21 could have addresses on different streets. The hotel could possibly have a Walnut Street address. If operations were being conducted

in both the office and hotel buildings, there could be confusion as to which side of the building is Side A. In this case, the IC should resolve the issue by designating Side A.

In **Figure 1-22**, geographic areas of responsibility are designated using street names. The chart in the BASF Chemical Company incident summary shows the deployment of companies for a fire that involves several buildings and operations occurring inside the complex between buildings. The normal alphabetic naming system would be difficult to apply in this situation where sides A, B, C, and D are not well defined.

It is best to use an alphanumeric naming system that names sides of the building A, B, C, and D and floors within the building by floor designation (e.g., Division 21 would be supervising multiple companies assigned to the 21st floor of a high-rise building). Naming tactical-level management units by floor number is intuitive and highly recommended. When one or more companies are operating on a floor

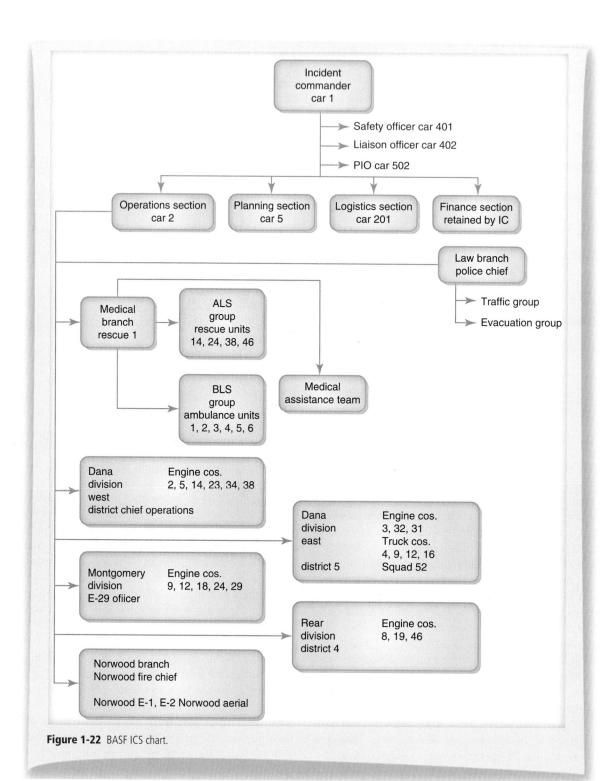

Figure 1-22 BASF ICS chart.

level, they retain their company name if a tactical-level management unit is not assigned. Division 21 indicates that one person is managing all operations on the 21st floor. When working with mutual aid departments, everyone should agree on a common system, include it in the SOPs, and use it during joint training exercises and tactical simulations as well as at the emergency scene. Whatever system you select, use it consistently and at every incident to make sure everyone is comfortable with the labeling/identifying system.

Communications

Communication is the lifeblood of any command system. It is impossible to coordinate and control a successful operation without effective communications.

During major emergencies, it is highly recommended that a chief officer with command experience and familiar with day-to-day operations at the dispatch center be assigned to manage operations and make resource allocation decisions at the dispatch center, such as:

- Reallocating companies to vacated stations
- Relocating and staffing extra apparatus
- Requesting mutual aid and outside agency assistance
- Recalling off-duty members
- Converting companies (e.g., truck companies to engine companies)
- Suggesting the dispatch of special equipment

Some general rules for communication at the incident scene include the following:

- Use face-to-face communication whenever possible.
- Provide mobile communication to units that are remote from the command post.
- Ensure that all operating units have some form of communication that ultimately relays information to the command post.
- Place representatives of agencies on different frequencies at the command post to handle communications within their agency.

- Follow the command organization structure, facilitating **unity of command**.
- Keep the number of radio channels used by any supervisor to an absolute minimum, preferably no more than two.
- Do not clutter radio channels with unnecessary transmissions.
- Use standard terminology.
- Use clear text; do not use 10-codes.

IC Rule of Engagement for Fire Fighter Safety

Maintain frequent two-way communications, and keep interior crews informed of changing conditions.

Reproduced from: © 2011, IAFC Safety, Health and Survival Section. All rights reserved.

Texture: © Eky Studio /Shutterstock, Inc; Steel: © Sharpshot/Dreamstime.com

Within the NIMS structure is a communications unit that reports to the logistics section chief. At large-scale incidents, establishing a communications unit is critical. This unit is responsible for establishing a communications plan and installing, procuring, and maintaining the communications equipment at the scene. At incidents where a formal communications unit is not established, preplanned resources should be used.

All radio communications at the incident should be in plain English. It is best to avoid codes because they can be easily misunderstood. Furthermore, not all personnel operating at large-scale incidents from different jurisdictions or agencies will be familiar with the codes.

All communications should be confined to essential messages. It is important to avoid clogging the airways with nonessential messages and to keep the frequency available for vital transmissions. Each transmission should be brief, not a long narrative of what is transpiring within the area.

Radios are the most common mobile communications tool for emergency operations, but other communications devices offer advantages and should be considered during long-term or communications-intensive situations. In addition to radios,

Figure 1-23 Mobile command vehicle exterior.

Courtesy of Bill Strite, Cincinnati, Ohio.

the following means of communication are often used at the scene:

- Mobile data terminals, mobile data computers, and/or tablet computers
- Face-to-face communication
- Messengers
- Telephones (cellular, satellite, and hardwired)
- Public address systems

Computers

General communication improvements in the private sector are being adapted to fire service use. Computers are now commonplace on fire apparatus and in some departments. Tablet computers are replacing laptop computers. Many departments now own or have access to field communications units that are often equipped with the latest innovations in communication **Figure 1-23** and **Figure 1-24**.

Face-to-Face Communication

The most effective form of communication is face to face; however, running an entire large-scale incident using only face-to-face communications would be all but impossible, as this means of communication is limited to those within earshot of the communicator. Face-to-face communication may be the best option when the information is sensitive and confidential, such as when discussing the fire cause or the death or serious injury of a fire fighter.

Figure 1-24 Mobile command vehicle interior (A and B).

Courtesy of Bill Strite, Cincinnati, Ohio.

Messengers

In some cases, during radio failure, messengers can be used to communicate orders and directions to units or tactical-level management unit supervisors who are in remote locations, but this approach will result in delayed communication. However, this can be an efficient way to relay messages between the command post and nearby section chiefs or to nearby units working under a division supervisor.

Cellular Telephones

Cellular telephones (cell phones) are often used to contact outside resources and are now being used by many departments as part of the fire-ground communications network. In some cases, cell phones have been found to be more reliable in large and high-rise buildings than radios. One problem with cell phones, however, is that the system can be overwhelmed or damaged during a major disaster. In some cases, media representatives and others at the scene tie up the available cell phone channels, locking in open lines by keeping their initial call active. During an extended incident, the cell phone companies might be able to provide additional channel capacity and resources. A fire department representative should communicate with cell phone service

providers to determine their emergency capabilities as part of the emergency planning process.

Satellite Telephones

Satellite phones are becoming more common in the fire service. Satellite service overcomes some of the shortcomings of local cell towers and repeaters. Signals are transmitted between the sender and receiver through a satellite as opposed to a cell tower. Cell towers can be damaged in catastrophic weather conditions and are subject to willful destruction by terrorists or others. Satellites are less vulnerable to these events; however, satellite service can be interrupted by storms.

Hardwired Communications Systems

One of the best examples of a hardwired system is a building emergency communication system. Over the years, this type of system has evolved from handsets, similar to a regular telephone, which can be carried by the fire fighter and plugged into phone jacks located throughout the building. Some newer systems have permanently mounted "fire phones" located in selected areas such as stairwells and elevator lobbies. These systems are

Incident Summary

After the Oklahoma City bombing in 1995, the local cell phone providers were able to provide additional capacity in the area of the Alfred P. Murrah Federal Building by bringing in portable cellular transmitters called cells on wheels. They also provided the many agencies operating at the scene with cell phones at no cost. These phones were programmed with priority numbers that could grab a channel. Several locations were established where telephone personnel were available around the clock to repair phones and provide charged batteries. Use of the cellular system was instrumental in coordinating the overwhelming logistics that this incident involved.

Data from: Ed Comeau and Stephen Foley, "Oklahoma City, April 19, 1995," NFPA Journal, July/August, 1995.

Flames: © Jag_cz/ShutterStock, Inc.; Paper: © silverjohn/ShutterStock, Inc.

designed primarily for use by fire fighters; however, they can also be part of an occupant emergency evacuation plan.

Building Communications Systems

Security and maintenance personnel often have radio systems designed to work within their building. Larger complexes may even have a cache of radios reserved for fire department use that are kept in the building's fire command center. In addition, a group assigned to work in different areas or floors can use installed landline telephones as a means of communication. As an example, the supervisor of a search and rescue group conducting searches on multiple floors could be stationed at a telephone or telephone bank in a safe location; units can report in and receive their next assignment via telephone, reducing radio traffic.

With the approval of the authority having jurisdiction, bidirectional antennas are finding their way into new high-rise buildings in lieu of hardwired systems. These radio systems are designed to improve radio reception inside the building. Local fire departments should test this type of system to ensure it works as designed before approving the system. Also consider the possibility of future radio replacements; will new radios be able to communicate effectively using the installed bidirectional antennas?

Public Address Systems

Public address systems are useful for mass notification within a building. As an example, in a high-rise incident, the public address system can be used to direct occupants to remain in place or to exit via a specific stairwell. Although rarely used, public address systems can also be effective as a means of communicating instructions to fire fighters.

Computer Systems

Computers provide nonverbal communication avenues that free "airtime," as well as provide additional information, especially in graphic form. Mobile data terminals, status message terminals, and tablet computers are systems commonly used to relay digital information. These terminal-based systems are dependent on a data network. Mobile data computers and other available devices combine dispatch communications with onboard data storage, thus reducing the dependence on a wireless network and the dispatch center. Use of on-scene computer systems in providing preincident plans is discussed further in the *Procedures, Preincident Planning, and Size-Up* chapter.

Radio Designation for the IC

For purposes of radio communications, the IC is simply referred to as "command," a designation that should be reserved solely for the IC. Some departments refer to each division, group, or branch using the term command, such as "rear command" or "interior command." This practice is strongly discouraged, as the use of multiple command designations results in confusion regarding who is in overall command of the incident.

The IC is designated as "command," whether this person is the chief of the department or a fire fighter working as an acting officer. Department SOPs must define the term command and specify how it is to be used and by whom. By using well-defined terminology, it is possible to eliminate confusion during command transfers. Operating units, divisions, and branches might not remember that Chief Jones assumed command, but they do recognize the importance of orders being issued by command and the need to communicate through channels to command.

Communications Network

It is crucial that the communications network support unity of command within the NIMS organization. **Figure 1-25** shows an example communications network following the NIMS structure. It is important to keep the communications network as simple as possible. It is not necessary to designate scores of radio channels even if you have equipment

Figure 1-25 Communications network following ICS organization.

capable of designating a large number of channels. However, there are sometimes good reasons to assign multiple channels to reduce interference among various units involved in the incident.

In Figure 1-25, the IC has retained control of the operations section and is using Channel 2 to communicate with all field units. Notice that the IC is using only two radio channels: one to communicate with dispatch and the other for field communications. The IC is communicating with the law branch by cell phone.

With the many divisions and fire companies using Channel 2, radio discipline would be imperative, and this single channel would be very crowded. If multiple channels were available, the communications network could be expanded, with each division using a separate channel to communicate with units under their supervision and Channel 2 or another channel to communicate with the Interior Branch. For example, command could communicate with the Interior Branch on Channel 2, and the Interior Branch could use Channel 4 to communicate with

Divisions 1, 2, 3, and 4 **Figure 1-26**. Each division could then be assigned a different channel. More airtime is now available for units communicating with the Interior Branch; however, the communications network is much more complex, and there are more communications layers between the IC and companies reporting to the divisions within the Interior Branch. Units subordinate to the Interior Branch can no longer listen to messages being transmitted by Command.

Interoperability is a concern at incidents involving multiple agencies or more than one jurisdiction. Technology provides a means of communicating on many radio channels. With proper planning and equipment, it is possible to communicate with all agencies at the scene, as well as agencies providing support from off site. Even if a department has the capability of communicating with all responding agencies, communicating via multiple radio channels is problematic. It is essential that the IC and all

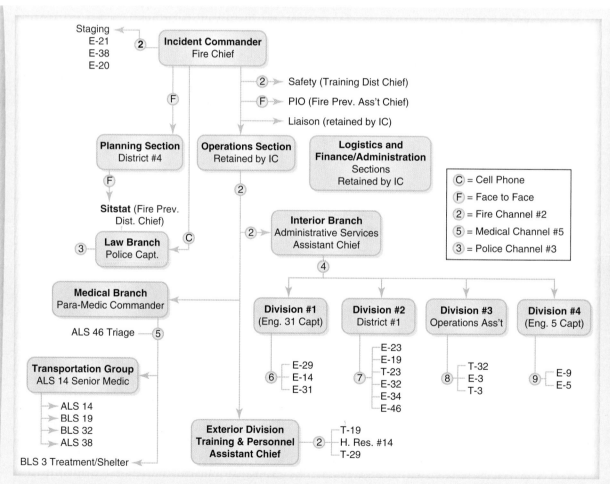

Figure 1-26 Revised communications network.

units that report to the operations section limit the number of radio channels used to reduce the possibility of missing critical messages from operating companies or crews.

Rule of Engagement for Fire Fighter Survival

Constantly monitor fire-ground communications for critical radio reports.

Reproduced from: © 2011, IAFC Safety, Health and Survival Section. All rights reserved.

Texture: © Eky Studio /Shutterstock, Inc; Steel: © Sharpshot/Dreamstime.com

Although it is possible to monitor several radio channels using radios as scanners, it is very difficult to conduct two-way communications on more than two channels. There are several ways to address this problem:

- Place a representative of the assisting agency with their communications equipment at the command post. The agency representative manages and communicates with responders from their agency.
- Assign a liaison officer to communicate with other agencies.

- Assign the logistics section to communicate with agencies that are providing equipment and supplies.
- Direct communications to technicians who retransmit critical messages to the IC. This role could be part of the communications unit within the logistics section.

Also consider using alternative communications methods such as computers, cell phones, or satellite telephones when communicating with other agencies.

Common terminology is the cornerstone of effective interoperability. It is essential that common terminology be established for any management system, especially one that will be used in joint operations by numerous and diverse participants under emergency conditions. This commonality of terms is particularly important in the areas of organizational functions, resources, and facilities.

Positions within the NIMS organizational framework are named, using the same names for the function regardless of the type of emergency. Line and staff positions remain the same, even though incident situations and objectives may differ significantly.

Chapter Summary

- The incident commander (IC) must manage resources, ensure fire fighter safety, and target priorities.
- The National Incident Management System (NIMS) enables the IC to delegate tasks and make decisions at both major and minor incidents.
- NIMS provides common terminology and operational assignments for all agencies. NIMS should be used by the entire response community, not only fire departments.
- When using a unified command structure, plan development is shared, but one operations chief should direct all field units.
- Strong command presence is critical to fire fighter safety.
- The initial report should confirm incident location, type, and conditions.
- Good communications and clear initial reporting are essential to command.
- When a chief officer assumes command, a stationary command post must be established.
- Officers assuming command must provide an updated status report based on observation and reconnaissance from units at the scene.
- When command is transferred to another officer, even another company officer, it should be formalized at a stationary command post.
- To avoid confusion, command should not be transferred between company officers unless there is a compelling reason to do so.
- The person assuming command must communicate with the previous IC to determine both situation and resource status of the operation in progress.
- The IC develops the incident strategy. Branch directors, division supervisors, and group supervisors develop tactics within the overall strategy.

- The command post is wherever the IC is located.
- Tactical-level management components should maintain a reasonable span of control and coordinate specific geographic and functional operations.
- Staging allows the IC to control access to an incident scene while deploying resources in a safe and effective manner.
- NIMS organizational structure develops in a modular fashion based on the type and size of the incident.
- IC is the one position that must always be staffed within a NIMS organization.
- A division is in charge of a geographic area, whereas a group is in charge of a functional area.
- Use geographic rather than functional assignments whenever possible to improve communications and eliminate the need for status reports within the division.
- Strike teams combine a set number of resources of the same type.
- Task forces are more flexible than strike teams and more likely to be formed at the time of the incident.
- Areas of responsibility are designated by using an alphanumeric system whenever possible.
- All radio communications should be in plain English, not codes, and confined to essential messages.
- Communications systems must support unity of command.

Key Terms

10-codes Numeric codes used to communicate predefined situations or conditions. For example, "10-4" generally means "finished

Flames: © Dni/Dreamstime.com; Steel Texture © Sharpshot/Dreamstime.com communicating." The use of 10-codes is discouraged.

branches The tactical-level management units immediately subordinate to sections in NIMS hierarchy. Units are subordinate to the logistics, finance/administration, and planning sections. Divisions and groups are subordinate to the operations section. Branches are used to reduce the span of control at very large operations or to manage a particular function/agency.

chain of command An organizational structure establishing a line of authority and responsibility along which orders and instructions are passed—e.g., incident commander to operations section to branch director to division supervisor to company officer to fire fighter, or a reply to a higher level in reverse order

division The tactical-level management unit in charge of a geographic area.

finance/administration section The section that tracks and provides financial and administrative services required to compensate people or organizations providing goods and services at the incident scene.

freelancing The act of performing tasks outside the incident's organizational structure.

group The tactical-level management unit in charge of a function.

incident commander (IC) The individual responsible for all incident activities, including the development of strategies and tactics and the ordering and the release of resources.

incident management system (IMS) A system that defines the roles and responsibilities to be assumed by responders and the standard operating procedures to be used in management and direction of emergency incidents and other functions.

incident management team (IMT) The incident commander and appropriate command and general staff personnel assigned to an incident.

safety officer A member of the command staff responsible for monitoring and assessing safety hazards and unsafe operations, and for developing measures for ensuring personnel safety.

liaison officer A member of the command staff, responsible for coordinating with representatives from cooperating and assisting agencies.

logistics section The section that obtains needed supplies, equipment, and facilities.

National Incident Management System (NIMS) A system mandated by HSPD-5 that provides a consistent, nationwide approach for federal, state, local and tribal governments; the private sector and nongovernmental organizations (NGOs) to work effectively and efficiently together to prepare for, respond to, and recover from domestic incidents, regardless of cause, size, or complexity.

operations section The section that manages all tactical units deployed at an incident scene.

planning section The section that gathers and evaluates information, assists the incident commander in developing the incident action plan, and tracks progress; also tracks resource status.

public information officer A member of the command staff responsible for interfacing with the public and media or with other agencies with incident related information requirements.

rational decision making (RDM) A form of decision making in which input is obtained from diverse sources and all options are carefully considered.

Flames: Dix/Dreamstime.com; Steel texture: © Sharpshot/Dreamstime.com

recognition-primed decision making (RPD) A form of decision making in which the incident commander must decide on the proper course of action with limited information available in a relatively short period of time.

single command A command structure in which one person is designated as the incident commander (IC). This person is responsible for the development and implementation of the incident action plan. The IC can delegate staff and command positions as needed to assist in command and control functions.

span of control The number of people reporting to a supervisor. The span of control should not exceed seven people reporting to a single supervisor under emergency conditions. As an example, a fire captain supervising an apparatus operator and three fire fighters would have a four-to-one span of control.

strike team A set number of resources of the same kind and type that have an established minimum number of personnel—e.g., five engine companies. Strike teams always have a leader (usually in a separate vehicle) and have common communications among resource elements.

task forces Any combination of resources that can be temporarily assembled for a specific mission. All resource elements within a task force must have common communications and a leader. Task forces should be established to meet specific tactical needs and should be demobilized as single resources. A typical task force would be two engine companies and a truck company under the supervision of a chief officer.

two-in/two-out rule A rule mandating that fire fighters working inside the hazard area work in crews of at least two people (two-in). The two people working in the hazard area must be backed up by at least two people outside the hazard area (two-out) who are properly equipped and immediately available to come to the aid of the inside crew. This rule applies to the first-arriving unit. As more units arrive, the two-out team is replaced by a rapid intervention crew.

unified command An application of the incident command system (ICS) that allows all agencies with jurisdictional responsibility for an incident or planned event, either geographical or functional, to manage an incident or planned event by establishing a common set of incident objectives and strategies.

unity of command A pyramidal command system ensuring that no one reports to more than one supervisor.

Flames: © Dra/Dreamstime.com; Steel texture: © Sharpshot/Dreamstime.com

Challenging Questions and Activities

1. Develop an organizational chart for a room-and-contents fire in a two-story, single-family, residential building. Assume that all first-alarm units are on the scene performing the following tasks:

   - *Engine 1*, per SOP, has secured a water supply via a forward lay using 5-in. (125-mm) hose. The four-person company is split into two crews. Crew 1 is the initial rapid intervention crew (IRIC) stationed outside the structure and consists of the hydrant fire fighter and apparatus operator. The officer and other fire fighter make up Crew 2 and are attacking the fire with a 1¾-in. (45-mm) hose line.
   - *Engine 2* relieves Engine 1's IRIC, with all four members on Engine 2 forming the rapid intervention crew (RIC). The Engine 1 hydrant fire fighter joins the officer and other fire fighter on the interior, and the apparatus operator continues operating the pump.

   - *Engine 3* secures a second source of water. The apparatus operator remains with the apparatus while the three remaining members advance a 1¾-in. (45-mm) hose line to the second floor.
   - *Truck 1* divides into two crews: Crew 1 is conducting a primary search in the second floor bedrooms; Crew 2 sets up positive-pressure ventilation then conducts a primary search of the first floor.
   - *Battalion Chief 1* assumes command at a command post on the A–D corner of the building after driving past the A–B side.

2. A large six-story warehouse is fully involved in fire, and the IC has decided on a defensive attack for the warehouse. There are occupied exposure buildings on each side of the warehouse, as shown in **Figure 1-27**. The IC has assigned companies to protect and evacuate the exposures on both sides of the warehouse. Develop an ICS organizational chart for this operation as it is presently being conducted.

Flames: Drx/Dreamstime.com; Steel texture: © Sharpshot/Dreamstime.com

Figure 1-27 Warehouse fire and exposures.

Endnotes

1. National Fire Protection Association. *NFPA 1561: Standard on Emergency Services Incident Management System and Command Safety*. Quincy, MA: 2014.

2. U.S. Department of Homeland Security. *National Incident Management System*. December 2008.

3. National Fire Protection Association. *NFPA 1500: Standard for Fire Department Occupational Safety and Health Program*. Quincy, MA: 2013

4. Code of Federal Regulations, Title 29, Section 1910.120, *Superfund Amendments and Reauthorization Act of 1986. Hazardous waste operations and emergency response*.

5. National Interagency Incident Management System. *Incident Command System: Operational System Description* (ICS 120-1). Stanford, CA: FIRESCOPE Program; 1981.

6. Isner, M. S., Klem, T. J. *NFPA Fire Investigation Report: World Trade Center Explosion and Fire*, New York, NY: National Fire Protection Association; February 26, 1993.

7. National Fire Protection Association. *NFPA 1521: Standard for Fire Department Safety Officer*. Quincy, MA: 2015.

8. U.S. Department of Homeland Security. *National Incident Management System*. December 2008.

Procedures, Preincident Planning, and Size-Up

LEARNING OBJECTIVES

- List the kinds of operations that should be covered by standard operating procedures (SOPs).
- Discuss the relationship between SOPs, preincident plans, and size-up.
- Examine the relationship between SOPs, equipment, and training.
- Compare standard operating *procedures* to standard operating *guidelines*, explaining the role of a "reasonable person" clause.
- Evaluate response district resources and challenges as part of the preincident planning process.
- Describe how enhanced security precautions affect response time and tactics.

Flames: © Dra/Dreamstime.com; Steel texture: © Snapshot/Dreamstime.com; Chapter opener photo: Courtesy of Scott Doran, ConocoPhillips Alaska

LEARNING OBJECTIVES (CONTINUED)

- Articulate the main components of preincident planning, identifying a step-by-step process used to gather, review, store, and retrieve preincident plan information.
- Recall major steps taken during size-up, and identify the order in which they will take place at an incident.
- Recognize the relationship between preplanning and construction characteristics common to a community.
- Analyze construction methods during everyday responses and while surveying buildings under construction and demolition.
- Examine compatibility issues and usefulness of computer programs in preincident planning.
- Construct a priority chart of buildings to be preplanned by occupancy type.
- List factors to be considered during size-up, and briefly define and explain the significance of each factor.
- Demonstrate (verbally and in writing) knowledge of fire behavior and the chemistry of fire.
- Recall the basics of building construction and how they interrelate to preincident planning and size-up.
- Define and explain the difference between *occupancy*, *occupant*, and *occupied*.
- Explain the size-up process in the chronological order in which information is received.
- Evaluate a specific fire department's SOPs.
- Prioritize occupancies to be preincident planned in a specific jurisdiction.
- Create a preincident plan drawing and narrative.
- Perform an initial size-up based on limited information.
- Apply size-up factors to a fire situation, and categorize factors as primary or secondary.

Case Study

Commercial laundry room in a nursing home.

Courtesy of Bernard J. Klaene

On April 29, 2008, a fire was reported in a difficult-to-access laundry room of a four-story skilled nursing home. The first-due company had previously conducted a preincident planned tour of the building and discussed possible scenarios. Knowledge of the building layout resulted in rapid ventilation and extinguishment. A defend-in-place tactic was successful in protecting occupants and avoided a difficult building evacuation.

1. How does a preincident plan improve size-up?
2. What effect does knowing the building have on response effectiveness?

Data from: NFPA 1620.B11 Nursing Home Fire.

Introduction

This chapter will discuss the importance of **preincident plans**, **standard operating procedures (SOPs)**, and their relationship to size-up. Every strategist recognizes the importance of gaining intelligence. Military officers go to great lengths to understand the factors that are involved in all operations, as well as the particulars of the specific battleground and enemy. In the fire service, factors that apply to fire-ground operations can and should be outlined in SOPs. The battle particulars are a result of preincident planning and analysis. As with military operations, the better the preincident plan, plan analysis, and SOPs, the fewer decisions that will need to be made in the heat of battle, which allows the incident commander (IC) to focus on important incident-related factors when developing an incident action plan.

SOPs, preincident plans, and incident-specific information are interrelated and are all important components of the size-up. The IC who is faced with a large or complex building and no preincident plan or applicable SOPs is at a great disadvantage.

SOPs and a good size-up are necessary prerequisites in the development of an incident action plan. For larger and more complex properties, preplanning is essential.

Important information about specific buildings can be obtained and analyzed in advance through preincident planning. This information is critical to the fire officer when making strategic and tactical decisions at the incident scene. Some things will be known only after the incident occurs, as is addressed in this chapter (see the Analyzing the Situation Through Size-Up section). Prefire information and incident-specific information are evaluated in terms of factors related to the operational priorities of life safety, extinguishment, and property conservation. Subtopics are structural conditions, resource needs, contents, occupancy type, access limitations, water supply, and special challenges.

The whole complex process of evaluating an incident and developing an incident action plan must take place in a few minutes. To accomplish this, it is necessary to focus on the major (primary) factors during the initial plan development process. As more information becomes available and the IC has time to reevaluate, the size-up information and the incident action plan should improve.

Developing Standard Operating Procedures

SOPs are general guidelines to be used at all structure fires or fires in similar occupancies. SOPs recognize similarities, covering areas such as those listed in Table 2-1 .

SOPs address any operation that can be handled using a standard approach. SOPs will vary among departments. The types of property to be protected, resources available, equipment, and training, among other factors, guide the promulgation of SOPs. This variability among departments does not mean fire departments should not share procedures. Instead, it is necessary to learn all that can be learned from the experiences of others. However, the final procedure must reflect department-specific needs and resources. As an example, a high-rise SOP would be of no value to a small rural department that does not respond to high-rise buildings within their jurisdiction or mutual aid response area. On the other hand, the SOP for an automobile fire would be similar in both a rural and a large urban fire department.

It is necessary to revise SOPs regularly. An automobile fire SOP written prior to the widespread use of alternate fuels, electric, and hybrid vehicles would be wholly inadequate. SOPs should be modified to include tactics related to recent research regarding fire behavior, venting, **flow path**, and staffing.

Large urban departments can quickly summon hundreds of fire fighters to the scene of a large structure fire. A smaller rural jurisdiction would, by necessity, handle a fire in the same size and type of structure differently, by calling on surrounding communities through mutual or automatic aid agreements. SOPs for handling the same type of fire may vary from department to department.

Table 2-1	Areas Covered in SOPs
Area	**Related Elements**
Command	• Implementation of the incident management system • Designation of person in charge • Establishment of command post • Command transfer • Communications • Development of an incident action plan • Rapid intervention crew • Accountability
Staging	• Exterior staging • Interior staging
Water supply	• Attack pumper • Forward or reverse hose lays • Pumper/hose wagon • Tender shuttle • Water relay
Special occupancy operations	• Places of assembly • Educational occupancies • Health care occupancies • Board and care occupancies • Hotels • Dormitories • Detention facilities • Mercantile • Business • Storage • Industrial • High-rise • Large-area buildings
Company operations	• Truck company • Engine company • Quint/quad company • Rescue company • Medical unit • Special functions (e.g., collapse, communications, hazardous materials, foam)
Special operations	• Technical rescues • Hazardous materials • Natural disasters • Electrical fires • Active shooter • Civil disturbances • Terrorism
Private fire protection	• Standpipe • Sprinkler • Non-water-based fire protection systems • Alarm systems

© Ely Studio/Shutterstock, Inc

Although SOPs must be written specifically for the department, there is a need for regional planning in writing procedures. It makes little sense to have specific procedures for the incident management or accountability systems limited to just one department. There is a good probability that these systems will be critical during large-scale incidents when other regional departments are on the scene providing mutual aid. Chapter 4 of NFPA 1720, *Standard for the Organization and Deployment of Fire Suppression Operations, Emergency Medical Operations, and Special Operations to the Public by Volunteer Fire Departments*,[1] requires procedures to ensure that an effective firefighting force can communicate with each other and conduct emergencies in a uniform manner.

Purpose of Standard Operating Procedures

Good department SOPs and preincident plans take the guesswork out of those first few precious moments on the fire ground. Company officers have predesignated assignments that allow them to take immediate action within the overall plan or, in some cases, to stage, awaiting orders from command.

In the absence of SOPs and preincident plans, the IC is so busy assigning companies that little time is left to collect the information necessary for the development of an incident action plan. Freelancing may also occur while the IC gathers the intelligence necessary for the formulation of an incident action plan, including the development of strategy and tactics. Either case results in inefficient and possibly dangerous operations.

With this in mind, it is best to provide specific SOPs for the first-arriving engine company. As an example, some departments require the first-arriving engine company to secure a source of water and advance a hose line to attack the fire. The SOP for the second-arriving unit could be specific or general in nature, depending on whether the first-arriving unit is directed to secure a source of water. If SOPs do not require the first-arriving engine company to secure a

water supply, the second-arriving engine company is generally assigned to supply water. Departments may assign the second-arriving engine to conduct staging, supply water, act as the rapid intervention crew, or provide general guidelines if the second-arriving company is not assigned specific tasks. The first-arriving truck company will generally choose from a menu of possible tasks, as enumerated in the *Company Operations* chapter. Truck company operations must be coordinated with the engine company attacking the fire.

During large-scale operations, the IC is faced with numerous, complex decisions. SOPs provide a structure for the decision-making process, including answering the questions of who makes what decisions, at what level of command, and from where? The National Incident Management System (NIMS) does an excellent job of establishing a command structure and describing the roles of various players at the incident scene and must be included in SOPs.

Relationship of Standard Operating Procedures to Training and Equipment

Regarding SOPs, NFPA 1500, *Standard on Fire Department Occupational Safety and Health Programs*,[2] states:

> **4.1.2** *The fire department shall prepare and maintain written policies and standard operating procedures that document the organization structure, membership, roles and responsibilities, expected functions, and training requirements, including the following:*
>
> (1) *The types of standard evolutions that are expected to be performed and the evolutions that must be performed simultaneously or in sequence for different types of situations*
> (2) *The minimum number of members who are required to perform each function or evolution and the manner in which the function is to be performed*

(3) *The number and types of apparatus and the number of personnel that will be dispatched to different types of incidents*

(4) *The procedures that will be employed to initiate and manage operations at the scene of an emergency incident*

5.1.2 *The fire department shall provide training, education, and professional development for all department members commensurate with the duties and functions that they are expected to perform.*

5.1.6 *The fire department shall provide all members with training and education on the fire department's written procedures.*

5.1.10 *Training programs for all members engaged in emergency operations shall include procedures for the safe exit and accountability of members during rapid evacuation, equipment failure, or other dangerous situations and events.*

NFPA 1500 makes a definite statement that training must be commensurate with SOPs. If new equipment is placed in service, the availability of the equipment makes a statement that fire fighters are expected to use the equipment safely. SOPs must outline how and under what circumstances the equipment is to be used.

When a new tactic or SOP is introduced, the equipment necessary to perform the new tasks must be procured and fire fighters must be trained to safely implement the tactic. Seldom-used equipment and tactics require frequent training, as fire fighters will lack field experience. For example, if a fire department decides to purchase wind control devices to counteract the effect of wind-driven fires in high-rise buildings, as discussed in the High-Rise Ventilation study conducted by the National Institute of Standards and Technology (NIST), it would be necessary to make inquiries regarding available wind control devices, purchase the selected equipment, then train all members who could conceivably be ordered to deploy the wind control device. Training should include the conditions under which the wind control device should be used along with actual hands-on training in deploying the device.

Recently, Underwriters Laboratories published research related to vertical ventilation.[3] The International Association of Fire Chiefs (IAFC) Safety, Health, and Survival Section and the International Association of Fire Service Instructors (IAFSI) issued suggested guidelines to be implemented based on the findings of this and other research.[4] One of their recommendations states, "Water should be applied to a fire as soon as possible and from the safest location." This tactic is often referred to as **softening the target.**

In this case, it is not necessary to purchase new equipment; you can employ the suggested tactics using existing nozzles and ventilation equipment. However, the suggested fire attack is much different from methods currently used by most fire departments. In many departments, this represents a radical departure from current fire attack methods. There could be considerable resistance to implementing these tactics, and there is a definite need for training that not only demonstrates the new method, but also a need to explain why the change is being made. Any department taking this advice to "attack the fire as soon as possible and from the safest location" should carefully study the associated research and note limitations prior to implementing these new tactics. After training members on the proper techniques, it is best to objectively evaluate fires where new procedures are implemented within the department as well as in other jurisdictions.

Consideration should be given to SOPs when developing new apparatus and equipment specifications. As an example, if department SOPs specify that the first-arriving pumper should pull past the main entrance to allow room for the first-arriving truck company, preconnected hose lines should be specified in locations that permit efficient deployment to the rear of the apparatus.

There is a relationship between SOPs, equipment, and training, as illustrated in **Figure 2-1** . Any time

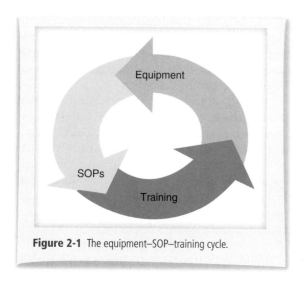

Figure 2-1 The equipment–SOP–training cycle.

new equipment is introduced or a new procedure is written, the entire cycle must be completed.

The Standard Operating Guidelines Controversy

Within the fire service, there is some discussion about whether to refer to standard operating *procedures* as standard operating *guidelines*, *general operating guidelines*, or another, similar name. Semantics can be important, but not in this case. It is far more important to have written procedures or guidelines than it is to argue about the name. Many in the fire service believe that calling procedures "standard operating guidelines" or "general operating guidelines" somehow changes their strength, in that a guideline may be considered less stringent than a procedure.

In reality, procedures are guidelines, and guidelines become procedures through practice. If a task is always performed the same way, that becomes the procedure for doing that task. Likewise, most procedures should be taken as general guidelines. Situations will arise in which the SOP does not fit the circumstance and the fire officer would be expected to modify actions accordingly. For example, an SOP that requires the first-arriving truck company to

ventilate would not be appropriate if the fire had self-vented, an attack hose line was not in position to control the fire and flow path, or there was an obvious defensive operation in progress. Some SOPs are always in effect, such as an SOP stating that fire fighters *must* wear self-contained breathing apparatus (SCBA) in contaminated atmospheres.

Because a degree of flexibility is sometimes required in following procedures, a "reasonable person" clause should be incorporated in the procedures manual. Essentially, a statement is made that procedures are to be followed but that fire fighters should follow a reasonable course of action when confronted with a situation in which modification of the procedure is appropriate. When a decision is made to modify procedures, the noncompliant unit or member is required to communicate the variance and be prepared to justify the modification.

Whether you call them procedures or guidelines, write them down, train to them, and use them consistently.

Evaluating Response District Resources and Challenges

The most valued fire suppression resource in any community is the fire fighter. A community with the best apparatus, equipment, and water supply is underprotected when the fire department is understaffed. Volunteer departments expend a good deal of time and effort in recruiting and retaining an adequate fire force. Paid departments struggle with budget issues that often result in understaffing. Determining the staffing level is based on many parameters. Over the years, many studies have been conducted in an attempt to establish the staffing needed for a structure fire. Prior to the advent of NFPA 1710, *Standard for the Organization and Deployment of Fire Suppression Operations, Emergency Medical Operations, and Special Operations to the Public by Career Fire Departments*,[5] there was no widely accepted staffing standard. NFPA 1710 specifically addresses staffing at residential fires. A study conducted by the National Institute of Standards and Technology verified the residential staffing requirements.[6] A more detailed discussion of staffing levels established in the NFPA 1710 standard can be found later in this chapter and in the *Fire Fighter Safety* chapter.

Apparatus requirements are also important when evaluating community fire protection. Apparatus needs vary depending on the water supply, height of buildings in the area, number of fire fighters per rig, and other factors.

Response Time

The ability to save lives and property is directly related to response time. Distance is but one factor affecting response time. A major highway with a higher speed limit will allow apparatus to travel faster compared to a winding two-lane road with numerous uncontrolled traffic signals. However, interstate highways may become extremely congested during rush hour. Likewise, flat terrain is easier to traverse with heavy apparatus than hilly topography. Railroad crossings, construction areas, weather, and other factors can also adversely affect response time. Response time will be discussed at length in the *Fire Fighter Safety* chapter.

The street layout, numbering system, and addresses are sometimes confusing. Natural and manmade barriers often break up street continuity. As an example, a street that previously continued for a considerable distance may be divided into different sections due to the construction of a limited-access highway. A hillside, river, or other obstruction may also result in an intermittent street system. When the street is divided into different sections, the address usually continues in the next street segment. Street address may end in the 1900 range at the bottom of a hill and continue in the 2100 range at the top of the hill. Continuous streets sometimes change name as they progress through different areas or jurisdictions. Likewise, streets laid out in a square pattern may have angular streets intersecting at various points. The street pattern in Washington DC has many such angular streets in what is basically a square street layout. A rapid response requires that fire fighters know the street layout and addresses in their response area. Having good street maps and/or a global positioning system (GPS) available is essential; however, accessing this information often results in a delayed response. Therefore, it is critical that training programs include knowledge of street addresses and response routes, as well as alternate routes. Fire stations in communities with confusing numbering systems or complicated street layouts often place large area maps in the fire station alarm room or apparatus floor to provide a quick reference.

Water Supply

Water supply is a critical part of community fire protection. SOPs and tactics must be modified to match the water supply. Fire fighters should know the capabilities of the water supply system available in their response area. In addition, it is necessary to have an alternate water supply strategy. Even the best water systems can be damaged due to an accident or natural disaster. Earthquakes have resulted in the loss of major sections of municipal water supplies, such as the damage in San Francisco during the Loma Prieta earthquake in 1989. Snow or floods can make hydrants inaccessible. In 2012, a conflagration occurred in Breezy Point, Queens, New York, during Hurricane Sandy after the storm surge had flooded streets and fire hydrants.

Many cities shut down mains or hydrant valves during floods to protect the potable water supply from contamination. During extremely cold weather, water mains have been known to freeze. Static sources of water can freeze, dry up, or be inaccessible. As water supplies improve, there is a tendency to abandon water tender or drafting operations even in areas with plentiful access to static sources of water. During Hurricane Sandy, pumpers drafted water from flooded streets to combat the Breezy Point conflagration. Preincident planning involves considering normal operations as well as the potential disruption of normal resources.

A well-maintained water distribution system with closely spaced fire hydrants and large flow capacity provides the ideal water supply. Commercial facilities that are remote from a reliable water supply may be required to store firefighting water on site. These on-site water supplies may be in the form of aboveground storage tanks or reservoirs. In most cases, these properties will be required to store an adequate water supply based on the hazard.

When all or part of the community is not serviced by a water distribution system with closely spaced fire hydrants, special provisions are necessary, such as:

- Working from the apparatus tank
- Setting up water relays
- Implementing tender shuttles

Evaluate these options during preincident planning by first determining the **rate of flow** (see the *Offensive Operations* chapter for rate-of-flow calculations). Once the rate of flow has been determined, the next step is to evaluate various sources of water and their accessibility and reliability.

Of the three options listed above, supplying water from the apparatus tank is the fastest and least labor intensive. The preferred strategy is an offensive attack whenever it can be carried out safely and effectively. Setting up water relays and shuttles requires considerable time and sometimes results in a delayed attack. Conducting an offensive attack when the risk-versus-benefit analysis indicates otherwise is dangerous and unnecessarily places fire fighters in jeopardy. Most residential fires are within the capacity of a 1¾-in. line and, if properly applied, can be extinguished, or at least controlled, with the water carried on the pumper.

Using the onboard water supply to attack the fire is sometimes referred to as an "attack pumper" tactic. Some departments apply this tactic at all residential fires. When an attack pumper deployment is used, the second-arriving pumper usually supplies the first-due pumper with water from the hydrant system or other source. This multiphase approach may be the best tactic when water supplies are located far away from a small structure. When an attack pumper deployment is used, fire fighters must be aware of the limited water supply provided by the apparatus tank and limit the time of attack to allow enough water to protect their retreat if necessary. This process should be addressed in the department training program. Whether the first-arriving pumper uses an attack pumper tactic or not, make provisions to ensure a reliable, adequate, and continuous water supply.

When the distance from the fire to the pumper supplying hose lines to attack the fire exceeds pump, hydraulic, or hose limitations, an alternative water supply layout is needed. Alternative layouts are more complicated and time consuming. If the water supply is from a static source, a drafting operation will also be required.

Relay pumping is one way to provide a continuous water supply when the water supply is a considerable distance from the fire. Relay pumping involves moving water from the source (static or hydrant) through multiple pumpers to the apparatus operating at the fire scene. This can be very effective in providing 1000 gallons per minute (GPM) (3785 liters per minute [L/min]) or more over a long distance when large-diameter hose or multiple relay lines are used. **Figure 2-2** shows a relay from a static source. Nearly every community has static sources of water, and the fire department should provide training and equipment necessary to set up a relay from a static water supply.

After evaluating the water supply during preincident planning and deciding on a water relay, many departments place permanent signs on the access road showing the position for each pumper in the relay **Figure 2-3**.

Communities sometimes rely on shuttling water via **tenders** (also referred to as a tanker or mobile water supply apparatus). High flow rates are also possible when using tenders, provided the water source can meet the demand, the source is relatively close to the fire scene, and several large-capacity tenders are available. However, moving the same amount of water the same distance, as in the water relay example shown in Figure 2-2, is generally more difficult when using tenders.

Figure 2-2 Water relay from a static source.

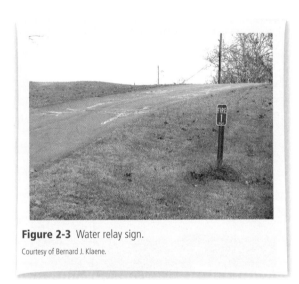

Figure 2-3 Water relay sign.

Courtesy of Bernard J. Klaene.

NFPA 1142, *Standard on Water Supplies for Suburban and Rural Fire Fighting*,[7] Appendix C-11, provides the following formula to determine the amount of water a tender can supply in GPM (L/min):

$$Q = \frac{V}{A + T_1 + T_2 + B} \times k$$

where:

Q = maximum continuous flow capabilities (GPM [L/min])

V = tank volume of the mobile supply apparatus in gallons (or liters)

A = time in minutes for the mobile water supply apparatus to drive 200 ft (61 m), dump water into the drop tank, and return 200 ft (61 m) to the starting point

T_1 = time in minutes for the mobile water supply apparatus to travel from the fire to the water source

T_2 = time in minutes for the same mobile water supply apparatus to travel from the water source back to the fire

B = time in minutes for the mobile water supply apparatus to drive 200 ft (61 m), fill the mobile water supply apparatus at the water source, and return 200 ft (61 m) to the starting point

k = 1 for pressure/vacuum water supply apparatus and 0.9 for other water supply apparatus

Setting up a tender operation using a single 3000-gallon (13,638-L) tender with a pressure/vacuum system transporting water 1 mile (1.6 kilometers [km]) from the source to the fire traveling at an average speed of 35 miles per hour (56 km/hr) yields:

V = 3000 gallons (13,638 L)
A = 3 minutes
T_1 and T_2 = 2.35 minutes each
B = 4
k = 1
Q = 256 GPM (1164 L/min)

Determine flow requirements and the best method to meet the calculated flow during preincident planning. Obviously, the calculations necessary to determine the flow rate from each tender and additional calculations to determine how many tenders are needed would be very difficult to carry out at the incident scene. It is best to try various tactics prior to the fire, as well as determining the time it takes to fill and empty available tenders. The A, B, V, and k variables in the formula can be known in advance. The T_1 and T_2 components can also be measured or calculated prior to the incident.

If a water shuttle operation is implemented, the IC should establish a water supply group. The water supply group supervisor would have the responsibility of determining the number of tenders needed to meet the flow requirement established by the IC. Adding more tenders will increase the continuous flow capacity. However, other variables, such as congestion at fill and dump sites, will affect the total flow. Available fill and dump sites, as well as routes to and from these sites, should be preplanned. If not preplanned, the water supply group supervisor will need to determine routes and dump and fill sites at the incident scene. When determining routes, it is generally best to set up separate pathways to and from the fire, taking traffic, road surfaces, bridge and road weight limits, and other factors into consideration.

Many serious accidents have occurred when vehicles not specifically designed as water tenders were used to transport water at the fire scene. Tank trucks that are not designed to transport water present unique safety hazards and should be inspected and approved before being pressed into service. This inspection should include, but not be limited to, adequate construction, vents, and discharges large enough to quickly dump and fill water, and the presence of baffle plates. The previous contents of the tank should be determined and the tank interior checked for residual product. An added consideration is the weight capacity of the tank truck. Water, at 8.3 pounds per gallon (1.0 kg/L), is heavier than many other liquids that are transported via tank truck. A full load of water could easily overload a tank truck that normally transports products lighter than water. An important safety consideration is the vehicle's center of gravity.

Construction Methods

Most fire districts have a predominance of buildings of similar construction. A great deal can be learned about properties in your jurisdiction during EMS or other responses. This is particularly true for buildings that are not formally preincident planned such as single-family residential buildings. In many cases, buildings in the same area were built as a subdivision and share common traits such as having inground basements, walk-out basements, cellars, or no basements. As an example, the 93 houses in the subdivision shown in **Figure 2-4** were built in the 1970s.

Figure 2-4 A subdivision in which the homes all have similar construction features.

Courtesy of Bernard J. Klaene.

Every house on the four streets that make up this subdivision has a basement, but none of the basements has direct access to the outside. Every house has a garage and lightweight **wood-truss** roof construction, but none was originally constructed using lightweight wood-truss or wooden I-beam floors support systems. At the time these homes were constructed, there was a shortage of natural gas, so all houses in the subdivision are "total electric," with no gas service.

Buildings constructed during a given period and by the same builder tend to be very similar. Frame buildings built prior to 1940 may be **balloon-frame construction**; whereas frame buildings built later will probably be **platform-frame construction**. Similarly, frame structures built prior to 1940 typically will have solid beam roof structures rather than lightweight truss roofs, unless the house has undergone major renovation or repair.

Fire companies should survey their district on a regular basis, paying particular attention to buildings being demolished and under construction. Buildings being demolished will show the common construction characteristics for an era and provide a look inside the building's skeleton. When a building is under construction, the interior framework and other features can be seen **Figure 2-5**. Fire

fighters often discover new construction methods and materials when examining a building under construction. This is a good time to take photographs and share discoveries with the other crews. Studying your response area by subdivisions and/or eras can reveal a lot about the residential and small commercial buildings within your jurisdiction.

Evaluating a Specific Property

During an official visit to a facility, it makes sense to gather information about all kinds of problems that may occur at a property. If the facility falls under the requirements of Title III of the **Superfund Amendments and Reauthorization Act (SARA)**, then hazardous materials planning is mandated by law.

Establishing SOPs is step 1 in the size-up. Preincident planning is step 2. After SOPs are developed, individual properties should be examined for specific hazards and characteristics. Preincident plans are a natural extension of SOPs. For example, a fire department should have a written procedure for operations conducted in buildings protected by sprinkler systems. These procedures would not be repeated in the preincident plan, but any deviations from the normal operation would be outlined. For example, if the water supply at a property is reliable and no off-site water supply exists, the preincident plan might modify the typical SOP at this sprinkler-protected building by not automatically requiring that the sprinkler system be supplied at the fire department connection.

Security Concerns

Security concerns are much greater now, resulting in building owners and management increasing security measures. Armed security personnel are commonplace, especially at government facilities. The World Trade Center attack of 2001 substantially increased awareness of the threat of terrorism. Schools and other properties are now taking precautions to protect occupants from violence. Recent

Figure 2-5 Building under construction.

Courtesy of Bernard J. Klaene.

"active shooter" incidents have resulted in extra security measures.

However, security measures often run contrary to fire safety. Locked doors, limited access, energized electrical fences, fences with razor or barbed wire, and other security measures can greatly affect fire department operations. Security barriers often delay fire department entry onto the property and hinder occupant evacuation.

Figure 2-6 shows an industrial property where entry to the property is blocked with large concrete barriers and a fence is topped with barbed wire. It is necessary to cooperate with building management and note any specific security concerns on the pre-incident plan.

Figure 2-6 Security features at an industrial complex.
Courtesy of Bernard J. Klaene.

Lockboxes

In the past, many fire departments carried keys to buildings or gates leading to complexes on fire apparatus. These keys were often misplaced or lost. Sometimes keys could no longer be identified or the property owner would change locks, failing to notify the fire department. This system resulted in a security breach for the property owner and in many cases did not provide the fire department access to the building or property. Many communities now require a lockbox system at specific commercial properties. Lockbox systems are usually placed at properties where:

- An alarm system is tied to a central station
- Chemicals are stored in reportable quantities according to the SARA act
- Properties require immediate access by the fire department

The local authority having jurisdiction decides which properties are required to have lockboxes and selects a standard key lock system that will be used throughout the community. A dedicated key opens all lockboxes in that community. The property owner procures the lockbox and contacts the local fire department to secure the key in the box after ensuring that the key placed in the lockbox will provide access to areas and buildings on the property. All properties requiring lockboxes according to the local code use the same dedicated key. This dedicated key is secured on fire apparatus and staff vehicles. It is important to secure the dedicated key that provides access to many major properties within the jurisdiction. Some departments use a key code system to access the apparatus lockbox that tracks when the dedicated key box was opened and by whom.

Lockboxes come in various sizes that can hold sets of keys, preincident plan information, or safety data sheets. In some cases, the dedicated key can also be used to unlock fire department connections to sprinkler or standpipe systems. The use of a key not only allows easy access to the building or facility, but it can also eliminate the need for forcible entry, thus reducing property damage. NFPA 1, *Fire Code*,[8] now requires that a standard elevator key be used throughout the jurisdiction. If a standard key cannot be used, property owners are required to place a nonstandard elevator key in a lockbox located near the elevator.

Safeguarding Preincident Plan Information

The information gathered while preincident planning could be invaluable to the would-be terrorist or thief. Historically, fire departments have been liberal in sharing information and somewhat lax in securing SOPs and preincident plans. Many times SOPs are posted on the Internet and, thus, are available to the general public. Preincident plans are often carried in three-ring binders inside unsecured vehicles. Lock-box and other keys are sometimes left in plain view within a command vehicle or apparatus.

It is recommended that fire departments take a more aggressive approach in protecting and securing preincident plans, department SOPs, and keys, including developing procedures for the disposal of computers, SOPs, preplans, and the medium on which they are stored. Procedures should be established for releasing SOP and preincident plan information.

Under the current climate of terrorism, the fire department would be open to criticism if they fail to be good stewards of information in their possession or neglect to take steps necessary to protect it.

Collecting Information and Preparing Preincident Plans

Preincident planning is vital to the safety of fire fighters and the efficient delivery of community fire protection. Therefore, it is imperative that preincident planning be approached in a positive manner. The information collected is essential to conducting safe and effective operations at large complex facilities or high-hazard properties.

Persons assigned to preincident planning must be properly trained and familiar with the preincident planning process used in their department. Some departments assign fire prevention staff to collect and store preincident plan information. Fire prevention personnel can often incorporate preincident planning into their inspection schedule and are generally well qualified. However, if possible, line personnel who are most likely to respond to an incident at the property should conduct the preincident plan survey. Whoever conducts the preincident survey should share their findings with other members who routinely respond to the premises, preferably in the form of an instructional program. There is a training

effect in actually surveying the property and also in sharing important findings with other responders. Once the information about a property is gathered, it may be best to have specially trained personnel enter the information into the department storage/retrieval system to maintain uniformity and avoid entry errors.

Preincident planning is a time-intensive activity that requires extensive data collection and data entry. A building's occupancy, layout, and contents may change frequently.

Preincident plans must be kept up to date. The entry gate shown in Figure 2-6 is blocked by concrete barriers. Formerly, this was the main entrance at this large industrial plant. A fire company arriving at the previous main entrance would not be able to proceed beyond the concrete barriers, and may not know where the new main entrance to this large industrial complex is located. Access to the plant via the new main entrance requires reentering the highway and proceeding approximately one-quarter mile (0.4 km) west, then reentering the complex by way of a newly constructed road. The old main entrance is only sporadically staffed by security personnel. At times there is no one at the gate to provide directions to the new entrance.

Changes to the interior of a building can greatly affect safety and tactics. In some cases, a building that contained a fairly light **fuel load** may change occupants, with a subsequent change in the life hazard and/or fuel load. As an example, a storage warehouse that previously stored noncombustible commodities could change occupants, with the new tenant storing rubber tires in open racks, greatly increasing the fuel load. If this storage warehouse is protected by a sprinkler system, the new fuel load could exceed the design limitations of the sprinkler system.

Keeping preincident plans current is at least as important as creating the initial preincident plan. Bad information can be worse than no information. The revision process is also time intensive, especially if plans must be completely redrawn and the information must be reentered. This process is when the computer can be used to great advantage; minor details can be changed and the information printed out or stored electronically with little effort. If computers are available at the incident scene, the process of finding the right preincident plan and the information needed to develop an incident action plan can be greatly simplified.

Imagine having hundreds of preincident plans, with thousands of pages and accompanying information (e.g., safety data sheets, hydrant maps), filed in three-ring binders. Finding the right property and safety data sheet could involve considerable time and effort. If a department has relatively few preincident plans, a three-ring binder or card system may be sufficient. However, if the department protects large numbers of properties requiring a preincident plan, then a computerized system, such as the one shown in **Figure 2-7**, is highly recommended.

A word of caution: The use of onboard computers is the preferred method of storing large numbers of preincident plans. However, before investing a great deal of time and effort in computerizing preincident plans, ensure that your computer system can handle all the required data, photographs, and drawings. It is essential that stored preincident plan and response information be quickly retrieved. Loading a CD or flash drive, waiting for the computer

to boot, or downloading your preincident plan can cause a considerable time delay. Preincident plan information that is not immediately available will be of limited use to the IC or company officers during initial operations.

The ultimate onboard computer system would remain in a ready mode and automatically load the incident address from dispatch into the vehicle computer while a global positioning system (GPS) would select the best route to the fire location and a graphic information system (GIS) would visually display the location of nearby fire hydrants, water mains, flood zones, underground pipelines, and so on.

GIS and other applications can also display aerial photographs or plot plans of buildings. If a preincident plan were available for the address, this ideal system would load the preincident plan, making it immediately available by "touch screen" or cursor navigation. In addition, the preincident plan drawing would be displayed on top with additional information available by accessing the narrative, via **hovering** over a drawing element or by **drilling down** for more text-based or graphic information. This "ultimate system" has one major drawback—it relies on information being received from a remote source (e.g., dispatch, telecommunications contractor). The reliability of the network must be verified via testing prior to making a final decision on system components and programs. As an additional precaution, preincident plan information should also be stored onboard fire vehicles as a backup when the communications network or terminal fails.

Many computer programs are available that allow easy navigation from page to page or provide information when the cursor is placed over the item. There are specific preplan software programs developed for fire service use, as well as common software programs such as PowerPoint that can be used. Easy navigation provides the means to quickly access needed information.

Probably the most difficult to construct and most important part of the preincident plan is the drawing(s). Most architectural drawings produced

Figure 2-7 Onboard computerized preincident plan.
Courtesy of Bernard J. Klaene.

today are drawn with AutoCad software. When considering new preplan software, it is best to purchase a program that can import and modify Auto-Cad drawings. If the property being preplanned has architectural drawings, you may be able to simply load the drawings and add your symbols. AutoCad and most other architectural drawing programs are done in layers, allowing you to eliminate unwanted drawing information and add layers with information important to the fire department. Some preincident plan drawings cannot be downloaded, or drawings may not be available. A powerful user-friendly drawing software program is an indispensable tool when developing and updating preincident plan drawings.

As an alternative to computerization or as an interim measure while preincident plans are being converted to onboard computer files, it may be a good idea to place printouts in the facility lockbox, provided the lockbox is large enough to accommodate the printout.

Preincident Plans

Formal preincident plans include both a narrative and drawings. Drawings are the most important part of the preincident plan. The IC and company officers do not have time to read pages of text-based information. However, some narrative information is needed, such as the address, occupants, emergency contacts, telephone numbers, and other general information.

Narratives are best written in outline form with extremely important information highlighted, color coded, or otherwise identified to draw attention to critical information. Much of the text-based information can be imbedded in graphic features, allowing the user to drill down or hover over the graphic feature to access additional graphic or text-based information. As an example, a symbol representing a standpipe can be used to show the location of the standpipe inside the building, with additional information such as pressure, size, flow rate, pressure regulator operation, and other pertinent details

revealed when the cursor is placed over the stand-pipe symbol.

Preincident plans can take various forms. A preincident plan that includes both a narrative and a

NOTE

There should be an SOP describing the preincident planning system.

Texture: © Ely Studio /Shutterstock, Inc; Steel: © Sharpshot/Dreamstime.com

drawing would be a formal preincident plan. On the other end of the spectrum is a building where a simple notation is made about a particular problem, such as holes in the floor due to a previous fire.

NFPA 1620, *Recommended Practice for Pre-Incident Planning*,[9] outlines the steps involved in developing, maintaining, and using a preincident plan.

Various occupancies are also defined in NFPA 1620, including:

- Assembly
- Educational
- Health care
- Detention and correctional
- Apartment buildings
- Dormitories
- Hotel
- Lodging and rooming house
- Residential board and care
- Mercantile
- Business
- Industrial
- Warehouse and storage

Appendix B of NFPA 1620 contains case histories of several fires where preincident planning led to success or where a lack of preincident planning led to failure. As an example, NFPA 1620.B.10 discusses a fire and explosion involving nitrocellulose where preincident planning resulted in a safer and more effective operation.

Incident Summary

Fire units responded to a fire alarm at a paint manufacturing facility. The department's hazmat team had just conducted a preplan at that facility the previous week. Shortly after arriving on scene, the IC requested additional resources, including the hazmat team, due to an explosion at the facility. The product involved was nitrocellulose, a DOT Class 4 flammable solid.

The first arriving companies attempted to gather information by looking the product up in the emergency response guide but were frustrated by the fact that there were six different entries for nitrocellulose. Due to the knowledge gained during the preplanning process, the hazmat unit was able to tell the companies exactly which entry to use until the hazardous materials team arrived. The hazardous materials team also knew the exact location of the product in the facility and the location of the deluge valve to turn off the suppression system. Because of the preplanning process, responding personnel had information on the product, its location, and the location of the pertinent fire protection systems without having to send a team into the hazard zone and without having to wait for property representatives to arrive on scene. Company officials were impressed with the fire department's knowledge base and the speed with which the incident was mitigated. While the incident could have been mitigated without a preplan, it certainly made the operation much faster and, more importantly, much safer.

Texture: © Eky Studio /Shutterstock, Inc; Steel: © Sharpshot/Dreamstime.com

Types of Preincident Plans

There are many ways to describe and categorize pre-incident plans. This text considers three levels or types of preincident plans:

1. **Complex preincident plan**. This plan is used when a property has more than three buildings or when it is necessary to show the layout of the premises and the relationship between buildings on the site. Complex pre-incident plans are used to identify building and fire protection features as well as hazards for each building. Fire companies and the IC cannot be expected to know the location of every building within a large complex. A street-level view of a large industrial complex is shown in **Figure 2-8**.

Figure 2-8 Industrial complex.
Courtesy of David J. Jones, Cincinnati, Ohio

Complex preincident plan drawings provide an overview of the complex to assist in locating buildings or areas within the complex. A preincident plan of this type may have several drawings of specific floor areas but also includes a drawing showing an overview of the complex and the position of various structures by name, number, letter, or other naming system, as shown in **Figure 2-9**. If "Building I" is given as the fire location, fire fighters and the IC use the overview drawing to locate Building I.

Responding units use the formal preincident plan for Buildings "H" and "I" to assist in sizing up the situation and formulating the incident action plan. Often, buildings are named according to their function or when they were constructed, rather than using a rational pattern that is apparent to the casual observer.

2. **Formal preincident plan**. A property with a substantial risk to life and/or property should

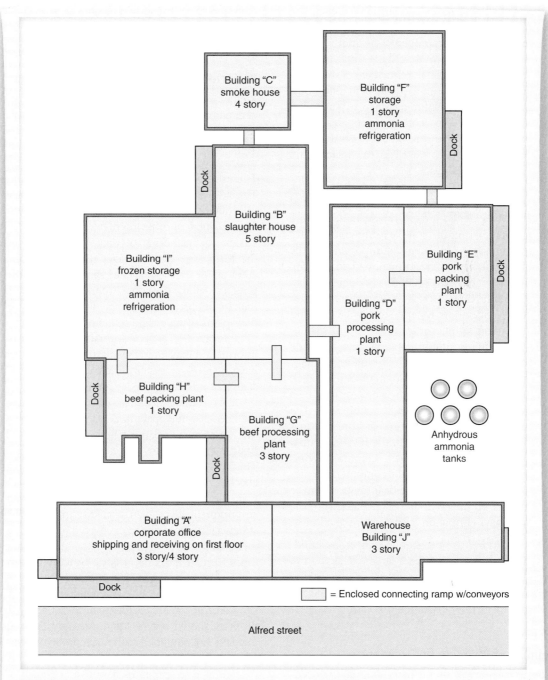

Figure 2-9 Complex preincident plan drawing.

be the subject of a formal preincident plan. This formal preincident plan would include a drawing of the property, specific floor layouts, and narrative describing important features. Several formal preincident plans could be included within a preincident plan for a single building within a complex preincident plan. One large, complex property could easily fill a three-ring binder with drawings of specific areas as well as the relationship between buildings.

The formal preincident plan drawing shows the interior layout for two attached buildings **Figure 2-10**. These buildings are located within the complex shown in Figures 2-8 and 2-9. Only items of importance to the IC are shown on the formal preincident plan. The narrative portion of this preincident plan would further describe the anhydrous ammonia refrigeration system, storage height, and any other significant features.

3. **Notation**. A simple notation may be made about the premises, such as building damage from a previous fire. Computer-aided dispatch systems usually have provisions for making a special notation for specific addresses. The dispatch center may be used to store information that is transmitted to the IC and companies at the incident scene. If communications applications are part of the computerized preincident plan system, this information may be shown on the computer screen as part of the graphic information available for the address.

Marking the exterior of dangerous buildings, such as painting a large "X" at the top level of a building that should not be entered because of structural problems, would be another form of notation preplan.

Preincident Planning by Occupancy

A fourth category of preincident planning by occupancy could be described as either a training issue or the topic of an SOP. For example, church fires have common traits that should be understood when conducting the size-up. (See the chapter titled *The Role of Occupancy* for more details about occupancies.)

Preincident Plan Checklist and Drawings

Preincident planning SOPs should identify special occupancies or specific types of buildings that are to be included in the preincident planning process. Some departments prefer to gather and distribute information using a standard preincident plan form. The use of forms has advantages and disadvantages. The advantages include being able to find specific information in a predictable location on each preincident plan and having defined categories, which ensure the gathering of data for items listed on the form. The disadvantages include having large amounts of "not applicable" space on the form and possibly failing to gather information about items that are not listed on the form. Preformatted preincident plans using checklists tend to be several pages long because of generic formats, and much of the information may be irrelevant. If a building does not have a sprinkler system, there is really no need to state that it is not sprinkler protected.

It is recommended that a detailed format based on the checklist in Appendix D of NFPA 1620 be used. The preincident plan narrative for Memory Lane Apartments lists items that can be known in advance. This information would typically be gathered using a checklist. To yield a more "user friendly" preincident plan, the information was then formatted to match the size-up/preincident plan checklist in **Figure 2-11**. When a factor is assumed or covered in SOPs, it is not included. As an example, the factor "Additional Hose Lines Needed" does not need to be addressed in this preincident plan narrative, as "nothing special" (SOPs call for a backup and line on the floor above the fire) should be adequate. Factors that do not apply to the property should not be listed. Memory Lane Apartments is not equipped with a standpipe or sprinkler system; therefore, no mention is made of a sprinkler or standpipe system in the narrative.

The size-up/preplan checklist includes 76 items (not counting titles where subtitles fully address the factor).

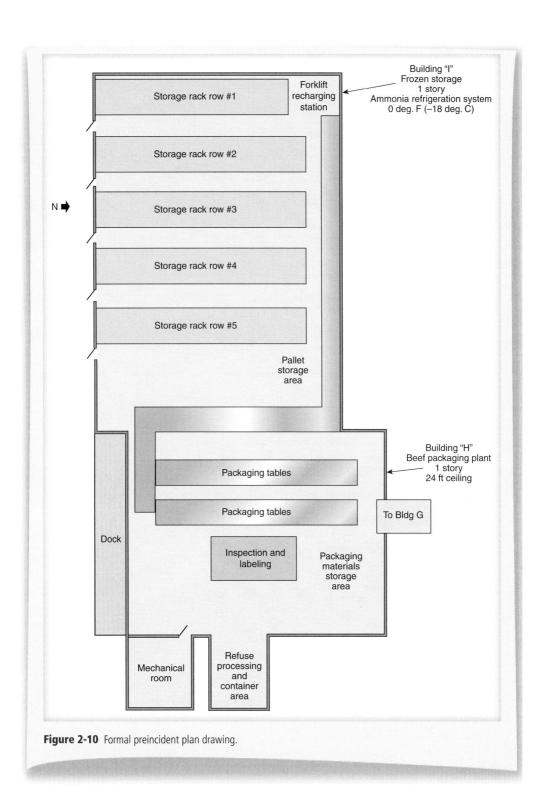

Figure 2-10 Formal preincident plan drawing.

Memory Lane Apartments
26 Memory Lane

Life Safety/Fire Fighter Safety
Occupancy Type

- 89-unit apartment building
- Most occupants are elderly, some with severe handicaps
- 7 small retail stores on first floor facing Main St.
- Video Rental Store (50 ft x 45 ft [15.2 m x 13.7 m])
- Joe's Barber Shop (24 ft x 45 ft [7.3 m x 13.7 m])
- Hair by Gloria Beauty Parlor (24 ft x 45 ft [7.3 m x 13.7 m])
- Old Tyme Antiques (12 ft x 45 ft [3.6 m x 13.7 m])
- Nail Boutique Manicure Salon (12 ft x 45 ft [3.6 m x 13.7 m])
- Song Shop CDs (24 ft x 45 ft [7.3 m x 13.7 m])
- Thrift Shop(24 ft x 45 ft [7.3 m x 13.7 m])
- Basement
- Bingo Hall (180 ft x 45 ft [55 m x 17.3 m])
    - Bingo hall posted for 500 maximum occupants
    - Exits for Bingo hall stairs to Memory Lane and Main St. sides

Estimated Number of Occupants

- Bingo hall could have 500 on Tuesday and Thursday evenings
- Occasionally used as a party hall
- Most apartments occupied by a single resident
- Shops vary, seldom more than 10 per shop

Primary and Alternative Egress Routes

- Three stairways and fire escape (see drawing)
- One-story roof to rear could provide emergency or secondary egress from second floor

Access to Building Exterior

- Access streets to front (Memory Lane) 50 ft (15 m) setback
- Right side (Main St)
- Driveway on left side
    - Narrow, but apparatus accessible
    - Too close for aerial use
    - U-shaped driveway to front (may not support apparatus without damage)
    - Rear – one-story roof – good access to second floor

Construction Type

- Ordinary construction (brick exterior wooden floor/ceiling assembly)

Roof Construction

- Built-up flat tar roof supported by 2^2 x 12^2

Condition

- Well maintained – no noted damage

Live and Dead Loads

- Average load for residential property

Enclosures and Fire Separations

- Only open areas in basement and stores
- Metal fire doors from apartments to hallways
- Hallway no separated, open from end to end

Extension Probability

- Stairways are not enclosed
- Utility openings penetrate floors

Concealed Spaces

- Inaccessible crawl space between top floor and roof

Age

- Built in 1932

Height and Area

- Four-story building
- Basement to rear of building above grade level
- 220 ft x 180 ft (67 m x 55 m) U-shaped (see drawing)

Figure 2-11 Sample preincident plan narrative.

Complexity and Layout
- U-shaped hallway with center hallway (see drawing)
- Access to basement down stairs
- Stores all face Main St. with direct access to street

Extinguishment
Probability of Extinguishment
- Basement only significant problem

External Exposures
- College of Design on left side (Side B) 15′ (4.6 m)

Internal Exposures
- Johnny's Restaurant attached to rear (Side C)
- Open stairways and floor to floor via walls/ceilings

Fuel Load
- Average residential
- Stores and bingo hall average

Calculated Rate of Flow Requirement
- Rate of flow for apartments and stores within the capacity of standard pre-connect and back-up line.
- Bingo hall 810 GPM (15 L/sec) required

Number and Size Hose Lines Needed for Extinguishment
- Standard pre-connect except bingo hall where 2-1/2″ (64 mm) lines may be required

Additional Hose Lines Needed
- Nothing special (SOP calls for back-up and line on floor above the fire)

Water Supply
- Hydrant located to front of building and on Main St.
- Additional hydrants on Main St.
- Hydrant flow on Memory Lane 1000 GPM (63 L/sec) per hydrant
- Hydrant flow on Main St. 1500 GPM (95 L/sec) per hydrant
- Water supply on grid system 8″ (203 mm) main on Memory Lane; 20″ (508 mm) main on Main St.
- System flow in area is approximately 12,000 GPM (757 L/sec)
- System can be cross-tied to two other supply systems for a total 30,000 GPM (1892 L/sec)

Property Conservation
- Ordinary household items in apartments
- Stores have property of moderate value

General Factors
Staffing/Apparatus on Alarm Card

Alarm	Engine Companies	Truck Companies	Others
First	23, 19, 32	23, 19	District Chief 1
Second	3, 9, 5	32, 3	District Chief 4
Third	46, 34, 29	29	H. Rescue 9, ALS 31, Deputy Chief
Fourth	14, 31, 12	None	Chief of Department
Fifth	21, 28, 20	None	Training Chief, Prevention Chief

Additional Company and Command Staff Available
- 11 engine companies
- 8 truck companies
- 1 heavy rescue company
- 4 ALS, 6 BLS units
- Mutual aid

Staging/Tactical Reserve
- Suggested staging area – parking lot off Main St. and Taft Road

Utilities: Water, Gas, Electric
- Gas, water, and electric shut-offs in basement utility room just left of Bingo hall

Figure 2-11 Sample preincident plan narrative. (Continued)

If you closely examine the Memory Lane Apartment preplan narrative and drawing, you can see that 27 of these factors are directly addressed, 5 are addressed in a general way, and 2 do not apply. Property conservation in this instance would follow SOPs and be typical; therefore, the 10 factors listed under property conservation are also addressed in a general way. In other words, information about 44 of 76 of the size-up/preincident plan factors can be known or partially known in advance. In some buildings, more or fewer factors may be known in advance. Incident-related factors, such as time and weather conditions, will not be known until the day of the fire or arrival at the fire scene.

In addition to addressing the size-up/preplan factors, the building name and address and the owner's/manager's/agent's name, with telephone numbers and emergency contact information, should be included in the preincident plan narrative.

The IC and responding units do not have time to read an encyclopedia about the property. They should be able to quickly find and easily read the needed information.

Notice the Maltese cross at the top of the Memory Lane Apartments preincident plan narrative. This is meant to be a quick first reference. This symbol could also be included in the drawing on a formal preincident plan. The Maltese cross marking is based on NFPA 1, *Fire Code*,[10] Annex E, which describes the Fire Fighter Safety Building Marking System. **Table 2-2** provides an overview of the marking system. This simple icon provides basic information regarding construction type, hazards of contents, sprinkler and standpipe systems, and occupancy or life hazards. The center of the cross is reserved for special hazards such as hazardous materials.

As an example, on the Maltese cross for the Memory Lane Apartments, the top of the cross is used to designate the construction method—"ORD" for ordinary construction. The right side addresses built-in fire protection features. The "N" signifies no protection; this building is not protected by a sprinkler or standpipe system. The bottom of the Maltese cross identifies the relative life hazard, which is rated as "H" (high) due to the number of occupants and their immobility. The left side is used to indicate the content hazard. The "M" indicates a moderate fire hazard regarding contents in this building.

The center of the Maltese cross is reserved for special hazards. Typically, this is used to identify hazardous chemicals using NFPA 704, *Standard System for the Identification of Hazardous Materials for Emergency Response*.[11] When hazardous materials are present and the department is using an electronic retrieval system, this or any other space on the Maltese cross can be used to navigate to additional information. In the case of hazardous materials, the NFPA 704 symbol could be linked to the chemical list, which could then be linked to another layer containing

Table 2-2	Maltese Cross - Fire Fighter Safety Building Marking System			
Top of Cross Construction Type	**Right Side Fire Protection**	**Bottom of Cross Life Hazard**	**Left Side Content Hazard**	**Center of Cross Special Designation**
FR = Fire Resistive	A= Full Sprinkler	L = Low	L = Low	At discretion of AHJ
NC = Non-Combustible	P= Partial Sprinkler	M = Medium	M = Medium	Haz Mats NFPA 704
OC = Ordinary	S = Standpipe	H = High	H = High	
HT = Heavy Timber	N = None			
C = Combustible				

© Eky Studio/Shutterstock, Inc.

chemical data sheets for each hazardous chemical. The center of the Maltese cross can also be used for other important information. The center is left blank on the Memory Lane Apartments preincident plan. NFPA 1 proposes using the Fire Fighter Safety Building Marking System on the exterior of a building. The authority having jurisdiction would need to pass a regulation, law, or ordinance to enforce the use of this marking system. However, using the system on preincident plans is at the discretion of the fire department and highly recommended.

It is said that a picture is worth a thousand words, and nowhere is this truer than on the incident scene. Including a drawing in a preincident plan is extremely useful. **Figure 2-12** is a sample of a preincident plan drawing.

Using intuitive drawing symbols on preincident plans that do not require a legend to understand is highly desirable. NFPA 170, *Standard for Fire Safety and Emergency Symbols*,[12] should be consulted when developing a symbol system for use in preincident planning. The symbols in Chapter 5 of NFPA 170 (Symbols for use by the Fire Service) are intuitive and particularly well suited to preincident plan drawings. The architectural and engineering drawings in Chapter 6 of the NFPA 170 standard use markings and symbols that are commonly used in professional drawings. Knowing these markings and symbols will assist in reading building drawings made available on scene by building management. If using a computer to retrieve and view drawings, note that many

Figure 2-12 Sample preincident plan drawing.

programs have the ability to describe the symbol or marking when the user places the cursor (hovers) over the drawing.

Drawings should include the following:

- Entry and exit doors
- Fences, gates, and other exterior security features
- Areas of safe refuge
- General floor layout
- Building and area dimensions
- Stairways
- Fire escapes
- Elevators
- Fire separations
- Utility shutoffs
- Alarm panels
- Lockbox
- Fire protection system and sectional control valves
- Fire protection system intakes
- Fire protection system manual actuation
- Fire protection system agent supply
- Hose outlets for standpipe systems
- Fire pumps
- Fire hydrants
- Potential staging areas
- Hazardous materials locations
- Emergency vents and controls
- Other information important to the IC

An important feature of a good preincident plan is its usefulness in the field. A hard-copy preincident plan consisting of more than two double-spaced printed pages in outline form plus a drawing page tends to be of little value during initial operations. The use of computerized preincident plans allows much more information to be accessed without cluttering the top layer of the drawing and narrative.

Simulating a fire during the preplanning process provides good training in preparing for fires in a building, but rarely should information about a specific fire scenario be part of the preincident plan. SOPs ensure that a company will supply the sprinkler intakes and that the incident management system will be implemented. Repeating SOP items unnecessarily

or including too many minor details on drawings or the narrative weakens the preincident plan. The IC does not always care about the location of bathrooms, plumbing, and other details included in architectural drawings. However, there are occasions when this information is useful. For example, to remove water from the upper floors of a building, crews can remove the toilets from the floors, creating a 4-in. (102-mm) drain. If a computerized preincident plan is properly designed, this information could be located on a level accessible by drilling down.

When preparing a preincident plan, one of the most common errors is preassigning companies to respond to specific locations. This is an acceptable practice up to a point. For example, in the absence of visual signs of fire, a company could be assigned to check the alarm system with other companies standing by, or a second truck company could be assigned to the rear of the building. However, it is usually poor practice to preselect a specific entry point or specific tactics or to make company assignments in advance. If the actual fire occurs in a location other than the assumed area, companies will be in the wrong locations. The preincident plan would then have made the situation worse. Including specific tactical assignments on the preincident plan should be avoided. Preassigning companies to specific tasks could result in vital tasks being unassigned or totally missed during the heat of battle.

What Structures Are Preplanned?

Just as some SOPs are department-specific, a decision regarding what properties to preincident plan depends on the jurisdiction being protected. Few communities have the resources necessary to preplan all buildings or even all commercial buildings. The operational priority list (life safety, extinguishment, property conservation) provides direction regarding which buildings may need to have preincident plans. If there is a high life hazard (including fire fighter safety issues), a particularly difficult extinguishment problem, or high-value property, then there is a need to prepare a preincident plan. Properties with high life hazards should be the highest priority for planning.

Preincident planning deals more with potential demand (properties with the potential for a large loss of life or property) than it does with realized demand (the actual number of fires in a building or type of building).

Buildings that present an extraordinary challenge in terms of life safety, extinguishment, and salvage should be preplanned. Other buildings that are generally preplanned, whether or not they present a high life safety, extinguishment, or property conservation problem, include the following:

- Buildings protected by fire protection systems (with the exception of small restaurants with a hood system)
- High-rise buildings
- Complexes with a large number of buildings

Modifying Standard Operating Procedures

SOPs describe a standardized method addressing predictable operational circumstances. Preplanning addresses what is different or unusual. Preincident plans are building specific, while SOPs are general. If the specific property fits the general category of operations, the SOP is assumed to be in effect; however, there may be special circumstances in which the SOP is not the most effective course of action.

Fallacy	Fact
All SOPs are always in effect.	SOPs are not etched in stone. They can be modified during preplanning or whenever situations dictate a different course of action.

Steel: © Sharpshot/Dreamstime.com

For example, department SOPs require the first-arriving engine company to secure a source of water. The specific building being planned is small and located a considerable distance from the nearest hydrant. Based on information obtained while preplanning the building, a recommendation is made to make an exception to the department water supply procedure. It is determined that the extra time involved in supplying water would result in complete destruction of the structure, and the rate of flow indicates that the apparatus tank has more than sufficient water capacity to conduct an offensive attack. Since the preplanned tactical consideration is different from the department SOP, it is very important that this recommended tactic be approved for this property and then shared with all other responding companies.

Estimating Life Safety Needs

As mentioned, any building that poses an unusually high risk to fire fighters or occupants must be included in the preplanning process. Occupancies with long-span roofs present a hazard to fire fighters but could also be preplanned due to the large rate of flow required to extinguish a fire. Heavy roof loading or unusual building features that could lead to partial or full collapse should be noted. A notation-type preincident plan may be enough for buildings that have previously been damaged by fire or weather.

Nursing homes, hospitals, places of assembly, schools, churches, and other places holding large numbers of people or a significant number of disabled people should be preplanned.

Estimating Extinguishment Needs

<u>Compartmentation</u> plays a major role in determining extinguishment needs. Most residential buildings are compartmented into a series of small rooms requiring much less water than in a large undivided warehouse. A building with compartments requiring more than two standard preconnected hose lines, as calculated by using the volume of the fire compartment divided by 100, should be preplanned. (Rate of flow will be covered in more detail in the *Offensive Operations* chapter.) If special or hazardous materials are present in quantity, such as flammable liquids, plastics, rubber tires, and idle wood pallets, smaller areas may require a preincident plan.

The estimated rate of flow should be included on the preincident plan when applicable. Where very large areas with heavy fuel loads exist, a decision may be made in advance of the fire regarding defensive or non-attack situations for fires beyond a certain percentage of involvement.

IC Rule of Engagement for Fire Fighter Safety

Do not risk fire fighter lives for lives or property that cannot be saved. Seriously consider a defensive strategy.

Reproduced from: © 2011, IAFC Safety, Health and Survival Section. All rights reserved.

Texture: © Eky Studio /Shutterstock, Inc; Steel: © Sharpshot/Dreamstime.com

Estimating Property Conservation Needs

Any number of items could be considered high-value contents. Jewelry, electronic equipment, and many other items could justify a preplanning effort. Identifying the location of floor drains is generally beyond the scope of a preincident plan. However, if information of this type would be of value to responding units and the IC, then include it in the preincident plan. When using drawing software, it is often possible to include information such as floor drains on a separate layer. The user can then turn on or drill down to the floor drain layer if this information is needed. Some property is more susceptible to damage, and the factors increasing its vulnerability should become part of the preincident plan. Stock that is piled on floors, rather than in racks or on pallets, is more likely to sustain water damage. Determining where the water from sprinkler activation or firefighting streams will go can be important.

As an example, a damaged sprinkler pipe at a high-rise building resulted in water flowing into a computer room that processed credit card transactions for the entire east coast of the United States. Immediate action by employees and fire fighters, who were able to stop the flow of water, prevented interruption of this vital service.

Relationship of Preplanning to Size-up

Size-up is a natural extension of the SOP and preplanning process. SOPs get the operation off to a predictable start while preplans provide specific information about the building in advance of the fire. Whenever possible, it is best to gather as much information in advance as possible; therefore the list of size-up factors is also used as a preincident plan list. Use the size-up list in Figure 2-13 , and the description of each factor, as a guideline when preplanning.

A brief explanation of what is meant by each of the size-up/preincident plan factors is provided in this chapter. Many of these factors are further explained in other chapters of this text.

Analyzing the Situation Through Size-up

Size-up factors, which may change from incident to incident, are difficult to categorize in terms of relative importance. Incident conditions will determine which size-up factors are most important. Factors related to life safety are most likely to be critical. These most important factors are known as **primary factors**. The initial size-up analysis is limited to evaluating primary factors. Less important factors are categorized as **secondary factors**. Incident conditions will dictate which conditions are primary. As an example, weather can be categorized as a critically important (primary) factor when at the extreme and as a less important (secondary) factor during moderate conditions. The fire officer must first sort through information and evaluate available data related to primary factors.

Size-up actually begins before the incident with the development of SOPs and preincident planning. When an alarm occurs, fire officers and the IC

Size-up/Pre-Plan Checklist Life Safety/Fire Fighter Safety

- ☐ Smoke and fire conditions
  - ☐ Fire location
  - ☐ Direction of travel
- ☐ Ventilation status
- ☐ Occupancy type*
- ☐ Occupant status
  - ☐ Estimated number of occupants*
  - ☐ Evacuation status
  - ☐ Occupant proximity to fire
  - ☐ Awareness of occupants*
  - ☐ Mobility of occupants*
  - ☐ Occupant familiarity with building*
  - ☐ Primary and alternative egress routes*
  - ☐ Medical status of occupants*
- ☐ Operational Status
  - ☐ Adherence to SOPs
  - ☐ Fire zone/perimeter
  - ☐ Accountability*
  - ☐ Rapid intervention*
  - ☐ Organization and coordination*
  - ☐ Rescue options*
  - ☐ Staffing needed to conduct primary search
  - ☐ Staffing needed to conduct secondary search
  - ☐ Staffing needed to assist in interior rescue/evacuation
  - ☐ Staffing needed for exterior rescue/evacuation
  - ☐ Apparatus and equipment needed for evacuation
  - ☐ Access to building exterior*
  - ☐ Access to building interior* (forcible entry)
- ☐ Structure
  - ☐ Signs of collapse
  - ☐ Collapse zone*
  - ☐ Construction type*
  - ☐ Roof construction*
  - ☐ Condition*
  - ☐ Live and dead loads*
  - ☐ Water load
  - ☐ Enclosures and fire separations*
  - ☐ Extension probability*
  - ☐ Concealed spaces*
  - ☐ Age*
  - ☐ Height and area*
  - ☐ Complexity and layout*

Extinguishment
- ☐ Probability of extinguishment*
- ☐ Offensive/defensive/non-attack

Figure 2-13 Size-up/preplan checklist. (Continues)

- ☐ Ventilation status
- ☐ External exposures*
- ☐ Internal exposures*
- ☐ Manual extinguishment
  - ☐ Fuel load*
  - ☐ Calculated rate of flow requirement*
  - ☐ Number and size hose lines needed for extinguishment*
  - ☐ Additional hose lines needed*
  - ☐ Staffing needed for hose lines*
  - ☐ Water supply*
  - ☐ Apparatus pump capacity*
  - ☐ Manual fire suppression system*
- ☐ Automatic fire suppression equipment*

Property Conservation
- ☐ Salvageable property*
- ☐ Location of salvageable property*
- ☐ Water damage
  - ☐ Probability of water damage*
  - ☐ Susceptibility of contents to water damage*
  - ☐ Water pathways to salvageable property*
  - ☐ Water removal methods available*
  - ☐ Water protective methods available*
- ☐ Smoke damage
  - ☐ Ventilation status
  - ☐ Probability of smoke damage*
- ☐ Susceptibility of contents to smoke damage*
- ☐ Damage from forcible entry and ventilation*

General Factors
- ☐ Total staffing available versus staffing needed*
- ☐ Total apparatus available versus apparatus needed*
- ☐ Staging/tactical reserve*
- ☐ Utilities (water, gas, electric)*
- ☐ Special resource needs*
- ☐ Time
  - ☐ Time of day
  - ☐ Day of week
  - ☐ Time of year
  - ☐ Special (e.g. holiday season)
- ☐ Weather
  - ☐ Temperature
  - ☐ Humidity
  - ☐ Precipitation
  - o Winds

* This factor can be at least partially known in advance of the fire through pre-planning.

Figure 2-13 Size-up/preplan checklist. (Continued)

consider what is already known about the specific property and generally about the type of property and area. Important factors such as weather conditions, time of the alarm, and day of the week will be known as the response begins. Other incident-related information begins with the dispatcher's information about the location of the fire, method of alarm, and other conditions reported to dispatch. Fire service personnel should maintain awareness about what is going on in their community. Are there any scheduled activities during their shift, such as protests, political activities, or religious holidays? Have there been any recent trends such as frequent arson fires?

In many situations, at least one fire company will arrive before the dispatched chief officer(s). Communications from the scene should add to or confirm dispatch information. Once on the scene, the responding chief officer will add to what is known through personal observation, communications with fire companies and building personnel, and reconnaissance. Once the on-scene information is processed, the IC must quickly evaluate the action that is taking place, fire stage, area of involvement, and condition of the building.

The amount and quality of information improve with time, and the IC has time to better evaluate the situation once initial assignments are made. Size-up continues throughout the incident and should continue through the overhaul phase.

Life Safety/Fire Fighter Safety

Smoke and Fire Conditions

Smoke and fire conditions are directly related to occupant survival and fire fighter safety and are primary factors at a structure fire. Heavy, dark pressurized smoke and visible fire conditions may necessitate a defensive attack, whereas light smoke with no fire evident indicates a high probability of saving occupants still inside the building. It must be remembered that reading smoke and fire is not an exact science and a hidden or ventilation-limited fire can rapidly increase when it breaks out of containment or when it receives additional air. Recent research by Underwriters Laboratories,[13] NIST, and others indicates that most fires in occupied residential buildings are oxygen limited and go through a decay phase where a decrease in temperature and smoke production could result in no visible signs of smoke on the exterior. A ventilation opening, including opening a door to gain access, could create a flow path that increases the oxygen concentration, resulting in a dramatic increase in visible smoke and flames. **Flashover** is a critical indicator at a structure fire. Occupants inside post-flashover compartments (rooms) have a very low probability of survival. Smoke and fire conditions can also provide a warning of an impending **backdraft**.

Experienced fire officers learn to evaluate pressure, smoke characteristics, and other factors in determining the intensity of the fire Figure 2-14 . However, interior reconnaissance is generally the best way to realistically determine fire intensity when an interior attack is possible.

Smoke volume, velocity, density, and color provide clues about the possibility of flashover, location of the fire, and survivability inside the burning building. Smoke and fire conditions will not be known until the time of the fire, but the compartment (room) size and tightness, as well as fuel content, provide prefire indicators of the relative fire potential.

Figure 2-14 Smoke and fire conditions.
Courtesy of Captain Nick Morgan, St. Louis Fire Department.

Fire Location

The fire's location determines the method and direction of attack. Extinguishment is the primary life safety tactic and an operational priority. In most cases, company officers will be able to determine the location of the fire on arrival. The information received from dispatch, alarm systems, information from occupants, and visual clues assist in finding the fire location. The first-in crews will typically be attacking or setting up to attack the fire when the chief officer arrives. There are times when the fire location is not obvious or when dispatch information is wrong.

When information is incomplete or incorrect, finding the fire can be very difficult, especially in a large building. High-rise buildings can hide a fairly large interior fire, and smoke alarms may be sounding far away from the actual fire location. A fairly small fire located at a window may appear to be much larger, whereas a large fire toward the core of the building may not be visible from the exterior. Locating the fire and its extension paths is critical to the development of the incident action plan.

Rule of Engagement for Fire Fighter Survival

Size up your tactical area of operation.

Reproduced from: © 2011, IAFC Safety, Health and Survival Section. All rights reserved.

Texture: © Eky Studio /Shutterstock, Inc; Steel: © Sharpshot/Dreamstime.com

Open flaming is a definite indicator of fire location, but the fire could originate in another location. The old adage "where there's smoke, there's fire" is true, but the primary goal here is to find the seat of the fire. Knowing the basic chemistry and physics of fire is helpful in learning to read smoke. All matter, including smoke and heat, naturally try to reach equilibrium. When openings are available, smoke will try to reach the same temperature as the temperature on the other side of the opening. In most cases, smoke will go to the nearest and largest opening unless wind conditions reverse the venting. Likewise, all gases, including fire gases, are governed by the physical properties of gases, which state that a given volume of gas will expand as the temperature of the gas rises. If the gas is contained under conditions in which expansion is not possible, the heated gas will undergo an increase in pressure. This increase can be calculated using the ideal gas law; however, the fire ground is no place for precise calculations. It is enough to understand that smoke rapidly flowing from an opening is a clue that the fire may be nearby. Fires develop flow paths based on air providing the oxygen necessary to sustain combustion flowing into the area and hot fire gases, smoke, and fire flowing out. Flow paths can be controlled by applying water and/or ventilation. It must be remembered that ventilation can occur either by fire fighters intentionally making ventilation openings or by the fire causing unintentional and uncontrolled venting.

Direction of Travel

Knowing where the fire is most likely to spread is important to life safety and extinguishment tactics. Search teams with hose lines need to protect and evacuate occupants in areas where the fire is likely to extend. Basic chemistry and physics of fire determine the direction of fire travel. The products of combustion travel upward until reaching a barrier. They will then travel horizontally to fill the top of the compartment and, if contained, will travel downward. Fire, heat, and smoke will therefore travel upward via the path of least resistance, before traveling horizontally or downward. Fire will most likely travel up stairways and other available openings to the floors above the fire. The upward path could also be through concealed wall areas (balloon-frame construction is particularly noted for this). Fire and smoke entering a wall will travel to the attic unless it meets a solid barrier. Nearly all buildings have vertical paths to the top of the building, although some are much better protected than others. Fire in an open wall, such as in balloon-frame construction, will travel unimpeded to the attic. In platform-frame construction and other types of construction with fire barriers, the path may not be as direct, but there are generally smaller pathways via utility openings for drains, water pipes, electrical wiring, and so on.

Although it is not possible to know the exact means of fire travel during preincident planning, it is possible to know the probability of fire travel. Rated and nonrated fire compartments tend to limit fire spread. Construction methods and alterations affect fire spread and can be known in advance.

Ventilation Status

Ventilation, whether intentional or unintentional, may change the direction of smoke and fire travel. Opening a door creates an air intake opening supplying oxygen to the fire and a low-pressure opening allowing smoke and fire gases to escape. Water application or other conditions influence the ventilation. As an example, wind-driven fires have been known to reverse venting and supply a superabundance of oxygen. Vent openings generally increase the available oxygen and often result in a rapid increase in fire intensity and size. Improper venting can spread and accelerate fire spread, change the ventilation flow path, and precipitate a backdraft or flashover, all of which increase the danger to occupants and fire fighters.

Notice that ventilation status is listed under all three operational priorities (*life safety, extinguishment,* and *property conservation*). Ventilation is a key factor during all phases of the operation. Venting for life safety involves moving the fire away from occupants and fire fighters. Understanding smoke and fire movement, as discussed previously under other factors, is essential to good venting. As mentioned earlier, the fire will follow the path of least resistance, upward then horizontally and finally downward. The vent opening should direct the flow path and fire away from occupants and fire fighters.

A common life safety venting tactic, included in the SOPs for many departments, is to remove a scuttle or other cover over a built-in opening at the top of a stairway in a multistory building.

The intended purpose of this tactic is to clear the stairway of smoke. However, this tactic can result in the unintended consequence of creating a flow path that extends fire, heat, and smoke into the stairway being vented. **Figure 2-15** , **Figure 2-16** , and **Figure 2-17** illustrate this unintended consequence. In this scenario, the first-arriving engine company is advancing a hose line to the fire floor, while the first truck company is split into two crews. One truck crew is gaining access to the roof, and the other is in the stairway making entry into the floor above the fire to check for extension and begin the primary search as well as facilitating evacuation (Figure 2-15).

Figure 2-15 Preparing to vent via scuttle.

When the engine company opens the door to attack the fire, an unvented fire will be released from confinement and will enter the stairway (Figure 2-16). This will tend to make smoke conditions in the stairway worse even if there is no vent opening at the top of the stairway. If an opening is made at the top of the stairway, fire and smoke could be pulled

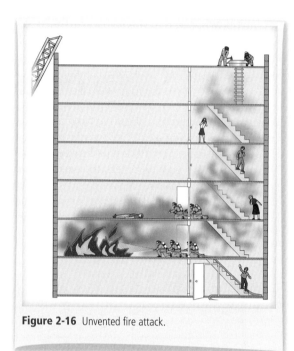

Figure 2-16 Unvented fire attack.

Figure 2-17 Fire vented at top of stairway. Fire is entering the stairway at the second floor, causing heavy smoke above.

into this "chimney-like" flow path of least resistance, making conditions much worse in the stairway (Figure 2-17). Opening the scuttle at the top of the stairs can be a good tactic if the fire is controlled but can cause serious injury or death if done under the wrong circumstances. Partially closing the door at the base of the stairs could reduce the unintended consequences of venting the stairs via the roof scuttle.

Ventilation possibilities can be known through preincident planning. Roof openings such as scuttles and vents may be present. The roof vents shown in Figure 2-18 could be opened to vent the area below.

Areas where aerial apparatus can be positioned to reach the roof can be known in advance. Department SOPs may determine what types of roofs are to be vented, as well as how and when positive pressure ventilation is to be done. When venting an area, it is essential that hose lines be in place to attack the fire.

In many situations it is advisable to "soften the target" by applying water prior to making entry. Earlier laboratory research conducted by Underwriters Laboratories was verified in cooperation with NIST

Figure 2-18 Fixed roof vents.
Courtesy of Bernard J. Klaene.

and the Fire Department of New York City (FDNY) on Governors Island New York during a series of live burns in residential buildings. This research further verified that limiting air intake and applying water to soften the target prior to making entry reduced temperatures at nearly every level of every room.[14]

Occupancy Type

The building's use will determine how likely the building is to be occupied at the time of the fire, number of occupants, fuel load, fuel type, value of the contents, and many other essential facts. Major occupancies can and should be identified in advance of the fire through preincident planning. Smaller occupancies can often be readily identified (e.g., a single-family detached residence). However, some buildings are used for purposes other than their original intended use (e.g., single-family residence being used as a real estate office). See the chapter titled *The Role of Occupancy* for a complete discussion of various occupancies and tactics related to occupancy.

> **NOTE**
>
> Occupancy: Building's primary use—e.g., school, hospital, nursing home
> Occupant: Person who could be in the building
> Occupied: Presence of occupant(s) in the building

Texture © Eky Studio /Shutterstock, Inc; Steel: © Snapshot/Dreamstime.com

Estimated Number of Occupants

Estimating the number of occupants in a large building is difficult at best. However, the building's intended use (occupancy) provides evidence of relative occupant density. NFPA 101, *Life Safety Code*,[15] establishes the maximum occupant load per square foot by occupancy type (e.g., one person per 100 ft^2 [9 m^2] in an office building [business]). In places of public assembly, the maximum number of

occupants may be posted. In hotels and apartment buildings, the number of available rooms/apartments provides an indicator of maximum occupant load. All of these details provide a starting point for estimating the number of people that could be in the building. Time factors, such as hours of operation, combined with the maximum occupant load and evacuation status help determine the number of people endangered by fire and the possible need for some level of assistance. Part of knowing your response district is being familiar with the location of local bars, restaurants, and other places that are most likely to be overcrowded. Some departments conduct special "life safety" inspections in places of assembly during times when overcrowding is most likely to occur.

Evacuation Status

Estimating the number of people still inside the building is the next logical step after estimating how many people may have occupied the building upon discovery of the fire. Some occupancies (e.g., primary and secondary schools) routinely conduct fire drills and account for occupants when a fire alarm sounds. The NFPA recommends that each home have an evacuation plan following the Exit Drills in the Home (EDITH) educational program. Many times family members account for other family members; unfortunately this is not always the case and information from occupants is not always accurate. Most buildings do not have an occupant accountability system.

Consider a fire in a large apartment building. Some of the residents probably would not be home when the fire occurred, others would evacuate and leave the scene, and some occupants may have guests. For all these reasons and more, the only way to be sure the building has been evacuated is to conduct a primary search, and even then there is a need to verify the search with a secondary search. Information from occupants who have escaped would be less reliable in a larger building without a formal accountability system than in a school or home.

If a preplanned building has an occupant accountability system, note the specifics of the plan on the preincident plan. Of particular interest is who will report to the fire department and where this person will be located (including an alternate site). Also note where occupants will be assembled. During preincident planning, be sure that occupant assembly areas do not conflict with expected fire department response routes or on-scene operations.

Occupant Proximity to Fire

Obviously, being close to a well-involved fire places occupants in imminent danger. The *Life Safety* chapter explains the life safety priorities for occupants on various levels of a multistory building. Knowing the location of occupied areas through preincident planning can be very useful during the size-up process.

Awareness of Occupants

People who are awake and alert are more likely to hear an alarm or sense the products of combustion and then take action to evacuate the building. Occupants who are asleep or mentally incapacitated may not be aware of the fire. Awareness is related to occupancy and will be further explained in the chapter titled *The Role of Occupancy*. Identifying the occupancy type during preincident planning provides an indication of occupant awareness (e.g., a nursing home fire at night where occupants are likely to be sleeping and possibly under the influence of medications).

Mobility of Occupants

Most large buildings will contain persons unable to fully evacuate on their own. As with other factors, knowing the occupancy of the building can be very helpful in determining if occupants are likely to successfully escape on their own.

Occupant Familiarity with Building

In most cases, the people inside a place of public assembly are not familiar with the building layout or alternate exit facilities. Many large loss-of-life fire reports in places of assembly address unfamiliarity

as a major problem. People who live or work in a building would be expected to know the building layout and location of exits. However, people will be most familiar with their regular ingress path. For example, a person working on an upper floor of a high-rise office building likely enters the lobby and takes an elevator to his or her office every workday. Unless the building management regularly conducts fire drills, many of the employees on upper floors will not use the stairway or even know its location.

Primary and Alternative Egress Routes

This factor is related to the mobility factor inasmuch as most occupants will escape unassisted if there is sufficient egress. Codes specify egress facilities, and preplan drawings should show the location of all exits. Some occupancies, such as hospitals, designate a defend-in-place strategy for occupants.

Many public buildings also have **areas of refuge** where immobile occupants wait for assistance. It is essential that preincident plans address these special facilities. Fire fighters must know the location, how many occupants could be at the area of refuge, and what the occupant and building managers expect to happen during an emergency. Often, fellow occupants or security personnel are designated to assist people in these locations. The fire department must always check areas of refuge during a fire emergency.

Medical Status of Occupants

This factor requires a twofold analysis: (1) What is the normal medical status of people occupying the building, and (2) what effect is the fire having on building occupants? Whenever people are still in the building upon arrival, emergency medical service units need to be summoned to the scene. A medical branch is needed if there are large numbers of potential victims.

Operational Status

SOPs provide a standard way of going to work at a fire scene. This allows the operation to get started, following a predictable plan. However, there are situations that require a different approach. The IC continually evaluates the safety and effectiveness of the standard operation and determines whether a nonstandard attack would be safer and/or more effective.

Safety is the IC's most important consideration. If the building is in danger of early collapse or the situation necessitates a transition from an offensive to a defensive operation for other reasons, the IC must order all personnel out of the building.

The effectiveness of the operation must constantly be evaluated. The IC must consider the following questions: Is the search being conducted in a systematic manner? Are the occupants who are in the most danger being rescued? Has the fire been properly vented to control fire spread? Is progress being made in controlling the fire? Are rate-of-flow requirements being met? Has salvage been considered?

It is critical that the IC continually reevaluate the situation in terms of risk management. The question that should regularly be asked is whether the desired benefit is worth placing fire fighters at risk.

Most of the operational status factors apply to an IC assuming command of an operation that is in progress.

Adherence to SOPs

This factor is evaluated when command is transferred. Are on-scene units following SOPs? Is the operation progressing as expected? Safety is a major part of this analysis. Is the operation being conducted safely? Are SCBAs and turnouts being used as prescribed? Is there a need to change from an offensive to a defensive strategy? This is one of the few size-up factors about which very little can be known in advance of the fire through preincident planning. However, SOPs that are developed in preparation for fire emergencies provide an advance indicator.

Hazard Control Zones and Perimeters

Fire zones and perimeters will be discussed in the *Fire Fighter Safety* chapter. Generally, the fire zone refers to an area where a specified level of protective clothing is required or possibly a safe area where no protective clothing is needed. The fire perimeter is set up to keep nonresponse people out of the area. The IC assuming command should evaluate the present zones/perimeter to determine if they are sufficient. General rules for fire zones/perimeters, including zones established during an active shooter incident, should be part of department SOPs. In addition, preincident plans for large secured properties may designate specific perimeters.

Accountability

The primary fire fighter accountability system is NIMS. A properly organized incident provides the first level of fire fighter accountability. NIMS is a safety system. It is important that freelancing be eliminated and that units working in the hazard area work as a unit (usually a fire company at a structure fire) within the command system. A formal accountability system is required by NFPA 1500, *Standard on Fire Department Occupational Safety and Health Programs*. The IC must ensure that the local system is being used. The *Fire Fighter Safety* chapter includes a detailed description of accountability systems.

> **NOTE**
>
> The fire department accountability system accounts for all fire fighters at the incident scene. It does not account for building occupants.

Texture: © Eky Studio /Shutterstock, Inc.; Steel: © Sharpshot/Dreamstime.com

Rapid Intervention

NFPA 1500, NFPA 1710, and NFPA 1720 require a **rapid intervention crew (RIC)**. A team of at least two fire fighters must be immediately available to rescue fellow fire fighters who need assistance. Although not required by regulations and standards, good practice dictates that at least an officer and three fire fighters be assigned to the RIC as soon as possible. The *Fire Fighter Safety* chapter also includes a detailed description of RICs.

Organization and Coordination

NIMS, as described in the chapter titled *Organizing, Coordinating, and Commanding Emergency Incidents* is the acceptable method of organizing an incident. All units on the scene must be included in the organizational structure. Activities of units must also be coordinated. The first-arriving company begins this process by organizing the initial attack and communicating orders to incoming units. When command is transferred, the person assuming command must be sure all units are working toward common tactical objectives within the overall strategy.

Rescue Options

The IC first evaluates the ways occupants can be removed from the building and then selects the safest and most efficient option. The preincident plan can identify primary means of egress, which are the preferred evacuation pathways, but should also address alternatives such as fire escapes and ladder positions. See the *Life Safety* chapter for a more detailed discussion of rescue options.

Staffing Needed to Conduct the Primary Search

The size of the area to be searched, smoke conditions, rescue methods available, and the condition of the occupants determine how many crews should be assigned to the primary search. In a single-family dwelling, a single company or possibly one company per floor conducts the primary search. In a large area with obstacles and poor visibility, several companies may be required to complete the primary search of a single floor. If victims must be physically removed, there will be a need for additional staffing.

Staffing Needed to Conduct the Secondary Search

Quite often the same teams that perform the primary search will conduct a secondary search in an area other than the area where they originally searched. It is recommended that a different company perform the secondary search, as they may look in areas that the first team overlooked. In most cases, the secondary search will not require increased staffing.

Staffing Needed for Interior Rescue/Evacuation

This task could involve the same staffing required for the primary search, or additional teams may be assigned to remove victims who are found during the primary search. In many cases, occupants will be attempting to evacuate on their own; staffing may be needed to assist and direct them to safety.

Staffing Needed for Exterior Rescue/Evacuation

If ladder or other exterior rescue methods are warranted, additional staffing will be needed. Most exterior rescues require more staffing per rescue than removal via the interior stairs.

Apparatus and Equipment Needed for Rescue/Evacuation

In most situations, an aerial apparatus or ground ladder will be the apparatus/equipment needed for rescue. There is other rescue equipment that is rarely used (e.g., rope rescue equipment).

Access to the Building's Exterior

When preincident planning a building, it is important to note apparatus entry points for secured properties. Quite often, special arrangements are made with building security for these properties. Street or road access around the structure should be included on the preincident plan drawing. Also note aerial access points and roadways that are not safe or not accessible to fire apparatus. When buildings are set back a considerable distance from the public roadway, companies may need to position apparatus on a private driveway to carry out rescues via aerials or to be within range of their attack hose.

Incident Summary

On September 3, 1991, a fire claimed the lives of 25 employees at a chicken processing plant in Hamlet, North Carolina. Locked exits were reported as the reason workers could not exit. These same locked exits could prevent fire fighters from exiting the building or delay the rapid intervention crew in gaining access.

Flames: © Jag_cz/ShutterStock, Inc.; Paper: © silverjohn/ShutterStock, Inc.

Access to the Building's Interior (Forcible Entry)

In response to the September 11, 2011, attacks and the increase in active shooter incidents, there is more security on the exterior and interior of buildings. Some buildings that were formally open to the public are now blockaded. School lockdowns often place students and staff in concealed secure locations. Be aware of changes to buildings and security measures in your jurisdiction. The need for forcible entry can significantly delay search and rescue operations as well as the initial fire attack.

Note the times when the building is secured or when special security features such as internal locks and gates impede entry. While performing preincident plan surveys, check the accessibility and security of doors while the building is open for business. Fire codes state that no special knowledge or key should be required to exit through a required exit.

Structure

A thorough risk-versus-benefit analysis determines whether the operation will be offensive or defensive. A major risk-versus-benefit consideration involves evaluating structural conditions. If the structure is in imminent danger of collapse, no expected benefit is worth the risk posed by entry into the structure. If the building is being damaged but adequate time

remains before a potential collapse, an offensive attack posture to save lives could be warranted. A decision may be made to contain the fire and rescue occupants, with a switch to a defensive attack once the primary search and rescue are complete. The effect of the initial operations should be reevaluated once occupants are removed from the structure.

IC Rule of Engagement for Fire Fighter Safety

Extend vigilant and measured risk to protect and rescue savable lives.

Texture: © Eky Studio /Shutterstock, Inc; Steel: © Sharpshot/Dreamstime.com

The necessary risk-versus-benefit analysis is somewhat subjective. However, experience and training can greatly improve its accuracy.

Signs of Collapse

Collapse can be catastrophic (failure of a supporting structure with complete or nearly complete destruction of the building). However, there are many partial collapses that injure and kill, such as a floor collapse where fire fighters are struck by the floor/ceiling assembly above or fall into a burning lower level. See the *Fire Fighter Safety* chapter for a list of possible collapse cues, but be aware that many collapses occur with no perceptible warning, especially in buildings constructed using lightweight materials such a trusses.

Collapse Zone

In the *Fire Fighter Safety* chapter, a collapse zone the height of the structure plus an allowance for debris scatter is recommended. In most cases, this translates to a collapse zone that is one and a half times the height of the building. It is very difficult to estimate the height of the building looking up from the street level. It is much easier to measure or estimate building height during preincident planning.

Other pragmatic decisions can be made during the preplanning tour. Determine if the width of the streets is less than the predicted collapse zone. In such a case, a decision could be made to avoid placing apparatus in the street in front of the building if there is a collapse hazard. Prior to the fire is also the time to determine the reach of master streams; can water be applied to the building from distances beyond the collapse zone? Determining a collapse zone provides a good example of how the same size-up factor can be primary or secondary. When confronted with a well-involved fire in a noncombustible structure, determining the collapse zone is a primary factor Figure 2-19 . In contrast, it would be a secondary factor for a minor fire in a fire-resistive building.

Construction Type

It is highly recommend that all members of fire departments become familiar with building types and common building problems. Few fire fighters would be able to conduct an engineering analysis to determine structural stability, but they can recognize potential structural problems during preincident planning and by studying past fires in different types of construction. *Building Construction for the Fire Service*[16] is a "must read" for anyone who might

take command at a structure fire, including the first-arriving officer.

During preincident planning, the building should be classified by construction type:

- **Type I construction**: Fire-resistive
- **Type II construction**: Noncombustible
- **Type III construction**: Ordinary
- **Type IV construction**: Heavy timber
- **Type V construction**: Frame

See NFPA 220, *Standard on Types of Building Construction*,[17] for specific fire-resistive requirements for each type of construction. Some buildings defy classification, particularly buildings that have been renovated. The original construction might be classified as fire-resistive construction, but removal of protective coatings on steel structural members could reduce the building to noncombustible status. NFPA 5000, *Building Construction and Safety Code*,[18] would classify a building so modified as the weakest form of construction (noncombustible in this example). There can also be different types of construction in various areas of a building. An older section of a building may be constructed using heavy wooden beams while a newer section may have wood or steel truss construction.

Code requirements change. Typically, the code enforced at the time of construction is the code that will apply to a building. For example, in some jurisdictions sprinklers were not required in high-rise buildings until 1972. Many older buildings remain unsprinklered, although many cities now require high-rise buildings to be retrofitted with sprinklers, regardless of when they were built.

The fire-resistive (Type I) building, as the name indicates, is superior to all other building types in regard to structural stability under fire conditions. Some might argue whether wood frame or noncombustible is most likely to collapse. Wood frame is generally of limited size, whereas noncombustible construction is used on very large buildings. Most large noncombustible buildings are sprinkler protected. The *Fire Fighter Safety* chapter discusses the relative safety of the five major types of construction.

Figure 2-19 Building in danger of imminent collapse.
© Glen E. Ellman.

Roof Construction

The roof covering can be important, particularly if it is combustible, such as wood shingle. However, the structure that supports the roof is generally most important to fire fighters in this era of nearly all non-combustible roof coverings. Roof collapse is a killing mechanism in its own right, but also a precursor to catastrophic collapse. The roof tends to tie in the walls of the building. When walls damaged by fire lose roof support, they often fail. Of particular concern are truss and other lightweight engineered supporting structures. Most modern buildings with large open areas will have a lightweight engineered roof support system. These long spans are particularly dangerous. The roof structure should be identified and noted during preincident planning. If a truss roof is present, especially a long span truss roof, it should be highlighted on the preplan narrative and a symbol added to the drawing indicating the truss roof. Truss roofs have been responsible for many fire fighter fatalities. The truss roof collapse mechanism and safety issues will be further discussed in the *Fire Fighter Safety* chapter.

Condition

A building in poor repair or one that is previously damaged presents an extra hazard for fire fighters. Abandoned buildings often have structural damage caused by weather, previous fires, or unauthorized occupants. Even if the building does not warrant formal preincident planning, a notation should warn responding fire fighters of structural damage or hazards such as holes in the floor. Some departments place a placard directly on the building to indicate structural damage or other structural problems. Even when a warning placard is affixed to the building, a backup notation system is recommended. The fire officer must consider serious prefire damage as a critical factor when deciding whether to take an offensive or a defensive approach.

Live and Dead Loads

The **dead load** is the load imposed on structural members by the building and permanent attachments.

The type of construction will be a major factor in determining the overall dead load. Of special consideration are heavy roof loads such as roof-mounted equipment, particularly if the roof is supported by unprotected truss construction. Heavy roof loads should be noted on the preincident plan.

Live loads are those loads produced by the building contents. The live load will vary from very light to extra hazard. Warehouse buildings tend to carry very heavy live loads, with a possible added impact load due to materials-handling equipment. If the materials being stored are combustible, they also add to the fuel load. Extraordinary live loads should be noted on preincident plans. Knowing a building's live load is important when developing incident-specific tactics.

Fire Suppression Water Load

The obvious water load is the weight of the water fire fighters discharge into the building during fire operations. Each gallon of water weighs 8.33 pounds (3.8 kg). Therefore, a 1000-GPM (63-L/sec) master stream operating into the building is adding 8330 pounds (3778 kg) to the building each minute. Given the fact that defensive operations often require several master streams operating over a prolonged period of time, it is easy to see how this water load could affect building integrity. Fortunately, most buildings will hold only a small percentage of the water from these master streams, as the water will run down vertical openings to the basement or outside. It is very difficult, if not impossible, to accurately determine how much water is being retained in any area of the building. Here again, preplan information can be invaluable. If stored materials are stacked directly on the floor and are absorbent, a large percentage of water will be retained. In contrast, storage on skids tends to allow water to run off. NFPA 1, *Fire Code 34.4.1.3*, addresses the hazard created by the weight of water:

> Where storing water-absorbent commodities, normal floor loads shall be reduced to take into account the added weight of water that can be absorbed during firefighting operations.

Water-absorbent materials are sometimes stacked against, or near, walls. As these materials absorb water from sprinklers or fire streams, they expand and push on the walls, sometimes resulting in an unexpected collapse. NFPA 1, *Fire Code,* requires that water-absorbent materials be stored no closer than 24-in. (0.6 m) to a wall. Other factors can also affect runoff. When a building is severely damaged, structural members, wall materials, and contents can fall on the floor, forming dams that retard water runoff. When pumping large quantities of water into a building, observe the runoff and consider the possibility that the weight of the water could facilitate structural failure.

Enclosures and Fire Separations

The type of construction and occupancy, both of which can be known before the fire, will be major clues in determining extension probability. Fire-resistive construction contains fire-rated compartments. However, the size of the compartments will vary by occupancy. A business occupancy such as the World Trade Center may have large, open office spaces, which will allow the fire to freely spread within the compartment. Residential high-rise buildings may have a few large common areas but will mainly consist of small rooms, which hinders fire extension.

Noncombustible (Type II) construction may also have large open areas, but the fire-resistive qualities between areas will generally be inferior to fire-resistive barriers.

The balloon-frame building is noted for a lack of compartmentation, having no rated fire assemblies to prevent fire spread. However, even nonrated doors and walls will retard fire spread for a period of time.

In a multistory building, the floor/ceiling assembly provides a barrier between floors. This barrier will be more effective in fire-resistive construction than in other types of construction. Floor/ceiling assemblies are seldom perfect barriers due to penetrations made for wiring, plumbing, air movement equipment, and other building services.

Extension Probability

Extension probability is directly related to the enclosures and fire separation factor just discussed and the concealed space factor discussed next. Fire spreads through concealed spaces and from area to area when enclosures are missing or weak. During size-up, the fire officer must consider how the fire spread will endanger fire fighters and occupants.

Concealed Spaces

Most buildings (except for heavy-timber, Type IV, construction) contain multiple concealed spaces. Of special note are attics, particularly when several buildings or units share a common attic. Fire separations are often damaged or removed in buildings that were originally constructed with attic fire separations. Sometimes the only separation between the attic and useable space below is a suspended ceiling. Many ICs have been surprised by a fire that breaks through the roof or extends to a distant apartment or store. If the concealed space involved in fire contains a truss roof or floor assembly, expect rapid collapse. Immediately communicate this finding to the IC.

In multistory buildings, suspended ceilings create a false space between the actual floor or roof assembly above and the ceiling tiles below. Flow paths created in this false space can result in fire cutting off fire fighter egress. As fire fighters make entry, they should use a thermal imaging camera (TIC) to determine whether the fire has penetrated the ceiling

NOTE

The thermal imaging camera is not actually a camera; therefore, NFPA documents, such as NFPA 1801, *Standard on Thermal Imagers for the Fire Service,* use the term "thermal imager." However, because the term "thermal imaging camera" is in widespread use within the fire service, this text uses the term "thermal imaging camera," or its abbreviation, "TIC."

Texture: © Eky Studio /Shutterstock, Inc; Steel: © Sharpshot/Dreamstime.com

and made its way into the false ceiling area. Under certain conditions, TICs are inaccurate; therefore, it is sometimes necessary to push up ceiling tiles to see if fire has extended into this space.

Fire fighters must use extreme caution when gaining access to this false space. Opening the ceiling could provide air to a ventilation-limited fire. A hose stream should always be immediately available before lifting ceiling tiles. The time to discover concealed spaces and truss roofs/floors is during preincident planning. Smoke conditions may not allow an examination during the fire.

Age

The age of the building can have positive and negative effects. Most older buildings will have heavier, more fire-resistive construction. However, buildings that have been renovated may contain new lightweight construction in the renovated areas. Not everyone who remodels secures a building permit, but renovations can be discovered as you tour your response area. It is more likely that a building permit will be secured when larger commercial buildings are remodeled. The fire department should be part of the permit approval process, or at the very least be notified when such construction is taking place. Renovated buildings using lighter construction can be an extra hazard because fire fighters expect heavy construction methods in older buildings. The negative side of the building's age is that the building may be getting weaker as it ages. This depends on the construction materials and building maintenance. The year the building was constructed should be noted on the preincident plan. In many counties, the auditor maintains a website that lists all properties in the county, including the date the building was constructed. Other useful information may also be available at the auditor's site, such as drawings or aerial photographs of properties.

Height and Area

It is impossible to have a large fire in a small building, thus the building's size will at least partially dictate the total volume of fire. Rate of flow formulas are based on the size of the fire compartment. Rate of flow is thoroughly discussed in the *Offensive Operations* chapter. The location and size of large undivided areas of buildings should be noted on preincident plans.

The height of the building determines the effectiveness of ground-based fire apparatus. When preincident planning a multistory building, determine the maximum reach of available aerial apparatus in terms of floors (e.g., aerial will reach the sixth floor). There may be areas around the building where aerial apparatus cannot be used due to obstructions or lack of access. Aerial access points should be identified in advance of the fire.

The height of the building will also affect the number of possible occupants, the type of construction, fuel load, and other factors. The height of the building should be included in the preincident plan. All high-rise buildings should be preplanned.

Complexity and Layout

The preincident plan should include a general floor layout for the building along with any information that might affect firefighting operations or fire fighter safety. The layout of some buildings is maze-like, particularly large buildings with multiple additions. **Figure 2-20** shows a large medical complex

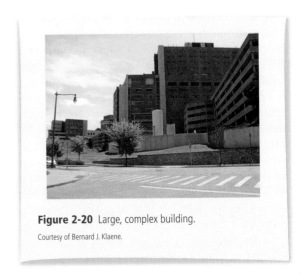

Figure 2-20 Large, complex building.
Courtesy of Bernard J. Klaene.

that has undergone multiple additions and demolitions. When visiting a large, complex building, take a moment to orient yourself. Try to determine your location in relation to the stairway, street, standpipe, and other critical landmarks. Could you find your way out under heavy smoke conditions? Storage occupancies may have aisles and cross-aisles that could be confusing under low-visibility conditions. Large open areas in business occupancies may have cubicles separating work areas. The layout of these cubicles can be extremely confusing and very difficult to search under fire conditions.

Green Buildings

The increased emphasis on energy conservation has resulted in an increase in "green energy technology." Many of the "green" concepts, such as solar panels, have been in use for decades. Solar panels have been placed on roofs for many years, but their use is now more commonplace. Standards regulating the installation of solar panels are evolving. Many existing solar panels do not meet current standards.[19] Green roofs designed to grow grass and other vegetation on the roof are a fairly new phenomenon. Both solar panels and green roofs add to the roof load and hinder vertical ventilation. Solar panels have proven difficult to deenergize,[20] and the wiring system is complex. The solar panels collect energy in the form of direct current electricity, which is then transformed into alternating current using an inverter. Disrupting power to the building at the inverter panel may not affect the energy level being emitted from the roof panels, which is usually 600 volts.

Fire departments should survey their response area to determine the location of "green roofs," massive arrays of solar panels, roof-mounted wind turbines, and other green construction. If such structures are found, contact the building department and request an analysis of the structural stability of the roof support system. The preincident plan should address whether roof operations should be conducted along with precautions to be taken.

Lightweight structural members and energy-efficient windows are considered green. The problems associated with lightweight construction under fire conditions are well known. Energy-efficient windows delay self-venting, which can be an advantage or disadvantage, depending on circumstances.

Extinguishment
Probability of Extinguishment

The probability of extinguishment is also important to life safety. When the fire is extinguished, the probability of death and injury are greatly diminished. Extinguishment is nearly always a primary factor at a working structure fire. Rate of flow is discussed later in this chapter, as well as in the *Offensive Operations* chapter. Flow requirements and the extent of fire are key factors in determining if a rapid, lifesaving extinguishment operation is possible.

Offensive/Defensive/Nonattack

Deciding the overall attack strategy is critical. The entire operation hinges on this early decision. Whenever command is transferred, the person assuming command must consider whether the present strategy is correct. Critical factors change as tactical objectives are achieved. The most important objectives are those related to life safety. Once the building is cleared of occupants, the reason for continuing an offensive attack is reduced; therefore a new risk-versus-benefit analysis should be conducted to determine if an offensive attack should continue.

IC Rule of Engagement for Fire Fighter Safety

If after completing the primary search, little or no progress toward fire control has been achieved, seriously consider a defensive strategy.

Reproduced from: © 2011, IAFC Safety, Health and Survival Section. All rights reserved.

Texture: © Eky Studio /Shutterstock, Inc; Steel: © Sharpshot/Dreamstime.com

Ventilation Status

Finding the fire location is much easier when the fire has self-vented in a building that is still safe to enter. However, a vented fire will generally progress toward full involvement much faster than a fire starving for oxygen (ventilation-controlled fire). If the fire is unvented, it is generally best to place hose streams in an attack position prior to venting.

Venting for extinguishment involves making openings to create a flow path that directs heat and smoke away from the hose crew to allow a quick and efficient fire attack. Part of the preincident planning process involves identifying available built-in vents and evaluating the roof structure for roof operations.

External Exposures

Are nearby structures (not connected), vehicles, and other property threatened by the fire? Protecting exposures can be done in several ways, as discussed in the *Offensive Operations* and *Defensive Operations* chapters. The preincident plan drawing should show the location of nearby exposures. The narrative part of the preplan should identify the occupancy and any special considerations regarding exposures. Many times, large liquefied petroleum gas tanks or tanks containing various hazardous materials are located at the rear of a building. A liquid oxygen tank can be seen in Figure 2-20. Check all sides of the building when noting the location of exposures.

Internal Exposures

Other parts of the fire building or connected structures are categorized as internal exposures. Identifying fire pathways is best done prior to the fire. At the time of the fire, the IC evaluates whether the fire is likely to follow one of the preidentified pathways. Tactics for protecting internal exposures are also discussed in the *Offensive Operations* chapter.

Manual Extinguishment

If there is no automatic fire suppression system or the system does not control the fire, manual extinguishment will be necessary. The *Offensive Operations* and *Defensive Operations* chapters both address this topic.

Fuel load

Fuel load varies as to quantity, type of fuel, geometric orientation, and other factors. Fuels identified as "extra hazard" in sprinkler calculations would generally require more water to extinguish than fires in ordinary combustibles. While fuel loads tend to change, in some buildings, the maximum and average fuel load can be evaluated during preincident planning. The Underwriters Laboratories study, Impact of Ventilation on Fire Behavior in Legacy and Contemporary Residential Construction, indicates that residential fuel loads found today are higher and burn faster than in the past.

Calculated Rate of Flow

The big question in calculating rate of flow is whether the standard attack line with a backup line can extinguish expected fires. It is impractical to calculate rate of flow at the time of the fire. If the fire compartment is larger than two standard preconnects can extinguish, or if the fuel load will create a fire situation beyond the control of two standard attack lines, the rate of flow should be precalculated for these areas and noted in the preincident plan. The *Offensive Operations* chapter explains the formulae used to calculate rate of flow.

Number of Hose Lines Needed for Extinguishment

The number of hose lines needed for extinguishment is an extension of rate of flow. Once the rate of flow and the flow provided by department hose lines are known, the required number of hose lines needed to extinguish an advanced fire can be estimated.

Additional Hose Lines/Master Streams Needed

Extinguishment is the primary purpose for using handheld hose lines and master streams, but there is also a need to cover internal and external exposures, provide a backup line, and protect critical egress routes.

Staffing Needed for Hose Lines

In most cases, a full fire company should be assigned to each hose line, except when the apparatus operator is needed at the pump panel (two members are assigned as the initial RIC) or when there are more than five people assigned to the apparatus. Departments should operate nozzles and hose lines under simulated fire conditions during training. This training should include advancing the hose up stairways, through doorways, and around obstacles to determine the actual staffing needs for each hose and nozzle combination. Keeping company members together also results in better accountability.

Water Supply

Once a fire occurs, it is too late to evaluate the water supply capability. Some areas are fortunate to be protected by a high-flow and reliable water supply. Most areas have weaknesses in the water supply system and others have no public water supply.

Growing communities often outstrip their water supply, resulting in reliability issues. These communities are likely to experience lower pressures and lower flows during business hours as demand for domestic water increases on a marginal water system. Even well-established high-volume water supplies will contain problem areas if hydrants are on dead-end or small mains. If the system pressure is supplied by gravity and the protected area is hilly, pressures and flows will be different at various locations, depending on the difference in height between the hydrant and the water supply tank as well as the distance from the water supply (friction loss).

Some public and nearly all private water systems have a limited supply. These limitations must be understood when deploying large-flow appliances. Private hydrants are likely to have a separate supply that may be connected to automatic fire suppression systems. Sometimes connecting to these hydrants will have a negative effect on the automatic fire suppression system.

Large municipal water systems are often supplied by multiple water systems. These multiple systems can usually be cross-tied to supply water from adjoining systems, thus increasing the total flow available at the scene of a major emergency. If the water system has this capability, members should know how to open valves (or how to get assistance in opening valves) connecting water systems. Many water utilities have the capability to increase water pressure when large flows are needed.

Some communities or sections of communities do not have a public water system and rely on other sources such as ponds, rivers, and pools as a water supply. As with the municipal water system, the reliability of the water source must be known in advance. Questions that should be asked include the following:

- Will the source freeze during cold months?
- Is there access for apparatus assigned to draft water?
- Are dry hydrants or other water supply equipment available?
- Are water sources dry during part of the year?
- When tender operations are anticipated, how many tenders will be required to maintain the required fire flow?

Even departments with a strong water supply should plan for situations in which the water supply could be reduced or unavailable. After Hurricane Katrina, water supplies were unavailable, requiring large municipal departments to use alternative water supplies. Some departments with reliable water supplies remove hard-suction hose from their pumpers and no longer train pump operators to draft water from static sources. As mentioned previously, FDNY found it necessary to draft water from flooded streets to suppress a conflagration during Hurricane Sandy. All departments should consider alternative sources of water as part of their disaster preparedness plan.

Apparatus Pump Capacity

Apparatus specifications should consider available water supply and fire hazards when pumps are specified. Individual apparatus pump capacity is a known factor. However, the total flow needed for a given situation is incident dependent. Seldom is the apparatus pump capacity challenged during an offensive attack. During defensive operations, the need for water could be greater than the apparatus pump capacity. In many cases, total pump capacity exceeds the available water supply.

Manual Fire Suppression Systems

In most cases, manual fire suppression system refers to a standpipe system. The fire officer must make a decision whether the system is to be used in a one-story or low-rise building or in the lower floor of a high-rise building. Usually, it is best to use the standpipe because its use reduces the work required to advance a hose line. However, some standpipe systems are supplied by the same water supply as the sprinkler system or they may be designed for occupant use having a minimal flow. There are three types of standpipe systems: Class 1, Class 2, and Class 3. Class 1 and 3 systems are designed to be used by fire fighters and have a required minimum flow rate. Sprinkler and standpipe systems will be discussed in more detail in the *Fire Protection Systems* chapter. The type of sprinkler and standpipe as well as the flow capacity should be noted in the preincident plan.

When there is a chance that the sprinkler system will be ineffective because hose lines are using water needed to supply the automatic sprinkler system, it is best to avoid using the standpipe. The water pressure and flow are preestablished for standpipe systems, although pressure and flow can usually be increased by pumping into the fire department connection. If the standpipe has a severely limited flow or very low pressures, it may be best to use hose taken directly from the apparatus to attack a fire. The reliance on the standpipe is directly related to the height of the fire floor; the higher in the building the fire, the greater the reliance on the standpipe.

Preincident plans should show the location of control valves (main and sectional), pumps, fire department connections, and hose outlets. Some systems have pressure or flow-reducing valves. The preincident plan should explain the use of these valves, whether they can be adjusted, and what equipment is needed to adjust the pressure/volume. SOPs and preincident plans should specify hose, nozzles, and

equipment needed during standpipe operations. Standpipe systems are discussed in the *Fire Protection Systems* and *High-Rise Buildings* chapters.

Automatic Fire Suppression Equipment

There may be more than one sprinkler system in a large building. These systems may or may not be interconnected. Just as with the standpipe system, the location of control valves (main and sectional), pumps, fire department connections, and hose outlets should be shown on the preincident plan for sprinkler-protected buildings. When a building is equipped with an automatic sprinkler system, the primary tactic involves letting the system do its job while fire fighters support the system and move in for final extinguishment. The sprinkler system becomes the first line of defense. It is essential that fire protection systems be properly maintained and that maintenance records be consulted when preincident planning a protected building.

Other automatic systems may be found in buildings; they must be addressed in the preincident plan. See the *Fire Protection Systems* chapter for more information on automatic fire suppression systems and tactics to be used at protected properties.

Property Conservation

Property conservation is the third operational priority. Seldom will factors related to property conservation take on the urgency associated with life safety and extinguishment. Information about property value and location within the building can usually be known in advance; preplanning property conservation measures can significantly reduce the overall loss. See the *Property Conservation* chapter for specific property conservation tactics.

Salvageable Property

Nearly every property has some form of salvageable property. During preincident planning, it is important to determine what there is to salvage and the value of the property. Remember that property might also have nonmonetary value. Most fires occur in residential occupancies that contain personal property, such as heirlooms and photographs that are considered invaluable by the occupants. Being careful and protecting these personal items can improve the public's perception of the fire department.

Location of Salvageable Property

In the residential occupancy, salvageable property can be found throughout the building. In other occupancies there may be concentrations of high-value property such as computer rooms in a school or business occupancy. Preincident plans should identify these locations and possibly provide guidance in handling the salvage effort (e.g., the location of accounts payable records, which are generally vital to business continuation). Make a note on pre-plans where critical files are kept or stored, and recommend to the owner that they protect these files.

Water Damage

The primary extinguishment tactic involves applying water to the burning material, thus there will be some quantity of water in the building. Water will migrate through various openings and possibly damage property on floors where water is discharged and below.

Probability of Water Damage

Some buildings tend to hold water while others have drains or pathways that allow water to harmlessly drain off. Property in water pathways below the fire will have a high probability of water damage.

Susceptibility of Contents to Water Damage

Susceptibility of contents to water damage has to do with how easily the property is damaged. Some property is highly vulnerable to water damage (e.g., paper files and electronic equipment).

Water Pathways to Salvageable Property

Water will flow downward through paths of least resistance. If the floor contains openings or holes,

water will flow through the openings or holes to the floor below. Otherwise, it will flow to stairways, elevators, drains, or other routes.

Water Removal Methods Available

Preferred methods involve using built-in features that drain the water out of the building. Otherwise, water removal will involve pushing or directing water to channels leading out of the building, or possibly removing it with water vacuums.

Protective Measures Available

When water cannot be directed away from salvageable property, the most common way to protect property from water damage is to place covers over the exposed contents, starting with the property that is in the water pathway and most susceptible to water damage. At times it is possible to move contents above floor level (e.g., placing furniture on skids or placing more valuable property on top of less valuable property). Water removal, as mentioned previously, is also a means of protecting property from flowing water.

Smoke Damage

Smoke can infiltrate the entire building, especially through ventilation systems. However, the most common pathway for smoke is upward; therefore, most smoke damage occurs on the fire floor and above.

Ventilation

The best way to reduce smoke damage is to ventilate the building. Some ventilation methods that are ill advised prior to full fire containment are acceptable after fire control or extinguishment. As an example, venting the stairway, as mentioned under Ventilation Status, could place fire fighters and occupants in danger. Later in the operation, this may be a good option for reducing smoke damage.

Property that is susceptible to smoke damage on the fire floor and above is most likely to be damaged. Materials that absorb smoke are more susceptible to damage as is sensitive electronic equipment.

Forcible entry and ventilation are often necessary to achieve tactical objectives, but they can be overdone. If a building is locked and no key is available, forcible entry will nearly always result in some damage, as will ventilating. Damage should be limited to what is needed to achieve tactical objectives.

General Factors

Total Staffing Available Versus Staffing Needed

The implementation of an incident action plan will require resources, and until the actual plan is developed, the exact number of resources needed is unknown. However, the IC needs to make approximations during the early size-up. Staffing is generally the most important and difficult resource to obtain when initiating an offensive attack.

NFPA 1710 establishes a minimum staffing level of 14 or 15 fire fighters to safely and efficiently combat a working structure fire in a two-story, single-family dwelling with no basement and no threatened exposures. As the size and complexity of the property increase, the required staffing will also increase, as demonstrated in the NIST High-Rise report.[21] If the calculated fire flow requires multiple 2½-in. (64-mm) attack lines, more staffing will be needed to meet the rate of flow. Staffing requirements enumerated in NFPA 1710 will be further discussed in later chapters. The *High-Rise Buildings* chapter further explains the NIST High-Rise staffing report.

Searching large areas, physically removing victims, meeting larger rate of flow requirements, or responding to areas beyond a fixed water supply will require additional staffing.

Available water resources become a high-priority issue in areas without hydrants or when the required rate of flow is large. Fire departments working in rural areas can sometimes summon large numbers of fire fighters to the scene, but the immediately available water supply is usually limited.

Offensive operations are typically labor intensive. Conversely, defensive operations require fewer fire fighters operating apparatus and equipment that require large-volume water supplies. In any case, the IC needs to estimate the staffing resources required.

Once an estimate of staffing requirements is made, the IC needs to consider how the responding resources match these requirements. Augmenting resources could take the form of requesting additional alarms or mutual aid. If resource capabilities cannot match the needs of an offensive attack, attempting an offensive operation will place fire fighters in extreme danger. If the needs of neither an offensive nor a defensive attack can be met, the IC's only safe option may be to assign available resources to exposure protection.

Total Apparatus Available Versus Apparatus Needed

Offensive operations are staffing intensive; the initial response generally provides an adequate number of apparatus. Defensive operations are apparatus intensive and require more apparatus to supply water for master streams, as well as master stream equipment requirements such as aerial apparatus and apparatus-mounted master streams. Each master stream will generally require a minimum of one apparatus. If the apparatus does not have pumps, such as an aerial truck without pumps, an additional apparatus will be needed to supply water.

Staging/Tactical Reserve

SOPs should address staging in a general way. Some properties, especially large complexes, may have specific staging opportunities that should be identified when preincident planning.

When all staffing or all apparatus are being used and the incident is not resolved, a tactical reserve is needed. For a small-scale incident, this reserve is usually at least one engine company and one truck company. Larger and more complex incidents require a larger tactical reserve. When confronted with a working fire, it is good practice to maintain a tactical reserve in staging, as forces can be rapidly deployed from a staging area. Waiting for resources to respond from distant fire stations can result in additional loss of life and property.

Utilities (Water, Gas, Electricity, Other)

Some departments routinely shut down the electrical supply in residential property. This can provide a safer environment and reduce chances of reignition. However, doing so can also be a very dangerous operation, even in a residence. Fire fighters should not attempt to disrupt the power supply in larger properties with high-voltage service. If the department routinely interrupts the power supply at residential properties, SOPs and preincident plans should specifically state that fire fighters should not disconnect power supplies when dealing with properties with high-energy service.

Residential gas or fuel supplies can be shut down if they are causing a problem or are leaking. Utility company personnel are better equipped to shut down electrical and piped gas supplies. Preincident plans should show the location of gas and electric shutoffs.

Special Resource Needs

Although not always considered a special resource, police and EMS are often needed at the scene. The IC should assess the need for these resources as part of the size-up. In many areas, police and EMS routinely respond to a reported structure fire. Other resources could also be required, such as air trucks, hazardous materials teams, the utility company, Red Cross, and others. Local Emergency Management Plans should include a list of resources available to the community with emergency contact information. The fire department should have access to this information.

Time

In examining the size-up/preincident plan factors, much can be known in advance of the fire through preincident planning. Time will not be known until a fire occurs. Consider the following time factors:

- Time of day: Time of day can determine the likelihood of people being in the building and their degree of awareness of the fire. An apartment building fire at noon generally presents a much different scenario than a fire in the same building at midnight. A fire in a Christian Church on Sunday morning could involve an extreme life hazard. A fire in the same church late on a Monday night might gain considerable headway before being discovered, but probably would not involve an extreme life hazard for occupants. Some properties (e.g., nursing homes) are continuously occupied.
- Day of week: As with time of day, occupancies are generally fully occupied on different days of the week. A school would normally be occupied during the week during the school year and minimally occupied at other times, except when a special event is taking place, such as an evening basketball game.
- Time of year: The time of year should also be considered. In the northern United States, winter is a time when people tend to stay inside more and use heating systems, which can serve as additional sources of ignition.
- Special times: Finally, special times that should be considered would include holidays. An office building early on Christmas day would contain few occupants, whereas this same building could house thousands of people on a normal workday. Likewise, a Christian church would be expected to contain a maximum number of occupants early on Christmas day.

Weather

The IC must consider the effect of the present weather conditions on the operation. The variety of weather conditions is nearly endless, but the IC should concentrate on extremes. High heat and humidity take a toll on fire fighters. Extreme cold also affects fire fighters and their equipment as well as victims who are being removed from a structure. High wind affects fire spread and ventilation, especially in high-rise structures or at large fuel fires, such as those at lumberyards. NIST has done extensive research on the subject of wind-driven fires, especially in high-rise buildings. However, wind-driven fires can be problematic in any building, as was experienced in a one-story residence in Houston, Texas, that resulted in two fire fighter on-duty fatalities.[22]

Fire departments in communities that experience little or no wind on a typical day can be caught off guard during rapidly changing weather conditions, such as thunderstorms, that can temporarily produce hurricane-force winds and change the dynamics of a fire. Improper ventilation tactics or sudden failure of windows can put fire fighters in peril before they realize what has happened. A vigilant IC, working with a planning and safety officer, should notice or anticipate changes in weather and adjust tactics accordingly.

The general kinds of weather that an area experiences can be known in advance, but the weather at the time of the incident is unknown until the alarm is transmitted. Fire fighters in southern Florida expect to deal with high temperatures and humidity several months a year, just as fire fighters in Alaska expect to deal with cold weather extremes; of course, a fire could occur in Alaska during the summer, with heat being a potential problem for fire fighters who are unaccustomed to working in warmer temperatures.

Extremely cold or hot weather conditions may require occupant shelter and fire fighter rehabilitation (rehab). Very cold weather can also result in frozen fire hydrants, frozen pumps on apparatus, and damaged apparatus equipment. During very cold or hot weather, most buildings will be tightly sealed to maintain internal temperatures, increasing the likelihood of ventilation-limited fires. Fires occurring during moderate weather conditions are more likely to be well ventilated via open windows and other openings.

Humidity is particular important during hot weather extremes and increases the fatigue factor. Humidity can also affect smoke movement. On a hot, humid day, smoke may stratify, creating heavy smoke conditions outside the building; this heavy smoke may require apparatus to be positioned farther away or require apparatus operators to don SCBA. High humidity and heat conditions can also cause smoke stratification within high-rise buildings.

Precipitation in the form of rain is less problematic than frozen precipitation, which slows response and makes it more dangerous, as well as creating hazardous on-scene operating conditions.

High winds can push the fire toward exposures or affect venting. High winds are particularly problematic during high-rise fire operations.

A list of size-up/preplan factors is included in Figure 2-13. An attempt has been made to include major factors affecting size-up, but invariably, one or more factors will be omitted from such a list. The three top priorities (life safety, extinguishment, and property conservation) are the basis of the size-up/preincident plan checklist. All of these factors are interrelated at some level. As an example, weather conditions have an effect on fire fighter and occupant safety. Weather also affects ventilation and thus extinguishment. Structural stability affects the attack method and almost every other tactic at the incident scene.

Size-up Chronology

Size-up follows a chronological sequence:

1. SOPs
2. Preincident plan
3. Shift/day/time
4. Alarm information
5. En route
6. Visual observations at the scene
7. Information gained during continuing operations
8. Overhaul

Size-up is a continuing and dynamic process. The on-scene evaluation should improve as observations and reports from working units provide additional and improved information.

Standard Operating Procedures and the Preincident Plan

SOPs and preincident planning activities are done well in advance of an alarm but play a significant role in the size-up when developing an incident action plan. They give the IC a head start on the size-up process and actually become part of the incident size-up.

Shift/Day/Time

As the fire officer reports for a shift, unusual weather conditions, day of the week, and related factors are considered. On occasion, special measures are taken, such as adding an extra company to certain locations because of snow-covered streets. The time of the alarm may also be very important. The life hazard in a high-rise office building will be much greater at 2:00 PM on a workday than at 2:00 AM on a weekend.

Special information may change from day to day. The water system may be out of service, a parade may be scheduled, or the area might have been experiencing extremely hot, dry weather for a prolonged period. These are all important factors to consider.

Alarm Information

When the alarm is received, the time of day factor is immediately processed along with dispatch information. Dispatch information can include the following:

- Building location/address
- Fire location
- Fire intensity (numerous calls usually mean big fires)
- Occupant status (e.g., report of an occupant trapped on the second floor)

Additional information may also be known about the area, or the specific building, or the dispatch file may contain a preincident plan notation.

While Responding

The dispatcher may be able to provide additional information to the responding units. An onboard computer could provide specific preincident plan information. Companies arriving on the scene should give status reports. For example, there may be visual indications of a working fire, such as visible smoke or flames. During the response, all previous information is reconsidered, such as what effect high winds will have on the operation. Initial information received from units on the scene is critical in the formulation of an incident action plan.

Visual Observations at the Scene

Many questions are answered as the IC gets the first look at the situation. When conditions permit, it is a good idea to get a view from several angles. By driving up to the building and going past the front prior to setting up a command post, the IC may be able to view three or even four sides of the building. However, this must be done quickly so that an incident action plan can be developed and implemented.

The combination of visual information and reconnaissance from companies working on the interior or other unseen locations provides the basis for the initial incident action plan. This may be the first opportunity to evaluate structural conditions and the effectiveness of action being taken. The big question is whether the current incident action plan will accomplish the desired objectives.

The quality and quantity of information will greatly increase with time. The IC goes from a dearth of information to what may be information overload. At this point, it is important to prioritize information as primary and secondary and spend time reevaluating the incident action plan on the basis of the information received.

Information Gained During Continuing Operations

If the operation is successful, good news such as "the primary search is complete and the fire is extinguished" will reach the IC. If on-scene resources are unable to accomplish tactical objectives, either bad news or no news will be communicated to the IC. Bad news requires an adjustment in plans. On occasion, the entire action plan is changed, such as shifting from an offensive attack to a defensive attack. After allowing time for companies to make entry and begin operations, "no news" is unacceptable. The IC must call for a status report when none is forthcoming.

Overhaul

Overhaul should be planned and should be deliberate. The emergency phase is over, and extreme caution should be taken to avoid injuring fire fighters. At this point, there is no justification for rushing to make decisions that could put fire fighters at risk.

Chapter Summary

- Standard operating procedures (SOPs), pre-incident plans, and incident-specific information are interrelated and are important components of the size-up.
- When preincident plans are available for a property, their use in evaluating the situation is crucial.
- A proper size-up is crucial to developing effective firefighting strategies and the incident action plan.
- Due to time constraints, the fire officer must focus on primary size-up factors when developing an initial incident action plan. Incident conditions will determine which size-up factors are more important (primary) or less important (secondary).
- The incident action plan should adapt to changes in circumstances or to additional critical information that indicates a need to change tactics.
- Structural firefighting SOPs are guidelines to be used at all structure fires except when circumstances indicate the need for a different approach.
- SOPs must address department-specific training and equipment but should also reflect regional procedures to facilitate operations with other jurisdictions and agencies.
- SOPs allow first-arriving units to take immediate action or stage while awaiting orders from command.
- SOPs specify action to be taken but should be flexible enough to allow fire fighters to follow a reasonable course of action when confronted with a situation in which modification of the procedure is appropriate.
- Preincident plans must be kept up to date.

- Preincident plans are needed when there is a high life hazard, a difficult extinguishment problem, or high-value property.
- Preincident plans should estimate the required rate of flow for large undivided areas, note any hazardous materials, and evaluate fuel loads.
- Building materials, roof construction, age and overall condition, fuel load, live loads, dead loads, and layout are key factors in determining likelihood of fire extension and building collapse.

Key Terms

area of refuge A floor area with at least two rooms separated by smoke-resisting partitions in a building protected by a sprinkler system, or a space located in an egress path that is separated from other building spaces.

backdraft A fire condition that occurs when oxygen (air) is introduced into a superheated, oxygen-deficient compartment charged with smoke and pyrolytic emissions, resulting in an explosive ignition.

balloon-frame construction An older type of wood-frame construction in which the wall studs extend vertically from the basement of the structure to the roof.

compartmentation Subdividing of a building into small areas (rooms) capable of limiting the spread of fire and the products of combustion.

complex preincident plan A plan used when a property has more than three buildings or when it is necessary to show the layout of the premises and relationship between buildings on the site.

Flames: Dix/Dreamstime.com; Steel texture: © Sharpshot/Dreamstime.com

Flames: Dizi/Dreamstime.com; Steel texture: © Sharpshot/Dreamstime.com

dead load The weight of a building; consists of the weight of all materials of construction incorporated into a building, including but not limited to floors, roofs, ceilings, stairways, built-in partitions, finishes, cladding, and other similarly incorporated architectural and structural items, as well as fixed service equipment.

drilling down Using a computer to navigate by pointing and clicking through a series of drop-down menus in a graphical user interface.

flashover An oxygen-sufficient condition in which room temperatures reach the ignition temperature of the suspended pyrolytic emissions, causing all combustible contents to suddenly ignite.

flow path The movement of heat and smoke from the higher pressure fire area toward lower pressure areas on the interior and exterior of the structure.

formal preincident plan A plan for a property with a substantial risk to life and/or property; includes a drawing of the property, specific floor layouts, and a narrative describing important features.

fuel load Fuels provided by a building's contents and combustible building materials; also called fire load.

hovering Accessing text or graphic items on a computer screen by placing the computer cursor over a specific area or by touching the screen in a touch-screen environment.

live loads The weight of the building's contents, people, or anything that is not part of or permanently attached to the structure.

notation A piece of information about the premises, such as damage to the building from a previous fire. This information may accompany a preincident plan or may be available when the building does not have a preincident plan.

platform-frame construction A construction technique using separate components to build the frame of a structure (one floor at a time). Each floor has a top and bottom plate that act as fire-stops.

preincident plans Written documents resulting from the gathering of general and detailed information to be used by public emergency response agencies and private industry for determining the response to reasonable anticipated emergency incidents at a specific facility.

primary factors The most important factors, assessed during size-up, which change from incident to incident and depend on specific incident conditions.

rapid intervention crew (RIC) A minimum of two fully equipped personnel on-site, for immediate rescue of injured or trapped fire fighters.

rate of flow The minimum water application rate required for extinguishment

secondary factors Less important factors at an incident, which change from incident to incident and depend on specific incident conditions.

softening the target Cooling hot fire gases using rapid, short-duration application (usually 15 to 60 seconds) from a straight or smooth bore fire stream aimed at a steep angle toward the ceiling from the safest effective location. When using this tactic, water often is applied from the exterior to the interior of a building.

standard operating procedures (SOPs) Written rules, policies, regulations, and procedures intended to organize operations in a predictable manner.

Superfund Amendments and Reauthorization Act (SARA) A federal law enacted in 1986 and also known as the Emergency Planning and Community Right to Know Act. Title III of this law requires businesses that handle or store hazardous chemicals in quantities above specific limits to report the location, quantity, and hazards of those chemicals to the State Emergency Response Commission (SERC), Local Emergency Planning Committee, and local fire department. Some of this information has been classified since the attack on the World Trade Center on September 11, 2001.

tenders NIMS term for vehicles that transport water from a water source to the fire scene; also referred to as a tanker or mobile water supply apparatus.

type I construction Construction method in which the structural members, including walls, columns, beams, girders, trusses, arches, floors, and roofs are of approved noncombustible or limited-combustible materials and have the highest level of fire-resistance ratings.

type II construction Construction method in which the structural members, including walls, columns, beams, girders, trusses, arches, floors, and roofs are of approved noncombustible or limited-combustible materials but the fire resistance rating does not meet the requirements for Type I construction.

type III construction Construction method in which exterior walls and structural members that are portions of exterior walls are of approved noncombustible or limited-combustible materials, and interior structural members, including walls, columns, beams, girders, trusses, arches, floors, and roofs, are entirely or partially of wood or smaller dimensions than those required for Type IV construction or of approved noncombustible, limited-combustible, or other approved combustible materials.

type IV construction Construction method in which exterior walls and structural members that are portions of exterior walls are of approved noncombustible or limited-combustible materials. Other interior structural members, including walls, columns, beams, girders, trusses, arches, floors, and roofs are of solid or laminated wood without concealed spaces. Wood columns supporting floor loads are not less than 8-in. (203 mm) in any dimension; wood columns supporting roof loads only are not less than 6-in. (152 mm) in the smallest dimension and not less than 8-in. (203 mm) in depth. Wood beams and girders supporting floor loads are not less than 6-in. (152 mm) in width and not less than 10-in. (254 mm) in depth. Wood beams and girders and other roof framing supporting roof loads only are not less than 4-in. (102 mm) in width and not less than 6-in. (152 mm) in depth. Specifics for other structural members are required to be large-dimension lumber as well.

type V construction Construction method in which exterior walls, bearing walls, columns, beams, girders, trusses, arches, floors, and roofs are entirely or partially of wood or other approved combustible material smaller than material required for Type IV construction.

wood truss An assembly made up of small-dimension lumber joined in a triangular configuration that can be used to support either roofs or floors.

Flames: Dno/Dreamstime.com; Steel Texture: © SharpShot/Dreamstime.com

Challenging Questions and Activities

1. Review your department's SOPs:

   A. Are they up to date? When was the last revision?

   B. Does the department have SOPs addressing:

   - Use of NIMS, including command transfer, establishing a command post, and specifying the IC?
   - Apparatus management—staging, positioning, collapse zones?
   - Structural firefighting, including water supply, initial attack, and company duties?
   - Operating in buildings with built-in fire equipment?
   - Special hazard operations, such as hazardous materials and technical rescues?
   - High-rise, large-area buildings and other special occupancies?

2. Select a department SOP that is out of date or otherwise needs revision. Rewrite the SOP.

3. Use the size-up/preplan checklist provided in Figure 2-13 to evaluate a fire in your jurisdiction. Assume the role of the first-arriving engine company officer responding to the fire shown in **Figure 2-21**. It is Sunday, September 4, at 7:00 AM; the temperature is 60°F (15.6°C). The address is 5442 Sunrise Avenue, which is located in an older (circa 1930) residential neighborhood. Upon arrival on the A side, you observe fire breaking through the roof. You then perform a 360-degree walk around. The conditions shown in Figure 2-21 are on the C side. No occupants are visible.

   A. Develop an initial report.

   B. Use the size-up/preincident plan check-list (Figure 2-13) to evaluate conditions.

   C. What is your command mode (fast attack, investigative, or command)?

   D. What would be your first action?

Figure 2-21 5442 Side C, Sunrise Avenue.

Courtesy of David J. Jones, Cincinanti, Ohio.

Flames: Dx/Dreamstime.com; Steel Texture: © Sharpshot/Dreamstime.com

WRAP-UP

End notes

1. National Fire Protection Association. *NFPA 1720: Standard for the Organization and Deployment of Fire Suppression Operations, Emergency Medical Operations and Special Operations to the Public by Volunteer Fire Departments.* Quincy, MA: 2010.

2. National Fire Protection Association. NFPA 1500, Standard on Fire Department Occupational Safety and Health Programs, Section 4.1.2, Copyright © 2014, National Fire Protection Association. This reprinted material is not the complete official position of the NFPA on the referenced subject, which is represented only by the standard in its entirety.

3. Kerber, S., UL Firefighter Safety Research Institute. Study of the effectiveness of fire service vertical ventilation and suppression tactics in single family homes. Northbrook, IL: Underwriters Laboratories; 2013.

4. International Association of Fire Chiefs (IAFC), Safety, Health and Survival Section; International Society of Fire Service Instructors. Firefighter safety call to action: New research informs need for updated procedures, policies. December 23, 2013. Retrieved from IAFC website: http://www.iafc.org/Media/articlePR.

5. National Fire Protection Association. *NFPA 1710: Standard for the Organization and Deployment of Fire Suppression Operations, Emergency Medical Operations, and Special Operations to the Public by Career Fire Departments.* Quincy, MA: 2010.

6. National Institute of Standards and Technology. Report on residential fireground field experiments. Gaithersburg, MD: 2010.

7. National Fire Protection Association. *NFPA 1142: Standard on Water Supplies for Suburban and Rural Fire Fighting.* Quincy, MA: 2012.

8. National Fire Protection Association. *NFPA 1: Fire Code.* Quincy, MA: 2012.

9. National Fire Protection Association. *NFPA 1620: Recommended Practice for Pre-incident Planning.* Quincy, MA: 2010.

10. National Fire Protection Association. *NFPA 1: Fire Code.* Quincy, MA: 2012.

11. National Fire Protection Association. *NFPA 704: Standard System for the Identification of Hazardous Materials for Emergency Response.* Quincy, MA: 2012.

12. National Fire Protection Association. *NFPA 170: Standard for Fire Safety and Emergency Symbols.* Quincy, MA: 2012.

13. Kerber, S., UL Firefighter Safety Research Institute. *Impact of Ventilation on Fire Behavior in Legacy and Contemporary Residential Construction.* Northbrook, IL: Underwriters Laboratories; 2010

14. Underwriters Laboratories, National Institute of Standards and Technology, Fire Department of New York City. Research Governors Island, NY. July 2012.

15. National Fire Protection Association. *NFPA 101: Life Safety Code.* Quincy MA: 2012.

16. Corbett, G. P., Brannigan, F. L. *Brannigan's Building Construction for the Fire Service,* 5th ed. Burlington, MA: Jones and Bartlett Learning; 2015.

17. National Fire Protection Association. *NFPA 220: Standard on Types of Building Construction.* Quincy, MA: 2012.

18. National Fire Protection Association. *NFPA 5000: Building Construction and Safety Code.* Quincy, MA: 2012.

19. The Fire Protection Research Foundation Forum on PV Panel Fire Risks. Las Vegas, NV. June 2014.

20. Durso, F., Jr. Easy being green. *NFPA Journal.* November/December 2012.

21. Averill, J., Moore-Merrell, L. NIST Report on high-rise fireground field experiments, NIST Technical Study Note 1797. Gaithersburg, MD: National Institute of Standards and Technology; 2010.

22. NIOSH Firefighter Fatality Investigation: Career probationary fire fighter and captain die as a result of rapid fire progression in a wind-driven residential structure fire—Texas 2009-11. April 8, 2010. Retrieved from http://www.cdc.gov/niosh/fire/reports/face200911.html.

Flames: Dra/Dreamstime.com; Steel texture: © Sharpshoot/Dreamstime.com

Developing an Incident Action Plan

LEARNING OBJECTIVES

- Explain the importance of deciding on an overall strategy in the development of an incident action plan.
- Describe how extinguishment is both an operational priority and tactical objective with an emphasis on the relationship between life safety and extinguishment.
- Evaluate conditions leading to an offensive or defensive operation.
- Compare probability of occupant survival to fire and building conditions.
- List situations when a written incident action plan is needed.
- Use a case study or actual fire to develop an incident action plan based on a risk-versus-benefit analysis.

Flames: © Dtx/Dreamstime.com; Steel texture: © Sharpshot/Dreamstime.com; Chapter opener photo: Tk © Sergey Toronto/ Shutter Stock, Inc.

Case Study

Fire fighters responding to a residential fire.

Courtesy of Bill Strite, Cincinnati Ohio.

Fire companies and a battalion chief arrived on the scene of a fire in a wood frame house. A man who was badly burned reported that he was repairing his automobile gas tank in the basement when the entire basement suddenly became engulfed in fire. His 5-year-old daughter was sleeping on the 2nd floor and he had not been able to locate her. The battalion chief assumed command and formulated an offensive incident action plan as the engine company attacked the fire and the truck company began a primary search per department SOPs. Fire was visible in the attic area and exposures appeared to be threatened. The second arriving truck company accessed the roof of the adjoining building without consulting the incident commander.

1. Explain the basic differences between an offensive and a defensive incident action plan.
2. What factors should be considered during a risk-versus-benefit analysis?
3. What tactics should be included in an offensive incident action plan when exposures are threatened?

Introduction

An **incident action plan** provides the central focus for operations. An offensive/defensive strategy decision based on a risk-versus-benefit analysis is the first step in developing an incident action plan. All tactical decisions and task assignments are based on the overall strategy. Everyone on the fire ground must be aware of the strategic mode. The offensive/defensive decision and overall incident action plan are derived from an analytical approach to information gained through preincident planning and size-up. The result should be a straightforward, easy-to-understand plan outlining major tactical objectives. All tactics should lead to completion of the major objectives identified in the incident action plan.

IC Rule of Engagement for Fire Fighter Safety

Conduct an initial risk assessment and implement a safe action plan.

Reproduced from: © 2011, IAFC Safety, Health and Survival Section. All rights reserved.

Texture: © Eky Studio /Shutterstock, inc; Steel: © Sharpshot/Dreamstime.com

Because conditions constantly change, size-up is a continuous process; therefore, the incident action plan must remain flexible. Often, as conditions change, the tactics necessary to accomplish a strategy are modified. Incident commanders (ICs) should develop alternative incident action plans that anticipate possible changes and be ready to react as necessary (i.e., change from an offensive to defensive strategy). A chief's aide or planning section chief can be of great assistance in developing a "plan B," allowing the IC to focus on the current operation.

The primary strategic considerations are always life safety, extinguishment, and property conservation. However, the three priorities are not mutually exclusive. When resources are limited—and they usually are in the beginning stages of an operation—the fire officer must take action related to life safety first, extinguishment second, and property conservation third. However, in most cases, extinguishment and life safety are closely related. If the fire is extinguished, rescue often takes care of itself, and the overall operation is much safer. Extinguishment is normally, but not always, the most important life safety tactic. Extinguishment is both an operational priority and a tactical objective.

Stocker photo background © Vitaliy Konovy/ShutterStock, Inc.; Paper: © silver-john/Shutterstock, Inc.

There are times when there are only a few easily rescued victims. When resources are limited, it may be best to rescue these victims before attempting to reach other possible victims. However, this is a very difficult decision. Victims that are in the most danger may not be visible and reports from occupants are often incorrect.

When developing an incident action plan, it is best to categorize tactical objectives in terms of their relative importance as primary and secondary (and possibly tertiary) objectives. As an example, extinguishing the fire would normally be a primary objective, as would search and rescue operations. Most property conservation measures would fall under secondary objectives. The idea is to provide an effective plan for deployment, with primary objective tasks and tactics being assigned prior to secondary objective tactics, when resources are limited.

Once there are available resources on the scene, all priorities can be handled simultaneously. If the fire operation will continue for more than a few minutes, the IC should request enough staffing and equipment to handle the immediate life safety, extinguishment, and property conservation activities plus an allowance for a tactical reserve. Anticipate the unexpected.

Remember, the key to a successful fire-ground operation is keeping it simple. Incident action plans should be simply stated and concise. This chapter emphasizes the development of an incident action plan leading to offensive or defensive tactics. Five scenarios (a dwelling, an apartment building, a high-rise residence, an industrial property, and a church) provide examples of incident action plan development.

Determining Life Safety Needs

The *Procedures, Preincident Planning, and Size-Up* chapter discusses the importance of life safety during preincident planning and size-up. The primary objective of structural firefighting is saving lives; therefore, life safety is the first consideration in developing an incident action plan. For this reason,

the IC may expose fire fighters to greater risk when savable lives are threatened.

IC Rule of Engagement for Fire Fighter Safety

Do not risk fire fighter lives for lives or property that cannot be saved. Seriously consider a defensive strategy.

Reproduced from: © 2011, IAFC Safety, Health and Survival Section. All rights reserved.

Texture: © Eky Studio /Shutterstock, Inc; Steel: © Sharpshot/Dreamstime.com

Evaluating Structural Conditions

Structural conditions bear heavily on the offensive/defensive decision. Even with sufficient resources, an interior attack should not be conducted in an unsafe building. Fire fighters should never enter a building that is in danger of imminent collapse and should maintain a collapse zone. However, rescue efforts may be justified in a building that is currently structurally sound but may ultimately fail as continuing fire damage undermines the structure. In this situation, conduct an offensive attack to assist occupants in exiting the building with a reevaluation of the structure and fire conditions once the building is evacuated. If extinguishment is being accomplished during the initial offensive attack, offensive tactics will normally be continued. However, if the fire has not been contained or extinguished during the initial offensive attack, after the rescue effort has been accomplished, the expected benefit has changed from life safety to saving property. There is less reason to stay in a building of questionable stability, as the benefit component of the **risk-versus-benefit analysis** has changed from life safety to saving property. Proficient ICs are always weighing risk against expected benefits. Staying on the right side of this equation requires the development of an alternative incident action plan. If the fire in a building of questionable structural integrity is not contained when all life safety tasks are completed, an alternative defensive plan (plan B) should be at hand and resources ready for implementation. It is a good idea to assign a planning section chief to develop defensive plans as soon as a transition from offensive to defensive is anticipated. A major part of this alternative plan consists of moving apparatus and fire fighters out of the collapse zone, as well as deciding the appropriate position and type of fire streams for the defensive attack.

Fallacy	Fact
The three tactical objectives are mutually exclusive and must be performed in order of priority: life safety, extinguishment, property conservation.	The three tactical objectives are interrelated; extinguishment is usually an important tactic in accomplishing the life safety objective. When resources are sufficient, all three objectives can be accomplished simultaneously.

Steel: © Sharpshot/Dreamstime.com

Estimating Resource Capability and Evaluating Resource Requirements

Comparing resource capability to incident requirements is part of the size-up process. The resources that are required to fight a fire offensively are different from those needed for a defensive battle. Generally, it takes more staffing to conduct an offensive operation. Lives and property are best saved by conducting an offensive attack. However, a lack of necessary resources could lead to a defensive decision, even when an offensive attack is clearly the better approach.

IC Rule of Engagement for Fire Fighter Safety

If you do not have the resources to safely support and protect fire fighters, seriously consider a defensive strategy.

Reproduced from: © 2011, IAFC Safety, Health and Survival Section. All rights reserved.

Texture: © Eky Studio /Shutterstock, Inc; Steel: © Sharpshot/Dreamstime.com

The IC must consider what will be needed to conduct an offensive attack early in the plan development process. In making the offensive-versus-defensive decision, the IC must apply sound risk management principles to ensure fire fighter safety. Figure 3-1 charts time/resources versus occupant probability of survival in terms of structural stability, temperature, and smoke layer. At the beginning of the incident, there is a very high probability of saving occupant lives and extinguishing the fire. As the temperature increases in an unvented structure, the smoke layer begins to fill the enclosure while the structure weakens; the probability of saving lives decreases, and extinguishment is more difficult. In other words, the probable benefit is lower and the risk is higher. In the early stages of the fire, no fire department personnel would be on the scene. As time passes, fire department resources become available, but the probable benefits of life safety and extinguishment decrease.

Each fire will be different; thus, Figure 3-1 is a general depiction rather than an actual fire model. Underwriters Laboratories conducted tests comparing time to flashover in a living room containing modern furniture versus a living room containing "legacy" furnishings. For purposes of the study, "legacy" was defined as the fire load typically found in a residential occupancy in the mid-twentieth century. With all variables other than the fire load held constant, the room with modern contents flashed in 3 minutes 40 seconds compared to 29 minutes 30 seconds for the legacy room in the one-story residential structure.[1] Even this very credible research is not universally applicable to home fires. Many factors determine the rate of progression to an untenable temperature or **immediately dangerous to life and health (IDLH)** toxicity level. It is important to point out that if the occupant immediately calls dispatch to report the fire, units responding within the guidelines of NFPA 1710 would usually arrive at or after flashover in a residential fire. Many factors affect fire dynamics and the progression to flashover, ventilation being a major factor, as pointed out in later research by Underwriters Laboratories.[2]

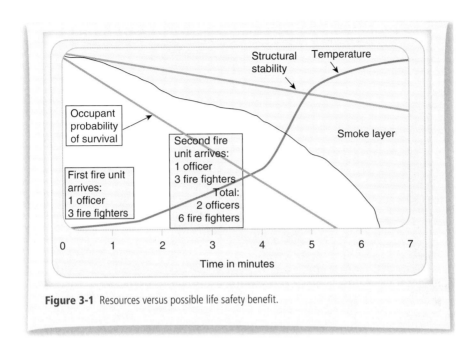

Figure 3-1 Resources versus possible life safety benefit.

Devising an Offensive or a Defensive Incident Action Plan

The entire operation is governed by the offensive/defensive decision. A casual observer ought to be able to determine whether the tactics being applied are interior (offensive) or exterior (defensive). However, it may not always be this clear. During the Underwriters Laboratories residential vertical ventilation research, an exterior hose stream was operated from the exterior through an opening for a short period (usually 15 seconds to 1 minute) to soften the target prior to making entry for an offensive attack.[2] Many fire service leaders are now advocating this tactic. Master streams are generally considered defensive tools. However, they can be used to support rescue efforts in large, complex structures with fire separations. Master streams are sometimes used to prevent fire from entering critical evacuation routes, such as a fire escape, or to cover exposures. When master streams are being used during an offensive operation, coordination through command

is critical. An operation that begins as an offensive attack is sometimes changed to defensive, but the actions taken during either attack must be coordinated as offensive or defensive; they must never be both. The *Offensive Operations* and *Defensive Operations* chapters discuss offensive and defensive attacks in detail. Whenever it is safe to do so, an offensive attack is preferred.

Formulating an Incident Action Plan

The IC sets the objectives, decides on the tactics necessary to achieve those objectives, and finally assigns units to complete the tasks associated with each objective. The incident action plan is the focus of the entire operation. All tactics are directed toward completing the objectives, and each objective is directed toward accomplishing the overall incident action plan. The incident action plan should be simple and understandable, as shown in Figure 3-2.

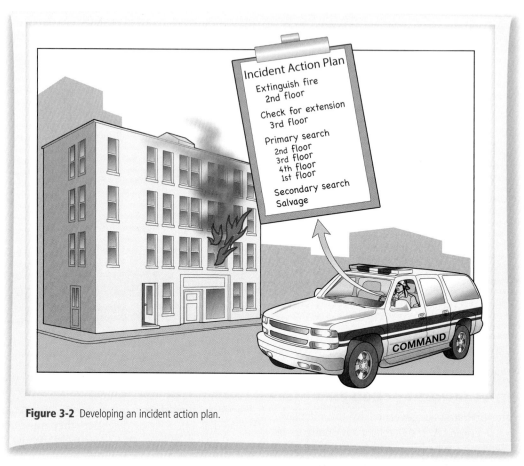

Figure 3-2 Developing an incident action plan.

Written incident action plans are required when an incident extends past a single operational period or if a unified command is established. Most structure fires do not require formal written incident action plans. However, critical aspects of an incident should be recorded to assist in the transfer of command and documenting the incident for the written report and critique. This could be a tactical worksheet or other media. At a minimum, the IC should track resources on scene, their current assignment, known hazards, and benchmarks completed. Including a sketch of the fire scene and an organizational chart is highly recommended. Typically, an assistant to the IC would maintain written documents. As the incident grows, the planning section chief would usually update information. When staffed, the planning section tracks progress and continuously develops the incident action plan and alternative plans. **Figure 3-3** shows a tactical worksheet.

The IC will establish objectives, making strategic determinations for the incident based on the requirements of the jurisdiction. In the case of a unified command, the incident objectives must adequately reflect the policies and needs of all the jurisdictional agencies. The incident action plan should cover all tactical and support activities required during the operational period.

An incident action plan becomes more important as the incident grows in size. However, there should be an incident action plan for every structure fire, and units operating on the scene should be coordinated using this plan.

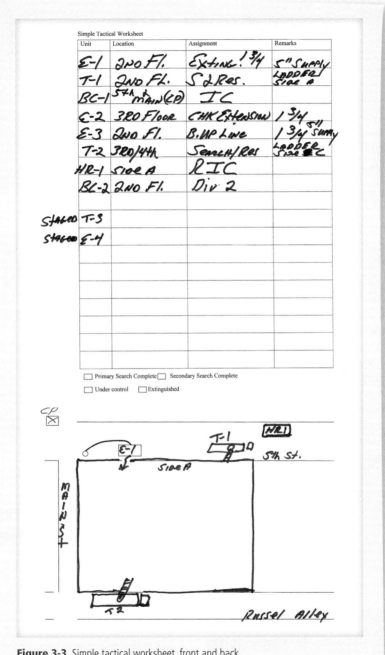

Figure 3-3 Simple tactical worksheet, front and back.
Courtesy of Bernard J. Klaene.

Incident Summary

At the Oklahoma City bombing site, the incident management team met twice daily to review the progress that had been made in the previous 12 hours and to establish plans for the next 12 hours. These plans were written, copied, and distributed immediately to everyone involved in the operation. By clearly stating the operational period's objectives, the plans let everyone know what the focus of their tasks were in relation to the overall objective.

Source: Comeau, E., Foley, S. Oklahoma City, April 19, 1995. NFPA Journal, July/August 1995.

Deployment

Writing a list of tactical objectives is useless unless sufficient resources are assigned to accomplish the objectives. Experienced ICs know that after making assignments, they must follow up and request status reports. Sometimes an assignment is made but not achieved. Experienced company-level officers provide status reports, as appropriate, without being prompted. However, when reports are not received in a timely manner and the assigned company has had ample time to complete the assignment, it is incumbent upon the IC to request a status report. If status reports are requested too frequently, the status report communications can interfere with efficient operations.

IC Rule of Engagement for Fire Fighter Safety

Obtain frequent progress reports and revise the action plan.

Reproduced from: © 2011, IAFC Safety, Health and Survival Section. All rights reserved.

Texture: © Eky Studio /Shutterstock, Inc; Steel: © Sharpshot/Dreamstime.com

NOTE

Ordering it done and getting it done are not always the same.

Reproduced from: © 2011, IAFC Safety, Health and Survival Section. All rights reserved.

Texture: © Eky Studio /Shutterstock, Inc; Steel: © Sharpshot/Dreamstime.com

The following five scenarios will illustrate the process involved in developing an incident action plan. In these scenarios, assume that all companies are staffed with the NFPA 1710 minimum of four personnel[3] (an officer, a driver, and two fire fighters). If your department operates fire companies with fewer than four responding members, two or more units should be combined under the supervision of a single officer. Likewise, more tasks may be accomplished if more than four members are assigned to a company. As an example, some departments staff quint apparatus with six or more fire fighters, allowing the company to simultaneously obtain a source of water, advance an attack line, and perform ventilation activities.

Scenario 1: Single-Family Detached Dwelling

Fires in the home make up the vast majority of fires in the United States. Sometimes department standard operating procedures (SOPs) spell out company duties for the entire first-alarm assignment. When this is the case, the IC need modify operations only when the SOPs are not being followed or when the procedures do not correspond to the circumstances. It could be said that the IC is on autopilot when SOPs are followed, but going on autopilot can be dangerous. No fire should ever be considered routine. On

Flames: © lag_cz/ShutterStock, Inc.; Paper © silver-john/ShutterStock, Inc.

the other hand, there is no need for unnecessary interference in an operation that is going well. Most ICs feel more comfortable handling a small fire than a larger, more complicated fire, yet the process that is used to develop an incident action plan for a small residential fire is the same as that for a very large and complex fire. Even when the fire operation is going well, the IC should focus on completing the tactics necessary to meet strategic objectives while simultaneously developing plan B as a precaution should the situation change.

It is 3:30 PM on Saturday, July 14. It is 90°F (32°C) and very humid. The fire was reported by a neighbor who noticed smoke coming from the roof. This residential area has a strong water supply system. The first-arriving engine company is making entry, and the truck company has placed their aerial on the roof. The status of the occupants is unknown. Conditions are as shown in **Figure 3-4**.

Figure 3-4 Scenario 1: Fire in a single-family detached dwelling.

Courtesy of Denny Baker, Cincinnati Fire Department.

Risk-Versus-Benefit Analysis
Expected Benefits

Saving the lives of the occupants and saving property. There is a chance that the occupants were not at home at the time of the fire, but this cannot be known until verified by a primary and secondary search. The exact location of the fire is unknown at this time. There appears to be a good chance of extinguishment.

NOTE

Benefit as used here is the expected or potential benefit to occupants or owners. Rescuing occupants would be a life safety benefit.

Texture: © Eky Studio /Shutterstock, Inc; Steel: © Sharpshot/Dreamstime.com

Expected Risk

Moderate. A life safety benefit is the primary reason for conducting an offensive operation. There is a risk associated with any offensive operation, but the risk to fire fighters is usually less in a single-family dwelling than it would be in a larger, more complex occupancy. The fire pictured in Figure 3-4 appears to be limited. This could be a ventilation-limited fire. The risk is somewhat increased due to the unknown location and size of the fire. There is a chance that the fire originated in the first floor or basement and spread to the attic.

NOTE

Risk, as used in the risk-versus-benefit analysis, refers to the risk to fire fighters, not the risk to occupants.

Texture: © Eky Studio /Shutterstock, Inc; Steel: © Sharpshot/Dreamstime.com

Structural Stability

Primary failure not expected until near full involvement. This is a frame structure. Frame buildings are fairly unstable structures and contribute significant fuel to the fire. However, the combustible characteristics of the building provide stability clues. Frame structures will normally have considerable fire involvement before collapse, with the exception of truss roof and engineered floor construction. For this reason, if the fire is in a modern residential building, extreme caution should be used if the IC decides that it is necessary to vent the roof. Whenever possible, it is a good

Figure 3-5 Venting a roof from an aerial ladder.
Courtesy of David J. Jones Cincinnati, Ohio.

idea to provide a stable platform using an aerial, an elevated platform, or a roof ladder **Figure 3-5**.

Offensive/Defensive Decision

Offensive. Given the possible life safety, as well as property saving, benefits with a moderate risk to fire fighters, a decision is made to initiate an offensive attack.

Resource Needs

Initial needs are finding the fire, advancing a standard preconnected attack line while controlling ventilation by minimizing the door opening, and simultaneously performing the primary search with coordinated ventilation. An initial rapid intervention crew (IRIC) should remain outside the building. In most operations, two or more fire fighters make entry with the hose line, with the IRIC remaining outside. In this case, one fire fighter and the pump operator are assigned as the IRIC. The truck company should begin the primary search and prepare to vent on orders of the IC. As the operation progresses, anticipate the need for an additional search and rescue crew, a backup hose line, a dedicated RIC, and tactical reserve. The size of the tactical reserve should take into account the need for rehabilitation due to the hot and humid weather conditions.

Incident Action Plan

- Make entry; find and extinguish the fire.
- Establish an IRIC followed by a dedicated rapid intervention crew (RIC).
- Search and rescue in basement, first floor, and attic (appears to be dormer rooms in attic area).
- Vent heat, smoke, and gases using horizontal and/or vertical venting.
- Advance a hose line to the attic and check for extension on second floor and attic.

Deployment

- *First engine.* Provide an initial status report and verify command. Complete a 360-degree size-up of the house. Establish a water supply, force entry, and attack the fire on the first floor with a 1¾-in. (45-mm) hose line. Two members remain outside the building as the IRIC. The fire fighter assigned to the IRIC could control the entry door as the attack team advances.
- *First truck.* Act as search and rescue teams on first floor, attic, and basement. Ventilate as needed, possibly using horizontal ventilation and/or opening the roof, depending on the location of the fire.
- *First battalion chief.* Transmits a status report, assumes command, and establishes a stationary command post as IC.
- *Second engine company.* Obtain a second source of water. Advance a 1¾-in. (45-mm) hose line into the building, covering the area not covered by the first engine; check for extension. The first-in truck company should assist in checking for extension.
- *Third engine company.* Function as RIC. A two-person RIC is the absolute minimum. While department SOPs vary regarding which company is assigned as the RIC, most departments assign at least four members (a four-person fire company) as the RIC.
- *Medic unit.*
- *Tactical reserve of at least two companies.*

NFPA 1710, 5.2.4.2.2, establishes a minimum staffing of 14 fire fighters for a fire in a 2000 ft² (186 m²) two-story, single-family structure with no basement and no exposures (15 minimum if an aerial device is used). A National Institute of Standards and Technology (NIST) staffing study[4] used a building constructed to meet the single-family structure described in NFPA 1710. This study proved the need for the staffing levels enumerated in NFPA 1710.

In this example, four fire companies and a chief officer—a total of 17 fire fighters—are deployed, plus a tactical reserve and medic unit. The additional staffing provides a four-person RIC. It is obvious that this deployment covers only the most basic tasks. If the first-arriving engine company is successful in extinguishing the fire, the entire operation is favorably changed, and a 15-person response will be adequate. This is the reason for placing "extinguish the fire" as the highest strategic priority, thus accomplishing the life safety objective by extinguishing the fire and ventilating the smoke, heat, and gases. A written incident action plan would not be required for this incident.

Scenario 2: Fourteen-Unit Apartment Building

Apartment building fires typically endanger more occupants than fires in single-family dwellings, and accounting for building occupants is generally impractical. The increased life safety problem is essentially found by multiplying the number of apartment units by the number of occupants typically found in a single-family house. Apartment buildings are second only to single- and two-family residential buildings in terms of the number of fires, fire deaths, and injuries. During the 5-year period of 2005 through 2009, there were an average of 109,360 fires in apartment buildings each year, resulting in 440 civilian fatalities per year.[5] Many apartment buildings (including this one) have non-fire-stopped attics that allow the fire to spread unimpeded throughout the attic. Shared basement areas used for tenant storage are the norm.

The attic and basement areas can require large flows of water for extinguishment. Most apartments comprise small individual compartments that are within the capabilities of preconnected attack lines carried by most fire departments.

This 14-unit apartment building has two basement-level walk-out apartments on the C side (Apartments 1 and 2). The remainder of the basement is undivided, with individual storage areas assigned to each tenant plus a laundry room. There are six apartments each on the first and second floors, connected by a hallway running the length of the building. Stairways to the second floor are located on each end of the building **Figure 3-6**.

It is 7:30 AM on Sunday, May 6. Temperature is 55°F (13°C), and winds are light and variable. Dispatch reports the call was received from Apartment 12 on the second floor reporting visible smoke.

Fire fighters on the first-arriving engine company encounter smoke issuing from the first floor hallway as they prepare to advance a 1¾-in. (45-mm) hose line through the hallway entrance on side A **Figure 3-7A**. Fire and heavy smoke can be seen issuing from side D **Figure 3-7B**. The first-arriving truck company is conducting a 360, venting windows from the exterior on side D, and checking all sides of the building for occupants.Figure 3-7B

Courtesy of Mike Carey and Jeff Neal, Cincinnati Fire Department.

Risk-Versus-Benefit Analysis
Expected Benefits
Saving the lives of the occupants and saving property. Given the day of the week (Sunday) and time of day (7:30 AM), there is a good chance that occupants will be at home and many will be sleeping. There is no report of anyone escaping or notifying occupants of the fire. Smoke conditions in the hallways may result in occupants seeking safe refuge inside their apartment. With no evidence of occupants escaping the building, a pessimistic assumption is made that all apartments are occupied. Information regarding fire extension is also unknown, but the presence of moderate smoke at the

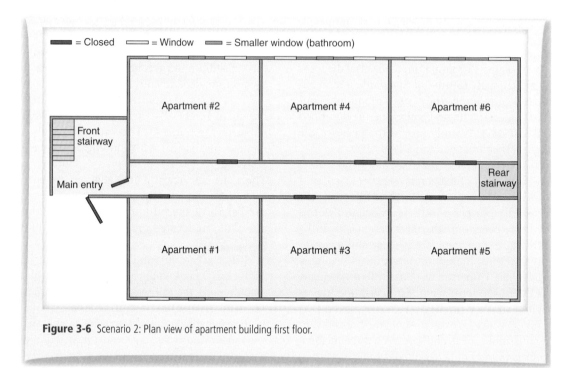

Figure 3-6 Scenario 2: Plan view of apartment building first floor.

A

B

Figure 3-7 A. Scenario 1: Fire in an apartment building, side A. **B.** Scenario 2: Fire in an apartment building, side D.

Courtesy of Mike Carey and Jeff Neal, Cincinnati Fire Department.

main entry indicates that smoke and possibly fire have migrated out of the apartment of origin to the hallway and possibly to other parts of the building. Heavy smoke and fire on side D indicate a working fire in one or more apartments on the first floor.

Expected Risk

Moderate. This fire presents a greater risk to fire fighters than the fire in Scenario 1 due to what appears to be a more advanced fire in a larger, more complex building. Premature and uncoordinated horizontal ventilation by fire fighters venting exterior windows on the D side may have provided air to a ventilation-limited fire, increasing fire intensity in some apartments and common areas.

Structural Stability

No signs of imminent structural collapse are evident, although this building probably has a wood truss roof structure. If fire extends into the undivided attic, rapid extension will occur, endangering residents on the second floor with a high probability of roof collapse. A roof collapse presents a danger to fire fighters working on the second floor, but intervening wall structures will limit the collapse area.

Offensive/Defensive Decision

Offensive. An offensive attack is clearly in progress, and this appears to be the correct decision.

Resource Needs

Three attack hose lines and possibly four are needed, with an initial and a backup water supply. A RIC is needed, as are search and rescue crews on all levels of the building. Additional ventilation may be required but must be coordinated with the fire attack. A tactical reserve and medical units are also required.

Incident Action Plan

- Extinguish fire in apartment of origin by coordinating a measured exterior attack with the interior attack to soften the target.

- Establish an IRIC and follow with a dedicated RIC.
- Extinguish fires in other apartments on the first floor.
- Extinguish fires on the second floor and attic.
- Search and rescue on first floor.
- Search and rescue on second floor.
- Search and rescue in basement.

Deployment

- *First battalion chief.* Transmits a status report, assumes command, and establishes a stationary command post.
- *First engine company.* Establish a water supply. Redirect initial attack line to the D side, applying water through open windows, then advance the fire line to the apartment of origin via the interior hallway and attack the fire on the first floor.
- *First truck company.* Continue conducting 360 of building, placing ground ladders as needed. Function as the IRIC and then all members are assigned as the dedicated RIC.
- *Second engine company.* Advance a 1¾-in. (45-mm) hose line into the first floor, and check for extension, extinguishing fires in other areas of the first floor. Obtain a second source of water.
- *Second truck company.* Conduct search and rescue operations on the first floor.
- *Third engine company.* Advance a 1¾-in. (45-mm) hose to the second floor, checking for fire extension in second floor apartments and the attic.
- *Third truck company.* Conduct a search and rescue operation on the second floor.
- *Fourth engine company.* Conduct search and rescue of basement and basement apartments while checking for fire extension.
- *Two medic units.* Provide medical treatment for fire fighters and occupants. Establish rehab in safe area; position medical transport vehicles out of harm's way and in position to transport injured.

- *Tactical reserve of at least three companies.* There is a good chance that the search and rescue effort will need to be augmented, especially if occupants are found who need assistance in evacuating the building.

Scenario 3: High-Rise Apartment Building

High-rise buildings pose extreme life hazards. Office buildings may contain thousands of people, and large residential buildings can house hundreds of families. The mere fact that the fire could be located several floors above grade level increases the complexity and difficulty in deploying fire streams. In addition, many floors above the fire could be occupied, making it necessary to conduct search and rescue operations on many different levels. The actual fire conditions could be similar to those in the single-family detached dwelling. However, the challenges that are presented will be much more complex. The *High-Rise Buildings* chapter is dedicated to the special strategies and tactics needed to combat high-rise fires.

Fires in residential high-rise buildings will typically be limited to a small area in the building unless the fire occurs in a common area or the apartment door is open or damaged. The smoke, however, can spread far from the fire, endangering a large number of residents. Most apartments in residential high-rise buildings have one-floor layouts. However, some high-end modern high-rise apartment buildings include apartments that span two floors with unprotected openings between floors. These are usually very large open-layout apartments that may require a higher rate of flow.

In this scenario, the fire occurs in the living room area on the sixth floor of a nine-story, fire-resistive apartment building housing seniors.

The fire is in Apartment 608 (X8), as illustrated in the floor plan in **Figure 3-8**. The building shown in **Figure 3-9** has 10 apartments per floor and primarily houses elderly occupants.

The building is equipped with a standpipe system, but is not protected by an automatic sprinkler

Figure 3-8 Scenario 3: Floor plan for high-rise apartment building.

Figure 3-9 Scenario 3: Fire in a high-rise apartment building.

Courtesy of Bernard J. Klaene.

system. According to the preincident plan, elevators are equipped with "fire fighter emergency operation." Using an elevator with a fire in progress is a calculated risk. Elevator safety requires that fire fighters dismount the elevator at least two floors below the fire. The risk associated with using elevators may not be warranted for a fire on the sixth floor. The *High-Rise Buildings* chapter offers a more complete discussion of the use of elevators by fire fighters.

The fire originates in the living room. It is 7:00 AM Saturday morning. The weather is mild and wind is calm.

Risk-Versus-Benefit Analysis
Expected Benefits
Saving the lives of the occupants and saving property. There is a good chance that most residents will be home and asleep. Occupants on the three floors above the fire and those on the fire floor are in imminent danger.

Expected Risk
Moderate to high. The risk in this building is somewhat greater than that in a smaller residential building, due to the number and location of occupants who are beyond the reach of aerial ladders and towers and the difficulty in ventilating the building or softening the target using an exterior stream. The layout of this building is fairly straightforward in comparison to some high-rise buildings. The fire-resistive construction and multiple stairs reduce the risk to some extent.

Structural Stability
Failure not expected. The structure is fire-resistive construction; therefore, it is unlikely that a moderate fire would seriously threaten the stability of this building.

Offensive/Defensive Decision
Offensive. Given the high likelihood of many occupants being endangered on four floor levels, a decision is made to initiate an offensive attack.

Resource Needs
Resource needs are much greater than those in low-rise residential properties. It will take longer to reach the fire floor, as confirmed in 2013 research conducted by the NIST. The floor layout is much larger, and there are three occupied floors above the fire.[6]

The backup hose line is also more critical in this situation. The early estimation is for three companies to be assigned to the fire floor, two to the floor above, and one each on the eighth and ninth floors. Property conservation, especially below the fire, is also important, but it is a secondary objective.

Incident Action Plan
- Force entry (if necessary) and extinguish the fire on the sixth floor.
- Search, rescue, and ventilate the sixth to ninth floors.
- Advance a hose line to the seventh floor and check for extension and smoke conditions.
- Evacuate occupants and arrange for shelter (as needed).
- Implement property conservation on the fifth floor.

Deployment
- *First engine and first truck.* Combine as one company. Establish water supply and connect to fire department connection. Engine officer to check alarm panel and establish command; is assigned as initial lobby control officer. Truck officer and five fire fighters walk the stairs, connect to the standpipe outlet on the fifth floor. The officer and three fire fighters advance into the sixth floor hallway with a hose line to attack the fire and start the initial search and rescue operation. Two fire fighters are assigned as the IRIC.
- *First battalion chief.* Takes command and reassigns first engine officer as the planning section chief.
- *Second engine.* Proceed to the fourth floor staging area as the RIC (two fire fighters formerly assigned as the IRIC advance a backup hose line into sixth floor).
- *Third engine.* Advance to seventh floor, connecting a hose line to the standpipe, then check for fire extension. The officer remains in the staging area as the staging officer.

The IC now orders all responding units to first report to the staging area for further assignment upon arrival.

- *Second truck.* Assigned to search and rescue/evacuation of seventh floor.
- *Second battalion chief.* Assigned as search and rescue group supervisor. Coordinates search/rescue/evacuation activities on all floors above and below the sixth floor.
- *Fourth engine.* Assigned to search and rescue group.
- *Fifth engine.* Assigned to search and rescue group.
- *Third truck.* Assigned to search and rescue group.
- *Heavy rescue.* Assigned to search and rescue group.

In addition to the assigned units, a tactical reserve should be established to relieve exhausted crews and/or augment extinguishment and life safety operations. If the first-arriving companies are successful in quickly extinguishing the fire and keeping smoke and toxic gases from extending to threaten occupants on other floors, some of the companies shown under deployment may not be needed. However, the IC should maintain a tactical reserve in the staging area to relieve fatigued members and to be prepared to handle unforeseen circumstances.

When confronted with a working fire in a high-rise building, additional resources should be requested on arrival.

High-rise residences come in all sizes, including buildings that are much larger than the one in this example. As the size and complexity of the building increase, the danger to both fire fighters and occupants also increases. Numerous other factors, such as extreme temperatures, strong winds, blocked stairways, unsafe elevators, and locked passageways further complicate operations. The danger is substantially decreased if the building has a properly installed, working sprinkler system.

Many high-rise residential buildings house elderly residents. This special population can be partially debilitated, making evacuation more labor intensive. Further, these elderly occupants are much more prone to injury and death from the products of combustion.

Scenario 4: Meat-Processing Plant Complex

The fire hazard challenge in large industrial complexes varies widely. Many of these facilities have tremendous fire loads, dangerous processes, and hazardous materials inside large, complicated buildings. Locating specific areas within a large complex is nearly impossible without a preincident plan drawing and/or assistance from employees. A working fire deep inside the complex may not be visible from the street entrance to the facility. In most cases, arrangements are made during preincident planning to have a representative from the facility meet the fire department and direct them to the reported fire area. However, it is not good practice to rely on plant personnel. At the BASF fire, discussed in the *Organizing, Coordinating, and Commanding Emergency Incidents* chapter, the plant safety supervisor was designated to meet the first-arriving fire company at a preselected area. On the day of the fire, he was seriously injured and found lying in the street in front of the plant. On other occasions, security personnel assigned to meet the fire department became involved in notifying occupants of the fire or using fire extinguishers. Preincident planning, including an overview of the property, is essential to successful operations in large industrial complexes.

The industrial complex used as an example of complex and formal preincident plans in the *Procedures, Preincident Planning, and Size-Up* chapter is the setting for this scenario **Figure 3-10**.

A fire is reported in Building H **Figure 3-11**. It is Wednesday, May 5, at 10:00 AM. Weather conditions are cool, dry, and calm. According to the preincident plan narrative, a maximum of 254 employees could be at the plant during normal working hours. There is no employee accountability system.

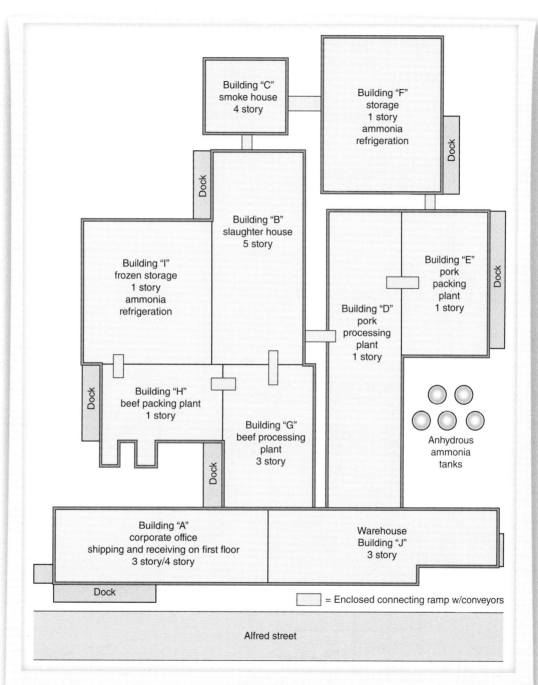

Figure 3-10 Scenario 4: Supreme Meat Packing Company.

Figure 3-11 Scenario 4: Buildings H and I.

First-arriving engine checks the preincident plan drawing, proceeds to Building H, and forward lays a large-diameter hose line to establish a water supply on the south side of the complex. First engine reports visible fire on the east side of the building and attempts to enter the building via the loading dock on the south side with a charged 2½-in. line. They are unable to advance on the fire and decide to back out of the building. As they retreat, an attempt is made to determine if any employees remain in the building by verbally calling into the building and doing a quick check using their thermal imaging camera. They are unsuccessful in locating anyone in Building H. First engine's officer provides a status report and recommends a defensive operation to the IC, who is just arriving at the scene. First engine repositions their 2½-in. (65-mm) hose line on the west side as shown in Figure 3-12 .

Figure 3-12 Scenario 4: West side of Building H.
Courtesy of David J. Jones Cincinnati, Ohio.

Risk-Versus-Benefit Analysis
Expected Benefit
Saving employee lives and saving property. At this time, there is a high probability that occupants remain inside buildings within the complex and possibly in Building H. The fire is likely to involve exposure buildings due to the close proximity of buildings and the access areas between buildings.

Expected Risk
- *Extreme for an offensive attack.* The fire is well advanced and the metal truss roof structure on Building H is being attacked by the fire.
- *Moderate for a defensive attack.* The space between Buildings A and H is limited, making it difficult to attack the fire from positions beyond the collapse zone. Large-exposure buildings combined with confusing layouts make search and rescue and fire operations hazardous.

Structural Stability
Expect collapse of Building H. Replace first engine's attack line on the east side with an unstaffed master stream or an elevated master stream located farther from the building. Evaluate all exposure buildings starting with Buildings G, I, and B.

Offensive/Defensive Decision
Defensive for Building H. Attempt entry and offensive operations in other buildings, prioritizing Buildings G, I, and B.

Resource Needs
A large contingent of fire companies is needed due to the large geographic area that must be searched and evacuated, as well as the many exposure and attack hose lines required to contain the fire to the building of origin.

Incident Action Plan
- Undertake search and rescue operations in the attached buildings.
- Evacuate all areas of the plant, accounting for as many employees as possible.
- Provide a large and uninterrupted water supply.
- Contain and limit the fire to Building H.
- Provide exterior master streams on all accessible sides of Building H.
- Advance hose lines into Buildings G, I, and B.
- Advance hose lines into strategic positions between buildings on all sides of the plant.

Deployment

- *First engine company.* Remain in position on east side of Building H, staying out of the collapse zone. Apparatus operator to remain at apparatus pumps.
- *First battalion chief.* Acts as IC.
- *Second engine company.* Establish a second large-diameter water supply on north side of complex, then assigned as the RIC.
- *First truck company.* Position apparatus for future elevated master stream protecting intersection between Buildings H and I. Crew enter Building I via the dock, conducting search and rescue operations. Evacuate occupants and evaluate fire spread into Building I.
- *Third engine company.* Advance 2½-in. (65-mm) hose line into Building G. Conduct search and rescue operations. Evaluate fire spread into Building G and protect wall exposed to Building H.
- *Second truck company.* Set up elevated master stream outside the collapse zone on east side of Building H.
- *Second battalion chief.* Assigned as west division supervisor (companies assigned to Buildings I, G, and B).
- *Department safety officer.* Assigned as incident safety officer.
- *Chief of department.* Assumes command, reassigning first battalion chief as planning section chief.
- *First engine.* Redeploy 2½-in. line to Building I to protect wall exposed to Building H.
- *Fourth engine.* Advance 2½-in. (65-mm) hose line into Building B to protect wall exposed to Buildings H and G; conduct search and rescue operation
- *First heavy rescue.* Assigned to Building G, conducting search and rescue operations. Evacuate occupants and evaluate fire spread into Building G.
- *Third truck.* Evacuate Building A, then move to and evacuate Building J.

- *Fifth engine.* Evacuate Building D, then move to and evacuate Building E.
- *Sixth engine.* Evacuate Building C, then move to and evacuate Building F.
- *Third battalion chief.* Assigned as evacuation group supervisor (companies assigned to Buildings A, J, D, E, F, and C).
- *Second heavy rescue.* Establish second RIC.
- *Seventh engine and fourth truck.* Establish large-diameter hose water supply; set up master streams on southeast corner and south side of Building H. Truck 4 officer assigned as exterior attack group supervisor.
- *Medical branch.* Treat and transport occupants as necessary. Establish safe area for evacuees.
- *Eighth and ninth engine companies.* Provide additional water supply on the south side of the complex via large-diameter hose. Assist in setting up master streams.
- *Four engine companies and two truck companies.* Staged as a tactical reserve. First company officer in staging assigned as staging manager.

Scenario 5: Church Fire

Older, gothic-style churches are characterized by very high ceilings, with huge concealed spaces above the main body of the church. Newer churches come in a variety of styles. Both old and new churches have a common characteristic: a large open area to accommodate the congregation and altar. This one factor means that total and catastrophic structural collapse of the roof is highly probable once a fire reaches the structure supporting the roof.

The large open area in the main body of the church requires an unusually high rate of flow because of the high ceiling. (Fire flow calculations are explained in the, *Offensive Operations* chapter.) It will be necessary to use 2½-in. (64-mm) or 3-in. (76-mm) hand lines or portable master stream appliances with solid stream nozzles to obtain the necessary reach if an interior attack is attempted.

Many people congregate during services, especially on religious holidays. During these times, life safety is a key tactical consideration. There are subsequent long periods when these buildings are unoccupied, allowing a fire to gain considerable headway before it is discovered. Few churches are protected by fire suppression systems or automatic alarms.

In this scenario, the twin bell towers in a gothic-style church are involved in fire **Figure 3-13** and **Figure 3-14**. It is Wednesday, October 5, at 9:00 PM. Weather is cool; winds are light and variable following a thunderstorm. The first report to the fire department is for a fire in the bell towers. The first-arriving engine company confirms a large volume of fire in the bell towers.

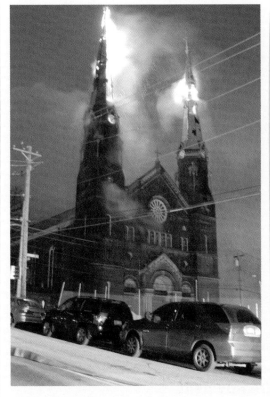

Figure 3-13 Scenario 5: Church fire.
Courtesy of David J. Jones Cincinnati, Ohio.

Risk-Versus-Benefit Analysis

Expected Benefits

Saving property only. The church is not having services at this time and the fire has gained considerable headway. There is a low probability that anyone is in the church. This historic gothic church has many stained glass windows and artwork that is considered priceless.

Expected Risk

Extreme. The very high bell towers combined with a large volume of fire make collapse of the towers highly probable. Fire extension to the interior of the church is likely but unknown at this time. Electrical wires close to the front of the church present a hazard.

Structural Stability

Partial failure inevitable. As previously stated, the bell towers should be expected to fail and collapse. The height of the bell towers would indicate a fairly large collapse zone that may be beyond the effective reach of exterior fire streams. If the fire has extended into the concealed space above the church interior, the church roof should also be expected to collapse.

Offensive/Defensive Decision

Defensive. Structural collapse potential is great; no one should enter the bell tower area. It may be possible to force entry to the rear of the church on side D to check for fire extension and make certain that everyone is out of the building. In checking for extension, evidence of fire in the roof area above the main body of the church may not be apparent. If fire is evident, it is usually too late to save the roof; therefore, it is preferable to make a defensive attack on the bell towers using streams placed beyond the collapse zone and away from electric wires in an attempt to prevent fire extension into the main body of the church.

Resource Needs

A large water supply and multiple elevated master streams are needed.

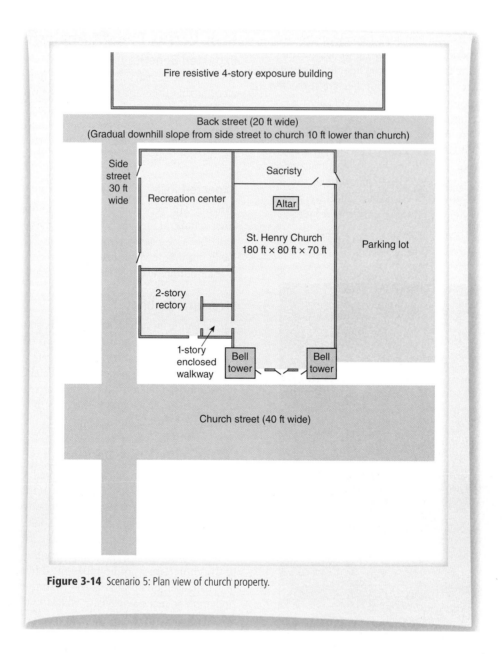

Figure 3-14 Scenario 5: Plan view of church property.

Incident Action Plan

- Extinguish the fire in the bell tower.
- Maintain a wide fire perimeter.
- Check the interior of the church via the rear door if this can be safely accomplished.

Deployment

- *First engine.* Officer takes command, establishing a command post; remaining company members supply water to the first truck's master stream.

- *First truck*. Divide into two crews:
  - *First truck, crew 1*: Set up an elevated master stream operating into the bell tower on the A–B corner of the church.
  - *First truck, crew 2*: Force entry to the rear of the church; check for visible fire extension; make certain the church is unoccupied, then return to A–B corner with crew 1.
- *First battalion chief*. Assumes command from first engine's officer. First engine company's officer reassigned as the planning section chief.
- *Second truck and second engine*. Set up a second elevated master stream operating into the bell tower on the A–B corner of the church. Elevated master stream supplied by the second engine company operating on the A–B corner of the church.
- *Second battalion chief*. Assigned as Division A–B supervisor.
- *First heavy rescue unit*. Advance a hose line into the one-story enclosed walkway between the church and rectory and check for extension along the wall separating the church and recreation center from the church.
- *Two ALS units*. Remain on standby at scene.
- *Third truck and third engine*. Set up a third elevated master stream operating into the bell tower on the A–D corner of the church supplied by the third engine company.
  - After master streams and water supplies are established, truck company officers from second and third truck companies supervise operation of master streams. First- and second-arriving engine company apparatus operators remain at pump panels. Second- and third-arriving engine company officers form two RICs using fire fighters from the second and third engine and truck companies who are not involved in supplying water or operating master streams.
- *Department safety officer*. Assigned as incident safety officer.
- *Fourth truck and fourth engine*. Set up a fourth elevated master stream operating into the bell tower on the A–D corner of the church supplied by the fourth engine company.
- *Third battalion chief*. Assigned as division A–D supervisor.
- *First battalion chief*. Still acting as planning section chief, develops a plan to set up two elevated master streams on the D side and an elevated master stream on Back Street as a contingency plan should the fire break through the roof of the main body of the church. The plan also includes advancing a 2½-in. hose line to the rear door of the church ready to enter the church to conduct a thorough search and check for extension after the bell towers are extinguished, provided structural conditions permit entry.
- *Three additional engine and ladder companies and two additional chief officers*. Called to the scene to implement this preemptive plan.
- *Two engines and two trucks*. Staged as a tactical reserve.

As previously stated, defensive operations require less staffing than interior operations. Once master streams have been set up, minimal personnel are needed to maintain the fire flow. However, they should remain on the scene as a tactical reserve or to complete extinguishment after the collapse danger has passed. Fire fighters on standby status should be assembled in a central area well away from the fire.

In the church fire example, the IC should direct the planning section to develop a plan changing from a defensive to an offensive operation if the bell towers are extinguished before the fire enters the main church structure. However, there is a real possibility that the fire has already entered the concealed spaces above the church, which will require a larger defensive operation with little hope of saving the building. A very limited degree of risk is acceptable as long as there is property to be saved. If a defensive operation is initiated for the entire church, the only savable property will be external exposures. Therefore, the IC should reduce the amount of risk to as near zero as possible. This may mean using unstaffed master streams or relocating staffed master streams farther from the church.

WRAP-UP

Chapter Summary

- An incident action plan outlines major tactical objectives and provides the central focus for operations.
- The incident action plan must be flexible, adapting to changing conditions.
- Primary strategic considerations are always life safety, extinguishment, and property conservation.
- Knowledge of the structure's components and the extent of fire involvement are essential in assessing structural conditions.
- Lives and property are best saved by conducting an offensive attack.
- A defensive attack should be used whenever the risk-versus-benefit analysis indicates the risks associated with an offensive operation outweigh the possible benefits (e.g., interior conditions are judged to present an extreme danger to fire fighters and there is no reasonable chance to save lives).
- A written action plan is needed when the incident extends beyond a single operational period. Rarely will written incident action plans be required on structure fires. Tactical work sheets will aid the incident commander in tracking resources and in the transfer of command.
- The incident action plan must be followed up with status reports, as changing conditions may require changes to the plan.

Key Terms

immediately dangerous to life and health (IDLH) Exposure to airborne contaminants that are likely to cause death or immediate or delayed permanent adverse health effects or prevent escape from such an environment. Examples include smoke or other poisonous gases at sufficiently high concentrations.

incident action plan The objectives for the overall incident strategy, tactics, risk management, and member safety that are developed by the incident commander and updated throughout the incident.

risk-versus-benefit analysis The process of weighing predicted risks to fire fighters against potential benefits for owners/occupants and making decisions based on the outcome of that analysis.

Challenging Questions and Activities

1. Critique a recent fire in your jurisdiction and evaluate the operation using the following criteria:

   A. Identify a reasonably expected benefit, and compare the actual benefit to the risks taken.
   B. Was the structure sound, or was it in danger of collapse?
   C. Was the operation offensive or defensive?
   D. Was this the correct strategy? Explain why.
   E. Were there sufficient resources to conduct a safe and effective offensive or defensive attack?
   F. List the objectives of the incident action plan in order of priority.

Flames: © Dex/Dreamstime.com; Steel texture: © Sharpshot/Dreamstime.com

2. Assume the role of a battalion chief responding as the IC from a fire station that also contains an engine and truck company. The fire department has 30 engine companies, 15 truck companies, 2 heavy rescue companies, 10 ALS units, and 5 on-duty battalion chiefs, off-duty battalion chiefs, the chief of department, five assistant chiefs and one safety chief are subject to recall. The water supply is from a city water supply system. Hydrants are closely spaced with high flows. The total first alarm response is three engine companies, two truck companies, a heavy rescue company, and two battalion chiefs. All companies are staffed with an officer, a driver, and two fire fighters.

It is Sunday, August 24, at 3:00 PM. It is partly cloudy and hot with a light wind from the east (B side). Conditions on arrival at the two-story house being renovated are as shown in Figure 3-15.

Apply the process used to evaluate the five scenarios in this chapter:

- Analyze the risks versus the benefits.
- Evaluate structural stability.
- Determine the strategy (offensive or defensive).
- Estimate resources needed.
- Develop an incident action plan.
- Deploy companies.

3. Develop a NIMS organizational chart based on the deployment in the five scenarios in this chapter.

Figure 3-15 Two-story house being renovated.
Courtesy of David J. Jones Cincinnati, Ohio.

Endnotes

1. Kerber, S. Impact of ventilation on fire behavior in legacy and contemporary residential construction. Northbrook, IL: Underwriters Laboratories; 2010.

2. Kerber, S., UL Firefighter Safety Research Institute. Study of the effectiveness of fire service vertical ventilation and suppression tactics in single family homes. Northbrook, IL: Underwriters Laboratories; 2013.

3. National Fire Protection Association. *NFPA 1710: Standard for the Organization and Deployment of Fire Suppression Operations, Emergency Medical Operations, and Special Operations to the Public by Career Fire Departments.* Quincy, MA: 2010.

4. National Institute of Standards and Technology. Report on residential fireground field experiments. Washington DC: 2010.

5. National Fire Protection Association. Fire analysis and research division home structure fires, home structure fires, Marty Ahrens. Quincy, MA: 2011.

6. National Institute of Standards and Technology. Report on high-rise fireground field experiments. Washington DC; 2013.

Flames: Dra/Dreamstime.com; Steel texture: © Sharpshot10-Dreamstime.com

Flames © Dx/Dreamstime.com; Steel texture © Sharpshot/Dreamstime.com; Chapter opener photo: SV/Luna/ShutterStock, Inc.

Company Operations

LEARNING OBJECTIVES

- Compare the division-of-labor concept as used in industry to company-level deployment.
- Describe structural firefighting functional assignments.
- Enumerate engine and ladder company fire-ground functions and tasks.
- Analyze tactics for the first-arriving fire company in relation to life safety.
- Apply engine and ladder company tasks to coordinating and controlling company-level deployment.
- List situations where splitting companies may be acceptable.
- Evaluate the positive and negative aspects of preassigning tasks, tools, and fire-ground positions.
- Assess proper and improper ventilation methods in regard to achieving the operational priorities of life safety, extinguishment, and property conservation.
- Given the fire and victim locations, determine the best vent location.
- Describe safe and efficient positioning of apparatus.
- Apply engine and ladder company tasks at a structure fire.
- Analyze company operations at a structure fire scenario.
- Develop an incident action plan using engine and ladder company functional assignments.

Case Study

Coordinated engine and truck company operations at a single-family residential house.

Courtesy of Bill Strite, Cincinnati Ohio

On August 5, 2014, fire fighters responded to a residential fire with closely spaced exposures. Fire was visible in the building of origin and spreading to D side exposures. The incident action plan called for an offensive strategy extinguishing the fire while simultaneously conducting search and rescue operations in the building of origin and exposure on the D side. After search and rescue activities were completed, fire was extending to two buildings on the D side. An alternative defensive incident action plan was briefly implemented to knock down the main body of fire in the fire building and D side exposures followed by interior operations for final knockdown.

1. What tasks should be conducted by engine companies during offensive operations?
2. What are the primary tasks conducted by truck companies during defensive operations?

Introduction

The primary responsibility of all companies working on the scene of an emergency is to work within the overall incident action plan. Freelancing cannot and should not be tolerated. Through the establishment of standard operating procedures (SOPs), first-in units can quickly go to work, allowing the incident commander (IC) time to evaluate options while maintaining control during those first few critical minutes on the fire ground. Assigning company-level operations should be included in the department's SOPs. This chapter focuses on identifying and coordinating engine and truck company tasks.

Companies and crews are supervised by a company officer, thus reducing the span of control by having one person report for the entire crew. In terms of both incident scene safety and operational efficiency, the importance of having a strong command system cannot be overstated.

The **division-of-labor principle** was the basis for the Industrial Revolution. It was found that developing job skills in a concentrated area allowed greater productivity, as compared to one individual developing a wide variety of skills. The problem implicit in the old saying "jack of all trades, master of none" is overcome by developing, within an individual, a high level of expertise in a limited number of tasks. In the fire service, the division-of-labor concept is applied by dividing tasks into the following functional areas:

- Command responsibilities
- Engine company tasks
- Truck company tasks
- Emergency medical services (EMS)
- Special operations
  - Hazardous materials teams
  - Technical rescue teams
  - Air operations
- Support services

The division-of-labor concept is not strictly adhered to in the fire service because fire fighters should be cross-trained to accomplish many different fire-ground

Flames: © Jag_cz/ShutterStock, Inc.; Steel texture: © Sharpshot/Dreamstime.com; Stacked photo background: © Vitaly Korovin/ShutterStock, Inc.; Paper: © silver-john/ShutterStock, Inc.

tasks. The division-of-labor concept stresses the need for preassigning duties (functions) at an emergency to ensure that, before the incident occurs, everyone is familiar with what will be expected of them during an emergency.

Members of hazardous materials teams, paramedics, and heavy rescue squads are required to complete specialized training and certification requirements that ensure proficiency in their particular disciplines. However, all fire fighters should have a general understanding of these special duties, just as members of the special teams understand the duties of line fire fighters. At a minimum, line fire fighters should be able to function as part of either an engine or a truck company crew and be trained in basic emergency medical and hazardous materials mitigation skills.

The incident action plan governs company-level operations. Engine company tasks during an offensive attack focus on the interior and involve "rescue as you go" tactics while applying water to extinguish the fire. During a defensive attack, the engine company sets up master streams and applies water to nearby buildings, deluges the main body of fire, or both.

Truck companies expend a great deal of effort ventilating, laddering, forcing entry, and conducting primary and secondary searches during an offensive attack. However, during defensive attacks, truck company duties are typically limited to setting up elevated master streams and support activities.

Simply put, engine company work involves applying water to the fire. Truck companies perform search and rescue activities, as well as assist engine companies by gaining entry, laddering, and controlling the fire through coordinated ventilation. Heavy rescue companies fill in wherever needed.

Engine Company Tasks

An engine company is the basic firefighting unit. On close examination, engine company operations can be dissected into the following functions:

- Performing rescue operations
- Establishing a water supply
- Advancing and operating hose lines

In most cases, an engine company is first to arrive at the scene. The first-in officer initiates the attack and provides a means of moving water from a water supply source to the fire.

Life safety—not only the safety of the victims, but also that of the fire fighters—is the first priority of everyone on the fire ground. Engine companies usually accomplish life safety objectives by placing attack lines in position to protect victims and rescuers and by providing safe evacuation routes.

Engine companies advancing hose through the main entrance of the structure sometimes encounter occupants attempting to exit the building or in some cases occupants who are unconscious. It may also be possible to conduct a cursory search and rescue as fire lines are being advanced into the fire area. Occupants may only be in need of direction and guidance, rather than actual physical assistance. In these cases, the engine crew will do whatever is necessary to rescue occupants.

When the location of victims is unknown, the determination by an engine company officer to position hose lines or to search for occupants is difficult. This tactical decision is based on the following concerns:

- Immediate danger to the occupants
- Available staffing and resources
- Time before additional resources arrive
- Extent of fire involvement
- Equipment available to perform the rescue

As an example, the fire shown in **Figure 4-1** is in a three-story apartment building with a reported basement fire. It is 9:00 AM on a Sunday morning. The first-due engine company and truck are located 2 miles from the fire scene. Both companies are staffed with an officer, a driver, and two fire fighters. Upon arrival, the engine company discovers that the basement fire has extended out of the basement and into the stairway. They also observe occupants on apartment balconies awaiting rescue. Fire walls divide the building into three separate fire areas. Each apartment has rated fire doors leading to the apartment and sliding glass doors leading to a patio. The truck company is not yet on the scene.

Figure 4-1 Apartment building fire with visible occupants.
Courtesy of Bernard J. Klaene

The engine company has several options, including the following:

- Rescue occupants on balconies using a ground ladder
- Conduct an interior search and evacuation
- Extinguish the fire or confine it to the stairway and basement

Of these three options, extinguishing the fire has the greatest potential to save lives. The occupants on the balconies are not in imminent danger. Occupants on the first floor of the fire building are in greatest danger, but the fire door will provide a measure of safety. If necessary, they can escape or be easily rescued via windows that are at grade level. Occupants on the second and third floor of the fire building will be safe for a time provided they take advantage of the fire door leading to their apartment from the hallway. Occupants in the two buildings beyond the fire wall have the protection of the fire wall and fire door, as well as the patio door. Fire and smoke will have to penetrate all three barriers before occupants are in imminent danger. The truck company will be on scene very soon. The basement fire may present a challenge, but containing the fire in the stairway should be well within the capabilities of a standard preconnected hose line.

If there is a potential rescue that can be made, life safety will always be the top priority. The incident action plan should direct all resources (either directly or indirectly) toward the successful evacuation of all endangered occupants. In the apartment building example, the engine company should save lives by extinguishment or fire containment—an indirect means of saving lives.

Truck Company Tasks

The terms "ladder company" and "truck company" are synonymous. It is not necessary to have an aerial ladder or elevated platform at every fire, or even in every fire department. However, it is necessary to have truck company *functions* assigned to a specific group of fire fighters on the fire ground. General truck company duties should be preassigned. Fire operations are more often unsuccessful because of poor or nonexistent truck company work than because of a lack of water.

The following tasks are normally assigned to truck companies:

- Conduct search (primary and secondary)
- Rescue trapped victims
- Ventilate
- Force entry
- Ladder the building
- Check for fire extension
- Access concealed spaces

As this list indicates, truck companies are responsible for a wide variety of tasks when an offensive attack is in progress; therefore, it is crucial to properly staff companies that are charged with carrying out truck company duties. Other than providing ladder pipe and other elevated master streams, truck company activities are very limited during defensive attacks. In low-staffing situations, the truck company may need to assist the engine company in initial placement of hose lines during an offensive attack. However, the tendency is for the truck company to continue handling hose lines, even when adequate engine company personnel

are available. This is not the most effective use of truck personnel.

A Note About Staffing

Many departments find it difficult to provide adequate fire company staffing. A standard engine or truck company should be staffed with at least four fire fighters to operate effectively. Some departments assign a driver to an elevated platform or aerial apparatus with no additional staffing. This apparatus provides a means of accessing upper levels of a building, but should not be considered a truck company unless additional staffing is provided. Additional staffing is often provided by members responding in ambulances, private automobiles, and other means.

Coordinating Company Operations

For engine companies, rescue is typically an indirect activity. Getting hose lines between the victims and the fire makes rescue or self-evacuation possible. Extinguishment or reducing heat production improves safety for fire fighters and occupants. For truck companies, rescue may be a more direct activity. Truck companies will generally ventilate to control the spread of heat, smoke, and fire while conducting the primary search and removing any victims. It is imperative that ventilation be coordinated with the engine company attacking the fire. The engine company should be in position and ready to attack the fire prior to venting. Venting will generally increase fire intensity and often changes the fire and smoke flow path. Truck companies will also ladder the building to establish alternative evacuation routes.

Figure 4-2 illustrates how an engine company and a truck company coordinate rescue and ventilation. It is imperative that engine and truck company rescue operations also be coordinated. Conducting a search and rescue operation without the protection of a hose line controlling the fire is extremely hazardous. As an example, the truck crew

Figure 4-2 Engine and truck company coordinated rescue.

operating on the third floor in Figure 4-2 would be in an extremely hazardous position if the engine company crew were not attacking the fire on the floor below. Likewise the outside venting at the second floor would tend to increase fire intensity if they vented the window prior to the fire attack team getting water on the fire. It is critical that all crews coordinate activities and keep the IC informed of their location and status.

The IC must coordinate all activities on the fire ground. Control and coordination are vital if the engine company is unable to control the fire or a decision to change from an offensive to a defensive attack is made. Truck companies will in many cases be working in areas above the fire while the engine company works at controlling the fire. If the fire is not controlled, truck companies working inside the structure need to be notified, and, if at all possible, their retreat should be protected by the engine company.

Safety and control dictate that operating units work as groups (companies or crews). The first-in truck company at an offensive operation is the exception to that rule. Given the variety of tasks and

the fact that many of these tasks can be performed by two-member teams, it is permissible for truck companies to split into separate crews. At a working structure fire, two members may go to the roof to ventilate, while a second two-member crew may conduct the primary search. In addition, once the first task is completed, these "truckies" may reposition to perform other ladder company tasks. For example, fire fighters who are assigned to the roof may also perform property conservation on the floor below the fire. Once these quick initial operations are accomplished, truck company members should be reunited as a company or full crew under the direction of the truck company officer. It is not freelancing when a company officer splits his or her crew into separate teams. Each crew should have a team leader who can communicate with the company officer. In some departments, the crew leader will be a higher-ranking company member (such as a sergeant); in others, the crew leader will be a fire apparatus operator or senior fire fighter.

Rule of Engagement for Fire Fighter Survival

Go in together, stay together, come out together.

Reproduced from: © 2011, IAFC Safety, Health and Survival Section. All rights reserved.

Texture: © Ely Studio /Shutterstock, Inc; Steel: © Sharpshot/Dreamstime.com

Other exceptions to the general rule of maintaining company unity under the direct control of the company officer include the first-arriving company splitting into an inside and outside crew in compliance with two-in/two-out rules.[1,2] When a company is split into two or more crews, the NIMS organization chart should account for each crew. There are also times when the apparatus operator should remain at the apparatus to operate pumps or an aerial device.

In Figure 4-2, the engine company is divided into inside and outside crews. The two members of the inside crew are advancing and operating the hose line to attack the fire. The outside crew is the **initial rapid intervention crew (IRIC)** and is staffed by the pump operator at the apparatus operating the pumps and another fire fighter controlling the door leading to the first floor.

All companies should remain together as a unit whenever possible. Maintaining company unity facilitates accountability because the officer can immediately verify the safety of the entire crew.

As has been pointed out, general truck company duties should be preassigned, usually to fire fighters arriving on aerial or elevated platform apparatus. This brings about a point of discussion: preassignment of specific tasks, tools, and positions. Many departments establish SOPs requiring truck company members to carry specific tools. For example, the officer carries a utility bar (e.g., K-tool, Hux bar), another member an ax, and still another member a salvage cover or rope rescue equipment. Are these good SOPs?

In a high-rise office building, is the same equipment needed for a fire on the first floor during working hours as is needed for a fire on the top floor when the building is unoccupied and secured? Forcible entry tools will be needed in many cases, but will the same tools be needed at every fire?

Considering that most jurisdictions have a wide variety of occupancies, from single-family dwellings to high-rise structures, carrying the same tool into every situation, or even the same tool for different fires within the same structure, may not prove to be a productive policy. The best that can be said for the practice of preassigning tools and equipment is that establishing default SOPs for tools has merit. If there is no reason to bring a different tool (e.g., nothing-showing situations), bring the assigned tool. This preassignment of tools should be based on experience and should be limited to one or two members, rather than the entire crew having preassigned tools. Company officers must be prepared to change tool assignments according to circumstances.

The same reasoning can be applied to preassigned tasks such as ventilation, forcible entry, and search and rescue. Ventilation may not be required or may have negative consequences under certain conditions. In many cases, the fire self-vents. When ventilation is warranted, its priority changes depending on fire conditions, the building, the location of the fire, the location of the occupants, and the ventilation options available. Primary search may require the entire truck crew or could be accomplished by the engine company as it advances.

Preassigning truck company positions, unless done in a very general way with company officers having the latitude to use discretion, may result in the inefficient use of resources. Assigning fire fighters to the roof when the fire is several stories below the roof simply places them in a useless—and often dangerous—position when they could be accomplishing more effective truck company activities. Few fire departments have sufficient staffing to place fire fighters in unnecessary positions when a working fire is in progress.

Fire officers and fire fighters should be highly trained, rational human beings. SOPs that limit their ability to make reasonable decisions are generally counterproductive. Fire fighters who are assigned to truck companies should bring tools into the structure, but the tools should be those needed for the specific operation. The roof assignment may be extremely important in many fires but useless in others. Some departments have discontinued the practice of roof ventilation on single-family dwelling fires. Forcible entry may be the single most important fire-ground task when victims are trapped in secured areas or when the fire attack crew cannot make entry. Other times, windows and doors may be open, making forcible entry an unnecessary task. Given the fact that timing ventilation and forcible entry can make the difference between success and failure, the preferred approach is to have "truckies" attending to necessary truck work while engine crews are securing a water supply and advancing hose lines.

Quint and Quad Companies

Some departments staff multifunction apparatus that are equipped to perform both engine and truck company operations. These are usually referred to as **quint** or **quad** companies. Some departments provide sufficient staffing on this single apparatus to achieve both engine and truck company objectives, while other departments do not provide the staffing necessary to perform both engine and truck company tasks. As an example, a quint staffed with an officer, driver, and four fighters could establish a two-person IRIC, advance a hose line into the building, and use the ladder either to rescue visible occupants or vent. Having a minimum of six fire fighters assigned to a quint company allows the versatility of using the pumps and ladder in many different ways. With four fire fighters assigned to a quint, the department via SOPs or the company officer must make a decision to be either a truck company or an engine company. It is essential that SOPs describe actions the quint company is to take when it is staffed with fewer than six fire fighters and is the first apparatus on the scene. In most cases, this first-arriving quint company will function as an engine company if staffing does not permit functioning as a fully staffed quint company. However, there are occasions when the aerial device can also be used to rescue visible occupants. Even if the aerial device is not being used on arrival, the quint apparatus should be positioned for future use of the aerial device. Fire fighters assigned to quint apparatus must be constantly aware that these apparatus are likely to be used as either an engine company or a truck company.

Ventilation

Ventilation is one of the IC's most important tactical considerations, requiring close coordination between fire fighters venting and attack crews. Proper ventilation can have a positive effect on all three fire-ground priorities (life safety, extinguishment, and property conservation). However, ventilation is a double-edged sword. Improper ventilation can adversely affect all three priorities. **Table 4-1** lists the effects of both proper and improper ventilation. **Figure 4-3** shows proper vent locations.

It must be remembered that any form of venting, including opening an exterior door, will increase the oxygen supply. In many cases, the fire is oxygen deficient, and this increase in oxygen can result in a dramatic increase in fire growth. This is yet another reason to have a charged hose line staffed and in place whenever interior operations are being conducted.

Apparatus Positioning

Apparatus positioning and the fire company's assignment are directly related. The positioning of

Table 4-1 The Effects of Proper and Improper Ventilation

	Proper Ventilation	Improper Ventilation
Life safety	• Pulls fire away from trapped occupants or their means of egress	• Provides air to a ventilation-limited fire, causing rapid increase in temperature • Draws fire toward victims or can extend fire through their exit path
Extinguishment	• Limits fire spread by channeling fire toward nearby openings and allows fire fighters to safely attack the fire	• Spreads the fire into previously undamaged areas • Can cause a backdraft or let the fire gain headway while lines are being advanced
Property conservation	• Limits smoke, heat, and water damage by allowing an interior attack and removing the products of combustion, which are often the leading cause of damage	• Causes excessive and unnecessary damage • Does not remove the damaging products of combustion, but rather spreads them throughout the building

© Eky Studio/Shutterstock, Inc.

Figure 4-3 Vent locations.

apparatus must be consistent with the company's objective. If aerial ladders or towers are needed, the first-arriving truck company will position for aerial or tower placement. All companies should make a conscious effort to keep the fire zone as accessible and safe as possible.

The company officer (or IC) should always think about positioning. If the first-in engine company will be used to lay attack lines, position the apparatus on the same side of the street as the fire and as close to the curb line as possible, with the side discharge gates aligned with the desired entry point, provided there is a very low probability of building collapse. In most residential fires, this will be directly in front of the house. If the apparatus is set up to use rear discharges for preconnected attack lines, then the pumper should be positioned close to the curb but forward of the desired entry point. Positioning of this first-arriving apparatus should be addressed in SOPs and, among other things, depends on the layout of the hose bed.

Department SOPs generally address actions to be taken by the second-in engine company. SOPs for this company may allow more discretion on the part of the company officer. Some departments specify that the first-in and second-in engine companies secure separate sources of water. Other departments permit an **attack pumper** configuration in which the first engine company goes directly

to the front of the building and the second engine company provides the water supply. It must be remembered that working from the limited water supply carried on a pumper increases the chance that the company attacking the fire will exhaust their water supply prior to extinguishment, and a continued operation is based on an assumption that the second-due engine company will arrive at the scene and supply water in a timely manner. Each department must decide what works best for its jurisdiction.

For purposes of discussion, when the second engine company is supplying the attack pumper from a hydrant, it should select the suction intake that will allow the apparatus operator to get close to the curb while keeping the intersection open.

If the first-in truck company will be using its aerial device, the truck must be positioned to obtain a safe operating angle and reach the desired point for roof access, rescue, ventilation, or other tasks. Whenever possible, the aerial should be placed in a position to access two sides of the building. If the aerial device is not being used, the truck should be parked out of the way but close to the building's entry point in a position that will allow the apparatus to be repositioned if the aerial device is needed. Getting tools, ladders, and other equipment to the building is a labor-intensive job if the truck is parked too far from the entry point. In positioning the truck, it is also important to remember that on some apparatus, ladders feed off the rear of the truck. When this is the case, be certain that access to the rear of the truck is not blocked.

Figure 4-4 shows an example of apparatus positioning for an attack pumper operation with the aerial ladder being used to gain access to the roof.

Some common errors when positioning apparatus include the following:

- Placing aerials under wires where they cannot be safely raised
- Placing aerials and platforms in unsafe and unstable positions
- Not allowing enough room to extend the outriggers
- Apparatus/staff cars blocking access to the fire area or front of the building

- Pumpers placed in positions where preconnected attack lines are difficult to lay
- Apparatus unnecessarily blocking streets or fire hydrants
- Unnecessarily blocking streets with large-diameter hose
- First-arriving companies not securing essential attack positions
- Positioning apparatus in dead ends or other locations where they cannot be quickly repositioned
- Companies assigned to staging responding through the fire area, causing unnecessary congestion (Staging operations must be defined in department SOPs.)

Other vehicles should be positioned well outside the operational area.

Figure 4-4 Coordinated attack pumper operation.

Chapter Summary

- At the company level, the span of control is maintained by having the company officer manage the assigned crew.
- The division-of-labor concept stresses the need for preassigning duties (functions) at an emergency scene.
- An engine company is the basic firefighting unit. Engine company operations include conducting rescue, establishing a water supply, and operating hose lines.
- General truck company duties include search and rescue, ventilation, forcible entry, laddering, and reconnaissance.
- Truck company activities are varied and numerous during offensive attacks but are limited during defensive attacks.
- Safety and control dictatate that operating units work as companies or crews.
- Maintaining company unity improves safety and accountability; however, it is permissible to split the first-arriving company with the two-in/two-out rule. Further, the first-arriving truck company is often split into two crews.
- Communication between company officers or crew leaders and the IC must be maintained.
- Preassigned tools, tasks, and operations should be adjusted to specific circumstances; strict adherence to tool and task SOPs can sometimes be counterproductive.
- Offensive firefighting involves a coordinated effort of hose placement and ventilation.

Key Terms

attack pumper The first-arriving engine company that goes directly to the fire building without securing a water supply.

division-of-labor principle The principle that an incident or task should be broken down into smaller, more manageable tasks and personnel assigned to complete those tasks (developing job skills in a concentrated area to allow for more productivity).

initial rapid intervention crew (IRIC) A stand-by team of at least two fire fighters located outside the hazard area available to provide assistance to fire fighters operating within the hazard area.

quad A multifunction apparatus that is equipped to provide for the following four functions: water, pumps, hose, and ground ladders. In other words, this is a quint company minus the aerial ladder.

quint A multifunction apparatus that is equipped to perform both engine and truck company operations. It is equipped to provide for the following five functions: water, pumps, hose, ground ladders, and an aerial ladder.

Challenging Questions and Activities

1. Using resources from your department or a department with which you are familiar, assign tasks to the first-arriving and subsequent-arriving companies for the scenario shown in **Figure 4-5** and described as follows:

- Time and day: 4:00 PM, Saturday afternoon
- Weather: Partly cloudy and warm
- Structure: One-story with basement, ordinary construction
- Occupancy: Print shop

Flames: © Dra/Dreamstime.com; Steel Texture: © Sharpshot/Dreamstime.com

Flames: Drx/Dreamstime.com; Steel texture: © Sharpshot/Dreamstime.com

- Exposures: Side A, restaurant; side B, medical supply distributor; side C, vacant lot with street beyond; side D, auto repair shop

- Conditions on arrival as shown in Figure 4-5
- Area water supply is very good and hydrants are spaced every 500 ft (152 m)

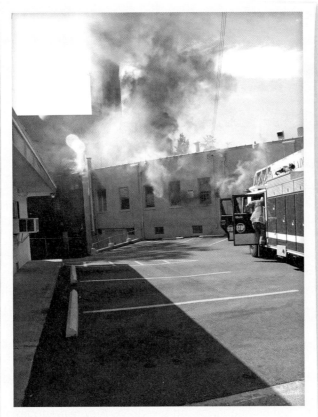

Figure 4-5 Print shop scenario.
Courtesy of Bill Strite, Cincinnati, Ohio

Endnotes

1. National Fire Protection Association. *NFPA 1500: Standard for Fire Department Occupational Safety and Health Program*. Quincy, MA: 2013.

2. Occupational Safety and Health Administration. *Code of Federal Regulations, 29 CFR 1910*. Washington, DC: U.S.

Flames: © Dra/Dreamstime.com; Steel texture: © SharpshotDreamstime.com; Chapter opener photo: © Johnny Habell/Shutter Stock, Inc.

Fire Fighter Safety

LEARNING OBJECTIVES

- Explain how applying a risk-versus-benefit analysis is the key to safe and effective fire-ground operations.
- Describe how employing effective fire-ground strategy and tactics affects safety on the fire ground.
- Identify and analyze the major causes involved in on-duty fire fighter fatalities related to health, wellness, fitness, and vehicle operations.
- Evaluate the effects of proper and improper ventilation during offensive fire attacks.
- Define the term "on-duty injury" as used by the National Fire Protection Association (NFPA).
- Analyze the trend in the number of fire fighter on-duty deaths and injuries over a 37-year period.
- Define frequency and severity as they relate to fire fighter injuries.
- Compare and contrast fire trends and fire fighter on-duty deaths.
- Describe the relative risk to fire fighters when combating fires in different occupancy types.
- Describe how "near miss" accidents improve the analysis of fire fighter injury statistics.
- Discuss risk management principles applied to the fire ground.
- List and compare the provisions in NFPA 1500, the *Offensive Operations* chapter, and the IAFC Rules of Engagement.

LEARNING OBJECTIVES (CONTINUED)

- Use a probability analysis to assess the occupied status of a building based on time, occupancy, and other factors.
- List the variables affecting the time to flashover.
- Estimate the collapse time based on burn time, fire intensity, content load, and construction type.
- List three important questions that must be answered prior to conducting roof operations at a structure fire.
- Describe the characteristics and possible components of a green building.
- Examine the difference between a managed retreat and an evacuation due to an imminent hazard.
- Evaluate the difference between modern lightweight and older structural components.
- Use experimental data and case studies to analyze the risk-versus-benefit analysis related to basement fires.
- Discuss and contrast prefire and fire conditions that contribute to structural collapse.
- Examine the hazards presented by suspended ceilings.
- Compare construction methods in terms of structural stability, fire extension, and fuel contribution.
- Develop zones and perimeters around a structure fire.
- Define and explain the five time segments from ignition to effective action.
- Evaluate the survivability, structural stability, and flashover from ignition to effective action.

- Compute the staffing necessary to achieve the tasks enumerated in NFPA 1710.
- Define and compare flashover and backdraft.
- Explain the relationship between the National Incident Management System (NIMS) and a fire fighter accountability system.
- List situations when a personal accountability report should be initiated.
- Explain the importance of alternative egress for fire fighters conducting an offensive attack.
- Define and describe the role of a rapid intervention crew (RIC).
- Determine the number of personnel to be assigned to the RIC based on the size and complexity of the building and incident.
- Construct an emergency message for a disoriented fire fighter needing assistance.
- Explain measures that can be taken to improve the chances of survival when fire fighters are lost and out of air in a large building, including supplying air to a firefighter who has exhausted his or her air supply.
- List tools that should be available to a RIC.
- Compare the advantages and disadvantages of a mobile RIC versus a stationary RIC.
- Evaluate hazards to fire fighters during overhaul operations.
- List factors the incident commander should consider when formulating an incident action plan to be used during overhaul.
- Describe hot and cold weather rehabilitation.

Case Study

Two career lieutenants killed and two career fire fighters injured following a flashover at an assembly hall fire.

Courtesy of the Bryan Fire Department.

At 8:13 PM, on February 15, 2013, two career fire officers were killed and two career fire fighters were injured combating an advanced fire in a 7400-ft² (687-m²) assembly hall. The fire was observed burning at the roof level on arrival. The parking lot was empty and there was no indication that the building was occupied. Two members of the first-arriving fire crew advanced a small-diameter hose line into the building in an unsuccessful attempt to extinguish the fire. No attempt was made to remove ceiling tiles or use a thermal imaging camera to determine if the fire visible at the roof was in the overhead false ceiling space.

Eventually, the attack crew was forced to retreat due to a low air supply. The fire fighter was able to follow the fire hose to safety, but for unknown reasons, the fire officer was unable to follow the fire hose out of the building. The trapped fire officer transmitted a call for help. A two-person rapid intervention crew (RIC) entered the building, following the hose line, and found the fire officer, who was responsive. The RIC and fire officer being rescued were caught in a flashover as they made their way toward the exit; the fire officer did not survive. The RIC officer later succumbed to his injuries while undergoing treatment at a burn center.

1. Explain the importance of a risk-versus-benefit analysis in regard to fire fighter safety.
2. How does ventilation affect fire fighter survival during structure fire?
3. What influence does compartment size have on fire fighter safety?
4. How can fire companies determine whether fire has entered a concealed area?

Data from: NIOSH Fire Fighter Fatality Investigation 2013-04. Two Career Lieutenants Killed and Two Career Fire ghters Injured Following a Flashover at an Assembly Hall Fire—Texas.

Introduction

Fire departments are dedicated to saving lives and property from the perils of fire. Saving lives is the highest tactical priority at the incident scene. The fire fighter's life is valued as being as important as any other life, and the fire fighter is the most valued resource of any fire department. Too often, fire fighters lose their lives in the process of saving the lives of others. Even more tragic are incidents where fire fighters lose their lives when there are no lives or property to be saved.

The guiding objective of this text is preparing the fire officer to take command of structure fires fully using available resources in a safe *and* effective way. Being safe *or* effective is relatively easy. The first-arriving company officer or chief officer assuming the role of incident commander (IC) concerned only with fire fighter safety would apply defensive streams from a distance beyond the collapse zone at all structure fires. In contrast, the fire officer interested only in being effective might routinely apply an aggressive offensive interior attack regardless of circumstances. Being safe *and* effective requires fire officers to carefully weigh the risk versus the benefit. Fire fighter safety is closely related to the risk-versus-benefit analysis discussed in the chapter titled *Developing an Incident Action Plan*. Probably the most important safety factor at the scene of a structure fire is a well-organized operation based on a solid risk-versus-benefit analysis using effective tactics.

In examining fire-ground incidents resulting in on-duty injuries and deaths, it can be seen that improper firefighting tactics were a contributing cause at many of these incidents. Extinguishing the fire improves safety for both occupants and fire fighters. In examining several cases, it was found that inadequate fire flows due to poor water supplies, hose kinks, or improper pump operation allowed the fire to gain in intensity resulting in injuries to fire

Stacked photo background: © Vitaliy Koroyin/ShutterStock, Inc.; Paper: © silver-john/ShutterStock, Inc.

fighters exposed to extreme heat and smoke conditions. In addition to extinguishing the fire, it is necessary to protect fire fighters by placing hose lines in close proximity to all areas where fire fighters are working within a structure. Proper ventilation can improve operational safety, but in many cases ventilation results in increased fire intensity unless hose lines are charged and in position to knock down the fire. Poor ventilation combined with a lack of protective hose lines is a deadly combination. A properly coordinated fire attack with proper ventilation greatly reduces the risk of fire fighter injury.

It is imperative that proper tactics be applied at structure fires to protect fire fighters. Proper training, combined with solid fire-ground procedures is an absolute necessity. Too often, a failure to complete basic fire-ground tasks, such as efficient pump operations, results in fire fighter injuries or deaths.

This chapter describes how NFPA 1500 is applied to assessing fire fighter safety at the incident scene. It also relates fire fighter safety to prefire and fire conditions—time, fire intensity, and structural stability—and the correlation between response time and progression to flashover. This chapter also discusses some of the factors relating to fire fighter safety, using statistics, as well as a series of case studies and established practices.

Fire Fighter Injuries and Fatalities

Identifying and analyzing how fire fighters are killed and injured is critical to reducing fire fighter injuries and deaths. The National Fire Protection Association (NFPA) has compiled and analyzed fire fighter on-duty fatality statistics since 1977. The annual On-Duty Firefighter Fatalities report for the previous calendar year is published each year in the July/August edition of the *NFPA Journal*. The annual On-Duty Firefighter Injury report is published in the November/December issue of the *NFPA Journal*. The charts and graphs used here are taken from data provided by the NFPA. A necessary part of any statistical study involves defining what is meant by various

terms and describing how data are gathered. Other organizations define fire fighter injuries and deaths differently and use other methodologies to gather data, thus yielding different results.

The NFPA defines the term "on-duty injury" as an injury occurring while at the scene of alarms, whether fire or nonfire; when responding to or returning from alarms; when performing other duties, including training, maintenance, public education, inspection, investigation, etc.; and while performing nonfire duties or official assignment station duty. If the on-duty injury proves fatal or any illness that was incurred as a result of actions while on duty proves fatal, then the death is counted as an on-duty fatality. On-duty fatalities are associated with specific on-duty activities and are reported as of the date of injury or onset. One of the real values in statistical analysis is defining trends. Statistical data occurring over a long period is more meaningful in determining trends than research limited to a few years. For this reason, this text uses NFPA statistics exclusively to evaluate fire fighter injuries and deaths.

It is essential that fire departments closely examine injuries, fatalities, and near miss accidents that occur in their department then take action to prevent future injuries. Several organizations publish fire investigative reports analyzing individual fires that resulted in fire fighter fatalities; such organizations include the NFPA, National Institute for Occupational Safety and Health (NIOSH), U.S. Fire Administration (USFA), and others. These reports provide information, which, if properly used and analyzed, can prevent future on-duty deaths and reduce injury **frequency** and **severity**.

In examining the charts in **Figure 5-1**, it is clear that the number of on-duty fire fighter deaths is trending downward over this 37-year period. Many measures have been taken to improve fire fighter safety since 1977. Notice that the 340 on-duty deaths at the World Trade Center fire in 2001 are not included. The graph would be off the chart in 2001 if these fatalities were included in the chart. Two very tragic fires resulted in 28 on-duty fatalities in 2013 (with the Yarnell, Arizona, wildfire leading to 19 deaths and the West, Texas, ammonium nitrate

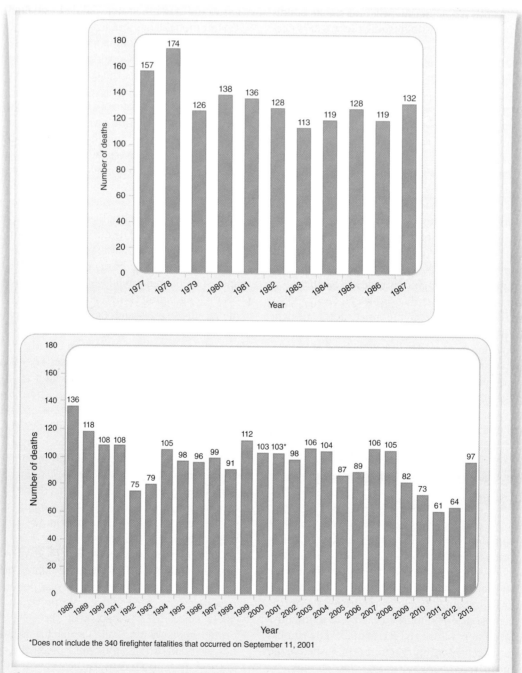

Figure 5-1 On-duty fire fighter deaths, 1977–2013.

Modified from: Fahy, Rita F. Firefighter Fatalities in the United States-2013. National Fire Protection Association.

explosion resulting in 9 deaths). These incidents will result in a higher number of fatalities than average number of fire fighter on-duty deaths during 2013.

Personal protective equipment and breathing apparatus used by fire fighters in the present day are vastly improved compared to protective gear used in 1977 when the NFPA began gathering data on fire fighter injuries and fatalities. In 1987, the first edition of NFPA 1500, *Standard on Fire Department Occupational Safety and Health Program*, was adopted, with subsequent revisions every 5 years. Following the provisions outlined in NFPA 1500[1] can substantially reduce injury frequency and severity.

Chapter 8 of NFPA 1500 provides the requirements for safety measures to be taken at the incident scene. These measures represent the absolute minimum. All fire fighters and officers must be familiar with this standard to be effective in protecting their most valued resource: the fire fighter. Several important fire-ground safety issues are enumerated in the NFPA 1500 standard:

- Risk management principles must be applied to fire-ground operations.
- The IC is responsible for overall safety at the scene.
- An incident management system must be used at all emergency scenes.
- The IC maintains command and control of all operating forces within a common strategy based on situation analysis. The situation analysis must be ongoing with changes in strategy commensurate with the changing situation.
- An adequate number of personnel must be available to implement the strategy.
- Preestablished standard operating procedures (SOPs) must be implemented.
- An accountability system must be used at all incidents.
- Rapid intervention crews (RICs) must be provided at all stages of the incident, beginning with the first-arriving unit using a two-in/two-out process.
- Inexperienced members must be directly supervised by more experienced members.

- Minimum basic and continued fire training must be provided in accordance with SOPs.
- Medical treatment and rehabilitation must be available as needed.
- Full personal protective clothing must be worn.
- Self-contained breathing apparatus (SCBA) must be worn in hazardous atmospheres, atmospheres that are suspected of being hazardous, or atmospheres that may rapidly become hazardous.
- Personal alert safety system (PASS) devices must be worn and activated before entering a hazardous area.
- Postincident analysis is necessary for significant incidents or those causing serious injury or death.

NFPA 1500 contains a detailed description of these minimum requirements. Standards issued by the Occupational Safety and Health Administration (OSHA) also address measures to be taken to protect fire fighters.[2]

There is no doubt that attention to safety and the attitude toward safe operations has greatly improved within the fire service. Unfortunately, even with these improvements the fire service is still experiencing a large number of on-duty deaths. Most troubling is the fact that the fire service continues to experience a large number of fire fighter deaths on the fire ground. **Figure 5-2** compares the number of noncardiac-related fire fighter fatalities at structure fires to the number of fires. Examining the data on noncardiac fire-ground deaths reveals that although fire fighters are responding to fewer fires, when operating at structure fires, they are dying at a higher rate per fire than in the late 1970s.

Considering all that has been done to improve safety, this is a very disturbing statistic. Could this fatality rate be due to a lack of fire experience and training? Fire fighters are now expected to perform many additional emergency tasks as compared to the 1970s. Each of these tasks necessitates extra training. As a result, many departments find it difficult to provide adequate structural firefighting training and education to make up for the lack of fire-ground experience.

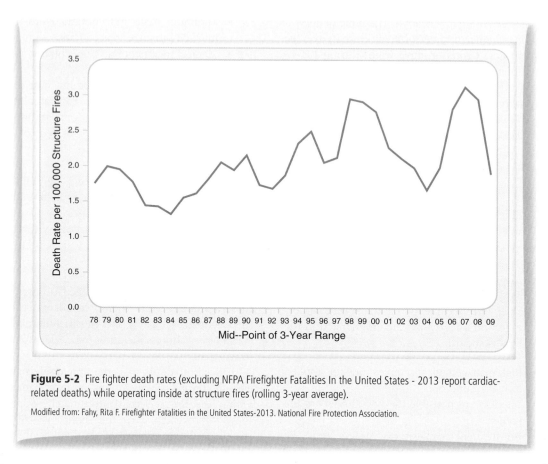

Figure 5-2 Fire fighter death rates (excluding NFPA Firefighter Fatalities In the United States - 2013 report cardiac-related deaths) while operating inside at structure fires (rolling 3-year average).

Modified from: Fahy, Rita F. Firefighter Fatalities in the United States-2013. National Fire Protection Association.

Nothing completely takes the place of experience, but task- and tactic-oriented fire training and education can improve safety during fire-ground operations.

The fire ground is becoming more hazardous. Lightweight construction combined with heavy fire loads (including large quantities of plastics) in larger buildings significantly increase the risk to fire fighters. Engineered lightweight roof and floor systems, present in residential and larger commercial buildings, are discussed later in this chapter. Lightweight construction methods create a substantial hazard to fire fighters.

Modern personal protective equipment provides a higher level of protection, thus permitting fire fighters to make a closer approach to the fire. Advancing farther inside the building and closer to a more intense fire places fire fighters at greater risk. The everyday home fire provides experience for larger fires, but different strategy and tactics are required to meet the challenges posed when a large, complex building is on fire. Occupancy plays a role in fire fighter safety. Never consider any fire routine. Most fire fighter fatalities occur in home (dwellings and apartments) fires because most fires occur in these occupancies. In 2011, 80% of all structure fires occurred in residential occupancies.[3] However, once a fire occurs, the risk to fire fighters is much greater in other occupancy types, as shown in **Figure 5-3**. The probability of fatality is much greater in a manufacturing occupancy than in a residence. Comparing the on-duty death rate of 18.9 deaths per 100,000 fires in manufacturing occupancies to the 2.9 deaths per 100,000 fires in residential occupancies indicates that the chances of a fire fighter fatality is 6.5 times greater when combating a manufacturing occupancy fire.

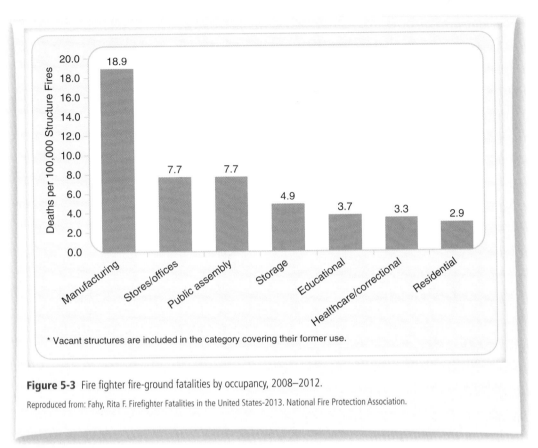

Figure 5-3 Fire fighter fire-ground fatalities by occupancy, 2008–2012.

Reproduced from: Fahy, Rita F. Firefighter Fatalities in the United States-2013. National Fire Protection Association.

When examining statistical information, it is important to consider the impact of a single fire where many lives are lost. As an example, if the 340 fire fighter on-duty deaths at the World Trade Center and the 9 fire fighter on-duty deaths that occurred at a fire in a Charleston, South Carolina, furniture store were added to the stores/offices category, the bar representing these occupancies would be off the chart. Vacant structures were once listed separately and represented the highest risk to fire fighters. The NFPA now statistically categorizes vacant structures according to their former use.

It is important to avoid a single-family residential mind-set. Escape routes are closer and easier to find in most single-family residential structures as compared to larger buildings where more time is needed to escape and a higher probability of disorientation exists. Depleting the SCBA air supply or an SCBA failure often results in survivable smoke inhalation injury in a one- or two-family dwelling fire. The loss of air supply is more likely to be fatal in a larger, more complex building. Even a defensive posture may not be sufficient to protect fire fighters when a building contains an extra-hazard fire load, as was the case in the West, Texas, ammonium nitrate storage facility.

Figure 5-4 shows the number of deaths by type of duty. Given the lower number of fires and increased activity in nonfire emergencies such as emergency medical services (EMS), the fire ground remains a very dangerous workplace.

In analyzing fire fighter on-duty deaths, it is also important to consider the nature and cause of the injury leading to fire fighter fatalities, as shown in **Figure 5-5** and **Figure 5-6**.

Figure 5-7 shows that cardiac events and asphyxiation are the two leading causes of on-duty

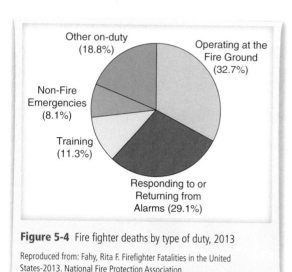

Figure 5-4 Fire fighter deaths by type of duty, 2013

Reproduced from: Fahy, Rita F. Firefighter Fatalities in the United States-2013. National Fire Protection Association.

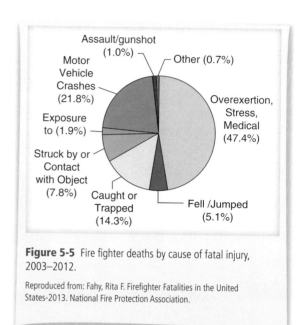

Figure 5-5 Fire fighter deaths by cause of fatal injury, 2003–2012.

Reproduced from: Fahy, Rita F. Firefighter Fatalities in the United States-2013. National Fire Protection Association.

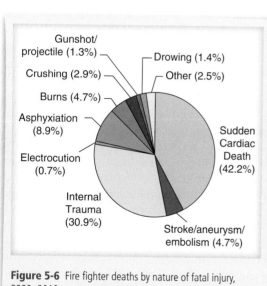

Figure 5-6 Fire fighter deaths by nature of fatal injury, 2003–2012.

Reproduced from: Fahy, Rita F. Firefighter Fatalities in the United States-2013. National Fire Protection Association.

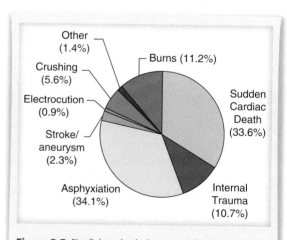

Figure 5-7 Fire fighter deaths by nature of injury—structure fire only, 2002-2011.

Reproduced from: Fahy, Rita F. Firefighter Fatalities in the United States-2013. National Fire Protection Association.

Incident Summary

On April 17, 2013, the West, Texas, volunteer fire department responded to a rapidly developing fire at a wood-frame warehouse storing approximately 60 tons of ammonium nitrate. The fire department made an effort to extinguish the fire while the staff of a nursing home located just 200 yards (183 meters) from the fire scene began moving residents to more protected areas inside the building.

About 20 minutes after the fire was reported, the ammonium nitrate stored inside the warehouse exploded, destroying the warehouse and sending large pieces of flying debris into the surrounding town (population 2800). The blast was heard 80 miles away and registered a magnitude 2.1 tremor that was recorded by the U.S. Geological Survey. At the nursing home, windows shattered and the roof collapsed, trapping many elderly residents. An apartment complex across the street from the nursing home was heavily damaged by the explosion. The middle and intermediate schools and nearly 200 homes were damaged due to the explosion. The ammonium nitrate explosion killed 15 people: 10 fire fighters, 2 civilians assisting the fire department, 2 residents from the apartment complex, and 1 nursing home resident. About 200 people were injured.

The U.S. Chemical Safety Board (CSB) estimates that total damages could exceed $230 million. The cause of the fire has not been determined; in its preliminary findings, the CSB said that "the explosion resulted from an intense fire that led to the detonation of" the ammonium nitrate. The Environmental Protection Agency estimates that 13,000 facilities similar to West Fertilizer pose threats to communities throughout the United States.

It is interesting to note that the largest loss of life in an ammonium nitrate explosion also occurred in Texas. On April 16, 1947, in Texas City, ammonium nitrate was being transferred from a chemical plant onto a ship when a fire erupted on board. A tremendous explosion killed hundreds, including 27 members of the Texas City Fire Department.

Data from: West's Wake, March/April 2014 NFPA Journal, Fred Durso

fire fighter fatalities at structure fires. Sudden cardiac deaths are responsible for slightly more than one third of fire fighter fatalities at structure fires. Many of these deaths are preventable through proper medical evaluation and wellness programs.

Asphyxiation accounts for an almost equal number of fire fighter on-duty deaths when working at structure fires. Deaths by asphyxiation require further study to determine common denominators. Much has been written about air management. Part of the problem is depleting the SCBA air supply while working on the interior at a structure fire. An air cylinder rated at 30 minutes usually provides an air supply of approximately 20 minutes under work conditions experienced on the interior at a working structure fire. Low air alarms should sound with approximately 5 to 7 minutes of escape time remaining. This notice is probably sufficient in the typical dwelling, provided the fire fighter immediately retreats when the alarm sounds. When a fire fighter is disoriented or working in a larger building, 5 to 7 minutes may not provide sufficient escape time. The accountability officer can track "on-air" time, notifying fire fighters of approximate time remaining to allow more escape time. Manually tracking on-air time is very difficult and may require the appointment of assistant accountability officers. Technology now available from breathing apparatus manufacturers shows promise in assisting in this task.

The advent of 45-minute and 60-minute open-circuit SCBAs provides additional time and increases the escape time. Unfortunately, higher-pressure and

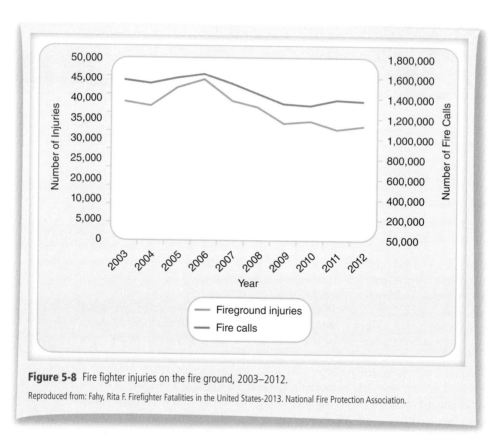

Figure 5-8 Fire fighter injuries on the fire ground, 2003–2012.

Reproduced from: Fahy, Rita F. Firefighter Fatalities in the United States-2013. National Fire Protection Association.

larger-volume cylinders weigh more and sometimes have a higher profile, which increases the workload and decreases mobility.

Evaluating fire fighter deaths in regard to situations and causes is essential to a continued decrease in the number of fire fighter fatalities. Moreover, the company officer and IC can also affect non-life-threatening injuries that often lead to pain, suffering, and disability. **Figure 5-8** shows a decrease in the number of fireground injuries per year for a 10-year period. Unlike the on-duty fatalities, the number of fire fighter injuries is closely tied to the number of fires. Over 45% of all injuries occur on the fire ground, and over 55% of on-duty injuries are strains and sprains.[4]

The statistical information presented here was gathered from actual fire fighter death and injury reports. Professional safety specialists have long realized that minor injury accidents are in the range of 10% of all accidents, and accidents resulting in lost

time are less than 1% of total accidents.[5] Most minor injuries and close calls (near misses) go unreported. Many of these near misses could have resulted in serious injuries or death with slightly different circumstances. An example of a near miss would be a fire fighter falling through a collapsed floor and being rescued without injury or failing to report a minor injury. These near misses are good indicators of common hazards and merit examination. With this in mind, the Department of Homeland Security, the International Association of Fire Chiefs (IAFC), and the International Association of Fire Fighters created a website (www.firefighternearmiss.com) that contains many reports of near misses and contributes to the overall understanding of how fire fighters are injured.

Fire administrators must be committed to fire fighter safety and provide the necessary procedures, training, and equipment. ICs have overall safety responsibility during an incident and must make the crucial offensive/defensive attack decision at structure fires, then monitor, organize, coordinate, and provide safety measures described throughout this text. Company officers have a responsibility to properly supervise members of their company. During the early stages of an operation, company officers are often the IC. Fire fighters must take personal responsibility for their safety and the safety of members working with them by following procedures, maintaining firefighting skills, and properly using the equipment provided. Sustaining a high level of physical fitness is important not only to the individual member but to the entire fire-ground operation.

Risk Management Applied to the Fire Ground

Probably the most important element of the incident safety program is applying risk management to fire-ground operations. NFPA 1500 specifically addresses risk management principles to be applied at the incident scene:

NFPA 1500-8.3.2 The concept of risk management shall be utilized on the basis of the following principles:

(1) Activities that present a significant risk to the safety of members shall be limited to situations where there is a potential to save endangered lives.

(2) Activities that are routinely employed to protect property shall be recognized as inherent risks to the safety of members, and actions shall be taken to reduce or avoid these risks.

(3) No risk to the safety of members shall be acceptable when there is no possibility to save lives or property.

(4) In situations where the risk to fire department members is excessive, activities shall be limited to defensive operations.

The military and law enforcement developed rules of engagement to provide soldiers and police officers with a checklist to be used when confronted with deadly force situations requiring an immediate decision. The IAFC developed rules of engagement for ICs and fire fighters faced with life-and-death decisions on the fire ground.[6] These rules were published along with objectives that briefly describe the implementation of each rule at the incident scene. The Rules of Engagement for Firefighter Survival are a list of precautions to be taken by company officers and fire fighters.

IAFC Rules of Engagement for Fire Fighter Survival

1. **Size-up your tactical area of operation.**
   Objective: To cause the company officer and firefighters to pause for a moment and look over their area of operation and evaluate their *individual* risk exposure and determine a safe approach to completing their assigned tactical objectives.

2. **Determine the occupant survival profile.**
   Objective: To cause the company officer and firefighter to consider fire conditions in relation to possible occupant survival of a *rescue event* as part of their initial and ongoing *individual risk assessment* and action plan development.

3. **DO NOT risk your life for lives or property that cannot be saved.**
   Objective: To prevent firefighters from engaging in high-risk search and rescue and firefighting operations which may harm them when fire conditions prevent occupant survival and significant or total destruction of the building is inevitable.

4. **Extend LIMITED risk to protect SAVABLE property.**
   Objective: To cause firefighters to limit risk exposure to a reasonable, cautious and conservative level when trying to save a building.

5. **Extend VIGILANT and MEASURED risk to protect and rescue SAVABLE lives.**
   Objective: To cause firefighters to manage search and rescue and supporting firefighting operations in a calculated, controlled and safe manner, *while remaining alert to changing conditions*, during high risk primary search and rescue operations where lives can be saved.

6. **Go in together, stay together, come out together.**

   Objective: To ensure that firefighters always enter a burning building as a team of two or more members and *no firefighter is allowed to be alone at any time* while entering, operating in or exiting a building.

7. **Maintain continuous awareness of your air supply, situation, location and fire conditions.**

   Objective: To cause all firefighters and company officers to maintain constant situational awareness of their SCBA air supply, where they are in the building and all that is happening in their area of operations and elsewhere on the fireground that may affect their risk and safety.

8. **Constantly monitor fireground communications for critical radio reports.**

   Objective: To cause all firefighters and company officers to maintain constant awareness of *all* fire ground radio communications on their assigned channel for progress reports, critical messages or other information that may affect their risk and safety.

9. **You are required to report unsafe practices or conditions that can harm you. Stop, evaluate, and decide.**

   Objective: To prevent company officers and firefighters from engaging in unsafe practices or exposure to unsafe conditions that can harm them and *allowing any member to raise an alert about a safety concern without penalty* and mandating the supervisor address the question to ensure safe operations.

10. **You are required to abandon your position and retreat before deteriorating conditions can harm you.**

    Objective: To cause firefighters and company officers to be aware of fire conditions and cause an early exit to a safe area when they are exposed to deteriorating conditions, unacceptable risk and a life threatening situation.

11. **Declare a May-Day as soon as you THINK you are in danger.**

    Objective: To ensure the firefighter is comfortable with, and there is no delay in, declaring a May Day when a firefighter is faced with a life threatening situation and the May Day is declared as soon as they THINK they are in trouble.

The Incident Commander's Rules of Engagement for Fire Fighter Safety are intended to be implemented by the Incident Commander. It is important to note that the first rule "conducting a 360" also applies to the first arriving company officer. In reality, the first arriving company officer or in some cases, the first arriving firefighter, is in fact the Incident Commander until another officer assumes command. Therefore, company officers should also know The Incident Commander's Rules of Engagement for Fire Fighter Safety and implement these rules when in command.

The Incident Commander's Rules of Engagement for Fire Fighter Safety

1. **Rapidly conduct, or obtain, a 360 degree situational size up of the incident.**

   Objective: To cause the incident commander to obtain an early 360 degree survey and risk assessment of the fireground in order to determine the safest approach to tactical operations as part the risk assessment and action plan development *and before firefighters are placed at substantial risk.*

2. **Determine the occupant survival profile.**

   Objective: To cause the incident commander to consider fire conditions in relation to possible occupant survival of a *rescue event* before committing firefighters to high risk search and rescue operations as part of the initial and ongoing *risk assessment* and action plan development.

3. **Conduct an initial risk assessment and implement a SAFE ACTION PLAN.**

   Objective: To cause the incident commander to develop a safe action plan by conducting a thorough size-up, assess the occupant survival profile and completing a risk assessment *before* firefighters are placed in high risk positions on the fireground.

4. **If you do not have the resources to safely support and protect firefighters, seriously consider a defensive strategy.**

Objective: To prevent the commitment of fire-fighters to high risk tactical objectives that cannot be accomplished safely due to inadequate resources on the scene.

5. **DO NOT risk firefighter lives for lives or property that can not be saved. Seriously consider a defensive strategy.**

Objective: To prevent the commitment of firefighters to high risk search and rescue and firefighting operations that may harm them when fire conditions prevent occupant survival and significant or total destruction of the building is inevitable.

6. **Extend LIMITED risk to protect SAVABLE property.**

Objective: To cause the incident commander to limit risk exposure to a reasonable, cautious and conservative level when trying to save a building that is believed, following a thorough size up, to be savable.

7. **Extend VIGILANT and MEASURED risk to protect and rescue SAVABLE lives.**

Objective: To cause the incident commander to manage search and rescue, and supporting firefighting operations, in a highly calculated, controlled, and cautious manner, *while remaining alert to changing conditions,* during high risk search and rescue operations where lives can be saved.

8. **Act upon reported unsafe practices and conditions that can harm firefighters. Stop, evaluate and decide.**

Objective: To prevent firefighters and supervisors from engaging in unsafe practices or exposure to unsafe conditions that will harm them and *allowing any member to raise an alert about a safety concern without penalty* and *mandating* the incident commander and command organization officers promptly address the question to insure safe operations.

9. **Maintain frequent two-way communications and keep interior crews informed of changing conditions.**

Objective: To ensure that the incident commander is obtaining frequent progress reports from command organization officers and all interior crews are kept informed of changing fire conditions observed from the exterior by the incident commander, or other command officers, that may affect crew safety.

10. **Obtain frequent progress reports and revise the action plan.**

Objective: To cause the incident commander, as well as all command organization officers, to obtain frequent progress reports, to continually assess fire conditions and any risk to firefighters, and to regularly adjust and revise the action plan to maintain safe operations.

11. **Ensure accurate accountability of every firefighter location and status**

Objective: To cause the incident commander, and command organization officers, to maintain a constant and accurate accountability of the location and status of all firefighters within a small geographic area of accuracy within the hazard zone and be aware of who is presently in or out of the building.

12. **If after completion of the primary search, little or no progress towards fire control has been achieved, seriously consider a defensive strategy.**

Objective: To cause a benchmark decision point, requiring the incident commander to determine if it's safe to continue offensive interior operations if there is no progress in controlling the fire and there are no lives to be saved following completion of the primary search.

13. **Always have a rapid intervention team in place at all working fires.**

Objective: To cause the incident commander to have a rapid intervention team in place ready to rescue firefighters at all working fires.

14. **Always have firefighter rehab services in place at all working fires.**

Objective: To ensure all firefighters who endured physically strenuous activity at a working fire are rehabilitated and medically evaluated for continued duty.

Some might wrongly interpret these rules to always justify a defensive operation. Actually, these Rules of Engagement require a risk-versus-benefit analysis that is consistent with NFPA 1500-8.3.2, described previously. This text continually stresses the importance of developing an incident action plan based on a risk-versus-benefit analysis and further describes how to apply NFPA 1500 as well as the IAFC Rules of Engagement for Fire Fighter Survival. These important guidelines must be applied to every situation. The single most important concept the fire officer must acquire is being skilled at recognizing the point at which the risk to fire fighters' lives outweighs the possible benefits of saving lives and property. Too many fire fighters are injured or killed while undertaking body recoveries or wetting down the ruins. The decision to employ a defensive approach based on the risks outweighing the possible benefits can be a difficult one. One of the clearest illustrations of this difficulty can be found in the Worcester, Massachusetts, warehouse fire, where the IC made a decision to abandon interior operations, knowing that six missing fire fighters who remained in the building would almost certainly perish.

The first statement in NFPA 1500-8.3.2 reads, "Activities that present a significant risk to the safety of members shall be limited to situations where there is a potential to save endangered lives." This

Incident Summary

In Worcester, Massachusetts, six career fire fighters died in a vacant, six-floor, maze-like, cold storage and warehouse building while searching for two homeless people who accidentally started the fire and escaped prior to the arrival of the fire department. Approximately 30 minutes after the first alarm was struck, two fire fighters searching for victims and checking for extension sounded an emergency message. A personal accountability report (PAR) confirmed two missing fire fighters. The search-and-rescue operation was expanded to find the two missing fire fighters and the homeless people thought to be in the building. Four additional fire fighters became disoriented during this part of the operation. The fire was believed to have 30 to 90 minutes of pre-burn time. After more than an hour on the scene, a company on the interior notified command of structural problems and an arson investigator on the exterior reported the fire was venting from the roof. A defensive attack was ordered at approximately 1 hour and 45 minutes after the initial alarm. The six missing fire fighters perished.

The combination of a massive fire load and the maze of rooms in this Worcester, MA warehouse contributed to the death of six fire fighters.

© Paul Connors/AP Images.

Data from: National Institute for Occupational Safety and Health (NIOSH), Fire Fighter Fatality Investigation 99 F-47, Six Career Fire Fighters Killed in Cold Storage Warehouse Building Fire–Massachusetts.

Flames: © Jag_cz/ShutterStock, Inc.; Paper: © silver-john/ShutterStock, Inc.

statement can be somewhat ambiguous and has been the subject of much discussion. The controversy heightens with the provision of an exception to the two-in/two-out rule in the face of an imminent life-threatening situation. The degree of risk associated with entering a burning building lacking the safety provided by the two-in/two-out provision is obviously higher than the risk taken when two fire fighters are outside the building ready to come to the assistance of the interior crew using the buddy system. Therefore, the perceived benefit would have to be more absolute to justify the risk of entry without an initial rapid intervention crew (IRIC).

It is relatively easy to visualize a situation with an imminent life-threatening situation. Or is it? An argument could be made for urgency in the late-night residential fire with an assumption that occupants are home asleep, possibly unaware of the fire, or already in serious trouble. This process could be continued with assumptions that the possibility of a threat to life always exists until a primary search verifies that no one is endangered. A warehouse or office building could also be occupied at night. Even vacant buildings can be occupied by the homeless or others. When a fire breaks out in a vacant structure, there is a possibility that the person who caused the fire is still in the building.

NOTE

Sometimes it is necessary to enter vacant structures to determine the life safety hazard or extinguish a minor fire. Fire officers must be extremely careful when confronted with a well-involved fire in a vacant structure.

Texture: © Eky Studio /Shutterstock, Inc; Steel: © Sharpshot/Dreamstime.com

The concept of probability is important to the risk management process. The first-arriving officer must consider the possibility of people being in the building not as a yes-or-no proposition, but rather as a degree of probability: How does the fire officer evaluate the probability that savable people are in the building?

The probability that a nursing home is occupied is very high at any time of the day or night, as is the potential for an occupied dwelling at 2:00 AM. Finding someone in an office complex at 2:00 AM on a Saturday night or Christmas Eve would be less likely but possible. Thus, occupancy type, time of day, day of week, and time of year are factors in determining the probability of the structure being occupied. The probability of the building being occupied is essential to analyzing risk versus benefit, applying the Rules of Engagement, and completing the size-up process.

The NFPA 1500 committee undoubtedly expects the fire officer to exercise judgment in regard to the issue of an imminent life-threatening situation.

Fallacy	Fact
Vacant buildings are always unoccupied. Apply a defensive attack.	While the probability that a vacant building is occupied is low, there is a chance that someone will be in the building. The only way to know for sure is to conduct a primary search.

Steel: © Sharpshot/Dreamstime.com

Fire Intensity

Time and occupancy factors are not the only considerations. Fire intensity and building construction are equally important in determining the probability of saving lives and the risk to fire fighters. The hazards to fire fighters entering a given area increase as the fire progresses toward flashover. The building is getting weaker throughout the process to the point of complete failure. No one can determine exactly when flashover will occur or when the building will fail. However, there are useful approximations. In post-flashover fires, the chance of occupant survival is minimal within the flashover compartment.

In the past, the time from ignition to flashover was given as 10 minutes. The actual time can vary significantly, depending on a number of variables, including the following:

- Compartment size
- Available air supply (ventilation)
- Ignition source
- Fuel type
- Fuel geometry
- Geometry of the enclosure
- Distance between fuel packages
- Location of the fuel

Recent studies by Underwriters Laboratories comparing modern furnishings to legacy (mid-20th century) furnishings indicate that flashover is likely to occur prior to the arrival the first-due fire company in residential occupancies.[7] Other studies by Underwriters Laboratories and the National Institute of Standards and Technology (NIST) indicate that the fire may become ventilation limited prior to flashover. In this case, fire fighters can expect a rapid acceleration when air is supplied via doors, windows, or other openings.

There are several computer programs that can provide rough estimates of when flashover will be reached in a compartment. However, given the wide range of variables that must be entered into these computer programs, it is very difficult to provide a reliable time frame that the fire officer can use on the fire ground. Larger-volume compartments generally take longer to flashover. However, ventilation can radically affect the time to flashover. A fire in an oxygen-starved room could go into the decay phase and not flash over; adding air to the room could result in a backdraft or rapid fire acceleration. Most fire fighters think of ventilation in terms of purposeful venting. Many times self-venting occurs when the fire breaches containment, such as when windows fracture. Unintentional venting also occurs when fire fighters open doors or windows to gain access.

Wind can have an adverse effect by providing an abundance of oxygen to rapidly increase fire intensity. Most studies of wind-driven fires have been conducted in high-rise buildings or simulated high-rise scenarios. However, wind can have an adverse impact on operations in a building of any size, and

the lessons learned from these studies apply to these situations, as well. Wind blowing toward a vent opening tends to reverse the normal flow path due to an increase in pressure at the opening. If a ventilation outlet is available on another side of the building, the fire and smoke will naturally flow toward the outlet; for example, wind blowing into the rear of the building through an open window when fire fighters open the front door to make entry. Higher wind speeds increase fire and smoke movement within the structure. Fire fighters in this ventilation stream are in great peril. This potential risk is one of many reasons why the company making initial entry should control the door and have a charged hose line before making entry.

Fuel Load

The fire or fuel load consists of fuels provided by the contents and combustible building materials. Most of the construction materials used in wood-frame buildings will burn, thus creating a large fuel load. Combustible building materials are very limited in fire-resistive and noncombustible construction. The primary fuel load for most structure fires is made up of the combustible contents. The combustible contents in commercial occupancies vary widely, even among occupancies of the same type. As an example, the fuel load in a warehouse storing plastic products would be very high compared to the fuel load in a warehouse containing metal parts in cardboard cartons. Sprinkler calculations take fuel load into account. Some residential properties contain excessive contents, but there is an average fuel load that can be anticipated. The residential fuel load has increased significantly due to people having more possessions and the amount of plastic and high-fuel-capacity contents.

Underwriters Laboratories[8] conducted a series of 15 full-scale experiments in residential properties to examine the effect of the increased fuel load, as well as the geometry and room size, in modern residential buildings. They found that the fuel load in modern buildings is much greater than in legacy residential buildings of the mid-20th century. This increase and more volatile fuel load dramatically reduce the time to

Incident Summary

On Sunday, April 12, 2009, shortly after midnight, a fire in a large one-story ranch residence in Houston, Texas, killed two fire fighters. Fire fighters entered through the front door and were advancing a hose line down the main entry hallway when a large rear window failed, accelerating the burn rate. There were sustained winds of 17 miles per hour (27 kilometers per hour [km/hr]) and wind gusts up to 26 miles per hour (42 km/hr) blowing from the rear of the house toward the front. The NIOSH report also addresses problems involving inadequate water supply for attack lines and ventilation problems as possible contributing factors.

Data from: NIOSH [2010]. Career Probationary Fire Fighter and Captain Die as a Result of Rapid Fire Progression in a Wind-Driven Residential Structure Fire – Texas: U.S. Department of Health and Human Services, Public Health Service, Centers for Disease Control and Prevention, National Institute for Occupational Safety and Health, Fatality Assessment and Control Evaluation (FACE) Report F2009-11.

Modified from: NIOSH [2010]. Career Probationary Fire Fighter and Captain Die as a Result of Rapid Fire Progression in a Wind-Driven Residential Structure Fire – Texas: U.S. Department of Health and Human Services, Public Health Service, Centers for Disease Control and Prevention, National Institute for Occupational Safety and Health, Fatality Assessment and Control Evaluation (FACE) Report F2009-11.

flashover. In Underwriters Laboratories experiments, the time to flashover in a living room furnished with modern furniture occurred in 3.5 minutes, compared to 29.5 minutes for the same room furnished with legacy furnishings. This study and later studies also examined the effects of ventilation, including tests to determine when windows and doors would fail due to fire exposure. Armed with this knowledge, it is incumbent on fire departments to adjust their tactics, as explained in the *Offensive Operations* chapter.

Building Design Loads

Loads imposed on a building are divided into live, dead, seismic, wind, snow, and ice loads. Building loads affect structural stability. Unusually high building loads can result in premature collapse. Loads placed on lightweight roof and lightweight floor structural assemblies are particularly dangerous under fire conditions. A fire fighter walking on a roof creates a live load that is concentrated on a fairly small area the size of his or her boot. Using a roof ladder distributes this live load over a much greater surface area, reducing the chance of penetrating a weakened section of roof. Working from an aerial ladder or platform to perform ventilation or other roof tasks is much safer. Likewise, placing a ladder on a stairway or floor that could be structurally compromised distributes live loads created by fire fighters over a larger area.

Structural Stability

Structural failure can occur at any time. Always consider structural stability when sizing up the fire scene. Fire intensity, burn time, content loads, and construction methods and materials all affect structural stability. An old rule of thumb called the "20-minute rule" was based on all of these factors except the content load. The **20-minute rule** states that when a heavy volume of fire is burning out of control on two or more floors for 20 minutes or longer, structural collapse should be anticipated. This rule is based on Type III (ordinary construction). A frame structure may be completely consumed in less than 20 minutes, and unprotected steel can fail in less than 20 minutes when subjected to a heavy volume of fire. Fire entering a concealed truss space causes failure in a relatively short time, as has happened in several fires that have killed fire fighters. Heavy-timber or fire-resistive structures subjected to the same volume of fire would be expected to withstand fire attack longer than a frame structure.

There is a subjective element in the 20-minute rule: heavy volume of fire. What is perceived as a heavy volume of fire by one fire officer is a moderate volume of fire to another. There is also reason to question the "two or more floors" part of the rule. A heavy volume of fire on a single floor could also result in collapse. Is 20 minutes an exact time? Absolutely not! Likewise, the length of burn time at heavy volume is subjective, as there is no way of knowing the exact time of ignition or the intensity of burn before arrival. However, time, intensity, and fire volume factors must be considered in determining whether it is safe to enter or be near a structure.

An IC confronted with a well-involved fire should "start the clock" when making a decision to conduct operations in the offensive mode. If, after a given period of time, the fire is still not being brought under control, the incident action plan should be reviewed and the operation possibly changed to a defensive attack.

Given the amount of information that an IC must assimilate, it is easy to lose track of how much time has passed on the fire ground. NFPA 1500 requires dispatch centers to notify command every 10 minutes until the fire is knocked down, the incident becomes static, or the IC cancels the notification. This notification helps the IC track elapsed time. It

is critical that burn time take into account the time the fire was burning prior to notification and while units were responding, not just the amount of time since they arrived on the fire scene.

Transitioning from an offensive to a defensive strategy is the ultimate test of command. This transition phase is a dangerous time on the fire ground. When changing from an offensive to a defensive operation, a managed retreat is best. Engine companies operating hose lines should provide protection while other companies, especially companies above the fire, evacuate the building. Planning and safety can be of great assistance to the IC in making the decision to go defensive and assist in managing a retreat. This is one reason this text recommends staffing these positions as soon as possible. Ultimately determining the need to retreat and managing the process is the IC's responsibility. Whenever signs of imminent collapse are observed, the offensive operation must immediately be abandoned. All units must be notified and immediately move to the exterior. Usually, an announcement is given over the radio followed by a preplanned signal such as several long blasts on an air horn (10 three-second blasts, for example). Forces operating on the outside must be moved back a safe distance from the building, usually a distance at least 1½ times the height of the building. In buildings with adequate fire walls and separate supporting structure, it may be possible to evacuate to or move operations to an adjoining compartment (e.g., a shopping center with fire walls separating stores).

Construction Methods and Materials

A wide variety of construction materials and methods have been employed in buildings over the years. These range from reinforced concrete to lightweight wood trusses. Because of the different materials, the behavior of buildings under fire attack will vary significantly. Broad assumptions can be made regarding structural stability based on construction methods and materials, such as a heavy-timber building will survive longer than a lightweight wood-truss structure. However, there are unknown factors, such as deterioration or repairs, that affect the reliability of assumptions based on construction type.

Fire spread inside of the building can occur through a variety of openings, either horizontally or vertically. Some buildings are constructed with large vertical shafts for utility lines or for other reasons. In the MGM Grand fire in Las Vegas in 1980, fire was able to spread vertically through seismic joints and stairways. In the NFPA report on this incident, a space 5/8 in. (1.6 cm) wide in one of the stairways that had not been properly sealed off was cited as one of the conditions that contributed to the spread of smoke and fire into the stairway.

What the fire officer should learn from these incidents and others is that a building's performance under fire conditions can be unpredictable, and exterior appearances may not indicate interior conditions. Renovations to a building can dramatically affect how it performs under fire conditions. Openings made in floors to run utilities are not always properly sealed.

No building is completely immune to structural failure, but some buildings will withstand a very large and intense fire without a catastrophic (total) collapse. Other structures are known to experience early collapse under intense fire conditions. When deciding on an offensive or defensive strategy or when placing companies for rescue or fire attack, the fire officer must take structural stability into account.

Modern construction methods conserve materials by using lighter-weight structural members that provide the same load-bearing capabilities as earlier construction methods that used massive structural members. The primary way this is accomplished is through various types of truss construction. Other forms of lightweight construction include lightweight metal C joists and engineered wooden I-beams. Lightweight structural systems take the place of large wood beams or steel I-beams. Lightweight construction methods are structurally sound under normal conditions. In fact, the added strength allows builders to place roof trusses 24 in. (61 cm)

on center, whereas the earlier methods required 16-in. (41-cm) on-center supports.

The problem is that these lightweight structural members are adversely affected by fire much sooner than the more massive building materials. Compounding the problem is the fact that once the truss loses its triangular configuration, it loses load-bearing capacity. Lightweight construction systems frequently fail under fire conditions, with little warning to the fire fighter. A well-involved fire in a truss space may not be detected until it breaks out of containment. Several fire reports mention that fire fighters working on the interior stated that there were few signs of a serious fire prior to collapse. NIOSH issued an alert identifying problems and solutions to the challenges presented by modern lightweight construction.[9]

Unprotected metal trusses used in noncombustible construction fail quickly when exposed to high heat **Figure 5-9**.

Lightweight wood-truss assemblies burn through sooner than structural assemblies that are built of more massive components. Metal-web wood trusses combine the problems of early failure and premature burn-through experienced in wood and metal trusses. Truss roofs above large open areas are particularly dangerous. A truss roof failure over

Figure 5-9 Fire attacking metal truss construction.
Courtesy of David J. Jones Cincinnati, Ohio

Incident Summary

In Hackensack, New Jersey, five fire fighters died in the collapse of a bow-string truss building. Fire conditions inside the building were relatively clear, but the concealed ceiling space was fully involved in fire.

Data from: Thomas J. Klem, "The Hackensack Fatalities," Fire Command, October 1988, p. 26.

This deadly fire at a Ford dealership killed five fire fighters when the building collapsed.

© Al Paglione/The Record

a compartmented area may be partially suspended by walls separating the rooms below. A fire entering an undivided attic area can extend unimpeded throughout the entire attic. Truss roof collapse generally results in a large area of damage. More fire-resistive buildings can suffer partial interior collapses that are not often noticed by the IC or companies working in other areas of the building.

Structural connections can play a critical role in a building under fire attack. A large I-beam supported by a single bolt is only as strong as the single bolt. Wood-truss construction typically uses gusset plates in place of nails **Figure 5-10 A**. Nails form a stronger connection under fire conditions, owing to the depth of their penetration into the wood. Gusset plates usually penetrate only a fraction of an inch and form a large surface area to collect heat. As soon as the gusset plate teeth lose their strength or the fire burns through the wood to the depth of the gusset plate teeth, the plate releases from the wood, and the wood truss loses its stability **Figure 5-10 B**.

Engineered wooden I-beams are now in use, using the same logic as a steel I-beam. A 2- × 4-in. (approximately 50- × 100-mm) board is placed on the top and bottom chords with a piece of plywood or oriented strand board (OSB) between, thus increasing the load-bearing strength. Light-gauge steel C-channels and other lightweight structures are also used to support floors and roofs. These methods are satisfactory and even superior in some cases under normal conditions, but they lack reliability and stability when exposed to fire and heat.

Many older buildings are modernized by suspending ceilings or by adding walls or paneling and installing a facade. These renovations provide additional concealed spaces for the fire and present additional dangers to fire fighters. As an example, a fire in Charleston, South Carolina, that killed nine fire fighters spread throughout a 5-ft (1.5-m) high concealed space between a suspended ceiling and the roof, and then suddenly broke out of containment, killing fire fighters working below.

Figure 5-10 A. Gusset plates. **B.** Fire-damaged gusset plates.

A. © David Huntley/ShutterStock, Inc. **B.** Reproduced from: NFPA Fire Investigative Summary, Church Fire — Lake Worth TX 2/15/99.

Understanding the dynamics associated with structural collapse is critical to the fire fighter. The IC and fire fighters working inside at a structure fire must pay close attention to signs of impending structural failure. Many decisions to conduct the attack in the defensive rather than offensive mode are based on structural stability. When a safety officer is on scene, he or she must pay close attention to signs of collapse. Once an engineered structural component is directly exposed to fire or high heat, little time remains to escape.

Roof Operations

Many departments automatically assign members to the roof for every fire. This can be a dangerous practice. A conscious decision should be made regarding roof safety. Lightweight roofs supported by trusses, C-channels, or other engineered construction methods are particularly dangerous. When a fire has self-vented, the roof is likely to be unsafe, and the need for roof operations is questionable. The same kind of risk-versus-benefit analysis that is made before deciding on interior operations should be made before placing fire fighters on the roof. The fire officer should ask four important questions before assigning fire fighters to roof operations:

- Are roof operations necessary?
- What do I hope to gain (benefit) by opening the roof? In many cases, roof ventilation is not necessary and sometimes has a negative impact on operations.
- How long has the fire been impinging on the roof components?
- What is the safest method of operating on this roof (e.g., aerial device, roof ladder)?

Also limit the number of people working on the roof. If two fire fighters can perform necessary roof operations, there is no reason to place more than two people on the roof. Some departments prohibit roof operations on truss roofs or at specific occupancies as part of their SOPs.

Truss roofs and/or concealed spaces have played a major role in fire fighter fatalities at structure fires.

Table 5-1 lists a number of fires where truss roofs and/or concealed spaces were contributing causes according to fire reports analyzing fire fighter on-duty deaths.

Table 5-1 Fire Fighter On-Duty Deaths with Roof Collapse and/or Concealed Spaces as Contributing Factors

Location	Occupancy	Year	Number of Fire Fighter Fatalities
New Jersey*	Automobile Dealership	1986	5
Tennessee***	Church	1992	2
Indiana	Hotel/Office	1992	2
Connecticut*	Carpet Store	1996	1
Virginia*	Auto Parts Store	1996	2
Georgia**	Church	1998	1
California**	Commercial	1998	1
Mississippi**	Strip Mall	1998	2
Texas*	Church	1999	3
Texas**	Restaurant	2000	2
Arkansas**	Church	2000	4
South Carolina**	Residential	2001	1
Wisconsin**	Commercial	2001	1
Oregon**	Auto Parts Store	2002	3
Tennessee**	Retail Store	2003	2
Texas**	Vacant Residential	2005	1
South Carolina**	Furniture Store	2007	9
Michigan**	Vacant Residential	2008	1
Illinois**	Vacant Commercial	2010	2
Indiana**	Church	2011	1
California **	Single Family Dwelling	2011	1
Illinois**	Vacant Commercial	2011	2
Tennessee**	Restaurant	2012	1
Pennsylvania**	Commercial	2012	1
Wisconsin**	Theater	2012	1

* NFPA Investigative Report available (www.nfpa.org)

** NIOSH Fire Fighter Fatality Investigation available (http://www2a.cdc.gov/NIOSH-fire-fighter-face/)

*** USFA Technical Report Series report available (https://apps.usfa.fema.gov/publications/display.cfm)

Incident Summary

In a 2011 apartment building fire in Texas, a fire lieutenant was fatally injured when he fell through a fire-weakened roof into the attic while performing vertical ventilation. The fire originated on the first floor of this two-story building and was rapidly contained. However, the fire had extended via the exterior walls into the attic, damaging the truss roof structure.

Data from: NIOSH Firefighter Fatality Investigation 2011-20. Career Lieutenant Dies After Being Trapped in the Attic After Falling Through a Roof While Conducting Ventilation – Texas

Incident Summary

On September 1, 2013, a roof fire in a 366,000-square-foot (34,003-square-meter) cold-storage food warehouse in Delanco, New Jersey, involved more than 7000 photovoltaic panels. The panels made it difficult to access the seat of the fire, necessitating a defensive attack that resulted in the total destruction of the building and contents.

Data from: The PV Problem, January/February 2014 NFPA Journal, Robert Duval

NOTE

Always assigning fire fighters to roof operations often places fire fighters in unnecessary peril. Never conducting roof operations denies the IC a valuable tactic.

Green Construction

Green construction is becoming more prevalent. Green roofs are being installed on new and existing buildings. The green roof consists of layers of materials supporting soil used to grow vegetation on a roof, which also provides roof insulation. This roof design not only makes roof ventilation more difficult or impossible, it adds substantial weight to the roof. Of particular concern is the additional roof load added to an existing roof that was not designed to carry the weight.

Solar panels have been used for many years but are becoming more common, with many photovoltaic (PV) solar arrays installed on roofs. Solar panels not only add to the roof load, they create a potential ignition source, make roof ventilation more difficult, and present an electrical shock hazard that is difficult to deenergize. Disconnecting the electrical service to a building does not deenergize the direct current roof system, which connects to an inverter that converts direct current from the solar panel to alternating current used in the building. Standards pertaining to installation and maintenance of PV panels are now being promulgated to reduce the fire hazard.[10] However, many existing solar panel installations do not meet these evolving standards. It is imperative that fire fighters familiarize themselves with the layout and operation of solar panels.

Other green energy materials such as alternative insulation materials, lightweight construction methods, and wind turbines also present special problems.

Floor Construction

The use of trusses and other lightweight support systems were once limited to roofs but are now commonplace in floor construction, creating some of the same premature collapse problems experienced

Incident Summary

On May 14, 2013, a "green building" in LaFarge, Wisconsin, suffered an estimated $13,000,000 loss as fire spread through denim wall insulation material. A large array of photovoltaic (PV) roof panels covering the metal roof made vertical ventilation nearly impossible. There was uncertainty regarding the stability of the roof due to fire entering the lightweight truss attic space. A major concern was the shock hazard presented by the PV solar panels. It was a partly cloudy day, meaning the solar panels were producing electrical energy. The local power utility measured 50 volts of direct current in the metal roof covering. This direct current voltage is high enough to cause a potentially lethal electrical shock. The roof showed signs of structural damage, and deformed solar panels were falling through the roof into the attic space below.

Data from: Perfect Storm, January/February 2014 NFPA Journal, Robert Duval

in lightweight roofs. Fire fighters must realize that lightweight building materials generally accelerate structural failure; however, older solid-beam construction materials are not immune to collapse. Of the 18 floor collapses, resulting in fire fighter fatalities documented by NIOSH since 1997, 11 were in floors constructed using dimensional lumber and 7 were in lightweight floor support systems.

The changes that have occurred in the density and dimensions of newer lumber have compounded the problem of structural failure in wood construction. Lumber manufactured today is less dense and smaller in size. Underwriters Laboratories compared old 2 × 8s to new 2 × 10s and found that the old 2 × 8s had a cross-sectional dimension of 1.75 in. by 7.56 in. (cross-sectional area of 13.23 in.). The new 2 × 10s had a cross-sectional dimension of 1.5 in. by 9.125 in. (cross-sectional area of 13.6875 in.). The new 2 × 10s had a cross-sectional area just slightly larger than that of the old 2 × 8s. The density of the new lumber was 32.5 lb/ft^3 compared to 36.9 lb/ft^3 for the older lumber. Underwriters Laboratories also conducted furnace tests on old and new dimensional lumber and found that older 2 × 8s did not collapse until after 18 minutes compared to 7 minutes for the new 2 × 10s.[11]

Basement Fires

Basement fires have always presented a difficult and dangerous problem, especially basements that are completely underground. Various tactics have been used to fight basement fires. Probably the most often used, and the most difficult and dangerous, is advancing a hose line down the interior stairway from the first floor to the basement. Opening the door at the top of the stairway leading to a basement fire creates a vertical vent opening. The fire and products of combustion will flow toward, into, and out of this chimney-like vent. If a basement fire has gained considerable headway, it may be impossible or ill advised to use the stairway.

Making an opening on the first floor to lower a cellar (distributor) nozzle also creates a vent, but fire fighters need not enter the vent opening or create a large hole to operate the nozzle. It is important to have the cellar (distributor) nozzle charged and ready before creating a hole in the floor to avoid placing fire fighters in a newly created flow path. The cellar nozzle does not provide adequate protection for fire fighters operating the nozzle; therefore, a standard attack line should also be available as a safety measure. Specially designed cellar nozzles have successfully extinguished below-grade fires but have been ineffective on other occasions. Cellar nozzles tend to have a limited stream range and are more effective on fires in smaller undivided areas. Fire fighters must also use caution when making an opening for a cellar nozzle, as they could cause floor damage that would hasten collapse of the floor. Basement fires are also likely to extend to other parts of the structure through utility openings in the floor between the basement and first floor.

If the basement is partially aboveground, other tactics may be available such as operating hose streams through a basement window to control and limit the fire by softening the target. This tactic may allow entry into the basement via the stairway for final extinguishment. Pulsing and indirect attacks are tactics that could be used to knock down the main body of fire in a basement that is partially out of the ground. The *Offensive Operations* chapter will explain these tactics.

Some basements are accessible at ground level. Usually, houses built on hillsides where the hill slopes downhill from front to back have the main entrance at grade level on the A side, but may also have a walk-out basement at grade level on the C side of the building. Garages at basement level or separate basement entrances can also provide direct access to the basement. Availability of a grade-level basement entrance allows a direct fire attack, but venting opportunities may be limited. Softening the target by applying water through a window or from the exterior through a grade-level basement entrance may provide the cooling effect needed to allow fire fighters to enter the basement for final extinguishment. Venting a basement fire through openings on the first floor is likely to spread the

Incident Summary

On April 4, 2008, a captain and fire fighter were killed at a basement fire in a two-story, 2050 ft² (190 m²) single-family residential building in Ohio. Companies were dispatched to an automatic alarm at 6:11 AM. The first-arriving engine company reported on scene at 6:23 AM. Response was delayed because the building was located in a wooded area 450 ft (137 m) from the roadway, making it difficult to locate. Upon arrival, moderate smoke was visible. Building occupants pointing the way to the basement stairs reported that the fire was in the basement and everyone was out of the house.

The company captain and a fire fighter donned their SCBA and entered the residence through the open front door with a 1¾-in. (45-mm) hose line. After making entry, the captain ordered the line charged and with the assistance of the fire fighter proceeded down the basement stairway. At the bottom of the stairs, they encountered heat and could see a glow to their left, but they needed additional hose to reach the fire area. Another fire fighter assisted in pulling additional hose at the top of the stairway and through the front door. Upon returning to the top of the stairway, this fire fighter witnessed the captain calling a mayday, but the captain was unable to communicate with units on the exterior. Smoke was now getting black and heavy and was pushing out the front door, whereupon the IC ordered an evacuation of the building. An engine company and the established rapid intervention crew (RIC) accessed the basement through the grade-level walk-out door in the rear of the building. The engine company was able to extinguish the fire, and the RIC found both victims under a collapsed floor area. The victims apparently fell through the floor from the first floor after backing out of the basement.

Data from: NIOSH [2009]. A Career Captain and a Part-time Fire Fighter Die in a Residential Floor Collapse—Ohio, (FACE) Report F2008-09.

Flames: © Jag_cz/ShutterStock, Inc.; Paper: © silver-john/ShutterStock, Inc.

fire to areas near the vent opening unless water is applied into the basement prior to making vent openings.

Performing a complete walk-around (360) can be a challenge when buildings are located on steep hillsides. However, every effort should be made to do a complete survey, especially when dealing with a suspected basement fire. Several of the basement fires listed in Table 5-2 had basement access at the rear. Unfortunately, in many of these cases, companies and the IC were unaware of the grade-level access to the basement.

A 2012 study by Underwriters Laboratories[12] examines fire fighter safety when combating basement fires. This comprehensive report addresses collapse time for various floor support systems commonly used in residential construction as well as the effect of floor load, fuel load, and venting on collapse time. As expected, lightweight support systems failed much earlier than the dimensional lumber. An earlier fire research project sponsored by the NFPA[13] reached the same conclusion.

In the Underwriters Laboratories study, collapse times were compared to estimated response times. It

Table 5-2 Fire Fighter On-Duty Deaths with Floor Collapse as a Contributing Factor

Location	Occupancy	Year	Number of Fire Fighter Fatalities	Fell Into Basement Due to Floor Collapse
Kentucky	Residential	1997	1	Yes
Pennsylvania	Residential	1997	2	No
Ohio	Residential	1998	2	No
District of Columbia	Residential	1999	2	No
Alabama	Residential	2000	1	Yes
Ohio	Residential	2001	1	Yes
Illinois	Residential	2001	2	No
North Carolina	Residential	2002	1	Yes
New York	Residential	2002	2	Yes
Massachusetts	Residential	2003	1	No
Pennsylvania	Residential	2004	1	Yes
New York	Residential	2005	1	No
Maryland	Residential	2006	1	No
New York	Mercantile	2006	2	Yes
Wisconsin	Residential	2006	1	Yes
Indiana	Residential	2006	1	Yes
Tennessee	Residential	2007	1	Yes
Ohio	Residential	2008	2	Yes
Illinois	Residential	2008	1	Yes
Pennsylvania	Residential	2008	1	No
New York	Mercantile/Residential	2009	2	Yes

© Eky Studio/Shutterstock, Inc.

was determined that collapse of all the floor systems tested would have occurred either before fire units arrived or while they were at the scene. Does this mean that fire fighters should never enter a building with a known basement fire? No! Instead it should be understood that under the conditions of the experiment with no fire attack, collapse should be expected within the estimated response times used during the research. Many basement fires are minor fires involving laundry appliances or small quantities of combustible materials. Quite often basement fires are confined to clothes dryers or heating ventilation and air conditioning (HVAC) equipment.

Each basement fire is different. Basement fires range from small fires involving limited combustibles or equipment typically found in a residential basement to a raging inferno involving the entire basement area of a large industrial building with an extra-hazard fuel load. Obviously, a small fire could extend to become a larger fire, but a small contained fire should not affect the structural stability of the floor above or require special tactics. A larger fire might be controllable, but special tactics such as softening the target prior to making entry should be employed.

Large commercial buildings should be preincident planned, allowing the fire department to make specific decisions in advance of a fire. As an example, a decision could be made not to enter the basement in a large industrial building when there are indications of a working fire. This preplanned attack could include using cellar nozzles at specific locations on the first floor or designated areas where water or high-expansion foam could be applied through an alternate opening leading to the basement. This plan would most likely be based on fuel content, size of the basement, and the floor support system.

Venting in the absence of effective suppression had a negative effect on structural stability during the Underwriters Laboratories test fires. Fires generally consume the available oxygen then go into decay when the fire is ventilation-limited. On its face, this would be an argument for staying out of the building because fire operations sometimes result in unintentional venting when gaining access to the fire area. The problem with this "no-entry" policy is that occupants remaining in the fire area would be at great risk due to the lack of oxygen and toxic gases and all basement fires would result in total destruction of the building. Enclosures with high-heat conditions and a lack of oxygen are ripe for a rapid intensification, flashover, or backdraft. After a careful risk-versus-benefit analysis, the fire officer should determine whether interior operations are reasonable. If a decision is made to conduct an offensive attack, fire fighters should enter the building very cautiously with a charged hose line, expecting a rapid increase in fire intensity when opening doors and windows that vent the basement.

According to the Underwriters Laboratories study, standard on-scene measures used by fire fighters to test a floor for structural stability (sounding the floor, thermal imaging, and floor sag) provide some indication of an impending collapse, but are *very* late indicators that are not entirely reliable. If any of these procedures reveal a weak flooring system, prompt and immediate withdrawal from the area should be compulsory. In several of the NIOSH investigations listed in Table 5-2, a "spongy" floor or roof was noted prior to collapse. These are *very* late indicators, and thermal images are not reliable indicators of impending floor collapse.

The worst-case scenario when fighting a basement fire is falling through the floor into a well-involved fire below. A close second would be opening a door leading from the first floor to the basement and being subjected to a superheated flow path. Fire reports and experiments confirm two basic fire dynamics concepts. First, heat and heated gases will travel upward, then horizontally, following the path of least resistance to reach equilibrium. Second, a ventilation-limited fire will rapidly increase in intensity when air is introduced. When there is a working basement fire, opening the front door and the door at the top of the stairs leading to the basement will allow fire, smoke, heat, and toxic gases to travel upward via the vertical opening (stairway), then horizontally out the front door **Figure 5-11**.

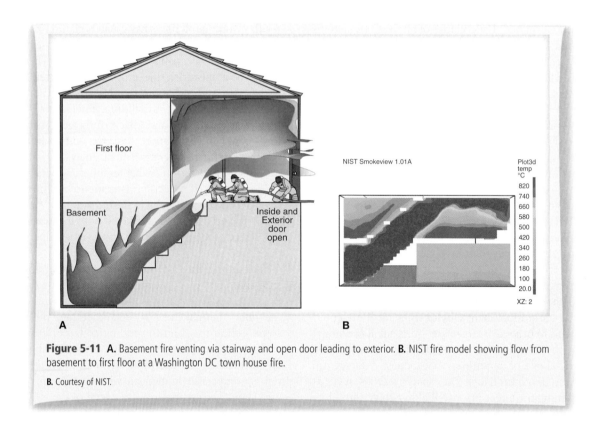

Figure 5-11 A. Basement fire venting via stairway and open door leading to exterior. **B.** NIST fire model showing flow from basement to first floor at a Washington DC town house fire.

B. Courtesy of NIST.

The standard practice of checking the door for heat and being ready to stop venting by controlling the door is especially important when opening doors leading to basements.

After two Washington DC fire fighters lost their lives at a basement fire in a town house, the NIST conducted research using fire models to determine heat conditions before and after opening windows and doors.[14] The model indicated that there was a distinct flow path from the basement up the stairway and out the front door. Temperatures on the basement stairway were as high as 1508°F (820°C). On the first floor, ceiling temperatures were in the same range as in the stairway. However, floor temperatures were 68°F (20°C).

Floor collapse is a major concern when combating a basement fire. Just as with roof operations, there is a danger that fire fighters working on the floor above will fall through the floor into an inferno below, or that when conducting a direct attack in the basement, a floor collapse from above could kill, injure, or entrap fire fighters below.

Based on the findings of the Underwriters Laboratories study and the number of on-duty fire fighter fatalities, some in the fire service are recommending that fire fighters never enter the first floor of a building with a working fire in the basement. This is a safe approach, but is it effective? Obviously when conditions indicate an impending collapse, staying out of the building is the only option. However, there are also good reasons to conduct operations inside the building on the first floor or basement. Several of the on-duty death reports from NIOSH listed in Table 5-2 reveal the following:

- There were savable occupants or reports of occupants in the fire building upon arrival.

- Fire fighters did not know the fire was in the basement. In some cases the initial size-up included smoke or fire visible at the roof when the point of origin was in the basement.
- Visual clues of fire were absent or indicated a minor fire.
- There was unknown direct access to the basement at grade level at the rear or side of the building.
- Floors were supported by dimensional wooden beams, or the construction method for the floor was not mentioned. There is no doubt that the percentage of lightweight flooring systems will increase; however, some contractors continue to use dimensional lumber to support floors.

As discussed in the chapter titled *Procedures, Pre-incident Planning, and Size-Up,* knowing common construction methods used in the department's response area is important to safe and effective fire operations. In some jurisdictions, building codes require that when the supporting members of the floor separating the basement from the first floor is of lightweight construction, it must be protected with a fire-rated basement ceiling assembly. Fire resistance ratings are most often accomplished by installing gypsum wallboard on the basement ceiling. Clearly, this requirement provides additional time for occupant escape or rescue, as well as extending the time available to conduct an offensive attack. The primary reason for using lightweight construction materials is to reduce construction costs. Applying gypsum wallboard to the ceiling increases the construction cost. Therefore, many contractors use dimensional lumber to avoid this additional expense. As a result, areas with fire-rated basement ceiling requirements will have fewer lightweight floor assemblies between the basement and first floor. It is critical to know your response area.

Finishing a basement is a common way to expand the living space within a residential property. Gypsum wallboard is sometimes applied to the ceiling in the finished basement area regardless of code requirements; however, finished living areas are more likely to be occupied by residents and often contain a substantial fire load.

Prefire Conditions and Fire Conditions

Several factors must be taken into consideration in evaluating the collapse potential for a building that is under fire attack. These factors fall into two general categories: prefire conditions of the building and fire conditions.

Prefire Conditions

Prefire conditions include the type of construction, as was mentioned previously. In addition, the following prefire conditions can contribute to a collapse in a building that is heavily involved in fire:

- Weight. Live and dead loads placed on floors, walls, and roofs can include the following:
  - Air-conditioning units
  - Green roofs
  - Solar panels
  - Tanks containing liquids
  - Large signs and marquees
  - False fronts (facades)
  - Cantilever appendages
  - Heavy machinery
- The type, location, and arrangement of combustible/flammable fuel loads inside of the building. Modern home furnishings create a much larger fuel load than legacy furnishings.[15]
- Damage to the building's structural support system from previous fires, weather, or partial collapse.
- Renovations to buildings, especially those that have had adjoining structures removed.
- Buildings or areas in poor repair, including vacant structures that may have deteriorated to create a hazard.
- Buildings with long spans, such as churches and warehouses. Such buildings should be carefully evaluated for collapse potential.
- Roofs and floors supported by trusses, wooden I-beams, C-channels, and other lightweight construction.

All of these factors can be known in advance through preincident planning.

Fire Conditions Leading to Structural Collapse

Building conditions leading to structural collapse are sometimes difficult to read, and structural failures have occurred without warning **Figure 5-12**. However, the proficient fire officer must recognize signs of imminent danger and maintain the span of control necessary to react quickly when fire fighters are endangered. As was mentioned previously, time and fire intensity are major factors. Signs of structural collapse include, but are not limited to, the following:

- Bulging, cracked, or unsupported walls.
- Walls leaking water or smoke.
- Falling bricks.

Figure 5-12 Fire conditions leading to structural collapse.

- Floors holding large volumes of water or stock soaked with water, thereby increasing the weight bearing on the floor. An indicator would be large quantities of water entering the building but little water draining out of the building.
- Movement in floors or the roof. If it feels as though it is unsafe to walk on, it probably is.
- Any other signs of structural movement, including unusual noises.
- Vertical structural members (e.g., columns, walls) that are out of plumb.

Fire Extension

Some buildings limit fire spread and contain the products of combustion better than others. As the fire enters concealed spaces, it can extend to remote locations. This **extension** can result in a sudden increase in heat intensity when the fire breaks out of containment, self-vents, or is vented by fire fighters. A sudden increase in temperature can occur at a location remote from the original fire and result in fire breaking out at multiple locations. Such extension is especially problematic when resources are limited and there is a need to attack the fire from several positions. Fire fighters performing search and rescue operations without a hose line are at particularly high risk, as are units working above the fire floor. The immediate availability of a charged hose line is more than a means of extinguishment; it is also an important safety factor. If there is any chance that the fire has spread into or is likely to advance into an area where fire fighters are working, a charged and staffed hose line is an imperative.

> **NOTE**
>
> Arsonists or terrorists may purposely set traps, shut off sprinkler systems, and use flammable liquids or explosives, resulting in unusually rapid fire spread and building collapse. Fire departments should track arson behaviors, and police agencies should communicate credible terrorist threats to the fire department.

Texture: © Eky Studio /Shutterstock, Inc; Steel: © Sharpshot/Dreamstime.com

Fire extension can cut off the primary means of egress for fire fighters. Fires that enter and are confined to concealed ceiling areas are particularly prone to getting behind the fire fighter **Figure 5-13**. Fire fighters must check concealed spaces in floors, ceilings, and walls to determine whether the fire is in these areas.

Thermal imaging cameras (TICs) can be used to find hotspots in concealed spaces, but these areas should be opened up if there is any doubt regarding fire extension. There is a high likelihood of a ventilation-limited fire in a concealed space. The fire may be in a decay phase, and as a result, it may not be producing high temperatures. Insulation, plaster, dry wall, finish flooring materials, or other building materials may also result in low temperature readings on the surfaces exposed to the view of the TIC. It is usually best to start by opening a small inspection hole to examine a concealed space, thus limiting oxygen to a possibly oxygen-starved fire.

Coordinated venting will direct the fire out of the building and protect fire fighters; improper venting pulls fire toward fire fighters or into areas behind fire fighters. Increasing the air supply to a ventilation-limited

Incident Summary

A fire fighter died and two other fire fighters were injured in a flashover at a residential fire as they entered the front door to attack the fire with an uncharged 1¾-in. (45-mm) hose line. The preconnected hose line was seven 50-ft (15-meter) sections in length, which is beyond the maximum length recommended by hose and nozzle manufacturers for 1¾-in. (45-mm) hose. The pump operator charged the hose line, but kinks in the hose prevented water from getting to the nozzle.

Data from: NIOSH [2005]. Career Fire Fighter Dies and Two Career Fire Fighters Injured in a Flashover During a House Fire—Ohio: U.S. Department of Health and Human Services, Public Health Service, Centers for Disease Control and Prevention, National Institute for Occupational Safety and Health, Fatality Assessment and Control Evaluation (FACE) Report F2003–12.

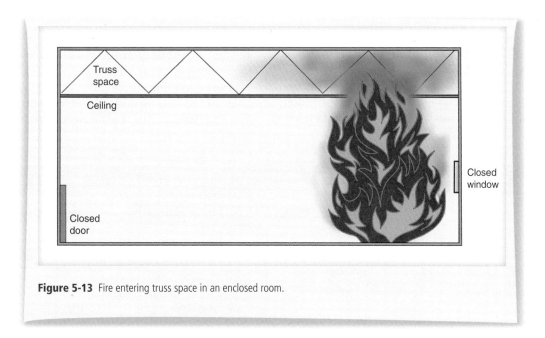

Figure 5-13 Fire entering truss space in an enclosed room.

fire will increase fire growth and intensity; therefore, a charged hose line must be in place to limit fire spread prior to opening concealed spaces.

Suspended ceilings are used to hide open construction methods such as trusses, thus creating a large, undivided concealed space. Most suspended ceilings are supported by wires attached to lightweight metal grid-work that is known to prematurely collapse under fire conditions. The area between the suspended ceiling and actual ceiling or roof provides a convenient location for lighting fixtures, ventilation equipment wiring, and other utilities. Heavy light fixtures, fans, and equipment are often located in this area, creating a serious falling-object hazard when the grid-work supporting the ceiling fails. Coaxial television cables, communication system wires, conduit, and the supporting grid-work also present an entanglement hazard.

The truss floor, like the truss roof, creates a large, concealed space and is less fire-resistive than heavier solid-beam construction Figure 5-14 .

Table 5-3 compares building construction methods in terms of structural stability, fire extension, and fuel contribution. It is important to remember that the five building categories listed do not represent all possible construction methods. As an example, non-combustible buildings are further subdivided into protected and nonprotected. Variations occur within all types of building categories. Buildings being renovated or remodeled may contain structures different from those prescribed by the original construction method. Additions and alterations sometimes result

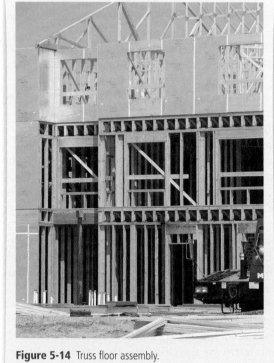

Figure 5-14 Truss floor assembly.
Courtesy of Captain David Jackson, Saginaw Township Fire Department

in what can best be described as a hybrid building (a combination of construction types).

The noncombustible building is sometimes mistaken for fire-resistive due to the nonburning characteristics of the structure. The differences are immense.

Table 5-3	Comparing Construction Types				
Construction	**Type**	**Structural Ability**	**Fire Extension Probability**	**Fuel Contribution**	
Fire resistive	I	Outstanding	Low	Low	
Non-combustive	II	Poor	Average	Low	
Ordinary	III	Average	Average	Average	
Heavy timber	IV	Excellent	Low	Average	
Frame	V	Poor	Average to high	High	

© Ely Studio/Shutterstock, Inc

Incident Summary

In a 1995 fire in a warehouse in Seattle, Washington, most of the building was built of heavy timber structural members. However, one critical support for the floors had been modified by using 2 x 4" (approximately 50 x 100 mm) wood studs. This support failed between 34 and 36 minutes after arrival on the scene, dropping four fire fighters into the fire below, killing them.

Data from: Edward R. Comeau, Fire Fighter Fatalities, Seattle, Washington, 1995, NFPA Fire Investigation Report. Quincy, MA: NFPA.

The noncombustible building will normally be masonry or metal on the exterior with lightweight metal trusses as a roof structure. These buildings often have large open areas with long structural spans. Once the fire enters the truss space, expect imminent roof collapse. Unfortunately, fire fighters fail to pay attention to this fact and are continuing to lose their lives in these buildings. Modern big-box retail stores, such as Walmart, Sam's Club, and Lowe's are generally of noncombustible construction with large open areas, but most of these buildings are protected by an automatic sprinkler system. A properly installed and operating sprinkler system will usually control fires in these stores. However, it is important to remember that many code variances or "trade-offs" may have been allowed because the building was constructed with sprinkler protection. If the sprinkler system is out of service or ineffective, these buildings should be considered non-code-compliant structures. Building management will sometimes arrange materials in a way that shields combustibles or results in a fire load beyond the designed capacity of the sprinkler system. If the sprinkler system is not controlling the fire in a building of this type, consider the hazards involved in entering this large-span truss space with a heavy fire load.

A complete discussion of structural collapse is beyond the scope of this text. Several textbooks discuss the different aspects of building design and how it can affect fire ground operations. *Brannigan's Building Construction for the Fire Service* by Glen Corbett and Frank Brannigan[16] is considered a companion text to this text. Fire officers should be familiar with the information contained in the Corbett–Brannigan text and how this information applies to their respective jurisdictions.

Incident Summary

On June 18, 2007, at 7:07 PM, the Charleston, South Carolina, Fire Department was dispatched to a fire in an enclosed loading dock behind a furniture store. The structure was a nonsprinkler-protected, one-story commercial furniture showroom and warehouse facility totaling over 51,500 ft² (4785 m²) that incorporated mixed construction types. The store/warehouse facility had been renovated and expanded a number of times. The original structure was constructed in the 1960s as a 17,500-ft² (1626-m²) grocery store with concrete block walls and lightweight metal bar joists (metal roof trusses) supporting the roof. After being converted to a furniture retail store, the original structure was expanded by adding a 6970-ft² (648-m²) addition on the right side (D side) and a 7020-ft² (652-m²) addition to the left (B side). Both additions were attached to the original exterior with large interior doors between buildings. A 15,600-ft² (1449-m²) warehouse was then added to the rear of the main showroom.

The main showroom measured 9 ft (2.7 m) from the floor to a suspended drop ceiling and approximately 14 ft (4.3 m) to the roof, creating almost 5 ft (1.5 m) of concealed space above the suspended ceiling. Using a thermal imaging camera or removing suspended ceiling tiles during initial entry into the furniture store might have revealed the presence of fire or heated products of combustion in this concealed area earlier in the fire, resulting in the use of different tactics.

The rear warehouse measured 29 ft (8.8 m) from the floor to the roof with rows of metal storage shelving that contained a variety of modern furniture items.

The fire department's initial response was three fire companies with three fire fighters per company and two chief officers (well below the response levels recommended in NFPA 1710). The first chief officer arrived at 7:09 PM and verified a fire behind the furniture store on the loading dock that connected a furniture store to the front and a warehouse to the rear. This first-arriving chief officer ordered an engine company to respond to that location with a 1½-in. (38-mm) line to combat the fire. The second-arriving chief officer checked the furniture store at the front and found no indications of fire spread into the store at that time. Later reports confirmed fires in the store and warehouse. The modern furniture provided a large fuel load. At approximately 7:26 PM (19 minutes after the original dispatch), a store employee called 911 reporting that he was trapped inside the building. This employee was rescued. At this time, three hose lines were inside the main showroom: an initial 1½-in. (38-mm) hose line, a 1-in. (25-mm) booster line, and a 2½-in. (65-mm) hose line. All three hose lines were pulled off a single pumper that was supplied by another pumper through a single 2½-in. (65-mm) supply line approximately 1850 ft (564 m) long.

Six fire companies working inside the furniture store became disoriented when heavy smoke obscured vision as the fire rapidly intensified. Several calls for help were received from interior crews along with at least one mayday. Fire conditions were so severe that rescue attempts by other fire companies were unsuccessful. Nine fire fighters remaining inside the building were caught in the rapid fire progression and were killed.

Data from: NIOSH [2009]. Nine Career Fire Fighters Die in Rapid Fire Progression at Commercial Furniture Showroom—South Carolina: U.S. Department of Health and Human Services, Public Health Service, Centers for Disease Control and Prevention, National Institute for Occupational Safety and Health, Fatality Assessment and Control Evaluation (FACE) Report F2007-18.

© Alexander Fox/Ap Images

Courtesy of NIOSH.

Hazard Control Zones and Fire Perimeter

Establishing a **collapse zone** is critical when structural stability is the reason for a defensive attack. Construction features combined with fire factors indicate the most probable type of structural failure. Given that the IC is always working with incomplete and imperfect information, it is impossible to accurately predict the type of collapse and resultant collapse zone. The only safe collapse zone is one that is equal to the height of the building plus an allowance for scattering debris. A good rule of thumb when determining collapse zone distances for most buildings is to establish an area 1½ times the height of the fire building. This sometimes presents a dilemma as the safe zone is beyond the street width, meaning that effective defensive positions are within the collapse zone. When a defensive stand is a reasonable alternative, positions at the corners of buildings are normally safer than those on the flat side of a wall. Consider using unstaffed ground monitors to reduce the risk of placing personnel in exposed positions.

A risk-versus-benefit analysis is essential. The crucial question that any IC must ask is, "What could I potentially save in relation to the risk being taken?" Obviously, no building is worth a fire fighter's life; therefore, imminent risk to a fire fighter's life to save a building is unacceptable. Also, remember that nothing should be risked to save what is already lost.

When total collapse is imminent, collapse zones represent **no-entry zones** that no one is permitted to enter, regardless of the level of protective clothing. No-entry zones can also exist within buildings, especially when roof or floor structures are suspect. In addition, no-entry zones would include other areas containing imminent hazards such as falling glass, areas containing atmospheres within or near the flammable range, and any other area that the IC or safety officer deems too hazardous to enter.

Collapse and no-entry zones are not the only safety considerations regarding access. The concept of limiting access to the fire scene is defined in a variety of ways. It seems appropriate to extend the lessons learned from hazardous materials responses regarding zones, as similar zones are possible at structure fires. In this text, working areas are called hazard control zones, whereas a wide area beyond the working zones is known as the **fire perimeter**. The fire perimeter is usually staffed by police, who keep unauthorized people away from the scene, as in an isolation area at a hazardous materials incident. Incident conditions must be considered when determining the dimensions of the fire perimeter. Two blocks in all directions beyond the building on fire is a good rule-of-thumb approximation for the fire perimeter.

Within the **hazard control zone**, there could be several subdivisions. In most cases there will be a **cold zone** where personal protective clothing is not required (similar to the cold zone at a hazardous materials incident). The command post, as well as other staff and command functions, would be based in this safe area. The cold zone would also include rehabilitation and medical treatment areas.

The **hot zone** would be an operating area, considered safe only for individuals wearing appropriate levels of personal protective clothing (much like the hot zone at a hazardous materials incident). The IC and safety officer have a responsibility to establish and enforce the hot zone. Everyone has a responsibility to abide by their decision.

It is not always necessary to establish a **warm zone** during a structure fire. A warm zone is established when different levels of protective clothing are needed for various areas. It is an intermediate zone (between the hot and cold zones). As an example, fire fighters working close to a structure (warm zone) may need to be in full protective clothing, but not breathing air from their SCBA. Members entering the structure (hot zone) would be required to don their SCBA face pieces. When the fire is in a larger building, donning the SCBA face piece may be delayed until the fire fighter reaches an area that is or could be an immediately dangerous to life and health (IDLH) atmosphere. Again, this intermediate area is referred to as a warm zone. As an example, the hot zone for a fire in a multistory building would normally include the fire floor, floors above the fire, and one or two floors below the fire. Lower floors would be the warm zone.

Figure 5-15 illustrates hazard control zones, with the building's interior being considered the hot zone. Fire fighters entering the building would be expected to be in full turnout gear, breathing from their SCBA. The area on all sides of the exterior would be the cold zone, and the fire perimeter would be maintained by police keeping unauthorized personnel away from the scene.

If the fire is not contained and an exterior (defensive) attack becomes necessary, the hot zone is moved far enough away from the structure to place fire fighters outside the collapse zone and in fresh air. The collapse zone then becomes a no-entry zone, as shown in Figure 5-16 .

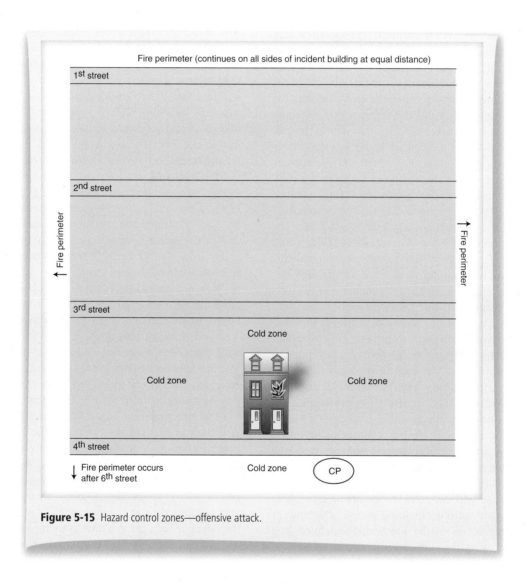

Figure 5-15 Hazard control zones—offensive attack.

Figure 5-16 Hazard control zones and fire perimeter—defensive attack.

Relationship of Time, Fire Intensity, and Structural Stability

As mentioned earlier in this chapter under *Fuel Load*, the Underwriters Laboratories study titled "Impact of Ventilation on Fire Behavior in Legacy and Contemporary Residential Construction" proves that modern synthetic fuels found in residential occupancies today burn faster and progress to flashover conditions much sooner than fuels found in mid-20th-century homes defined as "legacy" furnishings. The

Underwriters Laboratories flashover experiments were first conducted in two side-by-side rooms that were each 12' (3.7 m) by 12' by 8 ft (2.4 m) high. A second experiment was conducted in rooms 14 ft² (1.3 m²) larger than in the first experiment. In both cases, the fire compartment fuel load consisted of typical living room furnishings. One room was furnished with legacy furnishings and the other with modern furniture. For the most part, structural firefighting tactics in common use are based on materials found in homes 50 or more years ago. These furnishing no longer exist in the typical 21st-century home. An opening 8 ft (2.4 m) wide by 7 ft (2.1 m) high remained open during the experiments to allow better viewing of fire behavior. This opening provided more combustion air than average residential openings, where air is supplied by way of smaller window and/or door openings. As a result of these experiments Underwriters Laboratories concluded that:

> The modern and the legacy rooms demonstrated very different fire behavior. It was very clear that the natural materials in the legacy room burned slower and produced less energy and smoke than the fast burning synthetic furnished modern room. The times to flashover show that the flaming fire in a room with modern furnishings leaves significantly less time for occupants to escape the fire. It also demonstrates to the fire service that in most cases the fire has either transitioned to flashover prior to their arrival or became ventilation limited and is waiting for a ventilation opening to increase burning rate.

This is an extremely important research finding. Today fires in residential occupancies are indeed different than in the past. When the fire is well vented, there is little chance to save occupants remaining in the compartment of origin within a residential occupancy. With the 7 ft (2.1 m) by 8 ft (2.4 m) opening, the modern room experienced flashover in 3 minutes and 20 seconds and in 3 minutes and 30 seconds for the slightly larger room compared to a flashover time of 29 minutes and 30 seconds in the first burn in the legacy-furnished room. During the second burn, the legacy-furnished room did not reach flashover conditions. There is a high probability that these fires in a less-ventilated room would go into decay as the fire consumed available oxygen.

From a risk-versus-benefit analysis, the probability of saving lives in the fire compartment is greatly diminished when fire fighters arrive after flashover. However, occupants may be savable in areas separated by barriers (with closed interior doors) or

remote from the fire. Larger rooms with different furnishings (fuel load) may take longer to reach flashover conditions, allowing more time to conduct rescue activities on the interior.

Fire fighters arriving at a structure fire need to consider the effect of providing additional air by opening doors, windows, and roofs, which could lead to a rapid increase in combustion in a ventilation-limited fire compartment. Introducing air in such situations creates flow paths that endanger not only the remaining occupants but also fire fighters.

> **NOTE**
>
> The fuel load in the home of today burns much faster and contributes more fuel than in the past. This factor combined with lightweight construction places fire fighters and occupants at significantly greater risk than in the past.

Texture: © Eky Studio /Shutterstock, Inc; Steel: © Sharpshot/Dreamstime.com

Figure 5-17 A shows a relationship between response time, structural stability, and survivability of occupants in a well-ventilated legacy living room, **Figure 5-17 B** shows a well-ventilated modern living room, and **Figure 5-17 C** shows a ventilation-limited fire in a modern living room. These relationships are generally based on measurements taken during the Underwriters Laboratories research. However, the heat, toxicity, lack of oxygen, and response time are all highly dependent on a number of factors. During research, constants such as room size and ventilation are controlled to compare variables such as fire conditions produced by legacy furnishings versus modern furnishings. In actual fires there are few or no known constants and each variable has an effect on other variables. As an example, an uncontrolled fire in a room similar to the one used by Underwriters Laboratories during this research would probably be attached to other rooms and usually be less ventilated during the early stages of fire growth.

Structural stability and survivability were not tested during the Underwriters Laboratories

Figure 5-17 A. Flashover, survivability, and collapse versus response time— well-ventilated fire (legally living room).

Figure 5-17 B. Flashover, survivability, and collapse versus response time — well-ventilated fire (modern living room).

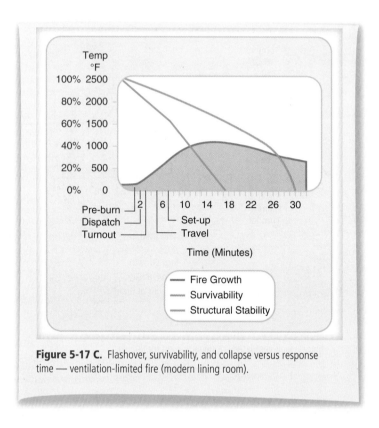

Figure 5-17 C. Flashover, survivability, and collapse versus response time — ventilation-limited fire (modern lining room).

experiments. However, we know that there is little chance of occupant survival under flashover conditions. Figure 5-17B shows the relationship between survivability and heat conditions in a well-ventilated room with modern furnishings. Therefore, the survivability part of the graph in Figure 5-17B would be considered accurate under the given conditions. In this well-ventilated fire involving modern furnishings, flashover occurs prior to the arrival of the fire department and there is little chance of rescuing the occupants who are in the fire compartment. However, a less-ventilated room may go into the **decay phase** due to a lack of oxygen or fuel (as illustrated in Figure 5-17C).

The building structure is only minimally affected during the **ignition phase** and early **growth phases** of the fire. Structural stability is compromised as the

fire continues to burn through the growth phase to the **fully developed phase**.

Structural stability is highly dependent on other factors such as construction methods and materials (e.g., wood truss versus an old dimensional timber supporting system). Another factor is construction materials that affect fire extension into walls, ceilings, and roofs.

The legacy fire in Figure 5-17A progresses slowly, and the probability that occupants will be able to safely escape is very high. In addition, fire fighters have time to rescue those who fail to escape. In the less-ventilated fire, which goes into the decay phase due to a lack of oxygen (air), there may well be savable occupants, although the effects of toxic gases from the burning synthetics and lack of oxygen will significantly reduce their chance of survival. The same building with a lighter fuel load (e.g., a vacant building) could burn out due to a lack of fuel.

There is also a chance that a fire would progress at a slower rate due to the ignition source and/or early detection. As an example, a fire ignited by a cigarette will take longer to develop into a flaming fire than the Underwriters Laboratories experimental fires using a flaming ignition source. This additional time could result in the alerting of occupants due to the smell of smoke or activating a smoke detector.

The longer the fire burns, the greater the risk to fire fighters and occupants. The probability of flashover, structural failure, low oxygen levels, high levels of toxic smoke/fire gases, or a combination of these factors increases over time. In some cases, the first-arriving fire company may arrive after flashover and be limited to saving lives and property in internal and external exposures. However, response time remains a major life safety and extinguishment factor, as a fire may progress at a slower rate due to a number of factors. As the fire continues to burn, the risk also increases in other parts of the fire building or adjoining buildings.

Structural collapse could occur before flashover, particularly in a large building where localized heating of the area above the fire could result in structural failure before the entire area reaches its flashover temperature. In Figure 5-17A, the green line marked "structural stability" starts at the top left corner of the chart indicating 100% structural strength. Downward deflection of the structural collapse line represents a weakening of the structure to the point of collapse where the line angles steeply downward. Structural stability will be different for every fire. Thus, this line represents a best guess. Newer structures with lightweight engineered-frame construction or in which fire enters a truss space are likely to collapse much sooner than well-maintained buildings made with dimensional lumber.

The blue line marked "survivability" is an estimation of the probability of occupant survival. As smoke and toxic gases reach the floor level, the probability of occupant survival is very low. Occupant survival is improbable in a post-flashover compartment, and even fully protected fire fighters will quickly succumb

under flashover conditions. The "survivability" line will always reach the bottom of the chart at or before flashover, as temperatures necessary for flashover are lethal. In ventilation-limited fires, heat could result in fatal injuries, but a lack of oxygen and toxic gases usually play a larger role than in fires that rapidly progress to flashover. Likewise, the probability of occupant survival is very low after a structural collapse due to fire. A different fuel configuration, a different ventilation profile, or a different type or size of structure could radically change the time value and curve configuration. The principle to keep in mind is that fire growth and time are critical factors.

Brannigan's Building Construction for the Fire Service makes an important distinction between building fires and structural fires. Fires involving the content of the building are categorized as building fires, whereas fires involving actual structural members are considered structure fires. This differentiation is not made in this text, but the concept is useful. According to Corbett and Brannigan's terminology, ignition and early growth stage fires are generally building fires, but somewhere during the progression to flashover, the fire generally finds its way into concealed spaces and begins to involve the structure. As mentioned previously, fires entering concealed spaces present many problems, including the potential to harm fire fighters by cutting off their egress. **Table 5-4** lists the time variables from ignition to when effective actions are taken on the fire ground. Each of the variables is subject to change depending on local resources and fire conditions.

Time: Ignition to Effective Actions

The time from ignition to effective actions is critical. The goal is to arrive prior to flashover and intervene, thus interrupting the fire's progression to flashover. Depending on conditions, progression to flashover in small enclosures will be rather fast and may well occur prior to the arrival of the fire department. The objective then is to contain the fire to the original flashover compartment and areas already involved in fire. In analyzing the time from ignition to effective action, it is necessary to consider the five components shown in **Table 5-4**. These times can vary depending on alarm systems, travel distance, proficiency of the attack team, and other factors.

Detection/Transmission

The first time segment shown in Figure 5-17 and Table 5-4 is preburn time, which is the time from ignition until the fire is reported to the dispatch center. Unless a supervised detection/alarm system is in the fire area, this time component is dependent upon a person discovering the fire and then calling in the alarm. This time element will vary greatly. A fire in an occupied area where everyone is awake will probably result in less detection/alarm time than one in an unoccupied building where the fire is not noticed until it shows on the exterior, which may be during the fully developed phase. When detection relies on a person detecting then transmitting an alarm, the detection/transmission time can be very long. There have been many cases in which a fire burned undetected until it reached sufficient intensity to break through a window or roof. Likewise, fire investigations of large uncontrolled fires often note a delayed alarm, resulting from occupants investigating the source of the fire

or trying to extinguish the fire before notifying the fire department. If the building is protected by a properly installed and operating supervised detection/alarm system, this time can be estimated based on experience. Otherwise this time component is an unknown.

Dispatch Time

The second time segment is the dispatch time. This includes the time for the dispatcher to take the call, select the units for the assignment, and then dispatch the companies. NFPA 1221, *Installation, Maintenance, and Use of Emergency Services Communications Systems, 2013 Edition*,[17] establishes time goals for dispatch centers. When the public safety answering point (PSAP) is at the communications center, the dispatcher is expected to answer the call from the public in 15 seconds then dispatch responders in 60 seconds for a total of 75 seconds (15 + 60). When the PSAP is not at the communications center, 30 seconds is added to both goals to allow time to answer the call at the remote PSAP then call the communications center. Every effort should be made to reduce dispatch time. Quite often, simple improvements in the way dispatchers handle calls can reduce dispatch time.

Table 5-4	Time Components: Ignition to Effective Operations	
	Definition	**Time Objective**
Pre-Burn Time	From ignition until the fire is reported to the Public Safety Answering Point (PSAP)	Unknown Variable No time objective
Dispatch Time	From public call until fire units are notified	NFPA 1221 75 seconds to 146 seconds
Turnout Time	From when fire units are notified until apparatus leaves the station	NFPA 1710 80 seconds
Travel Time	From when fire units leave the station until they arrive at the scene	NFPA 1710 240 seconds
Setup Time	From when fire units arrive at the scene until fire units take effective action	NFPA 1720 2 minutes

© Ely Studio/Shutterstock, Inc.

Turnout Time

The third segment is turnout time. This is the time from the receipt of the alarm by the fire department until the apparatus crosses the front door sill of the fire station or otherwise begins response toward the scene. Turnout time can differ greatly between fully staffed stations and on-call stations. NFPA 1710, *Standard for the Organization and Deployment of Fire Suppression Operations, Emergency Medical Operations, and Special Operations to the Public by Career Fire Departments, 2010 Edition*,[18] defines turnout time as follows:

> **3.3.53.8 Turnout Time.** *The time interval that begins when the emergency response facilities and emergency response units notification process begins either by an audible alarm or visual annunciation or both and ends at the beginning point of the travel time.*

NFPA 1710 sets a time objective of 1 minute and 20 seconds (80 seconds) turnout time for fire and special operations response (60 seconds for EMS response).

Travel Time

NFPA 1710.3.3.53.7 defines travel time as "the time interval that begins when a unit is enroute to the emergency incident and ends when a unit arrives at the scene." Travel time depends on distance, road conditions, terrain, traffic, and other factors. Average response times can be established by using computer models. If computer models are not available, reasonably accurate estimations can be determined by using empirical methods, such as the ISO (International Organization for Standardization) formula, which states that 1.7 × distance + 0.65 = travel time, or the RAND (Research ANd Development Corporation) average speed of 35 miles per hour (56 km/hr). Administering NFPA 1710 requires recordkeeping for turnout and travel time. These historical data can also be used to calculate response times within the response area.

NFPA 1710 establishes a travel time goal of 240 seconds (4 minutes) for the first-arriving engine company and 480 seconds (8 minutes) for the entire first alarm assignment. Career fire departments are expected to meet these goals on at least 90% of responses.

Setup Time

Setup time is the time necessary to position the apparatus, advance the first hose line into an attack position, and apply water. Staffing levels and training radically affect this part of the response time. The two-in/two-out rule is included in NFPA 1500 and has been adopted by OSHA.[19] This rule, in effect, changes the setup time. Now four people must be on the scene, with two of them positioned outside the hazard area.

NFPA 1710 does not address setup time. NFPA 1720, *Standard for the Organization and Deployment of Fire Suppression Operations, Emergency Medical Operations and Special Operations to the Public by Volunteer Fire Departments*, in 1720.4.3.3, states, "Upon assembling the necessary resources at the emergency scene, the fire department should have the capability to safely commence an initial attack within 2 minutes 90 percent of the time."

It is highly recommended that each fire department time its initial operations during training using its equipment and SOPs. In many cases, a significant amount of time can be saved through practicing, providing more accessible equipment, and revising SOPs. NFPA 1410, *Standard on Training for Initial Emergency Scene Operations*,[20] provides guidance on evaluating and timing a variety of fire-ground operations.

These five time segments are valid for the initial response only. Preburn time does not figure into on-scene calls for assistance, and the dispatch time is generally less when fire units are calling for help, as the dispatcher does not have to ask questions of the caller. Turnout and response times can also be reduced for subsequent calls for assistance by placing units on alert status, moving them into vacated stations, or placing them in staging. Setup time will change depending on the task assignment. These times are shown in Figure 5-17.

Adequate Number of Personnel

Setup time is related to staffing. If the initial company is staffed with less than four personnel and an imminent life-threatening situation does not exist, the attack should be delayed until additional personnel arrive. As was discussed previously, NFPA 1500 stipulates a minimum of four fire fighters as an initial crew at a working structure fire. The standard allows less than four in imminently life-threatening circumstances. According to NFPA 1500, 8.8.2.10:

> *Initial attack operations shall be organized to ensure that, if on arrival at the emergency scene, initial attack personnel find an imminent life-threatening situation where immediate action could prevent the loss of life or serious injury, such action shall be permitted with less than four personnel.*

From a commonsense point of view, a minimum crew size of four is logical, three being a compromise in situations of imminent danger. This is based on one person at the pump panel maintaining a reliable but limited water supply. The pump operator also acts as the outside safety backup during initial operations with a single crew on the scene. The buddy system is required whenever fire fighters enter a burning building; therefore two fire fighters would be required to make entry for fire suppression and/or rescue. Suppression is usually the best first action, and it takes at least two people to move and operate a standard attack line. When practical, one of the two fire fighters making entry (usually the company officer) would conduct a complete walk around the structure while the other fire fighter making entry would begin stretching the initial attack line toward the point of entry. When department SOPs direct the first-arriving engine company to secure a source of water, the fourth person would be the hydrant or water supply person. In most cases, the fire fighter assigned to the hydrant is the second person outside using a two-in/two-out rule. This person should also be able to control the entry door to limit air intake as the interior crew enters the building.

There are situations in which one or two fire fighters could start operations by setting up the pump, advancing a line toward the building, or securing a water supply. This would be acceptable under the NFPA 1500 safety standard, provided that entry was delayed until a team of at least two could enter the building with two remaining outside.

There is also a possibility of a first crew of one or two fire fighters raising ladders to rescue people from the exterior. This would meet the standard, as the fire fighters would not be making entry into the structure. However, this is a low-probability scenario. OSHA regulations require a two-in/two-out attack configuration for the initial attack when fire fighters enter a fire area but with an imminent life-threatening situation exception. The determination of imminent life-threatening situation quite often is somewhat subjective, thus inviting second guessing.

Assembling a minimum four-person team is highly recommended when conducting an offensive attack. Defensive attacks may require fewer people, but they are generally not effective in saving occupants and quite often result in a total loss of the building of origin.

The initial attack is designed to immediately save lives by extinguishing the fire or making a quick rescue. Many additional tasks are required to save lives and protect property. Some of these include the following (in no particular order):

- Ventilation
- Additional attack lines to meet flow requirements
- Attack line above the fire
- Attack line to concealed spaces
- Backup for the initial attack line
- Exposure protection
- Forced entry
- Laddering of the building
- Opening up of concealed spaces
- Salvage or property conservation
- Search and rescue of the area around the fire
- Search and rescue of the area immediately above the fire
- Search and rescue of other areas
- Utility control

Not all of these tasks will be needed at every working structure fire, and not all need to be performed by a separate team of two people. The exact number of fire fighters needed to perform these tasks will vary depending on conditions as well as the size and complexity of the building. It is obvious that many fire fighters will be needed to attack the fire, ventilate, perform the primary search, provide backup for inside crews, force entry, perform salvage, and, where necessary, protect exposure buildings.

NFPA 1710 establishes minimum staffing levels in terms of tasks to be accomplished and the number of personnel needed to accomplish specific tasks. NFPA 1710, 5.2.4.2.2, states the following:

The initial full alarm assignment to a structure fire in a typical 2000 ft^2 (186 m^2) two-story single family dwelling without basement and with no exposures shall provide for the following:

(1) *Establishment of incident command outside of the hazard area for the overall coordination and direction of the initial full alarm assignment with a minimum of one individual dedicated to this task.*

(2) *Establishment of an uninterrupted water supply of a minimum 400 gpm (1520 L/min) for 30 minutes with supply line(s) maintained by an operator.*

(3) *Establishment of an effective water flow application rate of 300 gpm (1140 L/min) from two handlines, each of which has a minimum flow rate of 100 gpm (380 L/min) with each handline operated by a minimum of two individuals to effectively and safely maintain the line.*

(4) *Provision of one support person for each attack and backup line deployed to provide hydrant hookup and to assist in the laying of hose lines, utility control, and forcible entry.*

(5) *Provision of at least one victim search and rescue team with each team consisting of a minimum of two individuals.*

(6) *Provision of at least one team, consisting of a minimum of two individuals, to raise ground ladders and perform ventilation.*

(7) *If an aerial device is used in operations, one person to function as an aerial operator and maintain primary control of the aerial device at all times.*

(8) *Establishment of an IRIC consisting of a minimum of two properly equipped and trained individuals.*

A minimum of 14 fire personnel would be required to staff the identified positions—15 if an aerial device is in use. This requirement is for what is termed a *low-hazard occupancy* in NFPA 1710. Exposures or more hazardous occupancies require additional staffing.

Many skeptics doubted the need for 14 or 15 fire fighters and a minimum 4-person crew at a typical residential fire with no exposures. In response, the NIST[21] built the 2000 ft^2 (186 m^2) two-story structure described in NFPA 1710 and conducted over 60 full-scale experiments in the structure to determine the effectiveness and safety of various crew sizes and different stagger times (interval between arriving crews). Technical fire service experts identified 22 separate tasks that should be conducted at this single-family residential fire. Each task and the overall time to completion were timed during the experiments. This scientific study proved the need for the staffing requirements identified in NFPA 1710.

The data for two-, three-, and four-person crews with a short interval between arriving companies showed a dramatic difference in the time to complete the 22 identified tasks. The two-person crew took 22 minutes and 16 seconds to complete the tasks, the three-person crew required 20 minutes and 30 seconds, and the four-person crew completed all 22 tasks in 15 minutes and 44 seconds.

Tactical Reserve

Planning ahead to obtain required resources is crucial. The number of fire fighters needed to safely and effectively combat a working structure fire can be significant. Tactical efficiency can reduce the number of people necessary to fight the fire. For example, immediate application of water with coordinated ventilation will assist the primary search team, allowing it to cover a much larger area with fewer people. This same tactic also permits the attack team to extinguish the fire and move to another task more rapidly.

After the IC staffs all of the attack and support positions, there is a need to establish a tactical reserve. The size of the reserve force depends on the stage of the incident and the number of units working, as well as the type of incident. It can be accurately said that if all units are assigned, you need additional personnel.

Correlation Between Elapsed Time and Progression to Flashover

The main point to be made in studying the flashover/survival/collapse charts is that time is a critical factor (Figure 5-17). The longer the fire has been burning, the less chance there will be to rescue occupants and the greater the chance of structural collapse. From a fire fighter safety viewpoint, the worst scenario is one in which the fire fighter arrives near the end of the buildup to flashover with occupant lives at risk in a large, undivided area. Once flashover occurs, the chance for survival is near zero in the flashover compartment. The risk to fire fighters has increased because of the fire intensity and the potential for significant structural damage. This is a situation in which risk management is critical to fire fighter safety and survival.

In general it can be said that the smaller the compartment, the sooner the progression to flashover or backdraft. Most of the fuel load in a structure fire will be in the form of solid materials. Ordinary solid materials must be preheated to the point at which vapor fuel is released, through a process called **pyrolysis**. The vapor fuel will ignite if there is an available and sufficient ignition source. However, as the fire progresses, it preheats solid fuels that release unignited pyrolytic vapors. If the fire area is enclosed, pyrolytic vapors and products of combustion such as smoke and carbon monoxide will accumulate. If there is sufficient oxygen available when the contained fuels reach ignition temperature, a flashover occurs. If there is a lack of oxygen, the fire could go into the decay phase. However, if air is introduced into this superheated and fuel-rich atmosphere, flashover or a backdraft explosion could occur. Fire fighters must use extreme caution when opening doors and windows when the fire is ventilation limited. The fire will almost surely gain intensity and possibly result in flashover or backdraft.

Small, enclosed rooms quickly consume available fuel and/or oxygen, creating potential flashover or backdraft conditions. In a small room, the fuel packages will tend to be close to each other, allowing heat radiation to raise their temperature toward ignition sooner. The volume of a smaller room allows the overall temperature to be raised more quickly than a larger room, decreasing the likely time to flashover or backdraft.

Observing smoke and temperature conditions is critical when entering small, enclosed areas. Sometimes these areas simply burn out before entry. Initially, flashover and backdraft tend to be the most critical problems when encountering a working fire in a small-volume compartment such as in the apartment building shown in Figure 5-18 . The chance of burning all available fuel or using up all the available oxygen is much less in larger areas.

Figure 5-18 Post-flashover fire in a single apartment.
Courtesy of Mike Carey and Jeff Neal, Cincinnati Fire Department

Flashover and backdraft will usually occur much later, if at all, in large fire compartments. However, the concealed space created between suspended ceilings and truss roofs may be a relatively small-volume area owing to the height of the area. In fact, many backdrafts occur in concealed spaces. One such incident on December 17, 1960, at the Parkmoor Bowling Alley in Louisville, Kentucky, resulted in the deaths of three fire fighters. Two fire fighters were fatally injured in an incident investigated by NIOSH due to a possible backdraft that contributed to a roof collapse and entrapment as ceiling tiles were removed from below in a 6670 ft^2 (620 m^2) retail store area.[22] This was also a problem in the previously discussed Charleston, South Carolina, Super Sofa fire.

Most buildings have more than one enclosed area or compartment. It may be possible to conduct search and rescue operations in other compartments after the compartment of origin has flashed over. As an example, suppose a fire occurs in the apartment building shown in Figure 5-18. The apartment of origin sustains flashover, but surrounding apartments are not yet involved. Survival chances for occupants remaining in the apartment of origin are near zero, but occupants in other areas near the fire and/or behind closed doors within the apartment of origin could be rescued. In this multistory building, there is a critical need to quickly get water on the fire and perform post-flashover rescue operations on the fire floor and on the floor above the fire.

Fire-Ground Operations
Communications

Communications are the lifeblood of any command organization. Without communications, the situation is at best perplexing, and at worst, chaotic. At a minimum, each crew working inside the structure, whether in the hot zone or the warm zone, must be provided with radio communications connecting them to the next level of command (e.g., division or group supervisor). Some departments issue every member a radio; this practice has been a documented lifesaver on several occasions. Radio discipline can prove to be a significant challenge, but it is imperative when everyone on the scene is assigned a radio. If everyone is talking simultaneously, it will be impossible to communicate assignments or emergency messages. Proper use of the radio in exchanging important information in a clear, calm manner is essential to good scene management and fire fighter safety.

Modern radios have the capacity to operate on a wide variety of channels. This is important to interoperability and can improve on-scene communications. However, it is important to keep the communications network simple so all units can be promptly notified of an emergency condition or critical need.

Frequent progress reports are essential to the IC, who should have a good overall view of the incident. Interior crews and crews working in areas not visible to the IC are the eyes and ears of the IC. Progress reports also provide everyone on the fire ground with the location of units and with information on other aspects of the fire that relate to their own particular operations.

Fire fighters in need of assistance transmit a mayday. A special radio alert tone indicating a mayday is an advisable enhancement. It is important that SOPs spell out the response to a mayday. The tendency is for all crews to stop what they are doing to go to the assistance of fire fighters who are calling the mayday. If everyone responds to the mayday call, however, there will be too many people in the area of the mayday. In most cases, the RIC will be assigned to rescue the fire fighters who are in danger. However, there are some commonsense exceptions, such as when another fire crew is in the immediate area of the problem.

Critical functions such as fire suppression will not be accomplished if all personnel focus on the fire fighter rescue operation. Controlling or extinguishing the fire provides the time needed for the rescue operation. Fire officers must understand the importance of remaining within the command system, and the IC must remain in total control at all times. Specific assignments should be made during Mayday rescue operations so that all tasks can be coordinated.

Departments should establish an emergency evacuation signal. One method is to use apparatus air horns to signal the retreat, such as 10 three-second blasts. Radio channels in use at the scene should also transmit an emergency evacuation message. Again, SOPs should be in place describing actions to be taken during an emergency evacuation. Just as the hose line controlling the fire buys critical time to accomplish a rescue, it can also provide additional time to evacuate and protect fire fighters above the fire. Only in rare cases is a "drop everything and run" evacuation warranted. The "drop everything and run" tactic is more often used during defensive operations. For offensive operations, an organized retreat is generally a better alternative.

Command and Control

Command and control was addressed in the chapter titled *Organizing, Coordinating, and Commanding Emergency Incidents*, but it is important to point out that the National Incident Management System (NIMS) is also a safety system. Taking measures to have everyone working toward a common goal in an organized fashion is the basis of a safe and effective operation. Conversely, freelancing leads to injuries and fatalities. NIMS is the foundation for accountability and rapid intervention procedures.

Accountability

An **accountability system** must be established on the fire ground for two purposes. The first is to ensure that everyone entering the area has a specific assignment. This is done to eliminate freelancing on the fire ground, which has led to several fire fighter fatalities.

The second purpose of an accountability system is to be able to track all personnel at the scene and to identify the location of any missing personnel if a catastrophic event should occur, such as a collapse. It also serves to identify whether any personnel or crews are overdue so that search and rescue operations can be initiated.

A good organizational structure accounts for all personnel operating at the scene and assigns responsibility for each crew with a reasonable span of control. Crew unity is an essential part of this process. Crew members should not be separated within the structure. The company officer or crew leader should be able to account for all members of his or her assigned crew working within the hazard area. Splitting crews inside the hazard area is inviting disaster. However, during offensive operations, it is acceptable to have an apparatus operator or other crew members working outside the hazard area while the officer and remaining crew members are inside the structure.

Incident Summary

In the fire in Seattle, Washington, described earlier in this chapter, the crews who had entered the building were not aware of a subbasement and did not realize that the main body of fire was below them. They thought they were making good progress on the fire and were in the mode of knocking down spot fires. Although every fire fighter was issued a radio, they did not communicate their actions over their radios, but instead provided a progress report to the IC face-to-face when they changed out bottles. If they had reported on the radios, the crew in the basement—which was facing a large fire but not applying any water because of the potential for opposing hose streams—would have realized that there was a significant mismatch between the two situations. Although face-to-face is often the most effective way to communicate information, it is also the IC's responsibility to ensure that everyone involved in the operation is aware of critical information.

Data from: Edward R. Comeau, Fire Fighter Fatalities, Seattle, Washington, 1995, NFPA Fire Investigation Report. Quincy, MA: NFPA.

Many department SOPs allow the first-arriving truck company to split into ventilation and search and rescue crews. This should be a temporary arrangement, with someone supervising each crew. Each of the split crews must have a radio to enable them to provide progress reports, receive assignments, and transmit a mayday if needed. Even then, the crews should reunite as a company as soon as possible. Except in situations of imminent life-threatening circumstances, splitting a crew to provide a two-out team during initial operations is required by NFPA 1500 and OSHA regulations when only one company or unit is on the scene and entry is made into a burning structure.

In addition to the accountability provided by NIMS, a separate system is needed to track and account for everyone at the scene **Figure 5-19**. Several systems are in use, and most are effective if they are properly used. High-tech tracking systems are available and in use. However, most accountability systems use roster sheets, personal tags, or labels. These accountability systems provide a means of gathering roster sheets or tags and establish a physical entry and exit location for the collection of accountability rosters or tags under the direction of an accountability officer(s). The success of the accountability system depends more on the consistent use of the system at every incident than on the type of system.

Department accountability procedures must address the initial accountability process used prior to the arrival of an assigned accountability officer and how the formal accountability officer includes units already on the scene. Personnel assigned as the accountability officer should receive appropriate training, including simulations in which the system is implemented. The IC may assign units that are en route to the scene or to staging. These units will not report to the command post. Provisions must be made to include all units at the scene in the accountability system.

If a situation deteriorates rapidly, the IC will call for a personnel accountability report (PAR). SOPs generally call for PARs in the following situations:

- Whenever the IC thinks it is necessary
- When the safety officer requests a PAR
- When the IC changes from an offensive to a defensive attack
- When sudden changes occur, such as backdraft, flashover, or collapse
- When the entire building has been searched
- When the fire is extinguished
- At prescribed times described in department SOPs

Routine requests for a PAR at a designated time or whenever there is a change in tactics reinforces the concept of personnel accountability and ensures that officers are aware of the location of the personnel assigned to them.

Because accountability procedures become more important as the incident increases in size and complexity, it is very likely that mutual aid resources will be at the scene. Therefore, a regional approach to an accountability system is the only logical choice.

Figure 5-19 Accountability officer at the command post.
Courtesy of Bill Strite Cincinnati, Ohio

The safety officer should not be the accountability officer. The safety officer position is mobile. The accountability officer should be at a fixed location. The safety officer must enter into the accountability process by keeping the accountability officer informed of his or her location. Likewise, the accountability officer is a valued informational resource for the IC, safety officer, and RIC.

Safety Officer

There is no exact time or situation that requires the handing off of the safety officer position at a structure fire. Each department should develop SOPs outlining when this staff position must be established. In general, the safety officer should be separately staffed whenever the IC believes that he or she can no longer effectively monitor safety at the scene. This situation usually happens fairly early, as the IC should be located at a fixed command post. The safety officer monitors all areas where fire fighters are working. The IC should be stationary, while the safety officer is mobile.

Alternative Egress

Most large buildings have multiple egress routes, as required by building and fire codes. When this is the case, the IC should send crews into each stairway for the purpose of managing the stairs. In high-rise fire operations, lobby control manages the stairs and elevators (See the *High-Rise Buildings* chapter for a complete explanation of lobby control).

This procedure will help to ensure that alternative routes that can be used during an emergency are identified and under control of the fire department. The lobby control function can also be useful in buildings that are not technically classified as high-rise structures.

The interior stairs are the preferred means of access and egress, but they are sometimes untenable or inadequate. When available, fire escapes provide additional means of escape and can also provide fire fighters with access to, and escape routes from, upper floors of the building. Fire escapes would be a second preferred choice of egress/access, because using fire escapes requires fewer fire fighters than positioning fire department ground ladders, and if properly maintained, they are safer. The building shown in Figure 5-20 through Figure 5-22 has two fire escapes and three stairways providing alternative egress from every location within the building. However, if the use of a fire escape is included in the preincident plan, the fire escape should be thoroughly inspected on a regular basis and its capacity should be noted. A proper inspection oftentimes requires more than a visual inspection, because the support structure may be imbedded in the building or otherwise hidden.

Figure 5-20 Apartment building with multiple means of egress—side A.

Courtesy of Bernard J. Klaene.

Figure 5-21 Apartment building with multiple means of egress—side D.

Courtesy of Bernard J. Klaene.

Figure 5-22 Apartment building with multiple means of egress—sides C and B.

Courtesy of Bernard J. Klaene.

A common practice in the past was for the first-due truck company to "claim the building" by placing a ladder somewhere (anywhere) on the building. This practice had a fundamental purpose of providing an alternative means of egress for occupants and fire fighters. Unfortunately, the purpose was often forgotten and the practice was eliminated as staffing on truck companies was reduced. Fortunately, this practice is making a comeback with many departments now "laddering the building" to provide an alternative means of escape for fire fighters working inside a burning building. Properly laddering the building is an important task that should be accomplished early in the operation. Placing a ladder to the one-story roof area shown in **Figure 5-23** would provide an alternative egress from several apartments on the second floor of this building.

SOPs should address laddering the building, possibly calling for ladders to be placed at predictable locations, such as the center window at the rear of the building or at the end of hallways. If ladders cannot be placed at a prescribed location or if SOPs do not address where ladders are to be placed, it is important for exterior crews placing ladders to communicate the location of ladders to the crews working on the interior.

Just as codes require two separate and distinct means of egress, there should be provision of a second way out for fire fighters. The alternative egress should be remote from the first. Access to the roof is generally by ladder, but there should also be a second means of egress from the roof, usually a second ladder placed remote from the first. Some departments assign laddering the building to the RIC. The

Figure 5-23 Apartment building with multiple means of egress—side C.

Courtesy of Bernard J. Klaene.

pros and cons of this and other RIC assignments will be discussed in the next section.

Rapid Intervention Crews

There is a tendency to delay the establishment of the RIC; assigning a later-arriving unit as the RIC results in limited RIC protection during the most hazardous early stages of the operation. Structural failure is more likely to occur later on the fire ground, but the Underwriters Laboratories experiments that examined floor structures during basement fires demonstrate that floor collapse often occurs early in the operation. Likewise, the Underwriters Laboratories study comparing modern furnishings to legacy furnishing established that flashover can be expected before or shortly after the arrival of the initial response. Water supply and hose line problems are most likely to occur during the beginning stages of the operation. When writing SOPs related to RIC operations, it is important to consider the importance of having a dedicated RIC available during the early stages of the operation. Waiting for a report of a confirmed working fire to dispatch a RIC company is unacceptable. Initial alarm staffing must include provisions for a RIC, and staffing should be considered insufficient until all safety and tactical provisions are covered, including a reserve force.

It is important to point out that the RIC is not a substitute for safe and effective operations. Formulating an incident action plan based on a risk-versus-benefit analysis, combined with good tactics and company-level attention to safety will reduce the need for emergency rescues. Tactics such as extinguishing the fire and venting the building can greatly reduce the possibility of fire fighters being trapped or disoriented.

Safe interior operations include:

- Maintaining crew integrity during interior operations with at least two members working together and in contact with one another
- Controlling and coordinating ventilation
- Providing hose line protection in areas where fire fighters are working

- Providing a means of communications to the exterior and among units working in an area
- Maintaining contact with the hose when operating a hose line
- Maintaining contact with the wall or rope when operating without a hose line
- Placing a fire fighter, signal light, or audible signal at the door leading to the room where fire fighters are working

Despite these measures, the unexpected can occur. Fire fighters must be taught self-survival techniques, including how to operate the emergency features on SCBA and PASS.

When a fire fighter or group of fire fighters suspect they are in trouble, and might need assistance, they should immediately transmit a mayday call reporting their situation. When requesting assistance, fire fighters should give their exact location, if possible; if they do not know their exact location, they should give the best estimate (e.g., We are located on the second floor on the C side."). Fire fighters should state the problem, such as "out of air," "disoriented," "trapped in collapse," "entangled," "fell through floor" or "fell through roof." They should also state the number of fire fighters needing assistance and company number. Transmitting a mayday with a location report should be practiced repetitively during training sessions until it becomes routine.

Even after requesting assistance, it is sometimes possible to continue self-rescue efforts or take measures that could extend survival time such as finding a source of outside air at a window or under a door. If fire fighters requesting assistance are merely disoriented and not in imminent danger, it may be best to activate their PASS and conserve air while awaiting assistance. In this case, isolation from the fire area (e.g., closing door to an area of safe refuge) can extend survival time.

The PASS device should be activated immediately after a "Mayday" is declared. It is possible to increase the efficiency of the PASS by directing it away from the fire fighter's body. Trapped members can sometimes indicate their location by tapping on doors, directing a light around the room, or otherwise signaling their location. Breaking or opening a window leading to the exterior can help fire fighters regain their orientation and call for assistance. However, the window becomes a vent opening and may draw fire and smoke toward the fire fighters unless the disoriented fire fighters have isolated their location by closing doors and other openings.

A team of fire fighters in Cincinnati depleted their air supply deep inside a smoke-filled warehouse/manufacturing building. They were able to extend their survival time by breathing air from a crack at the bottom of a metal overhead door that was locked on the exterior. They used a spanner wrench to bang on the door to indicate their location. They were eventually rescued and survived without serious injury.

In the initial stages of an operation, with only one crew operating at the scene, the pump operator and one additional fire fighter are generally assigned as the RIC. This outside crew must know the position of the fire fighters working inside the structure. Later in the operation, a minimum of two (preferably four or more) fire fighters with full turnouts are assigned as the RIC. This outside crew must know the location of fire fighters working inside the structure and be immediately available to rescue fire fighters. The officer in charge of the RIC is responsible for maintaining the crew's readiness at all times. As can be seen in **Figure 5-24** , *caught* or *trapped* is the cause of injury in more than 75% of all fire-ground fatalities. Many of these fire fighters could have been rescued by a properly staffed RIC.

The RIC should assemble necessary tools and equipment as prescribed in department SOPs. This tool list will be based on department operations, tools available, types of buildings protected, and

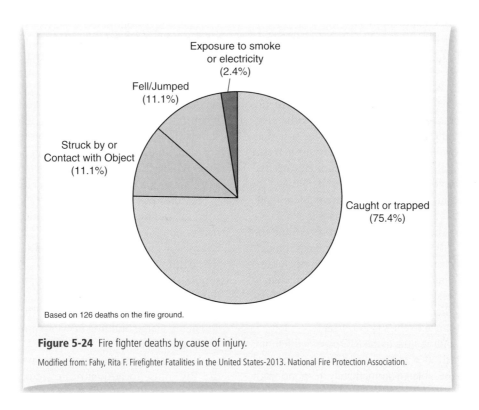

Based on 126 deaths on the fire ground.

Figure 5-24 Fire fighter deaths by cause of injury.

Modified from: Fahy, Rita F. Firefighter Fatalities in the United States-2013. National Fire Protection Association.

other factors. The RIC must have SCBA with connections to supply air as stated in NFPA 1500:

> **8.8.5** *At incidents where any SCBA being used is equipped with an RIC universal air connection (UAC), the RIC shall have the specialized rescue equipment necessary to complete the RIC UAC connection to a supplied air source.*

Tools commonly carried as part of the RIC cache include:

- Rescue ropes, search ropes, guideline ropes
- Thermal imaging camera (TIC)
- Patient carrier, webbing, or harness
- Portable ladders for above- and below-grade rescues
- Forcible entry tools
- Wire cutters and other hand tools
- Lighting equipment

A TIC is an indispensable tool when the exact location of the fire fighter(s) needing assistance is unknown. The RIC officer should determine if the situation calls for tools that may not be listed as part of standard RIC equipment. As an example, a building equipped with hardened security bars might require a special saw or cutting torch. The construction type, occupancy, fire location, and other factors will also indicate the type of rescue that may be needed as well as the tools necessary to perform the rescue. Additional tools and equipment needed because of building features could be known in advance and thus identified in the preincident plan.

A RIC that is assigned non-RIC tasks is at a severe disadvantage. The dedicated RIC can determine the best access locations, assemble tools and equipment, and generally preplan potential rescue operations according to conditions. They are also immediately available for assignment with full breathing apparatus. The dedicated RIC monitors conditions and provides assistance to the safety officer.

The availability status of the RIC poses a dilemma. A RIC stationed at a central location near the tool cache would be able to immediately respond with tools and equipment needed to rescue fire fighters calling for assistance. However, there is a benefit in having the RIC do a complete walk-around to determine the location of alternative means of ingress and egress and other building features that should be considered if a mayday occurs. While doing a walk-around, the RIC can also improve egress by forcing doors (being careful that the means of doing so does not create an unwanted vent opening), placing ladders, and removing obstacles. A mobile RIC can recon the building and provide valuable information to the IC. Mobile and stationary RICs both have advantages and disadvantages. Some departments opt for a combination of the two by assigning some RIC members to reconnaissance and improving egress while others remain at the equipment cache location. When writing RIC procedures, it is important to consider the advantages and disadvantages of each based on RIC staffing and other factors in the response area. If a preincident plan is available for the building, the RIC should have access to the preincident plan.

The USFA conducted a study[23] of fire department RIC operations. Eighty-three (mostly career) departments responded to the study. The number of members assigned to the RIC (when a RIC is established) and RIC operations vary greatly. The fire service is still learning the best procedures to use when rescuing our own. Even the name RIC is not standardized. Crews assigned to rescue fire fighters are sometimes titled RIT (rapid intervention team), RAT (rapid action team), FAST (fire fighter assist and search team), and go team, among other names.

Serious questions have been raised regarding the need for additional staffing for the RIC. Most departments now use the two-out team as an initial RIC but assign a full fire company as a RIC as the operation continues. Experience indicates that a two-person RIT is inadequate except in situations where fire fighters are easily located and able to escape with minimal assistance. A four-person RIC may be inadequate in a large building or where fire fighters need to be extricated.

When notified that a fire fighter needs assistance, the RIC must locate, extricate, remove/assist, and provide medical attention for fire fighters trapped inside a building. The sooner the RIC reaches trapped fire fighters and is able to get them to a place of safety, the better the chance of survival. In a very large structure, this could require establishing multiple RICs at various locations around the building.

Some departments categorize RIC operations as a "rapid removal" or an "extended operation." Rapid removal would normally consist of finding and assisting fire fighters to safety. In many cases, simply assisting a disoriented fire fighter to a safe area is all that is needed. Other times a fire fighter may be unconscious or debilitated, but located very near a safe area or near the exterior. In these cases, it may be best to quickly remove the fire fighter rather than trying to establish an air supply and performing other special rescue techniques. Extended operations may require two or more RICs. The first RIC

would make every effort to locate fire fighters needing assistance while a second team stands ready to extricate the fire fighters once they are found.

Fire departments may or may not require RICs to have a hose line as part of their standard procedure. Extending a hose line to the area where fire fighters need assistance can substantially delay search and extraction operations. However, it may be necessary to protect trapped fire fighters or to provide a path to the exit by extinguishing the fire. Some departments assign a truck or rescue company as the RIT and an engine company as the engine-RIC. The engine-RIC provides hose line protection. Other departments assign more members to the RIC to provide sufficient personnel to find and extract the fire fighter(s) requiring rescue and to protect all involved fire fighters. If the RIC deploys a hose line or another unit is assigned to protect RIC operations, it is best to

Incident Summary

In Phoenix, Arizona, a fire on the exterior of a large supermarket extended into a concealed space between the ceiling and roof in the storage area. A crew of fire fighters entering the store encountered worsening smoke conditions as they advanced. The low air supply warning sounded on two SCBAs, causing the crew to retreat. As they retreated, three members lost contact with the hose line that they had advanced into the store, and they became disoriented. One member was able to reestablish contact with the hose and find his way out. The other two fire fighters remained disoriented. One of the disoriented fire fighters heard another crew nearby and followed the sound to their location and ultimately to safety. The third member was found, but several crews were required to remove him, thus increasing his time in the hazard area. He was transported to the hospital, where he was pronounced dead.

Data from: Robert F. Duval and Stephen N. Foley, Supermarket Fire, Phoenix Arizona, March 14, 2001, 1 Firefighter Fatality. NFPA Fire Investigative Report. Quincy, MA: NFPA.

Debris fire fighters encountered while trying to find and rescue a missing fire fighter. Fire fighters needing rescue in large commercial structures may require more staffing and time to find and remove

Photograph by Robert Duval, © National Fire Protection Association, Inc. used with permission

Flames: © Jag_cz/ShutterStock, Inc.; Paper: © silver-john/ShutterStock, Inc.

provide a separate water supply for the RIC hose line, as hose line failure may be the reason for the call for assistance.

In a high-rise operation on an upper floor, the RIC and RIC equipment should be located one or more floors below the fire floor so RIC members do not expend their air supply while on standby, but are near areas where fire fighters are working. The IRIC should not be located in an area where they need to activate their SCBA while on standby.

It is critical to train and practice RIC procedures. Many times training under simulated fire conditions will point out deficiencies in procedures that can be changed to improve RIC operations. During training sessions, it is essential to develop a variety of scenarios in different buildings and use full protective clothing as you would while performing a rescue. RIC procedures should be practical under fire conditions and stress commonsense approaches. It is much easier to walk through various scenarios and practical rescue and self-rescue techniques during training than under the stress of an actual mayday.

Opposing Fire Streams

Hose streams or fire lines are important safety tools, but they can also do much harm when the flow path is changed, placing fire fighters or victims in danger. The Underwriters Laboratories study comparing the impact of ventilation on fire behavior in legacy versus modern furnishing[24] found that "while the fog stream 'pushed' steam along the flow path, there was no fire 'pushed.'" Earlier editions of this text included a description of hose lines, especially fog streams pushing fire. Hose streams do affect the flow path. The later Underwriters Laboratories study on vertical ventilation[25] further examined the "pushing" effect of fire streams, resulting in the following statement:

You cannot push fire with water. The previous [Underwriters Laboratories] study (Kerber S. 2010) discussed the concept of pushing fire in the data analysis. Since the release of that study there has been a lot of discussion about this and stories from well-respected fire service members where this did happen or was perceived to happen. The specific fires that were being recalled by the firefighters were discussed in detail. In many of these conversations, the firefighters were in the structure and in the flow path opposite the hose line. In most cases where firefighters experienced fire moving over their heads, fire attack crews were advancing on the inside and not applying water from the outside into a fully developed fire. All of the current experiments described in this report were designed to examine the operations and the impact of the initial arriving fire service units so we did not, do not, and will not suggest that firefighters should be in the position where they are in a flow path opposite the hose line. However, there are times when this may happen, so the experience of these firefighters should not be discounted. During our discussions, four events could have been witnessed which may have had the appearance of pushing fire:

(1) A flow path is changed with ventilation and not water application.
(2) A flow path is changed with water application.
(3) Turnout gear becomes saturated with energy and passes through to the firefighter.
(4) One room is extinguished, which allows air to entrain into another room, causing the second room to ignite or increase in burning.

Whenever possible all interior hose lines should attack from the same access point. If this is not possible or practical, communications between units attacking the fire is essential.

Master streams, especially fog streams, improperly operated on the exterior can change the flow path by disrupting the vent process. Offensive and defensive attacks should not be conducted at the same time in the same building. The IC assigns units so that everyone is working either offensively or in a defensive mode.

NOTE

The limited application of water from the exterior directed to the interior fire area to quickly reduce fire intensity in preparation for an offensive attack is not considered a defensive tactic.

Texture: © Eky Studio /Shutterstock, Inc; Steel: © Sharpshot/Dreamstime.com

Personal Protective Clothing

There have been great improvements in the level of protection provided by personal protective clothing since the NFPA first issued fire fighter on-duty death and injury reports in 1977. Yet there are still far too many inhalation, burn, and cutting injuries on the fire ground. Often, these injuries occur because fire fighters are not properly using personal protective clothing. However, there is also an attitude of invincibility due to the increased protective attributes of this clothing. Many departments now use longer-duration SCBAs that permit the fire fighter to stay in the IDLH atmosphere for a longer period, which increases the time available to escape; however, this equipment also adds to the fatigue factor. Positive pressure within the face piece eliminates many of the inhalation problems of the past, but can also result in a depletion of the air supply due to leakage. Likewise, the turnout ensemble has higher thermal ratings and is more cut resistant than in the past. Helmets and boots are improved, and the use of PASS and protective hoods is the norm.

These elements of protective clothing all have limitations. As an example, the minimum thermal protective performance (TPP) value is 35 for protective coats and trousers. Time in a heated environment gradually erodes the thermal protection depending on the temperature. At near flashover temperatures, a TPP of 35 only provides protection against second-degree burns for approximately 17.5 seconds. NFPA 1971, *Standard on Protective Ensembles for Structural Fire Fighting and Proximity Fire Fighting*,[26] provides specific test methods and requirements for each element of the protective clothing ensemble.

Many departments simply develop personal protective equipment specifications referencing the minimum levels of protection in the NFPA 1971 standard. A better approach is to specify greater protective levels if available when flexibility is not negatively affected.

The IC should establish the level of protective clothing necessary to enter hot and warm hazard control zones. The safety officer should be consulted on this issue.

NFPA 1500 addresses the use of SCBA in stating the following:

> **7.10.7** *When engaged in any operation where they could encounter atmospheres immediately dangerous to life or health (IDLH) or potentially IDLH or where the atmosphere is unknown, the fire department shall provide and require all members to use SCBA that has been certified as being compliant with NFPA 1981,* Standard on Open-Circuit Self-Contained Breathing Apparatus for the Fire Service.
>
> **7.10.8** *Members using SCBA shall not compromise the protective integrity of the SCBA for any reason when operating in IDLH, potentially IDLH, or unknown atmospheres by removing the facepiece or disconnecting any portion of the SCBA that would allow the ambient atmosphere to be breathed.*

Recent studies indicate that the SCBA face piece lens may fail under high-heat conditions. After reports of several face piece lens failures, NIST conducted tests subjecting the SCBA lens to high levels of heat that could be expected under high-heat conditions (572°F [300°C]) **Figure 5-25** .[27] One example of conditions leading to high heat cited in the report was opening the door to a room and causing a rapid increase in temperature as fire extends through the doorway. It is recommended that SCBA face piece lenses be checked for thermal damage on a regular basis.

Thermal ratings vary for different components of the protective clothing ensemble. Fire fighters must understand the limitations of all parts of their

Figure 5-25 Face piece lens damaged by high heat.
Courtesy of NIST.

personal protective equipment. When faced with an extremely hot interior fire, fire fighters can reduce the heat level by applying water prior to entering the fire compartment. In some cases, water is applied from the exterior prior to making entry.

The protective clothing performance ratings assume that the personal protective equipment is dry, properly maintained, and replaced when worn or damaged. Normal department maintenance is necessary, but each individual should check his or her equipment after every use. It is also recommended that protective clothing, including the SCBA face piece lens, be checked when members are in rehabilitation.

RIC and self-rescue methods sometimes involve making openings in interior or exterior walls. Typically, these are small openings such as the 16 in. (406 mm) between wall studs. Moving through this small space requires fire fighters to remove the SCBA backpack while leaving the face piece intact. This procedure does not necessarily result in the fire fighter breathing the ambient atmosphere, and at times, it may be necessary. However, consider making a larger opening, thus permitting a faster exit that does not potentially compromise any part of the protective clothing ensemble.

Overhaul

A tendency to "dress down" during the overhaul phase of an operation can lead to unnecessary injuries and long-term detrimental health effects. Even after airing out the structure, removing the SCBA face piece is a dangerous practice. Residual burning materials produce toxic gases, and concentrations of fire gases will be present within the building for a considerable time after fire control.

A 2010 Underwriters Laboratories report[28] provides further evidence of the toxic inhalation hazards faced by fire fighters when removing the SCBA during overhaul operations. Air monitoring was conducted at actual fire scenes during active firefighting and for 30 minutes following fire suppression. This Underwriters Laboratories research proves that fire fighters are exposed to toxic and potentially carcinogenic off-gases that are known to result in both acute and chronic adverse health effects. The burning contents found in structure fires today include more synthetic materials that produce an abundance of toxins. Some of the toxins emitted by common household furnishings and structural components during and after extinguishment are:

- Arsenic
- Benzene
- Carbon monoxide
- Formaldehyde
- Phenols
- Hydrochloric acid
- Hydrogen cyanide
- Formic acid

The study found that in addition to gaseous/vaporized toxins, the air and surfaces within a post-fire compartment contain smoke particles and condensed liquids that when inhaled pose a long-term health threat. Many of the smoke particles collected by Underwriters Laboratories during overhaul were too small to be visible by the naked eye, leading Underwriters Laboratories researchers to a conclusion that "clean" air was not really that clean. Even if open areas within the building were free of smoke and toxins, research conducted by NIOSH that this author took part in found that dead air spaces within a building contained toxic by-products of combustion long after the building was aired out.

Many fire fighters make a decision to remove their SCBA based on the observable presence or absence of smoke. Other departments take a more scientific approach, using portable gas-measuring instruments (usually a carbon monoxide meter) to determine the safety of the atmosphere. This approach, although more scientific, is no more reliable unless the only toxic material present is the one measurable by the available instrument. Even scientific studies, such as the 2010 Underwriters Laboratories study, often miss products that are not known to be present. Some in the fire service recommend that a lighter-weight respirator be used during overhaul. This was the subject of laboratory research conducted in 2007 that concluded that the respirators used during the testing do not fully protect during overhaul exposure.[29]

After a working structure fire and subsequent operations to gain access to concealed spaces, there are often many other hazards and dangers. Therefore, removing any part of the turnout ensemble inside the fire building is an unsafe practice.

The time between suppression and final **overhaul** is an ideal time to rotate personnel through rehabilitation while the building airs out. This recommended rehabilitation between active suppression and final overhaul is not meant to preclude the need for protective clothing, including the SCBA, or earlier rehabilitation when gaining control of the fire takes longer.

The IC should use the time between active firefighting and overhaul to develop an overhaul plan (a nonemergency incident action plan). Safety considerations include:

- Structural damage and structural stability
- Smoke and airborne contaminants
- Cutting hazards (broken glass, jagged metal, protruding nails)
- Holes in floors
- Damaged stairways
- Utility hazards (gas, oil, electric)

- Overhead hazards (damaged ceiling, loose materials)
- Visibility (provide lighting or wait until there is adequate natural light)

When the pre-overhaul inspection reveals imminent hazards or risks that cannot be determined, overhaul should be delayed. On some occasions, overhaul is conducted using heavy equipment. The building shown if **Figure 5-26** was heavily damaged by fire, and the structural stability could not be determined. Master streams remained in place for several days with the area being secured. Eventually, heavy equipment was used to raze the building.

When the pre-overhaul inspection finds no serious imminent hazards, the identified hazards should be discussed with crews assigned to that area. If possible, mark the hazardous conditions using barrier tape, ladders, and other means. See the *Property Conservation* chapter for more information on overhaul. Crews reentering the building should be on-air in full protective clothing.

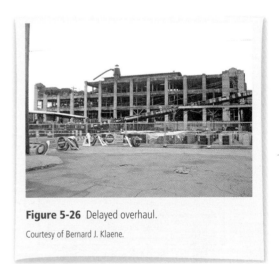

Figure 5-26 Delayed overhaul.
Courtesy of Bernard J. Klaene.

Rehabilitation

NFPA 1584, *Standard on the Rehabilitation Process for Members During Emergency Operations and Training Exercises, 2008 Edition*,[30] provides procedures for hot and cold weather rehabilitation.

Rehabilitation can be divided into three phases:

1. Preincident hydration and preparation
2. Incident rehabilitation
3. Postincident recovery

Thirst is a sign of mild dehydration; members should not wait until they are thirsty to ingest liquids. While awaiting calls, members should prehydrate by drinking 6 to 8 ounces of water every 6 hours plus liquids that are ingested with meals. This practice is especially important during hot and humid weather conditions. However, water or sports drinks should not be ingested in extreme quantities. Members should avoid activities that cause extreme fatigue or exhaustion while waiting for a response.

On-scene rehabilitation can be formal or informal depending on weather conditions, length of time on scene, and activity level. Working structure fires tend to require very high energy levels; thus they cause fatigue at a rapid rate. Informal rehabilitation usually takes place at the company apparatus. An apparatus used for rehabilitation should be in the cold zone so members can "dress down" while resting and rehydrating. Resting members should not be placed in areas where they will be breathing exhaust fumes. Formal rehabilitation involves setting up an area for rehabilitation. This area should provide shade and mechanical cooling equipment during hot weather and a warm location during cold weather. The formal rehabilitation area must have water or sports drinks available. If members are on the scene for an extended time, healthy food should also be provided. **Figure 5-27** shows the layout of a typical rehabilitation/treatment area.

NFPA 1584 delineates eight functions that are to be provided at the rehabilitation area. According to NFPA 1584.6.2, rehabilitation efforts shall include:

- Relief from climatic conditions
- Rest and recovery
- Active and/or passive cooling or warming as needed for the incident type and climate conditions
- Rehydration (fluid replacement)

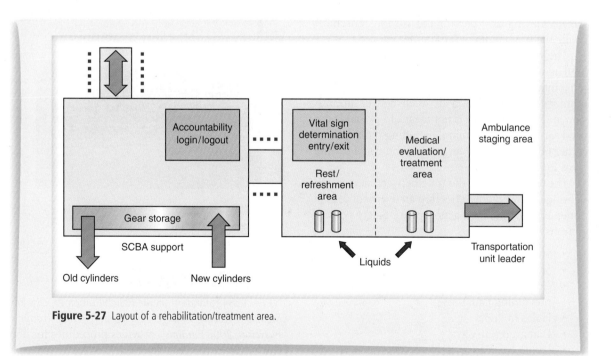

Figure 5-27 Layout of a rehabilitation/treatment area.

- Calorie and electrolyte replacement, as appropriate for longer-duration events
- Medical monitoring
- Member accountability
- Release

NFPA 1584 stipulates that rehabilitation should be provided after the second 30-minute cylinder, after one 45-minute or 60-minute cylinder, or after 40 minutes of strenuous work without an SCBA. The standard does allow these times to be adjusted depending on work or environmental conditions.

During temperature extremes, rehabilitation becomes a more critical issue, and longer rest periods may be needed. Hot weather rehabilitation could be outside using water mist and/or fans in a shaded area. Research regarding cooling techniques recommends active cooling techniques such as forearm immersion.[31] Cold weather requires that, at a minimum, blankets or other coverings be provided. A better option for hot or cold weather rehabilitation is an air-conditioned or heated area. This area can be a building, bus, or other climate-controlled facility.

An additional EMS unit, preferably an advanced life support (ALS) unit, should be assigned to rehabilitation when a formal rehabilitation is implemented to evaluate fire fighters entering and exiting rehabilitation Figure 5-28 . This unit's responsibility is to determine when members can return to action and to treat injuries and transport personnel when necessary. Rehabilitation protocols should be spelled out in department SOPs.

Before fire fighters are released from the rehabilitation area to return to active firefighting duties, their vital signs must be checked and be at safe levels. Vital signs to be checked can include pulse, respiration, and temperature. Consult the department physician when establishing reentry guidelines.

Maintaining company unity and the accountability system is important as members are cycled through rehabilitation. If a company member cannot be returned to duty, the accountability tracking system should be revised accordingly.

Figure 5-28 Rehabilitation area.
Courtesy of Captain Nick Morgan, St. Louis Fire Department.

WRAP-UP

Chapter Summary

- Fire fighter on-duty deaths have decreased dramatically in recent years.
- The number of fire fighter on-duty line-of-duty deaths per structure fire has not improved despite the reduction in fire fighter on-duty line-of-duty deaths.
- Minimum safety measures described in *NFPA 1500 and IAFC Rules of Engagement* should be familiar to all fire fighters and fire officers.
- NFPA 1500 requires a risk-versus-benefit analysis at all structure fires.
- Time from ignition to flashover can vary significantly, depending on a number of variables related to the compartment size, ventilation, and fuel load.
- The fuel load in modern residential occupancies tends to be much greater and faster burning than fuel loads in the past.
- Construction materials and methods affect structural stability under fire conditions.
- Green buildings present unique problems under fire conditions.
- Fire fighters must understand the dynamics associated with structural collapse and recognize the signs of impending structural failure.
- Fire entering concealed spaces can extend to remote locations and break out with a sudden increase in intensity.
- Fire extension can cut off the primary means of egress for fire fighters.
- The area between a suspended ceiling and the roof/ceiling is a common concealed space allowing fire to extend unnoticed.
- A safe collapse zone is equal to the height of the building, plus an allowance for debris scatter usually calculated as the distance equal to 1½ times the height of the fire building.
- Accountability for all persons entering the fire zone is essential.
- Structural stability decreases as time and fire intensity increase.
- Smaller compartments progress to flashover faster than larger compartments.
- Proper staffing requires that an adequate number of fire fighters be available to complete all necessary fire-ground tasks and to form a tactical reserve.
- Response to a mayday by the rapid intervention crew (RIC) must be orderly and organized. An emergency evacuation signal is crucial to warn of imminent collapse.
- Accountability should not be assigned to a safety officer; the safety officer must be mobile, while accountability is a stationary position.
- The safety officer monitors all areas where fire fighters are working.
- Safety is everyone's responsibility at all times at the fire scene, whether or not someone has been specifically assigned the role of safety officer.
- Fire fighters in doubt regarding the need to transmit a mayday should immediately declare a mayday.
- Removing any part of the turnout ensemble inside the fire building, even during overhaul, is an unsafe practice.
- On-scene rehabilitation is especially important in extremes of temperature, inclement weather, or extended incidents.

Flames: © Dax/Dreamstime.com; Steel texture: © Sharpshot/Dreamstime.com

Key Terms

20-minute rule A rule of thumb used for estimating the length of time until structural collapse occurs. It states that when a heavy volume of fire is burning out of control on two or more floors for 20 minutes or longer in a building of ordinary construction, structural collapse should be anticipated.

accountability system A system established on the fire ground to ensure that everyone entering the area has a specific assignment, to track all personnel at the scene, and to identify the location of any missing personnel if a catastrophic event should occur.

cold zone An area where personal protective clothing is not required; command post, rehabilitation, and medical treatment should be located in the cold zone.

collapse zone The area endangered by a potential building collapse; generally considered to be an area 1½ times the height of the involved building.

decay phase The phase of fire development in which the fire has consumed either the available fuel or oxygen to a point that the fire begins to diminish in intensity.

extension Fire that moves into areas not originally involved, including walls, ceilings, and attic spaces; also, the movement of fire into uninvolved areas of a structure.

fire perimeter A wide area beyond the hazard control zones, usually staffed by police to keep unauthorized people away from the scene.

frequency As related to fire fighter injuries, a measure of how often an injury occurs; for example, sprains and strains are the most frequent fire fighter injuries.

fully developed phase The phase of fire development at which the fire is free-burning and consuming much of the fuel.

growth phase The phase of fire development at which the fire is spreading beyond the point of origin and beginning to involve other fuels in the immediate fire area.

hazard control zone The area in which emergency responders are working; it can be subdivided into no-entry, hot, warm, and cold zones.

hot zone An operating area considered safe only for individuals wearing appropriate levels of personal protective clothing; established by the IC and safety officer.

ignition phase The phase of fire development at which the fire is limited to the immediate point of origin.

no-entry zone An area that is unsafe regardless of the level of personal protective equipment and that must be cleared of all personnel, including emergency response personnel.

overhaul Examination of all areas of the building and contents involved in a fire to ensure that the fire is completely extinguished.

prefire condition Factors that can contribute to a collapse in a building that is heavily involved in fire; these factors include construction types, weight, fuel loads, damage, renovations, deterioration, support systems, and related factors such as lightweight truss ceilings and floors.

pyrolysis The chemical decomposition of a compound into one or more other substances by heat alone; the process of heating solid materials until combustible vapors are emitted.

Flames: Dn/Dreamstime.com; Steel texture © Sharpshot/Dreamstime.com

Flames: Drx/Dreamstime.com; Steel texture: © Sharpshot/Dreamstime.com

rehabilitation The process of providing rest, rehydration, nourishment, and medical evaluation to members involved in extended incident scene operations and/or extreme weather conditions.

severity The extent of an injury's consequence, usually categorized as death, permanent disability, temporary disability, and minor.

warm zone An intermediate area between the hot and cold zones where personal protective equipment is required, but at a lower level than in the hot zone.

Challenging Questions and Activities

1. Use the most recent NFPA fire fighter fatality report to determine the type of duty as well as the nature and cause of injury most likely to result in an on-duty fire fighter fatality.

2. Use the most recent NFPA fire fighter injury report to determine the most common cause of injury. Using a scale of 1 to 10, where 10 is most severe, estimate the probable severity for various injury types (e.g., heart attack on the fire ground = 10).

3. Use a fire report from NFPA, NIOSH, USFA, or other source to evaluate the risk to fire fighters conducting an offensive attack. Classify the incident using the four risk/benefit categories from NFPA 1500-8.3.2 (listed earlier in this chapter).

4. Assume the role of incident safety officer for the fire shown in **Figure 5-29**. It is 4:00 PM. Weather conditions are 95°F (35°C) with high humidity and calm winds. The operation began as an offensive attack at 2:00 PM. An emergency evacuation was ordered at 2:45 PM. A PAR was conducted with all fire personnel accounted for at that time. The collapse shown in Figure 5-29 has just occurred. List your concerns and actions you would recommend to the IC.

Figure 5-29 Partial collapse of a large three-story commercial building.

5. Use the following scenarios to analyze and discuss the probability of the building described being occupied at the time of the alarm. Express the probability of occupancy as a percentage, where 0% means there is no chance anyone is inside the building and 100% indicates that the building is definitely

occupied by at least one person. Explain your answer.

A. Single-family residential building at 2:00 AM, Sunday, July 10, clear and mild weather conditions

B. Abandoned five-story building in the center of the city at 2:00 AM, Sunday, December 10, very cold and windy weather conditions

C. High-rise hotel in Michigan vacation resort area at 2:00 PM, Saturday, August 1, sunny and warm weather conditions

D. Nursing home at 1:00 PM, Wednesday, July 13, sunny and warm weather conditions

E. High school, 1:00 PM, Monday, October 2, mild and rainy weather conditions

F. High school, 1:00 PM, Monday, July 11, hot and rainy weather conditions

6. Use the scenarios in Question 5 to compute the staffing needed to safely and effectively conduct an offensive attack.

Endnotes

1. National Fire Protection Association. *NFPA 1500: Standard on Fire Department Occupational Safety and Health Program.* Copyright © 2013, National Fire Protection Association. This reprinted material is not the complete official position of the NFPA on the referenced subject, which is represented only by the standard in its entirety.

2. Code of Federal Regulations, 49 CFR 1910. Washington, DC: U.S. Government, April 18, 1998.

3. Karter, M. J., Jr. *Fire Loss in the United States During 2010.* Quincy, MA: National Fire Protection Association. September 2012.

4. Karter, M. J., Molis, J. Fire fighter injuries for 2012. *NFPA Journal.* November/December 2013:60–67.

5. Heinrich, H. W. *Industrial Accident Prevention* (4th ed.). New York, NY: McGraw-Hill; 1959.

6. International Association of Fire Chiefs. Rules of engagement for firefighter survival: Don't risk for what can't be saved; extend limited risk to save what can. *IAFC On Scene.* June 1, 2011.

7. Kerber, S., UL Firefighter Safety Research Institute. *Impact of Ventilation on Fire Behavior in Legacy and Contemporary Residential Construction.* Northbrook, IL: Underwriters Laboratories; 2010.

8. Kerber, S., UL Firefighter Safety Research Institute. *Impact of Ventilation on Fire Behavior in Legacy and Contemporary Residential Construction.* Northbrook, IL: Underwriters Laboratories; 2010.

9. National Institute for Occupational Safety and Health. *NIOSH Alert Preventing Injury and Deaths of Fire Fighters Due to Truss System Failures.* Atlanta, GA: 2005.

10. The Fire Protection Research Foundation. *Property Insurance Research Group Forum on PV Panel Fire Risk.* June 8, 2014.

11. Kerber, Stephen, Underwriters Laboratories. *Fire Service Collapse Hazard Floor Furnace Experiments.* Northbrook, IL: Underwriters Laboratories, 2012.

12. Kerber, Stephen, Underwriters Laboratories. *Improving Fire Safety by Understanding the Fire Performance of Engineered Floor Systems and Providing the Fire Service with Information and Tactical Decision Making.* Northbrook, IL: Underwriters Laboratories, 2012.

13. Grundahl, K. National Engineered Light Weight Construction Fire Research Technical Report. Quincy, MA: National Fire Protection Research Foundation; 1992.

14. Madrzykowski, D., Vettori, R. L. *NIST Building and Fire Research Laboratory, NISTIR 6510, Simulation of the Dynamics of the Fire at 3146 Cherry Road NE, Washington DC.* May 30, 1999, April 2000.

15. Kerber, S., UL Firefighter Safety Research Institute. *Impact of Ventilation on Fire Behavior in Legacy and Contemporary Residential Construction.* Northbrook, IL: Underwriters Laboratories; 2010.

Flames: Dx/Dreamstime.com; Steel texture © Sharpshot/Dreamstime.com

16. Corbett, G. P., Brannigan, F. L. *Brannigan's Building Construction for the Fire Service* (5th ed.). Burlington, MA: Jones and Bartlett Learning; 2015.

17. National Fire Protection Association. *NFPA 1221: Installation, Maintenance, and Use of Emergency Services Communications Systems.* Quincy, MA: 2013.

18. National Fire Protection Association. *NFPA 1710: Standard for the Organization and Deployment of Fire Suppression Operations, Emergency Medical Operations, and Special Operations to the Public by Career Fire Departments.* Copyright © 2013, National Fire Protection Association. This reprinted material is not the complete official position of the NFPA on the referenced subject, which is represented only by the standard in its entirety.

19. Code of Federal Regulations, 49 CFR 1910.

20. National Fire Protection Association. *NFPA 1410: Standard on Training for Initial Emergency Scene Operations.* Quincy, MA: 2010.

21. National Institute of Standards and Technology. *Technical Note 1661, Report on Residential Fireground Field Experiments.* Gaithersburg, MD: 2010

22. National Institute for Occupational Safety and Health. NIOSH Fire Fighter Fatality Investigation F-2003-18. Partial roof collapse in commercial structure fire claims the lives of two career fire fighters—Tennessee. Retrieved from http://www.cdc.gov/niosh/fire/reports/face200318.html

23. Williams, J., Stambaugh, H. *USFA-TR-123, Rapid Intervention Teams and How to Avoid Needing Them, Special Report.* U.S. Fire Administration; Emmitsburg, MD: March 2003.

24. Kerber, S., UL Firefighter Safety Research Institute. *Impact of Ventilation on Fire Behavior in Legacy and Contemporary Residential Construction.* Northbrook, IL: Underwriters Laboratories; 2010

25. Kerber, S., UL Firefighter Safety Research Institute. *Study of the Effectiveness of Fire Service Vertical Ventilation and Suppression Tactics in Single Family Homes.* Northbrook, IL: Underwriters Laboratories; 2013.

26. National Fire Protection Association. *NFPA 1971: Standard on Protective Ensembles for Structural Fire Fighting and Proximity Fire Fighting.* Quincy, MA: 2013.

27. Mensch, A. E., Cajaty, G. G., Braga, B., Bryner, N. P. *Fire Exposures of Fire Fighter Self-Contained Breathing Apparatus Facepiece Lenses.* Gaithersburg, MD: National Institute of Standards and Technology; 2011.

28. Fabian, T., Underwriters Laboratories, University of Cincinnati. *Firefighter Exposure to Smoke Particulates.* Cincinnati, OH: 2010.

29. Anthony, T. R., Joggerst P., James L., Burgess, J. L., Leonard, S. S., Shogren, E. S. Method development study for APR cartridge evaluation in fire overhaul exposures. *Annals of Occupational Hygiene.* 2007;51:703–716.

30. National Fire Protection Association. *NFPA 1584: Standard on the Rehabilitation Process for Members During Emergency Operations and Training Exercises.* Quincy, MA: 2008.

31. Selkirk, G. A., McLellan, T. M., Wong, J. Active versus passive cooling during work in warm environments while wearing firefighting protective clothing. *Journal of Occupational and Environmental Hygiene.* 2004:521–531.

Flames: Dra/Dreamstime.com; Steel texture: © SharpshotDreamstime.com

Flames © Dzr/Dreamstime.com; Steel texture © SharpshotDreamstime.com; Chapter opener photo: TK, ©Fotogra FFP/Shutter Stock, Inc.

Life Safety

LEARNING OBJECTIVES

- Explain the relationship between life safety and extinguishment.
- Determine the applicability of the vent, entry, isolate, search (VEIS) technique and exterior rescue tactics.
- Discuss the positive and negative aspects of ventilation in regard to life safety.
- Evaluate ventilation options as they relate to fire location, and given a scenario with several vent options, select the best option.
- List and evaluate rescue options.
- List rescue priorities in terms of occupant proximity to fire.
- Define a mass-casualty incident.
- Describe conditions that affect life safety staffing requirements.
- Evaluate tactics at a fire scenario in which a large number of occupants are in need of rescue.
- Use a scenario to evaluate basic life safety operations.
- Use a scenario to select and describe proper ventilation techniques.
- Use a scenario to describe and apply rescue options.
- Use a scenario to evaluate priorities as they relate to occupant proximity to the fire.
- Use a scenario to estimate staffing requirements at a fire in a structure occupied by a large number of people.
- Use scenarios to select proper life safety tactics when confronted with visual clues or reports of trapped occupants.

Case Study

Coordinated extinguishment, search, and rescue save two lives.

Courtesy of David J. Jones, Cincinanti, Ohio.

On November 5, 2010, at 11:07 PM, a one-alarm response of two engine companies, three truck companies (one assigned as a rapid intervention crew), a heavy rescue company, and two district chiefs responded to a reported structure fire. The first fire unit on the scene reported a "working fire" in a two-story house with heavy smoke and fire visible on the first floor. Two occupants (grandfather and granddaughter) were reported trapped on the second floor. Engine companies attacked the fire on the first floor, while truck companies used ground ladders to gain access to the second floor. The truck companies encountered heavy smoke conditions but were able to locate the two occupants and rescue them via ground ladders. Both victims were resuscitated at the scene and then transported to the hospital; both survived.

1. When confronted with a working fire and occupants trapped above the fire, what are some of the rescue options available?
2. What is the best first action for an engine company when confronted with a working fire where occupants remain in the building?
3. How would search and rescue operations change in a large building compared to a one- or two-family residence?

Introduction

This chapter discusses the most important of all fire-ground functions—life safety. Life safety presents the ultimate challenge to the fire officer. Decisions of whether to control the fire, remove the victims, or both might seem obvious. However, during the heat of battle, the fire officer faces numerous high-priority challenges. Deciding how best to assign initial limited resources at the incident scene is difficult. Initial strategic decisions at the incident scene must be made quickly based on the situation and the resources available to the incident commander (IC). Understanding the interrelationship of various life safety tactics discussed in this chapter is essential for an IC. The direct rescue of occupants is not the only way to save lives in a structure fire. Fire suppression and ventilation are also used to accomplish this most important mission.

Although fire extinguishment is generally considered the second priority, it is an essential part of most rescue operations. Fire fighter safety is a critical component of the life safety priority. The protection provided by fire streams, combined with proper ventilation, is an important part of protecting fire fighters as well as the occupants.

> **NOTE**
>
> In the vast majority of cases, the best life safety tactic is extinguishing or controlling the fire.

Texture: © Eky Studio /Shutterstock, Inc; Steel: © Sharpshot/Dreamstime.com

Evaluating the Probability of Extinguishment

Determining the probability of extinguishment is a major factor in life safety decisions. If the fire can be quickly extinguished, the top life safety priority is using hose streams to attack the fire. Only on rare

3rd red photo background: © Billy Kenny/Shutterstock, Inc; 1968: © Crystal Craig/Dreamstime LLC; ShutterStock, Inc; Photo: © Crystal Craig/Dreamstime LLC occasions is it best to rescue visible occupants rather than controlling or extinguishing the fire.

Fallacy	Fact
Occupants who are visible at windows are in the greatest danger and are always the first rescue priority.	People who are visible at windows might or might not be in grave danger. Occupants who are unable to reach a window are likely to be in more danger.

Steel: © Sharpshot/Dreamstime.com

Upon arrival at the fire scene, the fire officer must take several factors into account. A common tactical error is prioritizing victims according to visibility. Occupants who are still inside the building may be in grave danger, and it may be best to rescue people who are at windows via the interior stairs *after* the fire is under control or extinguished.

In some cases, it is possible to apply a **defend-in-place** tactic if the fire can be quickly controlled. Electing to leave people in a burning building is a calculated risk that the IC must be prepared to make based on fire conditions, available resources, and the threat to victims.

A fire may require a long lead time to suppress or even be impossible to contain when large-volume flows are needed. (Rate-of-flow calculations are explained in the *Offensive Operations* chapter.) Large interior flow requirements from hand-held lines translate into larger staffing needs, which, in turn, may require more time to assemble the needed resources. Seldom will large flows be required in residential occupancies. At times, the size of the fire or lack of resources results in a situation in which fire control cannot be achieved. In these cases, it may be possible to rescue occupants remote from the fire. However, the IC must keep in mind that the life of the fire fighter is just as valuable as any other life, and continuing with a futile effort that places fire fighters in imminent danger is unacceptable.

IC Rule of Engagement for Fire Fighter Safety

If after completing the primary search, little or no progress toward fire control has been achieved, seriously consider a defensive strategy.

Reproduced from: © 2011, IAFC Safety, Health and Survival Section. All rights reserved.

Texture: © Eky Studio /Shutterstock, Inc; Steel: © Sharpshot/Dreamstime.com

If the fire cannot be extinguished immediately, building features such as fire doors and fire walls can aid in extending the time available for evacuation by providing a barrier between the victims and the fire. Hose lines can be used to protect egress routes. It is critical to ensure that sufficient personnel and resources are available to provide the required water flow. Opening a fire door to apply water below the required rate of flow could have severe consequences. The opening that is made to make entry will allow the fire and smoke to extend into other noninvolved areas at an accelerated rate.

It is essential to get a charged line into position to extinguish the fire, which is usually accomplished by the first-arriving engine company crew. The crew advancing the hose line will assist victims they come upon but cannot be expected to perform a complete primary search of the area surrounding the fire. Their primary responsibility is extinguishment.

It is also critical to search the floor or floors above the fire. The number of crews needed for search and rescue depends on the number of floors to be searched, the size of the building, fire intensity, smoke conditions, and occupant status.

A single crew can search and evacuate an entire floor in a single-family detached dwelling where the occupants are able to assist themselves. If occupants are missing or unable to make their own way to safety, it may well take an entire crew to rescue one individual plus an emergency medical services (EMS) crew to provide treatment and transportation once the victim is removed from the building.

A high-rise building may have many floors above and below the fire to be searched. Floor areas in these buildings are generally large. More crews will be needed when confronted with large areas and/or when several floors are to be searched. There is sometimes a tendency to assign multiple companies to search a floor before getting at least one company in each exposed area above the fire to determine evacuation, fire, and smoke conditions and evacuation needs. In most situations, it is best to get at least one company into all areas on the fire floor and above as soon as possible. Once all critical areas have been assigned, additional units should be assigned to each area as needed to complete search and rescue activities.

Only in the most extreme circumstances should the first-in engine crew do anything other than advance an attack line. Generally, more lives will be saved by controlling the fire than by performing ladder rescues or knocking on doors to evacuate. Often, the only hope for occupants in the immediate fire area is fire control, especially if the fire is between them and their way out. What does all of this mean? The first-in engine crew *should not* be included in the staffing needed for rescue and evacuation. If multiple lines are needed to confine the fire, add the staffing needed for additional fire lines to the total staffing requirement.

Rescue Versus Fire Attack

In rare circumstances, the best tactical choice for the first-arriving unit may be to perform interior or exterior rescues prior to attacking the fire. However, this approach would be the exception rather than the normal procedure. An immediate fire attack is not an option when an apparatus without pumping capabilities (e.g., a heavy rescue or ladder truck without pumps) is first on the scene. If occupants requiring rescue are visible, placing a ladder or other exterior rescue method may be the best first action.

One technique that is sometimes used when victims are known to be inside the building but are not visible from the exterior is the vent, entry, search (VES) tactic. Information gained during the various Underwriters Laboratories studies has led to a modification of VES to VEIS (vent, entry, isolate, search). Indications for a VEIS approach include the following:

- Strong evidence exists that a savable victim remains inside the building.
- The location of the victim is known.
- The area where the victim is located is directly accessible through an exterior opening such as a window.
- The victim cannot be safely rescued via the interior stairway.

The *vent* part of VEIS involves opening the window or other exterior access point. Interior *entry* is made. The area is then *isolated* by closing doors leading to the area being searched. The final phase is *searching* for and rescuing the missing occupant. Operating without hose line protection is precarious, especially on the fire floor or above. The risk to fire fighters and occupants is greatly increased when a vent opening is made to enter the search area.

If the fire is in the room being entered, both the victim and fire fighters performing the rescue are in greater danger; opening the window will greatly intensify a ventilation-controlled fire. When this is the case and isolation is not possible, fire fighters entering the building should reverse the venting process, by closing the access window after making entry. Whenever possible, a fire fighter should remain at the point of entry to assist in victim removal. This fire fighter can control the access point and aid interior fire fighters in escaping.

In the majority of cases, a standard fire attack coordinated with ventilation and rescue is the best strategy when an engine company or quint company is first on the scene. In the event that extinguishment is known to be beyond the capabilities of initial responders, a VEIS operation may be the best choice. When occupants are known to be inside the building but are not visible (e.g., not at windows or on patios), initiating interior rescues without the protection of a hose line is more hazardous than when a quick entry and search are possible. The deeper within the interior of the building and the closer to the fire, the more dangerous the VEIS operation becomes.

Rule of Engagement for Fire Fighter Survival

Extend vigilant and measured risk to protect and rescue savable lives.

Texture: © Eky Studio /Shutterstock, Inc; Steel: © Sharpshot/Dreamstime.com

Assessing the Ventilation Profile

Ventilation has a positive component and a negative component. If fire fighters are working in an enclosed area when a window breaks, any fire not isolated from the vent opening will gain in intensity, but the products of combustion will be vented out of the area. For this reason, ventilation must be coordinated with the placement of attack lines to control the fire. In most cases, ventilation makes the job of finding the fire and victims much easier. However, under certain conditions, improper venting could result in redirecting the flow path toward victims, accelerate fire growth, or produce a backdraft. The fire fighter making entry or occupants could be seriously injured or killed by the resultant rapid fire growth. Further, a vent opening made between fire fighters or victims and their path to egress could prove fatal. Any opening in, or directly connected to, the fire area will provide ventilation (e.g., opening a door to gain entry to the interior of the building). Multiple openings often result in air flowing into one opening and smoke and fire exiting the other vent. As an example, if wind is blowing against the rear of a building when high heat fractures rear windows, the flow path could be reversed. Fire fighters opening the front door would then provide a flow path, causing fire, heat, and smoke to exit the front door they are entering.

Ventilation is done to relieve the products of combustion, allowing fire fighters to advance on the fire. When venting to support extinguishment, it is important to coordinate hose line placement with ventilation. Close coordination of venting means that the hose line is ready to quickly overcome the increase in combustion that will likely occur. In many cases, water can be applied through an exterior opening for a short time prior to making entry, thus reducing the heat being produced by the fire. Limiting the door opening reduces air entering or exiting through the door; applying water through the door often reduces heat production in the interior, allowing a closer approach to the fire and buying time for occupants still inside the building. This tactic is particularly useful for small structures such as single-family residential properties.

Venting can be a very effective life safety procedure. When venting for life safety purposes, the principle is to pull the fire, heat, smoke, and toxic gases away from victims, stairs, and other egress routes.

A common life safety venting tactic involves opening a scuttle at the top of a stairway. Doing so can be an effective means of venting the stairs because the smoke will generally rise, due to the natural stack effect. However, there is a negative component to this method of vertical venting: If a door to the fire area is opened, the vent opening will pull the fire and smoke from the fire compartment into the stairs (see Figures 2-15 through 2-17 in the chapter titled *Procedures, Preincident Planning, and Size-Up*).

Vent openings made in the wrong location may pull fire, heat, and toxic gases into areas containing fire fighters and occupants. **Figure 6-1** is a plan view of a one-story warehouse building with overhead doors identified as A, B, C, D, and E. Roof vents are labeled 1, 2, 3, and 4. Given the location of the fire, the best possible vertical vent would be the roof vent directly over the fire (roof vent 1). However, the roof could be weakened in this area or the fire intensity below could cause injury when the roof is opened. Vent 2 would be an acceptable alternative, although

Figure 6-1 Plan view of a one-story warehouse.

it would spread the fire into unburned areas. Cutting open the roof in the area near the fire would also be acceptable, but cutting through the roof of this commercial building would cause additional damage and could be significantly more difficult than opening a roof in a residential building. Opening roof vents 3 or 4 would be unacceptable because they would redirect the flow path toward fire fighters operating the hose line. Ventilation through door C would be best if horizontal venting were chosen, with the understanding that opening this door might be hazardous due to the sudden release of fire through the door. Door D would be the next best horizontal opening. Doors A and B could be used, but opening either one would not be an ideal choice, as venting through these doors would spread the fire. Opening door E could have disastrous effects because it would redirect the flow path toward fire fighters advancing on the fire and through a large portion of the warehouse. When venting, consider wind direction and velocity, as well as the location of the fire.

Strategically placed hose lines can offset the negative effects of improper venting, but creating a flow path that results in fire, heat, and smoke flowing toward occupants and fire crews is *never* a good tactic. At times, particularly when a large rate of flow is required, the hose line will not be able to stop the forward progress of an improperly vented fire.

Building features can retard fire growth and prevent extension into occupied areas if they remain closed. For this reason, it is usually best to avoid opening doors and windows in the fire area before charged hose lines are in place.

Fire fighters may break windows as they conduct their primary search. This venting tactic is often necessary; however, when possible, it is better to open the windows rather than breaking them. Reversible venting is the preferred method, as once a window or door is shattered, the vent opening cannot be easily closed. If the windows or door are opened so that they can again be tightly closed, unintended negative consequences caused by the vent opening can be reversed by closing the window or door.

In making openings for ventilation, it is important to consider the fact that the fire and heated gases will naturally move upward, horizontally, and then downward. Flow paths will be from low pressure to high pressure and from hot to cold.

Positive-pressure ventilation has become popular and is a useful tool. The key to successful positive-pressure venting is to control outlet openings. If too many doors and windows are opened, positive-pressure venting will prove ineffective. As with all other ventilation methods, there is a negative side to positive-pressure venting: The fire can be directed toward victims, toward their escape routes, or into unburned areas.

Analyzing the Available Rescue Options

Occupants should be rescued by the safest, shortest, easiest, and most direct route. Victims who are in danger above or below grade level should be evacuated by using the interior stairways whenever possible.

A defend-in-place concept is often used in occupancies such as high-rise buildings and health care facilities where occupants are moved away from the fire area or protected in place while remaining in the structure. Many large buildings are constructed in a manner that would allow a defend-in-place strategy. Some buildings, particularly high-rise structures, evacuate occupants sequentially to better manage travel in stairways. As an example, with a fire on the 15th floor of a 30-story building, occupants on the 14th, 15th, 16th, and 17th floors would be directed to evacuate the building or go to a lower floor in the building. All other occupants would remain in place. If the fire were not immediately controlled, other occupants would be directed to evacuate the building.

A defend-in-place or sequential evacuation requires the cooperation of building occupants. Many occupants of the World Trade Center fire on September 11, 2001, survived by acting contrary to the evacuation plan. People are aware of this incident and may be more likely to disobey evacuation instructions.

The decision to leave people inside a burning building is particularly perplexing on floors above the fire that are not yet threatened, especially if stairwell capacity is needed for evacuating more endangered occupants or for moving fire control forces and equipment to the fire area.

When deciding on an evacuation, the following question arises: Would the occupants be safer remaining in their present locations, which have not been invaded by the products of combustion, or would it be better to evacuate them through smoke-filled corridors and stairways?

NOTE

Making a decision to defend in place, when the fire is not yet under control, is a calculated risk taken by the IC with the lives of occupants.

Texture: © Eky Studio /Shutterstock, Inc; Steel: © Sharpshot/Dreamstime.com

The following sections cover some of the most common means of building evacuation from upper floors. They are listed in preferential order; for example, *well-maintained* fire escapes are generally a better rescue option than ground ladders. The overriding factors in determining the best means of evacuation are based on safety and efficiency. Large numbers of occupants can be directed down the interior stairs with little or no assistance from fire fighters. All of the other listed methods are more difficult, require more staffing, and are more hazardous. However, when fire or smoke enters the interior stairway, alternative egress may be required. Rescue via an aerial device is usually safer and requires fewer fire fighters than rescue using ground ladders.

Interior Stairs

Interior stairs are preferred over all other means of egress from upper floors when they are tenable. The safest and easiest way to move occupants is by using the stairs, which are intended for that purpose. Any other method is a less desirable alternative, and the use of alternative methods should be justified by special circumstances.

Flames © Jag_cz/ShutterStock, Inc.; Paper © silver-john/ShutterStock, Inc.

Incident Summary

In several fatal fires, victims were found in the corridors and stairways, but people who remained in their rooms were not injured. In a fatal high-rise fire in North York, Ontario, six victims were found in the stairways in the upper stories of a 29-story building. The fire occurred on the fifth floor. In the MGM Grand fire in Las Vegas, Nevada, almost 60% of the 85 fatalities in the high-rise tower were found in corridors, stairs, and elevators.

The door to the stairway where victims were found in the North York, Ontario fire.

© 1995, National Research Council Canada

Data from: Residential High Rise, North York, Ontario, NFPA Fire Investigation Report. Quincy, MA: NFPA.

The MGM Grand fire caused by an electrical short killed 83 people and injured hundreds more.

© AP Photos

Fire Escapes

Fire escapes are poor substitutes for interior stairs, and their structural integrity is sometimes questionable, especially when large numbers of people are attempting to use the fire escape. However, when properly maintained, they are the second best option. If the fire escape is structurally sound, it is preferred over methods other than the interior stairway. The *NFPA Building and Exits Code* (predecessor of *NFPA 101 Life Safety Code*) has not recognized fire escapes as an acceptable means of egress in new construction since 1927[1]. However, not every jurisdiction immediately prohibited fire escapes in new construction and many older buildings increased egress capacity by adding fire escapes to existing buildings.

In many cases, occupants choose to use the fire escape prior to the arrival of the fire department. Once occupants are on the fire escape, fire fighters must facilitate their movement to safety. Often, this involves assisting or directing people who are impeding movement on the fire escape.

Fire or smoke pushing out of window or door openings leading to the fire escape will slow or stop movement **Figure 6-2**. If fire fighters are inside the building, they may be able to close windows emitting smoke or close interior doors to keep smoke and fire from entering rooms with openings to the fire escape. In some cases, it may be necessary to apply water from the exterior to knock down the fire threatening the fire escape. In this case, every attempt should be made to advance on the fire and/or isolate the room. If fire impinging on the fire escape cannot be contained, it may be necessary to remove occupants from the fire escape via ladder or provide assistance to the interior stairway.

Drop ladders are typically difficult to climb or descend and sometimes will not release due to rust or paint. If the drop ladder or counterbalanced stair is in the up position, fire crews should attempt to pull it to the down position being cognizant that these ladders may drop suddenly. Placing ground ladders in position next to the fire escape at the second floor level provides an alternative or second means of escape from the fire escape. It is best to

Figure 6-2 Aerial ladder being placed in position to provide alternative egress from fire escape.

Courtesy of David J. Jones, Cincinnati, Ohio.

place the ground ladder on the side opposite the fire escape ladder.

Fire fighters can support self-evacuation via the fire escape by directing hose streams at visible fire threatening the fire escape, by providing ladders at strategic locations, and by directing or assisting people who are impeding movement.

Aerial Ladders and Elevated Platforms

Following interior stairs and fire escapes, the next rescue preference is for aerial ladders or elevated platforms. As noted previously, aerial ladders and platforms can be used in conjunction with evacuations via the fire escape. There are two schools of thought when using aerial devices for rescue. Aerials can move larger numbers of people from a single area in a shorter period of time. However, elevated platforms provide a more secure, less stressful rescue for the occupant, and articulated platforms can reach areas that an aerial would have difficulty reaching. Fire departments should survey hazard areas before purchasing apparatus, matching apparatus to identified needs.

Ground Ladders

Ground ladders follow aerials and elevated platforms on the rescue preference list. Ground ladders are generally less stable, do not have the reach of most aerial ladders, and require more personnel to place in position. However, in certain circumstances, the use of aerial devices may not be practical, such as when buildings are set back from the street or there is no access road to the side of the fire where occupants are in need of rescue. Electric wires and other obstructions may also impede rescues via aerial devices. When two or three fire fighters can quickly access the second or third floor, ground ladders may be the best option. As noted under fire escapes, ground ladders can also provide an alternative pathway to the ground level from a fire escape.

When fire and/or smoke are between the victim, the interior stairs, and other building egress facilities, such as fire escapes, aerial ladders, elevated platforms, and ground ladders are the best alternative when a quick knockdown is not possible or when extinguishment tactics further endanger occupants. Again, remember that if the stairways and corridors leading to the stairways are relatively clear of smoke, use the stairs.

When the fire threat seems real to the occupants, they will sometimes go to a window or other visible position to await rescue. They may or may not have checked the stairs before deciding to wait at the window. Do not attempt a ladder rescue just because people are at the window. Reassure them that you are there to help, but determine first whether they must be moved and second whether stairways are available. Whenever possible, it is best to send fire fighters

into the building to check the stairways and evaluate the need for ground ladder rescues. Fire fighters on the interior are in the best position to determine the need for evacuation and the best rescue tactics and to evaluate which occupants are in greatest danger.

Raising and positioning ground ladders may require several fire fighters. A ground ladder rescue will involve at least one fire fighter assisting the occupant and probably another trying to assist the person onto the ladder with a third fire fighter at the foot of the ladder.

Elevator Rescues

An interesting question arises regarding the use of elevators for rescue. Fire safety professionals warn against occupants using the elevators as a means of escape; however, they could be safely used when under the fire department's control. Under certain conditions, the use of elevators for evacuation is justified in buildings that are subdivided with good fire-resistive construction. If an elevator is remote and separated from the fire area with an auxiliary power supply, using it for rescue purposes, especially to evacuate immobile occupants, may be the best tactic. The use of elevators in the immediate fire area is hazardous to everyone, including fire fighters. Engineers and code officials are examining the possibility of improving elevator reliability and permitting occupants to use elevators for evacuation during a fire emergency. These systems are still in the exploratory phase. When code officials approve

NOTE

Rescues and evacuations via interior stairs, fire escapes, aerial apparatus, and ground ladders are considered acceptable, and even the preferred means of egress under certain circumstances. Elevators, scaling ladders, ropes, helicopters, and air bags are all questionable rescue methods that are potentially hazardous.

Texture: © Eky Studio /Shutterstock, Inc; Steel: © Sharpshot/Dreamstime.com

the use of elevators during a fire emergency, the local fire department should be part of the decision-making process and include this information in the preincident plan. The *High-Rise Buildings* chapter further discusses the safe use of elevators.

Rope Rescues

The use of ropes for rescue purposes can be justified only in extreme cases in which the victims are beyond the reach of aerials and elevated platforms. Before resorting to this last-option rescue technique, fire fighters should make every effort to perform an interior rescue. Rope rescues are extremely slow and dangerous and require specialized equipment and expertise. There is an added hazard to the rescuer and the victim because of the inherent danger of technical rescue operations. This rescue method is a tactic of near last resort.

Helicopter Rescues

Before deciding on rooftop rescues, the IC should ask whether the occupants who have reached the roof are safe. In some cases, it is possible for occupants to wait out the fire on the roof. Most roofs provide a difficult operating platform for helicopters, owing to obstructions that include roof top facades, antennas, satellite dishes, and ventilation equipment Figure 6-3 . These obstructions make it impossible for the helicopter to land on the roof. Even if the roof is clear of obstructions, it may not be a good landing platform because of the hot air currents caused by an intense fire.

Because of the limited number of people who can be safely carried in a helicopter, the choice of who will be rescued and who will stay behind must also be made. Victims may think they are in immediate danger and believe they cannot survive long enough to wait for another trip. They may attempt to climb into the helicopter, overloading it beyond its safe operating capacity. The helicopter has been used to advantage in some rescue situations, but its usefulness has been overstated. At many fires, dramatic helicopter rescues were unnecessary.

Some departments use helicopters to place fire fighters on the roof. These fire fighters then assist and reassure occupants on the roof or descend the stairway to perform rescues. They could also use the standpipe for extinguishment. Helicopters can provide quick and less fatiguing access to upper floors in a high-rise building. However, fire fighters typically use the path of ingress as their egress. A helicopter that drops fire fighters onto the roof may not be able to return later and remove them from a position above the fire floor. If your department anticipates the use of helicopters, continuous training of both fire personnel and helicopter crews is essential. Most private helicopter owners will resist getting involved in rescue work because they lack rescue equipment, training, and insurance coverage for performing rescues.

Figure 6-3 Rooftop obstructions.

Courtesy of Bernard J. Klaene.

Classifying Evacuation Status

During the early stages of an operation, the IC is forced to make staffing estimates based on incomplete or even inaccurate information. As companies are deployed, crews will give status reports, request assistance, or possibly release fire fighters for other assignments. Information becomes more complete and reliable as fire crews provide reconnaissance after entering endangered areas.

Fires with a significant life safety challenge reinforce the need to build a command structure from the very start of the incident so that a systematic search-and-rescue operation can be made without duplication of effort. When searching a large building with many rooms, a marking system is highly recommended. Many departments use plastic door hangers or other marking procedures to indicate rooms that have been searched. Even a simple chalk-marking system can be effective. For example, use chalk or another medium to mark an "X" on the door leading to an enclosed area to indicate the area has been searched. A "half X" denotes that the search is underway in that compartment. An "X" on a door leading to a floor or opposite the elevator indicates that the entire floor has been searched **Figure 6-4** . Whatever method is chosen, fire fighters must know the system and have the necessary materials immediately available to carry out the marking system. Adopting a marking system as an SOP is highly recommended.

Occupants who escape on their own may need direction, emergency medical treatment, a place of safe refuge, or other assistance. The care of these victims is usually a nonemergency function that can be delayed if necessary until personnel have verified that no one is in immediate danger.

Other victims may not be aware of the fire or of egress routes. They may require direction to safety, or fire fighters may need to provide a safe egress route for them. These victims will then be able to escape with little assistance. It may not be necessary to escort them all the way to the outside or, for that matter, to a place of safe refuge. When the number of available emergency personnel is limited, these victims should be provided direction and permitted to evacuate on their own whenever possible, allowing available fire fighters to perform the more crucial search and rescue tasks.

Sometimes it is possible to designate specific stairways as evacuation routes and other stairways for fire operations. This is a good tactic because fire fighters advancing hose lines through the stairs to the fire area will allow smoke to enter the stairs through the partially open door and impede egress because of the hose laying on the stairs. Six occupants of the Cook County Administration Building in Chicago perished while attempting to exit via the stairway being used by fire fighters to attack the fire. The Cook County Administration Building fire is discussed further in the *High-Rise Buildings* chapter.

A practical problem occurs in attempting to channel the flow of occupants into a designated stairway. In most cases, fire forces will not be in total control of the evacuation until they reach each floor to direct escaping occupants to the stairway of choice. There may be confusion regarding which stairway is to be used unless doors leading to stairways are plainly marked (e.g., stairway "A"). Simply designating the stairway is not enough; specific instructions must be given to both fire fighters and occupants. Some large buildings have voice annunciation systems or public address systems that permit the IC to relay messages to evacuees.

Figure 6-4 Door marking system.

Incident Summary

A fire occurred in a high-rise building with elderly occupants in St. Louis, Missouri. An attempt was made to attack the fire from both stairwells, but because of the number of elderly people coming down one of the stairwells, this approach was not possible. Whenever the door to the fire floor was opened, the smoke and heat entered the stairway, endangering the occupants. Once the flow of residents was stopped, it was possible to enter the floor from the second stairway to fight the fire and rescue an injured fire fighter. However, this operation was delayed because of the danger in which the occupants were placed whenever the door was opened.

Data from: Edward R. Comeau, personal correspondence following the investigation.

Immobile or unconscious victims will place additional requirements on the rescuers. In most cases these victims must be physically removed to the outside or to a place of safe refuge. This effort usually requires a minimum of two fire fighters (often more) to rescue one victim.

A commonly held fallacy is that the occupants will be able to provide rescuers with reliable information concerning occupant status and that the IC will be able to account for the building's occupants.

Fallacy	Fact
It is possible to account for all the occupants of the building without actually conducting a thorough and complete primary search.	The only way to be sure the building is completely evacuated is to conduct a primary search throughout the entire structure.

Steel: © Sharpshot/Dreamstime.com

The *only* reliable way to verify evacuation status and fire conditions is for fire fighters to enter the structure and systematically check every room. Occupancies such as primary schools and health care facilities conduct regular fire drills, and staff should have control over students and patients. But even in these occupancies, the staff may not account for visitors in the building. Some industrial settings have evacuation plans that include accountability. However, the mobility of employees within the facility may make these plans less effective than would be possible in a school or health care facility where the location of occupants is known. In the typical apartment building, some of the occupants may not be home when the fire occurs, and others may have escaped and then left the scene.

Whenever possible, the first-arriving fire company should conduct a 360-degree survey of the fire building's exterior, then take action to save the largest number of occupants with the least risk to fire crews. If the companies arriving prior to the IC could not obtain a full assessment of the exterior, the IC should personally conduct a complete exterior survey of the fire building or assign someone to recon the building exterior to determine the safest and most effective tactics.

Rule of Engagement for Fire Fighter Survival

Size up your tactical area of operation.

Texture: © Eky Studio /Shutterstock, Inc; Steel: © Sharpshot/Dreamstime.com

Incident Summary

Flames: © Jag_cz/ShutterStock, Inc.; Paper: © silver-john/ShutterStock, Inc.

The fire chief in Chapel Hill, North Carolina, faced problems accounting for occupants when a fraternity building caught fire. The fire occurred on a Sunday morning that was also graduation day. A number of the residents had already moved out for the semester, but no one was sure who or how many or whether any visitors were in the building. It was originally feared that as many as 25 people were inside the building. Ultimately, it was determined that five people had been trapped and killed by the fire.

Data from: Michael Isner, Fraternity Fire, Chapel Hill, North Carolina, NFPA Fire Investigation Report. Quincy, MA: NFPA.

Fire fighters entering the fraternity building in Chapel Hill, North Carolina.

© The News & Observer

IC Rule of Engagement for Fire Fighter Safety

Rapidly conduct, or obtain, a 360-degree situational size-up of the incident.

Reproduced from: © 2011, IAFC Safety, Health and Survival Section. All rights reserved.

Texture: © Eky Studio /Shutterstock, Inc; Steel: © Sharpshot/Dreamstime.com

Reports from occupants may be unreliable, but they should not be ignored. In many cases, occupants will be able to specifically direct rescuers to the location of a victim.

Flashover is the critical landmark. Before flashover, rescue is possible; after flashover, rescue is highly improbable within the flashover compartment. Recognizing the fire stage is important to the life safety effort. In the very early stages of a fire (ignition and early growth phases), occupants tend to rescue themselves once they are aware of the fire. Later, as the fire becomes larger and produces greater quantities of smoke and toxic gases, occupants may be conscious but unable to exit until the fire is extinguished. The most difficult victim to rescue is one who has been felled by the smoke, heat, or gases in an unknown location.

Fire conditions are related to many factors, such as the time that elapsed before the fire was detected, the time it takes to notify the fire department, and the response time of the units. If a fire is extending through a structure, the threat to occupants is directly related to their awareness of the fire, their ability to escape, the construction of the building, and the provisions that have been made for egress. Once a fire is in progress, the IC has little control over these factors.

IC Rule of Engagement for Fire Fighter Safety

Determine the occupant survival profile.

Reproduced from: © 2011, IAFC Safety, Health and Survival Section. All rights reserved.

Texture: © Eky Studio /Shutterstock, Inc; Steel: © Sharpshot/Dreamstime.com

Critical time (the time available until the structure becomes untenable) will vary depending on numerous building factors, as well as the building's fuel load. No exact figures can be given, but experience shows that fires spread rapidly upward in balloon-frame construction or, for that matter, in any construction that allows vertical fire spread to occur unimpeded. The time from ignition to full involvement in a balloon-frame structure will be very short in comparison to a fire in a fire-resistive building. The fire-resistive building allows fire fighters or occupants to confine the fire by using built-in features such as fire doors. An enclosed fire-resistive area can quickly become untenable because the products of combustion, including heat, are not relieved as they would be in a less compartmentalized structure.

The Underwriters Laboratories study of modern fuel loads in homes indicates that fire progression to flashover will occur much more rapidly than in the past.[2] If the IC determines that time does not permit the rescue of all the occupants, victim triage should be established to rescue as many victims as possible. The importance of fire control to life safety cannot be overstated. Buying time with fire control tactics is a critical strategy that must be implemented whenever possible. If the fire is extinguished, or at least diminished, fire fighters and victims within the building have a much better chance to evacuate safely.

Estimating the Number of People Needing Assistance

Once the IC has considered the preincident plan, visual, and reconnaissance information, he or she should be able to evaluate life safety requirements in terms of staffing and equipment.

Fire fighters assigned to search and rescue tasks should be equipped with forcible entry tools. Heavy metal doors with locks will extend the time and effort required for the primary search. In forcibly entering an interior area, it is important to consider the value of the opening being forced. A fire door or even a nonrated door will act as a barrier to fire intrusion. A fire door in particular can contain a fire for a significant period of time or keep fire from extending to an exposed area. If the fire extends to threaten the hallway while the search is being conducted, the door can isolate fire fighters and victims from the fire. Whenever possible, doors should be closed after the search is complete to limit fire spread and damage.

The fire attack crew should have hose lines in place and charged before forcing the door to the room or area of origin. In buildings with strong compartmentalization, conditions beyond the door are unknown, and extreme caution must be exercised when making entry. If the door is hot to the touch or a thermal imaging camera (TIC) indicates high temperature and evacuation is not complete, it may be best for the crews to stand ready at the door and allow the door to contain the fire while other crews complete the evacuation.

By making return trips to the fire area, fire fighters can rescue more than one person even in situations where carry-and-drag rescues are necessary; however, this is very demanding work and will quickly exhaust the available rescuers. In sizing up the personnel requirements, estimate the physical condition of the occupants, the number of occupants on the fire floor, and the travel distance to safety, as well as fire growth. Be proactive in requesting additional companies when faced with a potential large-scale rescue effort.

After estimating the numbers of victims requiring rescue, consider the method of rescue and the victims' condition. Immobile or unconscious victims will require at least one rescuer for each victim or, more typically, two rescuers per victim.

Few, if any, fire departments could successfully evacuate a high-rise building with 1000 or more occupants in need of assistance. Further, if all of the occupants of a high-rise building are placed in the exit facilities at the same time, more people might be placed in danger. Trade-offs must be made. An evacuation must be prioritized by removing the most endangered occupants first. Meanwhile, fire fighters can plan ahead to move potential victims in less immediate danger before they are felled by the products of combustion, allowing rescuers to conserve strength by directing or assisting victims to a safe area while they are still mobile.

Surveying Floor Layout and Size

The size of the area to be searched and evacuated and the location of stairways, halls, and fire escapes are key factors in organizing the search effort. Quite often, the primary and secondary searches are done by assigning floors to each search team. Large buildings may be further subdivided by wing or building side.

Many buildings have mazelike designs, like the one for the Beverly Hills Supper Club **Figure 6-5**. This is particularly true of buildings with multiple additions or office buildings that use cubicles for workspaces. In some cases, building additions result in confusing egress paths.

Whenever possible, fire fighters working in multistory buildings should survey an uninvolved floor

before ascending to the fire floor or floors above. The layout may be different from floor to floor, but some building features tend to remain constant, such as stairs, elevators, and standpipes. However, the general layout of the space can be different on each floor. One floor may be wide open, while the next is subdivided into small offices or suites.

Another problem is consistency in floor numbering. Buildings that are constructed on a grade may have unusual floor numbers. Sometimes what appears to be the first floor of a building from one side is actually identified as a higher or lower floor. For example, the first floor main entrance to Good Samaritan Hospital in Cincinnati is the sixth floor. The terrain slopes away from the front of this hospital, and a section that was added to the building has a grade-level entrance six floors below the main entrance. To avoid mixing floor numbers at the same level, the entire hospital floor numbering system was changed to be consistent with the new addition. This same situation can exist in single-family detached dwellings that are built on a grade. In high-rise buildings, there may be more than one set of windows on the first floor. The mezzanine might not be counted as a floor, and these buildings

Figure 6-5 Floor plan of the Beverly Hills Supper Club in Southgate, Kentucky, where 165 people died in a 1977 fire.

sometimes do not have a thirteenth floor. Each of these factors adds to the confusion. The grade-level sixth floor at Good Samaritan Hospital or the absence of a thirteenth floor can be known in advance through preincident planning.

In smaller buildings, it is often possible to get a good idea of the interior layout from the exterior of the building. Large windows indicate areas that are most likely common areas (living or family rooms), whereas bedrooms tend to have smaller windows. Some buildings have windows at stairway landings that can be seen between floor levels, thus indicating the location of the stairway. Buildings of similar construction in an area will typically have similar floor layouts. For instance, the locations of stairs to the second floor of a single-family detached dwelling tend to be near the main entry.

Even a small building that has been renovated or one that is at various grade levels on different sides can be difficult to search. These buildings can be deadly to the fire fighters who go inside to conduct offensive operations.

Larger buildings, particularly large open buildings, present an additional hazard to fire fighters and occupants when smoke conditions obscure vision. Most fire fighters have experience in home and apartment building fires, where most fires occur. The same tactics and procedures used in small, compartmented buildings could be fatal when used in a large open structure with low visibility. Searching a small room using a right- or left-hand search can be safe and effective. A larger open area would require the use of a rope or other guideline and different search techniques. A TIC can be used to quickly scan a small room prior to entry. Larger rooms may require the fire fighters conducting the search to advance with the TIC as they check around obstacles and advance throughout the compartment. A very large room may require substantial escape time—more than the time available when the low pressure alarm on the SCBA sounds. The person in charge of the search may need to assign someone as timekeeper. New technology also allows the accountability officer to monitor air supply from a remote location. As one crew depletes their air supply, a second crew will need to move forward to continue the search. Large compartments can produce large overwhelming fires that can overrun a hose crew.

Rule of Engagement for Fire Fighter Survival

Maintain continuous awareness of your air supply, situation, location, and fire conditions.

Reproduced from: © 2011, IAFC Safety, Health and Survival Section. All rights reserved.

Texture: © Eky Studio /Shutterstock, Inc; Steel: © Sharpshot/Dreamstime.com

Incident Summary

At two fires—one in Seattle, Washington, and another in Pittsburgh, Pennsylvania—a total of seven fire fighters were killed in buildings that were built on sloping grades. Because of the confusing topography, it was not clear to everyone on the fire ground how many floors the buildings had or the exact location of the crews operating inside of the buildings in relation to the fire.

Data from: Edward R. Comeau, Warehouse Fire, Seattle, Washington, NFPA Fire Investigation Report. Quincy, MA: NFPA. Michael Isner, Residential Fire, Pittsburgh, Pennsylvania, NFPA Investigation Report. Quincy, MA: NFPA.

Flames: © Jag_cz/ShutterStock, Inc.; Paper: © silver-john/ShutterStock, Inc.

Incident Summary

A situation similar to the tragedies in Seattle and Pittsburg occurred in a California fire in a residence that appeared to be two stories at the front, but had a basement and sub-basement that were not visible from the front of the building. The fire originated in the basement. Two fire fighters attacking the fire at the top of the basement stairway perished.

Data from: Fire Fighter Fatality Investigation Reports. NIOSH 2011-13 A career lieutenant and fire fighter/paramedic die in a hillside residential house fire—California

Side C showing basement fire.

Reproduced from: Fire Fighter Fatality Investigation Reports, NIOSH 2011-13

Side view of building showing basement and sub-basement.

Reproduced from: Fire Fighter Fatality Investigation Reports, NIOSH 2011-13

Flames: © Jag_cz/ShutterStock, Inc.; Paper: © silverjohn/ShutterStock, Inc.

Prioritizing Rescues by Location/Proximity

After determining the total number of occupants and developing a strategy for a full or partial evacuation, establish the search and rescue priority; this is determined by deciding who is in the greatest danger. Search and rescue is then prioritized on the basis of rescuing those in the greatest danger first. The priority list is as follows:

1. People on the fire floor nearest the immediate fire area
2. People in proximity to the fire area on the same level as the fire
3. People on the floor above the fire, especially immediately over the fire area
4. People on the top floor (unless fire conditions result in smoke stratification, as discussed in the *High-Rise Buildings* chapter)

5. People on the floors between the floor above the fire and the top floor
6. People on the floors below the fire
7. People in nearby buildings
8. People outside (in the collapse or falling glass zones)

Number 8 could be placed anywhere in the priority list, depending on the structural integrity of the building, falling glass, and so on.

Figure 6-6 shows rescue priorities. Those in the greatest danger are rescued first. An exception to this priority list would be when it is not possible to save everyone. In these situations, the IC should opt to save the largest number of people possible. Doing so will require committing resources to areas where the largest number of occupants can be rescued. Another possible exception exists if one or more of the floors are known to be unoccupied. For example, given a fire at midnight in the office building

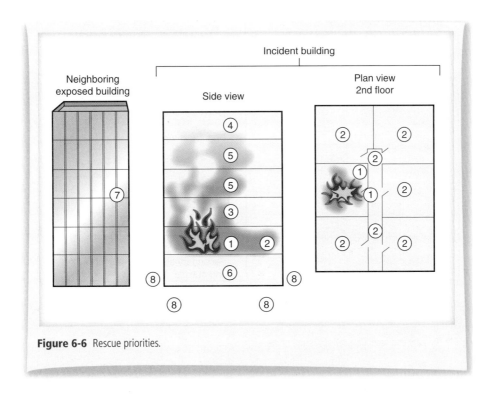

Figure 6-6 Rescue priorities.

shown in Figure 6-6, the only occupants would be security personnel on the first floor and a night club on the top floor. Even in this case, the first-due engine company should be assigned to attack the second floor fire.

The key to successful search operations is to be *systematic*. In most cases, the IC should follow the priorities listed above in assigning primary search areas. Crews should use a consistent method of marking and recording areas that have been searched so that all threatened areas are checked. The primary search is a quick but thorough search of the area. The secondary search ensures that no one was missed the first time through. If conditions and resources permit, a secondary search should be conducted as soon as the primary search is complete.

To ensure complete coverage during the secondary search, good practice dictates using a different crew. For example, if the first truck crew conducted the primary search of the fire floor and the second truck crew checked the floor above during the primary search, the secondary search assignments should be reversed (with the second truck crew assigned to the fire floor and the first truck crew to the floor above). Some fires require multiple search assignments, making this reassignment more complex. The officer in charge of the search and rescue group would manage the reassignment from primary to secondary search. **Figure 6-7** provides an example reassignment for the secondary search, with five companies being assigned to primary searches in different areas of the building and in an exposure building.

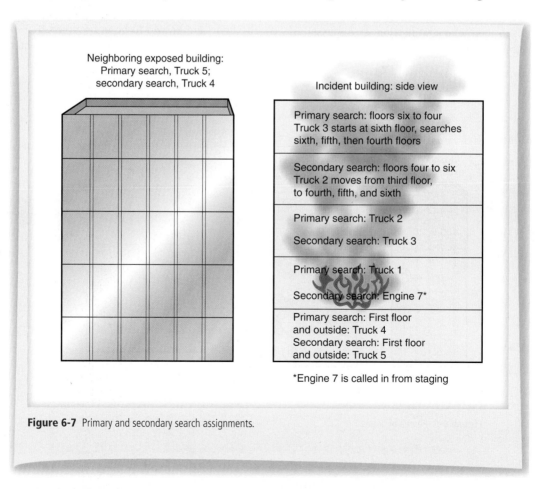

Neighboring exposed building:
Primary search, Truck 5;
secondary search, Truck 4

Incident building: side view

Primary search: floors six to four
Truck 3 starts at sixth floor, searches
sixth, fifth, then fourth floors

Secondary search: floors four to six
Truck 2 moves from third floor,
to fourth, fifth, and sixth

Primary search: Truck 2

Secondary search: Truck 3

Primary search: Truck 1

Secondary search: Engine 7*

Primary search: First floor
and outside: Truck 4
Secondary search: First floor
and outside: Truck 5

*Engine 7 is called in from staging

Figure 6-7 Primary and secondary search assignments.

The time and effort expended conducting the primary search dictate whether these same companies should be reassigned to secondary search or whether additional companies should be assigned to perform the secondary search. When necessary, additional secondary search companies can be summoned to the scene, but having a tactical reserve in staging ready for immediate assignment would be preferred.

The fire officer managing the search and rescue effort should assign specific areas of responsibility and then track these assignments and progress reports on a status board. When a large evacuation effort is required, the units involved in search and rescue will rely on communications and possibly a marking system, such as the one shown in Figure 6-4, to ensure that all areas are searched and to avoid duplication of effort. If at all possible, the search group should be assigned a separate radio channel or use telephones specifically designated for this purpose. It may also be possible to use the existing telephone system within the building, as long as it is not affected by the fire.

Evaluating the Medical Status of Victims

The IC must have wide discretion in calling medical assistance to the scene. Medical assistance can be requested to treat and transport injured fire fighters or occupants. The relative hazard, number of potential victims, and type of incident will dictate the need for EMS at the scene. A general rule for calling assistance applies here: If you think you need help, you do. It is much better to have medical assistance standing by at the scene than to have injured fire fighters and occupants waiting for treatment and transportation while EMS is responding.

For most fire incidents, medical units should be set up at one or more designated locations within the cold zone or beyond. Medical personnel should remain with their ambulance, ready to receive the injured. A fire in a multistory building may warrant moving medical personnel and equipment to a safe area inside the building. It may be a good idea to set up the medical triage/treatment area near the rehabilitation area. However, sufficient room should be available to physically separate the two areas.

Evaluating Victims in Mass-Casualty Incidents

Mass-casualty situations are the result of specific incidents, such as transportation accidents, hazardous materials releases, or natural disasters. Fires in buildings with large numbers of occupants have the potential to be mass-casualty incidents. Providing a system that includes triage, treatment, and transportation is a proven method of managing the medical component when large numbers of victims are encountered, regardless of the type of incident.

Mass casualty is defined in terms of department and community capabilities. A mass-casualty incident occurs when one or more of the following situations exist:

- The number of victims and the nature of their injuries make the normal level of stabilization and care unattainable.
- The number of immediately available trained personnel and transportation vehicles is insufficient.
- Hospital capabilities are insufficient to handle all of the victims requiring care.

Triage, Prioritizing, and Transport

Triage is the first medical priority in managing a medical disaster. The first-arriving EMS personnel should not leave the scene until they are relieved of triage responsibilities. Emergency medical teams are accustomed to treating one or more individuals, followed by immediate transport. Categorizing victims without treatment and transportation runs contrary to everyday operations in which emergency medical crews stabilize the injured person then immediately transport the victim to a hospital. The tendency to immediately treat and transport must be overcome by rigid enforcement of triage procedures.

When staffing permits, it is best to physically separate the dead from living patients. The walking wounded should be directed to another location for treatment. The walking wounded could be located in a separate area within the shelter being used for evacuated occupants.

Once triage teams have prioritized the victims based on their injuries, treatment teams will follow. Treatment teams first treat those needing immediate care (priority 1) and arrange for their transportation. After this, priority 2 and 3 patients can be treated. If sufficient personnel are on the scene, multiple teams can be formed to treat as many patients as possible. When available, paramedics or doctors should be assigned to treat the most seriously injured; EMTs and first responders can be assigned to take care of the remaining patients.

Providing transportation for victims is an important function of the medical transportation officer. A drive-through arrangement for ambulances will keep traffic lanes open.

At times, it is beneficial to use helicopters to transport patients. In using helicopters, the landing zones must be in safe areas, far enough away from treatment and triage areas that they will not interfere with those activities. Helicopters that are used to transport victims should not be flown over the fire area or command post. The noise and effect of the downdraft from the helicopter rotors can be extremely disruptive and can create a hazardous situation.

Staging and categorizing transport capabilities of medical units are essential in managing transportation needs. Incoming EMS transportation vehicles should be directed to an established staging area during mass-casualty operations and requested as needed. It may be possible to transport several priority 2 or 3 victims in the same vehicle. Priority 1 victims often require advanced life support (ALS) personnel on board the transport vehicle. Minor care (priority 2 or 3) victims are usually staged near the site for later treatment. Depending on the nature of the injuries, it may be possible to transport priority 2 and 3 patients in buses or other non-ambulance vehicles.

Communication with hospital facilities is necessary to determine how many patients and what types of injuries can be treated at each facility. Hospitals determine their limits based on personnel, staff expertise, and space. The hospitals must be prepared to exceed their normal patient capacity; however, they must recognize when they have reached their emergency capacity.

Patients should be directed to hospitals and trauma care centers according to hospital capabilities. The medical planning officer (similar to the incident planning officer, but assigned to the medical branch or medical group) or scribe should document where and when each patient is transported. The number and severity of injuries may require that only priority 1 patients be transported. Field hospitals can be used to treat priority 2 and 3 patients. Even with transportation available, it is generally not a good idea to flood hospitals with the walking wounded.

During normal operations, EMS personnel can treat individual fire fighters and occupants. When the number of potential victims exceeds normal capacity for individual treatment, the IC is well advised to staff a medical branch or medical group to coordinate medical activities. When the number and/or condition of the victims approach mass-casualty status, separate staffing of a medical branch is essential. **Figure 6-8** is a flowchart for a mass-casualty incident showing victims needing various levels of care. It should be read from the top down. The objective is to provide the best possible medical assistance within the limitations of available resources.

Evaluating the Need for Shelter

During weather extremes, there is a need to rehabilitate fire fighters and provide shelter for occupants. Occupants of nursing homes, hospitals, and other places where special needs exist will require shelter even during normal weather conditions.

In many cases, evacuees will find their own shelter by leaving the scene to go home or to the home of a friend. When shelter is required, a nearby

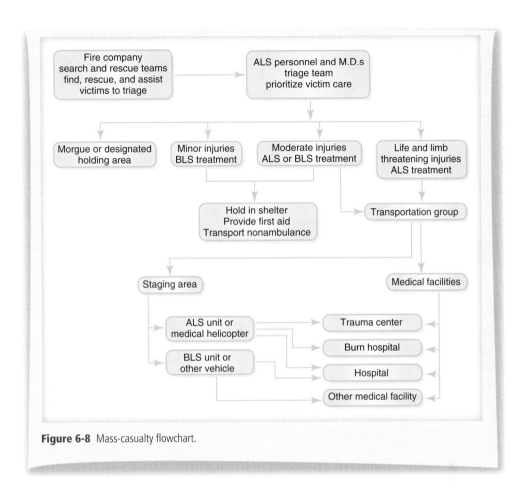

Figure 6-8 Mass-casualty flowchart.

building, well outside the fire zone, should be used. Agencies such as the Red Cross generally have personnel available to assist in setting up and managing shelters. However, their response times may be fairly long, making it necessary to assign fire, police, or EMS crews to this task until other agencies arrive.

Estimating Life Safety Staffing Requirements

The IC must have sufficient staffing to extinguish or at least control the fire, conduct search and rescue operations, treat and transport the injured, and remove victims to a place of safe refuge, while preventing reentry. The number of fire fighters needed depends on several factors:

- Number of victims
- Rescue methods used
- Condition of victims
- Fire conditions
- Smoke conditions
- Victim mobility
- Weather conditions
- Accessibility (need to force entry)

Additional staffing will be needed in the following situations:

- Victims are close to the immediate fire area.
- Victims have existing or fire-related physical impairments.
- The fire occurs during weather extremes, especially freezing weather.
- Evacuation routes other than the interior stairs must be used.
- It is necessary to force entry to rooms or hallways.

Evacuation status is determined by the following:

- Preincident planning information
- Occupant information
- Visual observation
- Reconnaissance
- Alarm information

Each fire department must develop SOPs based on local conditions and resources. Throughout most of the United States, crews, units, or companies are assembled according to apparatus type and function. Many different tasks are required on the fire ground; therefore, a division-of-labor approach is common practice. Fire department duties normally include the following:

- Water supply and application (engine company duties)
- Ventilation, entry, search and rescue, rapid intervention, and property conservation (ladder company duties)
- Search, rescue, and rapid intervention (rescue company duties)
- Triage, treatment, transportation, and rehabilitation (emergency medical duties)
- Planning, organizing, coordinating, and establishing command (duties of the IC)

This division-of-labor concept is important. Without the preassignment of general duties, actions are delayed and/or duplicated. In a rescue scenario with occupants showing at windows, there would be a tendency to assist the visible occupants at the expense of addressing more critical tasks. In other cases, fire fighters might neglect ventilation, entry, and search in favor of water application.

Many references are available that complement the information presented here or describe actual tasks necessary to fulfill the objectives of life safety. Suggested activities at the end of this chapter are designed to extend the learning experience by providing an opportunity to apply the material presented.

Chapter Summary

- Life safety is the first incident priority.
- Suppressing or extinguishing the fire is often the best way to protect lives.
- The fire floor and floor(s) above the fire are critical areas and searching them must be a high priority.
- Ventilation must be coordinated with the placement of attack lines to control the fire.
- Improper venting could produce a backdraft; vents opened between fire fighters or victims and their path to egress could be fatal.
- Ventilation strategy must consider the fact that the fire and heated gases will naturally move first upward, then laterally, and finally downward.
- Whenever possible, occupants should be rescued by the shortest, easiest, and most direct route using the interior stairways for above- or below-grade rescues.
- A defend-in-place strategy may be a better option when occupants are located far from the fire zone or are not able to evacuate rapidly.
- Interior stairs are the best means of egress from above grade, but fire escapes, aerial ladders, elevated platforms, ground ladders, and other less-desirable alternatives may be used as circumstances require.
- Ground ladders are less desirable than aerial ladders and elevated platforms but can provide quicker access in some situations.
- Elevators should not be used for rescue except when they are verified as safe to use or when they are remote and separated from the fire area with an auxiliary power supply.
- Rope rescues are among the least desirable rescue options, as this tactic is often slow and dangerous.
- When considering helicopter rescues, remember that they are often unnecessary; landing zones may be obstructed and air currents above a burning structure may be hazardous to helicopter operations.
- The *only* reliable way to verify evacuation status is for fire fighters to enter the structure and systematically check every room.
- Before flashover, rescue is possible; after flashover, rescue is highly improbable within the flashover compartment.
- Fire fighters working in multistory buildings should survey an uninvolved floor before ascending to the fire floor or floors above, as some building features will be constant from floor to floor.
- Establish search-and-rescue priority by deciding who is in the greatest danger from the fire.
- The key to successful search operations is to be systematic.
- A system to expedite triage, treatment, and transportation is essential in mass-casualty situations.
- During weather extremes, shelter is required for rehabilitating fire fighters and evacuated occupants.
- Agencies such as the Red Cross generally assist in setting up and managing shelters, but fire fighters sometimes must begin the process until they arrive.
- The staff needed for life safety depends on the number of victims, available rescue options, condition of victims, and fire and smoke.

Flames: Dny/Dreamstime.com; Steel texture: © Sharpshot/Dreamstime.com

Key Terms

critical time The time available until the structure becomes untenable.

defend-in-place A tactic utilized during a structure fire when it is very difficult to remove occupants from the building. Occupants are either protected at their present location or moved to a safe location within the building.

Challenging Questions and Activities

1. It is 6:30 AM on Wednesday, November 5. Winds are light and variable, with clear skies and moderate temperatures. Assume the role of the engine company officer. All responding companies are assigned an officer, driver, and two fire fighters. The two fire fighters assigned to your company have just graduated from the fire training academy. A truck company is also assigned to your station. You and the truck company are dispatched to a one-alarm fire at 4321 Southern Avenue. While en route, dispatch advises that a woman caller reports that her husband is still inside the building and there is "lots of fire." You arrive on the scene of the fire shown in **Figure 6-9**.

A woman meets you at the curb reporting that her husband is still inside their house on the second floor. The truck company will be arriving soon.

The next closest engine company is expected to arrive in 4 minutes. The full response to this alarm is three engine companies, two truck companies, and a battalion chief. All companies are staffed with an officer, a driver, and two fire fighters.

A. Provide an initial report and assume command.

B. Would you establish a command post and be part of the initial rapid intervention crew or would you make entry?

C. Explain your water supply and initial actions.

D. What would be the truck company's initial assignment?

Figure 6-9 One-alarm fire at 4321 Southern Avenue.

Flames: Drx/Dreamstime.com; Steel texture: © Sharpshot/Dreamstime.com

2. Assume the role of a truck company officer in Question 1. The truck is a straight stick aerial and is not equipped with hose or pumps. The engine company is out of service on an EMS response. You arrive on the scene of the fire shown in Figure 6-9. A woman meets you at the curb reporting that her husband is still inside their house on the second floor, pointing to the windows on the D side of the house.

   The next closest engine company is expected to arrive in 4 minutes. Provide an initial report and assume command.

   A. Is this a vent, entry, isolate, search (VEIS) situation?
   B. Explain your answer and explain how you would rescue the husband or what other actions you would take prior to the arrival of the first engine company.

3. A tenant using a charcoal grill on the first floor patio applied additional charcoal lighter fluid to hot coals, resulting in a large fireball that ignited the balcony above and spread to the third floor balcony. Fire has entered the first floor apartment where the patio fire originated (Figure 6-10). It is July 4 at 3:00 PM. It is hot and humid with calm winds.

   A. Where would you place the first hose line?
   B. Prioritize life safety activities by area (e.g., third priority—third floor above the apartment of origin).

Figure 6-10 Balcony fire that originated on the ground-level patio.

Courtesy of Bill Strite, Cincinnati, Ohio

   C. Where would you assign the first search and rescue crew?
   D. What tactics would you use to rescue occupants from various areas of the building?
   E. How would you rate the degree of danger for the first, second, and third floor apartments in the immediate fire area, as well as the apartments on either side of these apartments on each level. Use a scale of 0 to 10, where 0 means the fire poses no danger and 10 indicates that immediate life safety tactics are required.

Endnotes

1. Baldassarra, Carl, F, Safety Feature/Safety Hazard. *NFPA Journal*. September/October 2014:40–45.
2. Kerber, S., UL Firefighter Safety Research Institute. *Impact of Ventilation on Fire Behavior in Legacy and Contemporary Residential Construction*. Northbrook, IL: Underwriters Laboratories; 2010.

Fire Protection Systems

LEARNING OBJECTIVES

- List information related to fire protection systems that should be included in a preincident plan for a protected building.
- Compare residential sprinkler systems to commercial sprinkler systems.
- Explain why it is important to "let the system do its job" when conducting operations in a building protected by an automatic fire suppression system.
- Recognize the differences between wet pipe, dry pipe, and deluge sprinkler systems.
- Compare the reliability of a wet pipe sprinkler system to dry pipe and deluge sprinkler systems.
- Compare and contrast operations at a sprinkler-protected building with and without obvious signs of a fire or system operation.
- Describe fire department operations at a building protected by a deluge system.
- Identify, classify, and describe different types of standpipe systems.
- Describe fire department operations at a building equipped with a standpipe system.
- Explain discharge pressure differences in standpipe systems and how these differences affect operations.
- Describe the advantages and disadvantages of solid-bore and automatic nozzles when operating from a standpipe.
- Compute the pump discharge pressure needed to supply a fire line in a high-rise building equipped with a standpipe.

Flames: © Drx/Dreamstime.com; Steel texture: © Sharpshout/Dreamstime.com; Chapter opener photo: © Jones & Bartlett Learning. Photographed by Glen E. Ellman.

LEARNING OBJECTIVES (CONTINUED)

- Develop a list of standard standpipe equipment.
- List and describe fire protection systems other than sprinkler or standpipe systems.
- Explain fire department operations at facilities protected by a foam, carbon dioxide, dry chemical, and clean agent systems.
- Define the term interlock and provide an example of an interlock on a carbon dioxide system.

- As it relates to company responses, explain the possible problems with habitual false alarm system activations.
- Preplan buildings protected by sprinkler, standpipe, foam, or non-water-based systems.
- Evaluate operations at a fire in a building protected by a fire protection system.

Case Study

Aftermath of fraternity house fire—University of North Carolina at Chapel Hill.

© The News & Observer.

On May 12, 1996, a fire in an off-campus fraternity house at the University of North Carolina at Chapel Hill resulted in five fatalities and $475,000 in property loss. Less than a month later, an arson fire struck another fraternity house. On June 19, 1999, an ordinance was passed requiring all fraternity and sorority houses to be sprinkler protected.

1. Why should an automatic sprinkler system be installed in residential occupancies?
2. What would be the predicted outcome had the Chapel Hill fraternity house been protected by an automatic sprinkler system?
3. How does a sprinkler system change fire department operations and reduce fire fighter injuries?

Data from: Michael S. Isner, NFPA Fire Investigation Report, Fraternity House Fire, Chapel Hill NC, May 12, 1996.

Stacked photo background: © Vitaly Korovin/ShutterStock, Inc.; Paper: © silver-john/ ShutterStock, Inc.

Introduction

When fire protection systems are available, the offensive attack takes on an entirely different character. By properly using an installed fire protection system, the incident commander's (IC's) job can be made much easier, and the risk to fire fighters and civilians can be significantly reduced. This chapter covers operations at properties protected by the various types of fire protection systems.

Fire protection systems can greatly simplify fire-fighting efforts. When automatic fire protection systems are present, a considerable part of the offensive attack strategy involves properly supporting or using the system. If the building or area that is on fire is equipped with an automatic fire suppression system, the primary tactic is to support the system and let it do its job.

It is much safer to let the system handle the fire than to expose fire fighters to positions deep inside the building. Efforts should be directed toward maintaining the system in a fully operational status, while laying backup lines for final extinguishment.

Preincident planning is essential for any building that is protected by an automatic fire suppression system, with the possible exception of a building containing a range hood system as the only fire protection system.

During preincident planning, fire fighters need to familiarize themselves with the general layout of the building. Emergency contact telephone numbers of owners and building managers should be available to responders. A lockbox is an ideal place to keep this information. If a lockbox is not available, contact information should be kept on first-alarm fire apparatus and at the dispatch center. The people included on the emergency contact lists can then be called for assistance in gaining access to all parts of the building, restoring the system to service, and having employees assist in salvage operations. The location and operation of various water supply components should be known in advance, including the following:

- Main control valves
- Divisional control valves
- Fire pump (electric, diesel, other)

- Fire department connections (FDCs)
- Water supply (gravity tank, pump, public water system)
- Hydrant water supply (same as sprinkler supply, independent supply)

System limitations and peculiarities should also be addressed during preincident planning.

Buildings equipped with fire protection systems, particularly sprinkler systems, are often permitted to reduce fire protection measures in other areas such as increasing the fire load or reducing the exit capacity. If the system is out of service, these buildings no longer meet minimum code requirements and present a greater danger to occupants and fire fighters. Buildings under construction or demolition may have the fire protection systems removed, disabled, or shut off. The fire department should receive notification when a system is out of service and revise tactics accordingly.

Fallacy	**Fact**
Manually applying water is the best way to control a fire.	In a building equipped with a properly operating automatic fire suppression system, the best tactic is to conduct preplanned support activities and let the system do its job.

Steel: © Sharpshot/Dreamstime.com

Sprinkler Systems

Residential sprinkler system installations, as described in NFPA 13R, *Standard for the Installation of Sprinkler Systems in Low-Rise Residential Occupancies* and NFPA 13D, *Standard for the Installation of Sprinkler Systems in One- and Two-Family Dwellings and Manufactured Homes*,[1,2] are still fairly rare. Only 4.6% of homes, including apartment buildings, had sprinkler systems in 2009. However, this represents an increase from 3.9% of homes equipped with sprinklers in 2007. Home sprinkler systems have a proven track record in saving lives and property. The fire death rate in homes protected by wet pipe sprinkler systems is 83% lower than in nonsprinkler-protected homes. The fire damage in homes with wet pipe sprinkler systems is 69% lower than in nonsprinkler-protected homes.[3] The primary purpose of these life safety systems is to allow additional escape time. The hardware requirements for life safety systems are different from those of a sprinkler system in an industrial or office occupancy. For example, fire pumps are not generally included in the 13R and 13D systems. The tactics described next apply to all buildings that are protected by sprinkler systems; however, some modifications are necessary because of the differences in the sprinkler standards applicable to homes.

Commercial-type automatic sprinkler systems meeting the requirements of NFPA 13, *Standard for Sprinkler Systems*,[4] have an exceptional record in controlling fires. Large losses of life and property are practically nonexistent in buildings equipped with a properly designed, maintained, and operating sprinkler system **Figure 7-1**. When large-loss fires do occur in these properties, some degree of human error is generally involved **Figure 7-2** and **Figure 7-3**.

NFPA 13E, *Recommended Practices for Fire Department Operations in Properties Protected by Sprinkler and Standpipe Systems*,[5] lists some frequent causes of sprinkler system failure that fire fighters should be prepared to resolve, stating the following:

> **4.1.1** *Fire department personnel should be knowledgeable of and prepared to deal with the following three principal causes of unsatisfactory sprinkler performance:*
>
> (1) *A closed valve in the water supply line*
> (2) *The delivery of an inadequate water supply to the sprinkler system*
> (3) *Occupancy changes that render the installed system unsuitable*

Although rare, large losses do occur in sprinkler-protected buildings. There are times when an

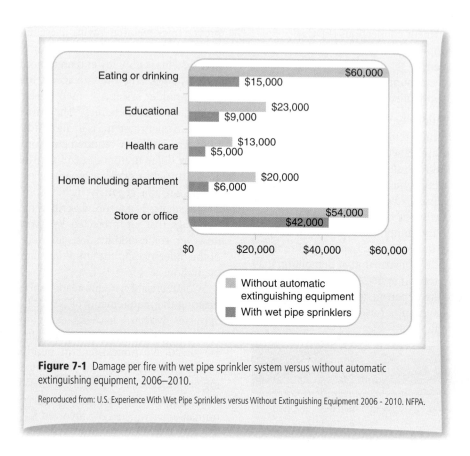

Figure 7-1 Damage per fire with wet pipe sprinkler system versus without automatic extinguishing equipment, 2006–2010.

Reproduced from: U.S. Experience With Wet Pipe Sprinklers versus Without Extinguishing Equipment 2006 - 2010. NFPA.

explosion or extremely fast-moving fire overwhelms a sprinkler system. Sometimes improper design, such as a building feature constructed below the sprinkler, may shield the fire from sprinkler water. Losses also occur when the sprinkler system is shut down or otherwise out of service. When the system or part of the system is out of service, the system should be tagged as specified in NFPA 25, *Standard for the Inspection, Testing, and Maintenance of Water-Based Fire Protection Systems*,[6] which states the following:

> *15.3.1 A tag shall be used to indicate that a system, or part thereof, has been removed from service.*

> *15.3.2 The tag shall be posted at each fire department connection and system control valve, and other locations required by the authority having jurisdiction indicating which system, or part thereof, has been removed from service.*

Proficient fire officers are familiar with the general operation of fire protection systems. Department standard operating procedures (SOPs) should address working in structures that are protected by fire protection systems.

The system that is most often encountered is the **wet pipe sprinkler system**, which is very effective and the most reliable fire suppression system. The

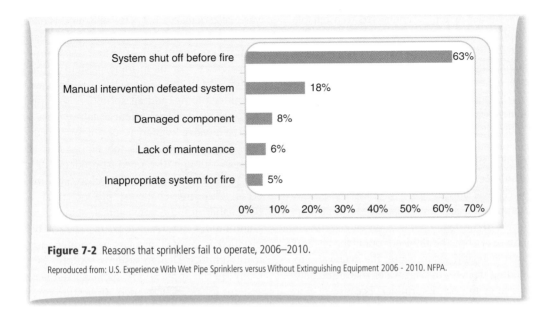

Figure 7-2 Reasons that sprinklers fail to operate, 2006–2010.

Reproduced from: U.S. Experience With Wet Pipe Sprinklers versus Without Extinguishing Equipment 2006 - 2010. NFPA.

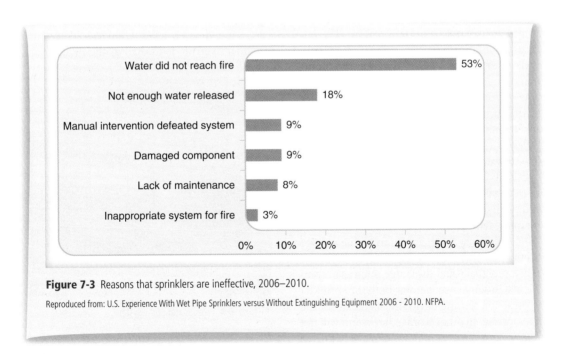

Figure 7-3 Reasons that sprinklers are ineffective, 2006–2010.

Reproduced from: U.S. Experience With Wet Pipe Sprinklers versus Without Extinguishing Equipment 2006 - 2010. NFPA.

wet pipe sprinkler system is connected to a reliable water supply such as a water main or tank (or both). The water is distributed throughout the protected structure and applied to the fire through sprinklers. Individual sprinklers are self-contained detection/application devices, which account for their reliability. Valves control water distribution to sprinkler piping, and fire pumps may be needed to provide the necessary water pressure and volume to the system.

A **dry pipe sprinkler system** is used in areas that might be subject to freezing temperatures. The difference between wet and dry pipe sprinkler systems is that the piping in a dry system is filled with air that is under pressure instead of water. When a sprinkler opens because of a fire, the air bleeds out of the system. This reduction in air pressure causes the main valve to open, flooding the system with water that ultimately discharges from the open sprinkler. Because of this design, water may take longer to reach the fire than in a wet pipe system. Sometimes a wet pipe system is protected using antifreeze to avoid the delayed operation and installation of additional equipment.

A **preaction sprinkler system** is also filled with air that may or may not be under pressure. The one difference is that a sensing device, such as a smoke or heat detector, opens a valve, flooding the piping with water. If a sprinkler has also fused, then water will come out of the sprinkler onto the fire. In some systems the air in the sprinkler piping will be under pressure. If there is a loss of air pressure, as when a sprinkler fuses, the valve will open, flooding the system with water.

In both systems, if the capacity of the piping exceeds 500 gallons (1893 L), the system must be equipped with an accelerator or exhauster. These devices rapidly remove the air from the system, reducing the time necessary for the water to fill the piping.

Preaction systems can be used when a warning is desired before actual water discharge, such as in computer rooms. In some installations, both the sprinkler and the preaction sensing device must actuate for the system to operate. In other words, the system will not discharge water when a sprinkler is damaged or fused unless the secondary sensing device controlling the valve also detects a fire. This safeguard is used where incidents of vandalism require a backup alarm, such as in parking garages.

Another type of sprinkler system is the **deluge system**. With this system there is no water in the sprinkler piping, and all sprinklers (or applicators) are open. When the system is activated, a detector-operated valve (which is normally closed) opens, releasing water that fills the piping and then discharges through the open sprinklers (or applicators) into the protected area. Deluge systems protect areas with high-challenge fire potential, such as flammable liquids, conveyors moving combustible commodities, and transformers. They may also be installed in aircraft hangars as combination water/foam systems. Deluge systems can also be used in many special applications, such as fixed water spray systems **Figure 7-4**. Such applications are used to protect aboveground liquefied petroleum gas tanks from fires, dilute flammable liquids, disperse flammable gases, and protect tanks from exposure fires.

Deluge systems are also equipped with manually operated override valves that can be used if the detection devices fail to operate. Many times the manual override is located at the deluge valve, but

Figure 7-4 Fixed water spray system.
Courtesy of Veltre Engineering.

it could be located elsewhere. Fire fighters should know where these control valves are located and how to manually activate the system. An exterior sprinkler system, designed as a water curtain to protect a building from exposure fires, may also be available. Some of these systems will require manual actuation, and it may also be necessary to supply or augment the water supply to the exposure protection system. This is yet another reason for preincident planning. Many deluge and exterior sprinkler systems will have operating instructions printed near the manual control valve.

The most common operational error at properties protected by automatic sprinkler systems is shutting down the system prematurely. Fire fighters working at the scene must be sure the fire is totally under control before shutting down the system. When the system is shut down, a fire fighter should be assigned to stand by the valve in case it is necessary to reopen it quickly. This fire fighter should be equipped with a radio or other reliable means of communication.

In many cases the sprinkler system will control the fire but may not completely extinguish fires that are shielded from direct water contact **Figure 7-5**. In a

Figure 7-5 Burning materials shielded by obstructions.

warehouse, commodities may be stored in shelves that block the spray pattern and do not allow the sprinkler system to apply water directly on the burning stock. Hand lines must be in place, ready to extinguish any remaining fires, and the fire fighter who is assigned to the sprinkler valve must stand by, ready to reactivate the system should it become necessary.

In managing fire department operations at a property protected by an automatic sprinkler system, the IC should keep in mind the potential to deprive the sprinkler system of water. A properly designed, installed, and maintained sprinkler system should have a calculated water requirement that includes enough water to support hose streams. However, the IC must be extremely careful to avoid depleting the sprinkler system of the water pressure and volume necessary to properly support operating sprinklers. The best practice is to connect attack lines to a water supply that is separate from the one supplying the sprinkler system. If a decision must be made between using hose lines and properly supplying the sprinkler system, it is usually best to supply the sprinkler system. If it becomes obvious that the sprinkler system is not achieving the desired result because of damaged piping or related problems, then the IC may elect to redirect water from the sprinkler system to support fire department hose lines.

Efforts should be made to obtain additional water supplies from sources that will not affect the sprinkler system operation. This can be done by identifying nearby water mains that are not part of the system that is supplying the building. Water can then be pumped from the remote water mains to the incident to either augment the sprinkler system or supply apparatus on the scene. The local water department may also be able to bring additional water pumps on-line and provide further volume and pressure to the area.

In rare cases, usually where sprinkler piping is damaged, it may be necessary to shut the system down so that enough water is available for manual suppression. If a sprinkler system is having difficulty controlling the fire, it is much better to have hose lines at strategic locations and secure additional off-site water supplies than to shut down the system if the sprinkler piping is still in place.

Following are general guidelines that could be incorporated into department SOPs for operations at protected properties.

Working at a Sprinkler-Protected Building with No Signs of Fire

Listed here are tasks that should be performed at buildings that are equipped with a sprinkler system when no signs of fire or system operation are evident from the outside. It is assumed that there was some report of an unusual condition or a fire alarm system was activated that caused the fire department to be notified.

Gaining Entry

If a key to the building is contained in a lockbox, it may not be necessary to force entry. Likewise, if someone who can provide access is responding and will arrive within a few minutes of the fire department's arrival, it may be advisable to wait for assistance. Conversely, if there will be a considerable delay in entering the building, it may not be appropriate to wait for assistance. The potential fire or water damage generally outweighs the damage done by forcible entry. When it is necessary to force entry, crews should do so with consideration to property damage. Many times, ladders can be used to gain entry through upper story windows. Guidance concerning forcible entry should be outlined in SOPs. Preincident plans should identify the location of lockboxes, identify if keys are available, and include information concerning potential entry locations.

Checking the Main Control Valve

A fire fighter equipped with a radio should be sent to the sprinkler system riser (main control valve). Depending on available staffing, the fire fighter may be assigned to remain at the valve throughout the entire operation. This fire fighter determines whether the system is flowing and checks the valve to ensure that it is in the fully open position. One method of checking for water flow, if the riser is not equipped with a device such as a water motor gong, is to place an ear against the riser and listen for water flowing through the pipe. If this area has the potential to become a hazard area, then the buddy system would apply, with two fire fighters assigned to the control valve.

Several types of main valves are used to open and close sprinkler systems. The two most commonly used are the outside stem and yolk valve **Figure 7-6** and the post indicator valve **Figure 7-7**. In preparing preincident plans, the IC should note the sprinkler valve type, location, and operation. In most situations, the sprinkler valve will be locked in the open position. It may be necessary to cut or break the lock to shut down the system.

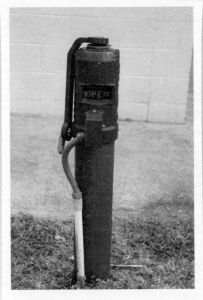

Figure 7-7 Post indicator valve (PIV).

Courtesy of Bernard J. Klaene.

Figure 7-6 Outside stem and yolk (OS&Y) valve.

© Jones & Bartlett Learning. Photographed by Glen E. Ellman.

Checking the Fire Pumps

When the sprinkler system water supply is augmented by a fire pump(s), a fire fighter (or two fire fighters if conditions warrant) equipped with a radio should be assigned to check the fire pumps. If the main pump is operating, there is a good chance that the system is discharging water, either by accident or onto a fire. If a fire is detected but the pumps are not operating, the fire fighter can manually start the pumps at the direction of the IC. It is a poor practice to rely on remote annunciator panels to determine whether the fire pumps are running. An actual physical check of the pumps is needed.

The fire fighter who is assigned to the main control valve may also be in a position to monitor the fire pump. In many cases the fire pumps are located near the main control valve. This is an important position that should not be relegated to a part-time task, requiring the fire fighter to walk a considerable distance between the riser and pumps. A good rule of thumb is to separate the tasks if the pumps cannot be seen or heard by someone standing at the main control valve.

Checking the Building for Fire and/or Sprinkler Operation

Fire fighters should be assigned to conduct a systematic check of the entire building. If there was a reason to notify the fire department because of an unusual condition, then it must be verified that nothing is amiss. There is a tendency to approach a sprinkler alarm with no visible sign of fire as a nonemergency situation, making entry without full protective clothing and the equipment needed to handle a fire situation. This is an unsafe procedure that could result in serious injury if a working fire is encountered while checking out the building.

Supplying the Fire Department Connection

A pumper connected to an adequate off-site water supply should connect two 2½-in. (65-mm) or two 3-in. (77-mm) hose lines (large-diameter hose if the system is so equipped) to the FDC. The water supply for this pumper should be large-diameter hose, two 2½-in. (65-mm) or 3-in. (77-mm) hose lines, or a direct hydrant connection. A single 2½-in. (65-mm) or 3-in. (77-mm) supply line is inadequate. During preincident planning, the IC should determine off-site water sources. Department SOPs should assign an engine company the task of supplying water to the FDC.

Large buildings or complexes could have several FDCs supplying different sprinkler or standpipe systems. FDCs are often interconnected; supplying any of the FDCs provides supplemental water to the entire building or complex . However, other buildings or complexes are equipped with separate FDCs supplying a standpipe and/or sprinkler system protecting a single area or building within a complex. Preplans must address this issue. If the intakes are not interconnected, the area covered by each intake

Figure 7-8 Fire department connection with signage showing area covered by system.

Courtesy of Bernard J. Klaene.

should be specified on the preplan and a marking system used to identify the area protected by each FDC.

The fire department should have access to plans for new structures within their jurisdiction and be provided an opportunity to have input regarding the best location for FDCs, fire hydrants, and other fire protection equipment. Response personnel reviewing these plans should make recommendations that coincide with current SOPs and facilitate effective operations. As an example, most codes require the FDC to be placed at the front of the building, but do not necessarily address how close to the front of the building. Many department SOPs direct the first-arriving engine company driver to secure a water supply and be prepared to pump into the FDC while the officer and remaining crew members make entry. The fire department intake shown in **Figure 7-9** is on the main road at the front of the building but well away from the entrance to the protected building. Following an SOP that places the pumper near the FDC would result in the apparatus and driver being located a considerable distance from the building. The driver could not be an effective member of a two-out team, and the apparatus, hose, and tools would be located far from the building. This situation would require a change in the SOP, or special provisions would need to be made in the preincident plan for this specific property.

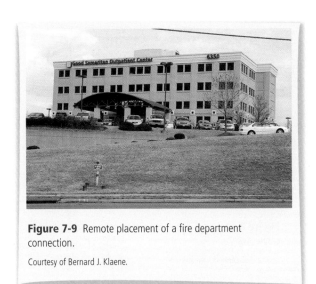

Figure 7-9 Remote placement of a fire department connection.

Courtesy of Bernard J. Klaene.

Working at a Sprinkler-Protected Building with Evidence of Fire Showing from the Exterior

If there are signs of a fire and/or the sprinkler system is operating, the main objective is to support the system while ensuring that occupants are safe.

Gaining Entry

Again, gain entry to the building using the minimum force necessary. However, with a fire in progress or a system operating, the time spent gaining entry will increase the risk to the occupants and will cause additional fire, water, and smoke damage; therefore, a more forceful entry is justified.

Checking the Main Control Valve and Fire Pump

Fire fighters should be assigned to the main valve and pump as when there is no visible sign of fire. However, when there is a confirmed fire, their primary responsibilities are to ensure continued operation of the system and to provide rapid shutdown when appropriate. These positions are critical and should be staffed throughout the operation.

Large systems will be equipped with control valves on portions of the system. These valves are not as prone to accidental closing as main control valves, but fire crews should check them as soon as possible.

Supplying the Fire Department Connection

When there is a confirmed working fire, an engine company must supply water by pumping into the FDCs, and the pump operator is required to remain at the apparatus. The department SOPs should identify minimum water supplies and pump pressures. Most departments provide a pressure of 150 pounds per square inch (psi) (1034 kilopascals [kPa]) to the FDCs. It is important to avoid surges when initially charging the hoses connected to the FDC, as many sprinkler systems are designed and tested to a maximum 175 psi (1207 kPa). As was previously noted, the preincident plan modifies SOPs whenever necessary according to the characteristics of the specific property. If a pressure higher than specified in the SOPs is required to adequately supply the sprinkler system or the standard pressure is likely to cause damage, this should be noted on the preplan.

When the building is supplied by a public water system, a fire hydrant will normally be located close by. The pumper assigned to supply the FDC has several options:

- Use soft suction to connect to the fire hydrant and hand lay hose to the FDCs (usually two 50-ft (15-m) sections from pumper to each side of a siamese FDC) or a single section of large-diameter hose to a large-diameter hose connection.
- Forward lay large-diameter hose and connect to the FDC.
- Place one pumper at the hydrant and another close to the FDC.

Each of these layouts has advantages and disadvantages. Placing a pumper near the building may result in the driver and apparatus being within the collapse or falling glass zone, but doing so does slightly reduce the friction loss. Using two pumpers requires two pump operators to remain at their apparatus, reducing initial staffing. However, placing a second pumper at the hydrant provides redundancy should one apparatus experience mechanical problems.

The typical older FDC will be two female 2½-in. (65-mm) intakes **Figure 7-10**. Two 2½-in. (65-mm) or preferably two 3-in. (77-mm) hose lines should be connected to the intakes. Large-diameter hose 4-in. (100-mm) or 5-in. (125-mm) connections are now being substituted for 2½-in. (45 mm) intakes **Figure 7-11**. Large-diameter hose provides large quantities of water and greatly reduces friction loss. However, there are several potential problems to address when using large-diameter hose to supply a sprinkler or standpipe system:

- The weight of the hose and water bearing on the intake is significant.
- Kinking is more likely.
- The unisex connection can accidentally disconnect (requires safety latch).

Figure 7-11 Large-diameter hose intake to fire department connection.

Courtesy of Bernard J. Klaene.

- There is no redundancy. (If a single section of large-diameter hose ruptures, the total supply to the FDC is lost.)
- Standard large-diameter hose is generally tested at a lower pressure* than 2½-in. (65-mm) or 3-in. hose (77-mm).

*At the time of this writing, large-diameter hose with higher-pressure ratings is available.

Large-diameter hose is typically tested to 200 psi. This rating is adequate for most sprinkler operations, but high pressures may be required for standpipe operations discussed later in this chapter.

Department training sessions should include overcoming potential difficulties in connecting to the FDC, such as:

- Damaged threads
- Threads different than those used by the jurisdiction
- Debris and other obstructions (missing intake caps are a clue)
- Damage to intakes or female connections

On occasion, male threads could be encountered at what is believed to be an FDC **Figure 7-12**. Fire and building codes require female connections on

Figure 7-10 Supplying fire department connection.

© Jones & Bartlett Learning. Photographed by Glen E. Ellman.

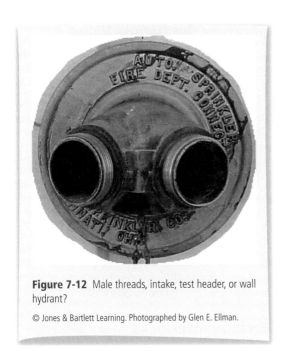

Figure 7-12 Male threads, intake, test header, or wall hydrant?

© Jones & Bartlett Learning. Photographed by Glen E. Ellman.

standpipes. Male threads are commonly used to test the system (test header) or as a wall hydrant. Preincident plans should be developed for all buildings protected by a fire protection system. Fire companies should routinely familiarize themselves with FDCs throughout their response area and check intakes for damage, operability, or unusual features.

Letting the System Do Its Job

It is better to shut down a sprinkler system too late than too early. The sprinkler system should be permitted to operate until the IC is certain that the fire is under control. When the sprinkler system is shut down, the only fire remaining should be small spot fires. Hose lines should then be in place to complete extinguishment.

Backing Up the System

Prepare for an offensive attack and overhaul by positioning hose lines. The hose lines should be staffed by fire fighters in full protective clothing, including self-contained breathing apparatus (SCBA). Hose lines should be operated only to perform rescue operations and limit fire spread or to complete overhaul operations after the sprinkler system has been shut down. However, in the rare case that the sprinkler system is ineffective due to damaged piping, malfunction, or inadequacy, hose lines may take priority.

Ventilating

When it is safe to do so, crews should make ventilation openings above the fire. Proper ventilation will channel the fire and limit its extension. The cooling effect of sprinkler water can inhibit upward smoke movement, possibly making it more difficult to ventilate. Recommended ventilation tactics should be part of the preincident plan.

Some large-area, sprinkler-protected buildings are equipped with automatic or manual roof vents and **draft curtains** that are designed to limit fire spread. Overhaul will most certainly require ventilation, usually positive- or negative-pressure mechanical ventilation. Fire fighters should not shut down the system to locate the fire; they should ventilate.

Performing Property Conservation Tasks

When possible, property conservation tasks should be accomplished while extinguishment is in progress. Extinguishment takes priority, but water damage often dictates that the IC summon enough assistance to start property conservation simultaneously.

Placing the System Back in Service

If fire crews are properly trained to do so, they should place the sprinkler system back into service by replacing fused sprinklers and reopening valves. Most codes require that a supply of spare sprinklers, of the type used in the system, be kept on the premises for this purpose. When replacing sprinklers, it is critical that the proper sprinkler be used. In some buildings, the same type of sprinkler will be used throughout the building. In other buildings, pendant, upright, sidewall, and other types of sprinklers may be used. Sprinklers designed to operate at different temperatures may also be in use. The sprinkler being replaced must be of the same type and temperature rating. If the system is too complex or the sprinklers are not available, the owner or manager should restore the system. If the system is equipped with division control valves, restoring the unaffected part of the system may be possible even if sprinklers are not replaced in the fire area. Some department SOPs prohibit reactivating the system, owing to possible legal implications; others believe that it is important to place the system back in service as soon as possible. The property owner should contact a licensed sprinkler contractor to inspect the system after it has operated during a fire. There is a possibility that sprinklers or piping could be damaged. Department policy and guidance on this subject should be included in the department's SOPs.

Working at a Property Protected by a Deluge System

The tasks required at a sprinkler-protected building are basically the same for wet, dry, or preaction systems. The deluge system presents at least one additional consideration: manual operation of the deluge valve. Many of these systems are located outside of buildings. Usually, the hazard protected by deluge systems can create extreme risks for fire fighters who are attempting to manually suppress the fire. System operation will be obvious, thus negating the need for the thorough investigation required in sprinkler-protected buildings. Following are general steps to be taken at these properties. However, preincident planning is the key to a successful operation.

Checking the Control Valve and Fire Pump

Just as with the wet and dry pipe systems, it is important to ensure that valves are open and that the pumps are operating properly.

Operating the Deluge Valve

It is possible, though improbable, that a fire would be in progress in an area protected by a deluge system that failed to operate properly. If the system is needed for fire control, the deluge valve should be activated manually provided it is in a safe location. However, it is more likely that an exposure fire would threaten an area protected by a deluge system. It may be possible to cover these protected exposures with the deluge system by operating the deluge valve if operation of the system does not create a safety hazard or cause additional damage. Consideration must be given to the water supply requirements for these systems when they are being used for exposure protection. A deluge system may well deplete a private water supply system. In the case of a transformer fire, it is best to stay out of secured areas and allow the system to do its job.

Checking Interlocks

Deluge systems often trip interlocking devices when activated. For example, system operation may deenergize electric transformers, shut down conveyor belts, or shut off a fuel supply. In most cases, there is a means of manually activating the interlock. If these interlocks have failed to operate and it is possible to safely shut down a fuel supply, conveyor belts, or other processes that contribute to the spread of fire, this should be done. However, transformers should not be shut down manually. Usually, the deluge system will control the fire even if the interlocks do not function. Sometimes, it may be advisable to wait for plant personnel to shut off fuel supplies or perform other related activities.

Letting the System Do Its Job

As with the wet and dry pipe systems, it is better to shut down the deluge system too late than too early. A determination must be made that the fire is completely under control before shutting down. This system will be flowing large quantities of water, increasing the temptation to shut down prematurely. Remember, even with hose lines in place, it may be impossible to safely and effectively apply the quantity of water necessary to control the fire if the system is shut down too soon. Many facilities that are protected by deluge systems are such that total loss has already occurred in the area of operation; the deluge system is merely preventing extension. For example, a transformer that catches fire, setting off a deluge system, has probably already sustained the maximum fire loss. In this case, the deluge system is usually designed to protect exposures.

Backing Up the System

Hose lines staffed by fully protected fire fighters are required at strategic locations for fires such as those involving conveyors. However, manually operated hose lines create a substantial safety hazard if the deluge system is protecting high-voltage transformers. The fire officer must evaluate this situation carefully, knowing that charged hose lines offer a dangerous temptation. Most of the time, the protected transformer could completely burn out without endangering lives or additional property.

Working at a Building Equipped with a Standpipe System

A **standpipe system** is not an automatic fire suppression system. Unlike sprinkler systems, standpipe systems cannot operate without human intervention. A properly operating and maintained building standpipe system is helpful in conducting offensive attacks. In high-rise structures, it may well be impossible to conduct a safe and effective interior operation on upper stories if the standpipe system is inoperative.

NFPA 14, *Standard for the Installation of Standpipe and Hose Systems*,[7] lists six major types of standpipe systems:

- *Automatic dry*. In this system, piping is filled with pressurized air or nitrogen but is connected to a water supply that automatically admits water into the system when a discharge is opened.

- *Automatic wet*. In this system, piping is filled with water, and water is provided when the discharge is opened. This is the most common standpipe system.

- *Semiautomatic dry*. In this system, a dry standpipe admits water into the system piping upon activation of a remote-control device located at a hose connection.

- *Manual dry*. This system is filled with air and does not have a water supply; the system relies entirely on water provided via the FDC to supply hose lines.

- *Manual wet*. This system is filled with water connected to a water supply that maintains water in the system, but it is not capable of providing water for firefighting purposes unless it is supplied by a fire department pumper.

- *Combined system*. This system supplies both sprinklers and standpipes.

- *Wet standpipe system*. A standpipe system having piping containing water at all times.

NFPA 14 also defines three types of standpipes (Class I, Class II, and Class III) as follows:

- The Class I standpipe system provides a 2½-in. (65-mm) hose connection to supply water for fire department use.

- The Class II standpipe system provides 1½-in. (38-mm) hose connections and hose stations to supply water for use primarily by trained personnel or by the fire department during initial response. The hose station may be provided with hose as small as 1-in. (25-mm) diameter.

- The Class III standpipe system provides 1½-in. (38-mm) hose stations to supply water for use by trained personnel and 2½-in. (65-mm) hose connections to supply a larger volume of water for use by the fire department.

The standpipe can be independent from or connected to the sprinkler system (combined system), which may have different flow requirements than the three classes of standpipes. As mentioned previously, it is important to evaluate the effect of diverting water from the sprinkler system to manually operated hose streams when connected to a combined system. Expect high-rise buildings to be equipped with standpipes, but other buildings may also have standpipes. Most new "big box" mercantile properties will have a drop from the sprinkler system with a hose connection.

Code requirements change. In most cases the standpipe system will meet the requirements of the code in effect when the building was originally built. One major change that occurred was the minimum pressure requirement at the standpipe outlet. Older codes generally specify 65 psi (448 kPa) at the standpipe outlet, whereas modern codes usually require 100 psi (689 kPa). Pressure and flow regulators have changed in design. Given the different types and classes of systems, as well as variations in pressure and design, it is imperative that preplan information be available for all buildings equipped with a standpipe system. At a minimum, preincident plan information about the standpipe should include:

- Location of FDC and hydrant closest to the FDC
- Special hydraulic considerations (maximum pressure, internal pump operating pressure, pressure zones, unusual pipe configuration)
- Location of fire pump and valves
- Maximum and minimum pressures
- Pressure-limiting valves (location of special tools, how to adjust/defeat)
- Pressure at standpipe outlets
- Maximum flow

When the standpipe system is connected to a fire pump, it is important to determine whether the building's fire pump is providing sufficient pressure and volume to support firefighting operations. If not, it may be necessary for the fire department to either supplement the water supply or provide all of the water needed through the FDC.

Checking Fire Pumps and Main Control Valves

As with a sprinkler system, it is important to check that fire pumps are operating and the main control valve is open. A properly operating system is crucial when a fire occurs in the upper stories of a high-rise building.

Supplying Fire Department Connections

The water supply requirements for standpipe operations are much the same as those for a sprinkler-protected building: two 2½-in. (65-mm) or 3-in. (77-mm) hose lines connected to the FDC (large-diameter hose when the connection is so equipped), with the pumper being supplied from a hydrant by large-diameter hose or soft suction connection. Some FDCs have more than two intakes. In most cases, this indicates that a flow greater than 500 gallons per minute (GPM) (1892 liters per minute [L/min]) may be required. It is best to supply all intake connections to the system being used.

Some standpipe systems will not have a fixed water supply and will rely entirely on fire department pumpers to provide water to the system. Manual wet and dry standpipes will require water to be supplied through the FDC. Manual dry standpipe systems are sometimes unserviceable due to valves being left open, caps missing, pipe damage, and other problems. With a manual dry system, these problems might not be known until the fire department attempts to use the system. An automatic or wet manual standpipe system will flow water if damaged or otherwise compromised, bringing immediate attention to the problem, unless the main control valve has been shut down.

Pressure requirements for a standpipe system include elevation loss, friction loss in the hose, friction loss in the standpipe piping system, and nozzle pressure. Standard fire-ground hydraulic calculations should be adequate, with a slight allowance for system piping. In most operations, an allowance of 10 to 15 psi (69 to 103 kPa) of friction loss is adequate for the standpipe system piping. It is not necessary to determine the exact friction loss in the standpipe system when a fire occurs. However, it should be noted if additional information regarding friction loss in the system piping is available through preplanning, especially if it is appreciably more than 15 psi (103 kPa). Communications between the pump operator and the company on the fire floor are critical so pump pressure can be regulated to meet rate-of-flow requirements. Some departments carry a pressure regulator with gauge as part of the stand-pipe equipment. Doing so is highly recommended because it allows the fire attack crew to control the outlet pressure at the point of use. When using 2½-in. (65-mm) smooth-bore nozzles, controlling the nozzle pressure can be critical, as overpressurization can result in an extreme nozzle reaction force.

In very large high-rise buildings, there is greater reliance on interior pumps. The laws of hydraulics tell us that there is 0.434 psi of **backpressure** for every 1 ft of elevation (3 kPa of backpressure for every 305 mm of elevation). As an example, in the 60-story high-rise building shown in Figure 7-13, each floor is 10 ft (3 m) high, yielding a backpressure of 256 psi (1765 kPa) at the base of the building. In this example, a pump pressure of 358 psi (2468 kPa) is required at the FDC, not including the friction loss in the hose from the pumper to the FDC. This exceeds the maximum test pressure for most hose lines. The pumper supplying the FDC could augment the water supply, but there is a great reliance on the interior pumping system. A building of this height would probably be designed with multiple pressure zones or pumps in series.

Providing Standpipe Equipment

It is poor practice to use the hose lines that are pre-connected to the building's standpipe system. Pre-connected hose supplied in a hose cabinet is often unlined and seldom tested or properly maintained. Fire fighters should bring their own hose, nozzles, and adapters into the building.

At a minimum, standpipe equipment should consist of the following:

1. First-arriving engine company:
   – Three 50-ft (15-m) lengths of 2½-in. (65-mm) hose
   – One 50-ft (15-m) section of 1¾-in. (45-mm) hose (If staffing is insufficient, the section of 1¾-in. [45-mm] hose can be brought to the fire floor by the second-arriving engine or the truck company.)

Standpipe

59th Floor

Standpipe

Elevation loss = 256 psi
Nozzle pressure = 50 psi
Friction loss in 150 ft of 2 1/2 in. hose = 37 psi
Friction loss in standpipe 15 psi
Total pressure required = 358 psi

Figure 7-13 Calculating pressure needed to supply standpipe operations.

– Smooth-bore, broken-stream, or combination nozzle capable of providing sufficient flow at low pressures
– Spanner wrenches
– 30-degree standpipe outlet elbow
– Pressure gauge
– Set of 2½-in. (65-mm) male and female adapters
– Standpipe hand wheel
– 18-in. (457-mm) pipe wrench
– Channellock and/or vise-grip pliers
– Set of Allen wrenches
– Screwdriver
– Wire brush
– Door chocks
– 2½-in. (65-mm) gated wye

2. Second-arriving engine company:
   – Three 50-ft (15-m) lengths of 2½-in. (65-mm) hose
   – 1¼-in. (32-mm) or 1⅛-in. (28-mm) smooth-bore nozzle (If a combination nozzle is preferred, it must be capable of providing sufficient flow at low pressures and equipped with an adapter that allows it to be converted to a smooth bore.)

3. Truck company:
   – Thermal imaging camera
   – Search and rescue equipment, including search rope as well as chalk (preferably sidewalk chalk), tags, or other marking materials to identify areas searched
   – Forcible entry, ventilation, and salvage equipment, as required
   – Extra air cylinders

Whenever possible these first three arriving companies should work together as a task force or division on the fire floor. Even with the standard four-person staffing, the first-arriving engine company will not be able to safely and effectively begin operations on the fire floor of a multistory building. SOPs generally assign the apparatus operator to supply the FDC, leaving a three-person crew to proceed to the fire floor, connect hose at the standpipe outlet on the floor below the fire, and advance a hose line up

the stairway through a door to the fire floor. In most cases, it will be necessary to advance hose through one or more doorways or around other obstacles to attack the fire. Maneuvering a hose line through multiple obstacles often requires fire fighters to be stationed at doorways or places where the hose has to make a sharp bend. Placing a fire fighter at the door leading to the hallway and rooms allows the door to be controlled to limit ventilation. With a four-person crew and the apparatus operator outside, even if a two-person crew can reach the fire area, only one person remains to staff the initial rapid intervention crew for fires above grade level or in large buildings.

Many departments use 1¾-in. (45-mm) or 2-in. (50-mm) hose lines as the initial attack line from a standpipe. A 150-ft (30-m) length of 1¾-in. (45-mm) hose with a 7/8-in. (22-mm) or 15/16-in. (24-mm) solid stream nozzle or with a low-constant-flow nozzle should provide a flow capable of containing most fires in buildings with compartments that do not require a flow exceeding 100 GPM (379 L/min). Using the formula of volume divided by 100 to calculate the size of the area that could be extinguished with 100 GPM (379 L/min) computes to an enclosure that does not exceed 10,000 ft^3 (283 m^3). Seldom will rooms in residential buildings exceed this size. However, there have been situations in which very low pressures were encountered at the standpipe outlet or where fires involved compartments exceeding 10,000 ft^3 (283 m^3). In these situations, larger-diameter hose reduces the friction loss and allows the use of larger-volume nozzles, thus increasing the available rate of flow at low pressures.

Likewise, the minimum 150 ft (46 m) of hose should be enough to reach the seat of the fire, even when it becomes necessary to maneuver around multiple obstructions or doorways to reach the fire area. The 1¾-in. (45-mm) hose line is much easier to advance than larger-diameter hose lines and is particularly beneficial when there is limited staffing on the fire floor. However, larger-diameter hose increases the chance of successfully extinguishing the fire; therefore, it should be the primary attack line. The recommended standpipe equipment listed previously allows the attack team to add a 50-ft (15-m) section of 1¾-in. (45-mm) hose at the nozzle tip. At 100 GPM (379 L/min), the one additional section of 1¾-in. (45-mm) hose will add approximately 5 psi (34 kPa) of friction loss to the hose layout.

What diameter and length of hose are best? This question has to be answered by each fire department. If standpipes are spaced such that 150 ft (30 m) of hose is sufficient to reach all areas in the standpipe-equipped buildings in your jurisdiction, three sections of hose may be the best choice. If your department responds to multiple buildings with low-pressure standpipes or with large compartments, the availability of 2½-in. (65-mm) hose is a must.

Preincident plans should be consulted in making decisions regarding hose diameter and length. Preincident plans could also identify the need for additional, specialized equipment. Rate-of-flow determinations, as described in the *Offensive Operations* chapter, are as essential in standpipe operations as they are in any other manual fire suppression operation.

Figure 7-14 shows standpipe equipment that would be placed in two quick-opening carrying cases with each fire fighter carrying a hose pack. This arrangement allows fire fighters to share the equipment load and provides an efficient means of carrying the standpipe equipment into the building and deploying it at the standpipe outlet. It is extremely important to select the correct standpipe equipment and hose for your jurisdiction. It is equally important to train on the proper use of this equipment under conditions likely to be found in buildings equipped with standpipes. As an example, if 2½-in. (65-mm) hose is selected for use in standpipe operations, training should simulate safely advancing a charged line with the minimum crew size.

Some buildings have pressure or flow reduction valves in the standpipe system. These valves can cause problems for fire fighters using the system, as was experienced at the Meridian Plaza fire in Philadelphia[8] that killed three fire fighters. In this case, pressure-reducing valves were improperly set to reduce the pressure to less than 60 psi (414 kPa) at the valve. Steps have been taken to eliminate this problem, but good preincident planning and routine inspections will reduce the possibility of this occurring. Many pressure- or flow-reducing valves are field adjustable. Some field adjustments are as simple as removing a fitting that reduces the size of the discharge opening. Other valves are more difficult to adjust and may require special tools to regulate the pressure. Instructions for increasing the flow and/or pressure should be included in preplans, and members must be shown how to change the pressure

A.

B.

Figure 7-14 A. Standpipe equipment. B. Fire fighters with standpipe equipment.

Courtesy of Bernard J. Klaene.

setting during training sessions. The preplanning survey should include arrangements to obtain special standpipe adjustment tools. The pressure-regulating tool should be added to the standpipe equipment before ascending to the fire floor. Pressure-reducing valves must be installed and maintained properly to ensure that they can provide the required volume and pressure for firefighting operations. With proper preplanning and maintenance, it should not be necessary to field-adjust pressure-reducing valves.

Automatic nozzles do not provide good flows at pressures below their design parameters. Some automatic nozzles require a nozzle pressure of 100 psi (689 kPa) and will not provide any flow at very low pressures. These nozzles should not be used when connected to a standpipe. Conversely, smooth-bore nozzles can produce an adequate stream at low discharge pressures and are often the best choice for standpipe operations. A 2½-in. (65-mm) line with 50 psi (345 kPa) of nozzle pressure and a 1¼-in. (32-mm) tip will flow approximately 326 GPM (1234 L/min). At 30 psi (207 kPa) of nozzle pressure, a 1¼-in. (32-mm) smooth-bore nozzle will flow approximately 253 GPM (958 L/min).

At the other extreme, it is possible to find standpipe discharge pressures above 130 psi (896 kPa). A 2½-in. (65-mm) line with some variable-stream nozzles will produce a large flow at this pressure. The 1¼-in. (32-mm) smooth-bore nozzle is not designed to operate at such high pressures. After discounting a friction loss of 55 psi (379 kPa), the 1¼-in. (32-mm) tip will be flowing nearly 400 GPM (1514 L/min), with a nozzle reaction force of nearly 220 psi (1517 kPa). In this situation, an in-line regulator/gauge placed on the standpipe outlet would be an invaluable tool. The solution is having the right tools at the right place, as well as knowing how and when to use them.

Nozzle manufacturers are now designing nozzles for standpipe use that can be changed from variable to solid bore without changing the tip. When deciding on standpipe hose and nozzles, it is important to determine the actual flow capabilities at the pressures available at standpipe discharges. Some nozzles are significantly more efficient than others. When changing from a variable-stream to a smooth-bore nozzle, it is important to recognize that there may be a significant increase in flow with a resultant increase in nozzle reaction (see the *Offensive Operations* chapter for more information regarding rate of flow and nozzle reaction).

Department training should include the use of the various types of nozzles at low and high pressures. It is essential that members of the department be familiar with and operate standpipe equipment under conditions similar to those found in your community and at the various pressures anticipated. These training sessions should be in full personal protective equipment, including standard-issue gloves. Many tasks that are easily performed in street clothing with clear visibility are very difficult to carry out under fire conditions wearing full fire gear.

Departments should also conduct flow tests (see the *Offensive Operations* chapter) to determine the actual flow using various hose layouts at expected pressures. The time to safely experiment and learn how to properly use firefighting equipment is during fire department training sessions. Trial and error on the fire ground can produce deadly results for both civilians and fire fighters.

Connecting to the Standpipe Discharge

Most departments have standing orders requiring connection to the standpipe one floor below the fire. Excess hose in the stairway may prove to be a problem. In the book *Firefighting Principles and Practices*,[9] William E. Clark presents a method of dealing with the excess hose. Hose is laid up the stairway above the fire floor before fire fighters enter the fire floor. This allows them to enter the fire floor pulling the hose *down* the steps rather than up, while keeping excess hose out of the way. In using this tactic, it is important to lay the hose up the steps before making entry to avoid being above the fire when the door to the fire floor is opened.

Recognize that fire hose in the stairway will impede occupant egress and that smoke conditions in the stairway may place occupants at greater risk. Whenever possible, it is best to dedicate a stairway for fire operations and other stairs for occupant evacuation. In many cases, only one stairway has standpipe outlets, making it the obvious choice as the fire operations stairway. Every effort should be made to keep occupants out of the fire operations stairway. See the *High-Rise Buildings* chapter for a more in-depth discussion regarding protecting occupant egress in high-rise buildings.

In conducting preincident planning, the standpipe outlets must be identified and special operational requirements noted. As with sprinkler operations, the preincident plan may identify operational needs that are different from department SOPs.

Non-Water-Based Extinguishing Systems

In addition to water-based systems, there are many other fixed extinguishing systems that use different agents. Following is a list of fire suppression agents and types of systems that use them:

- Foam
  - Surface application
  - Subsurface application
  - Deluge
- Halon (and other clean agents)
  - Total flooding
  - Local application
- Carbon dioxide
  - Total flooding
  - Local application
  - Extended discharge
  - Hand lines
- Dry chemical
  - Local application
  - Hand lines
- Other inerting systems (using inert gases to extinguish or contain a fire)

There is an extensive variety of systems and system components that can be used, depending on the hazard being protected. Each of these specialized systems will influence the development of a strategic plan. Therefore, fire officers should have a good

working knowledge of all systems. In most cases, the systems will probably have activated before the fire department's arrival. In atypical situations, the IC may need to order that the system be activated manually.

The proper use of in-place suppression devices is yet another reason for preincident planning. The preincident plan should include a drawing showing the location of system components (risers, shutoffs, pumps, and agent supply containers).

Some systems can create hazards for the occupants or fire fighters. For instance, carbon dioxide can suffocate anyone inside the protected enclosure. Dry chemical can cause physical harm and, at the very least, obscure vision.

Foam Systems

High-expansion foam systems are designed to protect buildings; however, these systems are rare. They are designed to fill an area, such as a basement, with foam, thereby smothering the fire.

Low-expansion foam systems are usually found at properties storing large quantities of flammable and combustible liquids. Some low-expansion foam systems are automatic; others require fire department support. Even fully automatic foam systems have provisions for manual operation.

Refineries and petroleum storage depots normally protect aboveground storage tanks with a foam system. A **foam house** will be located on the property or nearby, containing additional quantities of foam agent and a means of manually operating the system. System operation will be different at each facility, and the responding fire department should be familiar with the hazards being protected and the operation of the foam system. During the preplanning survey, consider the proximity of the foam house to the flammable or combustible liquid hazard being protected. If the foam house is located too close to the hazard, it may be unsafe to enter the area when the fire is in close proximity to the foam house. A department protecting such a facility that fails to prepare a preincident plan is rightfully open to public criticism.

Fire departments must train on operating the foam system but should also insist that written operating instructions be posted within the foam house. Foam is the number one defense against flammable liquid fires but is nearly useless on pressurized liquids or gases. Remember, for the automatic system, let the system do its job and support it with additional water supply and foam as needed. For the manual system, gain access to the foam house and operate the system as required.

At bulk storage facilities, the quantity of foam required to suppress a fire may be very large. The quantity should be determined either by the property owner or during preincident planning. If the required volume is not immediately available on site, then part of the preincident plan should be directed toward protecting exposures until additional quantities of foam can be obtained. Without the required quantity of foam on site, it may be counterproductive to launch a fire attack that will ultimately fail. Furthermore, foam will not suppress a three-dimensional fire, such as may occur with leaking fuel.

<u>Class A foam</u> increases the effectiveness of water and adheres well to exposed structures. Therefore, at incidents where these are important operational objectives, the use of Class A foam should be considered. Class A foam systems are rare.

Carbon Dioxide Systems

There are two types of automatic carbon dioxide systems: total flooding and local application. In addition, some occupancies will provide carbon dioxide hose reels for manual application of the agent. Carbon dioxide is used in areas where preventing water damage is a prime objective or where the extinguishing agent is more effective than water or dry chemical. Carbon dioxide is not a conductor of electricity and is also used near electrical equipment. Carbon dioxide systems rely on detectors for their activation. Storage of carbon dioxide is limited; therefore, the system has a limited discharge time and coverage.

Total flooding systems are dependent on agent containment for a period of time (soak time) to be effective. If ventilation systems are not shut down or the compartment is opened, the carbon dioxide will quickly dissipate and the fire may rekindle. Carbon

dioxide extinguishes by depleting the oxygen supply, and a carbon dioxide system generally requires carbon dioxide concentrations ranging from 34% to 75%. A room flooded with carbon dioxide may appear normal while actually being oxygen deficient. Carbon dioxide is very cold when discharged, as well as being heavier than air, and may accumulate in low or remote locations. Fire fighters entering a room where a carbon dioxide system has discharged must wear SCBA.

Halon and Other Clean Agents

Many large computer installations are protected by sprinklers because of the reliability factor. Clean agent systems are also used because of their ability to react quickly and to suppress a fire in its beginning stages, without damaging sensitive equipment. The quick extinguishing capabilities of Halon have long been recognized in explosion-suppression systems in which deflagrations are actually suppressed before pressure builds up.

Halogenated agents were greeted by many as the extinguishing agent of the future. They were considered the ultimate agents for all situations because they were believed to be nontoxic, nondamaging, are easily applied, and thought to leave the compartment safe for human habitation during discharge. The excitement that surrounded the use of halogenated agents in the early days has disappeared. We now know that halogenated agents can damage equipment, and although some do not cause immediate death, a health risk may be associated with staying in the compartment after the agent has discharged. Halon is not considered toxic, but the products of decomposition (hydrogen fluoride, hydrogen chloride, and hydrogen bromide) are harmful to humans and might damage some electronic components.

Further, Halon and other clean agents are expensive, and studies now indicate that Halon causes environmental harm by destroying the ozone layer. What was once thought to be a miracle agent is on its way to oblivion. Other clean agents are now being used, and halogenated agents are being phased out.

Fixed clean agent systems rely on smoke detectors for activation. They have a limited supply of agent that must be discharged into a confined room or area. Most clean agents do not require suffocating quantities to extinguish the fire; a low concentration is usually sufficient. However, to be effective, the room must remain closed, and the ventilation system must automatically shut down on discharge.

Clean agents are generally considered nontoxic to humans in the concentrations that are found in computer rooms and other installations. This is a time to be skeptical of current beliefs. Insist that fire fighters entering a room where a discharge has occurred wear SCBA until the room has been completely ventilated.

Dry and Wet Chemical Systems

Dry chemical systems are used in a number of different applications. One of the most common is in kitchen hoods in restaurants. Because of the ability of dry chemical systems to suppress fires in cooking appliances, ductwork, and other related areas, it is widely used for this purpose. Dry chemical systems are used in many other applications, such as dip tanks and gasoline-dispensing facilities. However, kitchen hood systems are the predominant application.

A kitchen hood system can be activated either automatically or manually at a pull station. For automatic operation, fusible links are usually located over the area being protected and/or in the ductwork. In the event of a fire, the link will fuse, releasing the tension on a cable that, in turn, will cause the dry chemical agent to be discharged through a series of nozzles.

Wet chemical agent systems, similar in design to dry chemical systems, are also found in kitchen hood applications. Wet chemical agents react with hot grease or oil to form a foam blanket that suppresses the release of combustible vapors. Property owners often prefer wet agent systems because cleaning up after the system discharges is much easier than with dry chemical agents.

As with the other specialized systems, these systems will probably have already discharged before the fire department's arrival.

Working in Areas Protected by Total Flooding Carbon Dioxide or Clean Agent Systems

Letting the System Do Its Job

If the system is controlling the fire, maintain the chemical concentration by keeping the doors closed. Unless occupants failed to escape, there is no need to enter the area if the fire is being controlled. Entering the area will allow the carbon dioxide or clean agent to dissipate, thereby reducing its effectiveness. Unlike the sprinkler and standpipe systems, carbon dioxide and clean agent systems have a very limited supply of extinguishing agent.

Final Extinguishment and Rescue

If it becomes necessary to enter the room to perform a rescue or for final extinguishment, members must wear full protective clothing and must don their face pieces before entering. After a carbon dioxide release, the area may appear to be completely clear yet pose a serious hazard due to a lack of oxygen. Clean agents may also deplete the oxygen levels, and they may pose a threat because of corrosive decomposition gases. Overhaul operations must be completed, especially where Class A materials are involved. Otherwise, rekindles may occur as the agent concentration is diluted over time.

Manual Activation

These systems will generally be equipped with a manual actuation device that can be operated in the event of fire where automatic detection/activation sensors fail. Some systems are equipped with an abort switch, which can be held to prevent agent discharge. Sometimes the abort switch is located outside the protected area with no provisions to determine what is going on inside the protected area. Employees may decide to keep the agent from

Incident Summary

An early morning fire occurred in a restaurant in Boston, Massachusetts. The cleaning crew was attempting to remove grease from the burners by covering them with aluminum foil and then turning the burners on high. A fire occurred and spread into the ductwork above the stove. The dry chemical system did not automatically discharge, and the fire spread upward, inside the ductwork, until it reached a mechanical room on the fifth (top) floor. From that point, the fire was able to ignite the roof structure. The roof on this historic building was completely burned off as a result of a kitchen fire five stories below it.

Data from: Edward R. Comeau, unpublished NFPA fire investigation

discharging for a number of reasons. The first-arriving company must determine whether preventing system discharge is justified.

Checking Interlocks

Fire suppression systems often trip interlocking devices when they are activated. At a minimum, there will be provisions for shutting down the ventilation system when using carbon dioxide and clean agents. Carbon dioxide systems and many clean agent systems have preaction alarms interlocked to the system, allowing occupants time to escape. In most cases there is a means of manually activating the interlock. If ventilation systems are operating, it is imperative that they be closed, especially with carbon dioxide total flood application. However, if the ventilation system has been running for a period of time after discharge of either carbon dioxide or clean agents, then the concentration of agent within the room may be insufficient to suppress a fire, at which time manual hose streams may be necessary.

Checking Agent Supply

Occasionally, a cylinder or other container holding clean agents or carbon dioxide may be accidentally shut off and not reopened. There may be extra containers of agent that can be connected to the system once the original supply has been depleted. A fire fighter should be assigned the task of checking supply valves, or if extra supplies of agent are available, an entire company may be needed to connect the additional supply. However, this task may require specialized knowledge, and by the time an assessment is made, the fire may have grown too large for the system to control. Manual suppression may be required, and the IC should be prepared for this eventuality.

System Restoration

System restoration will, by necessity, be left to the property owner and/or a contractor who is capable of recharging and resetting the system.

Working in Areas Protected by Local-Application Carbon Dioxide, Clean Agents, Dry Chemical, or Other Special Extinguishing Agents
Letting the System Do Its Job

Make sure valves are fully open and do not interfere with system operation.

Checking the Interlocks

Interlocks may shut off fuel supplies or deenergize equipment. Manual operation of the interlocks may be possible, or employees may be able to shut down equipment as needed.

Manual Activation

Support the system by activating manual devices when necessary. These systems are generally protecting Class B or Class C hazards and do not depend on an enclosure.

Backing Up the System

Be prepared with backup equipment, such as hose lines, foam lines, or portable extinguishers as required to augment the system and/or complete overhaul.

In kitchen hood systems, it is common for dry chemical systems also to protect the ductwork leading from the cooking appliances to where it exhausts outside of the building. It is important to inspect the entire ductwork and exhaust system to verify that the fire did not spread beyond the cooking appliance.

System Restoration

System restoration should be left to the property owner or a contractor who is capable of recharging and resetting the system.

Responses to Building Fire Alarm Systems

False alarms transmitted by fire suppression or fire alarm systems are common. In some areas they have become so frequent that special responses are sent to fire alarm activations. Although the number of fire alarms can be a nuisance, it is important to remember that a percentage of these alarms will be for actual fires. The high number of false alarms causes apathy, and apathy lulls fire forces into complacency. Then, when least expected, the inevitable occurs and lives and property are lost—lives and property that could have been saved had the alarm been heeded.

False alarms are often the result of poor system maintenance or improper installation. Unfortunately, no one has all of the answers to this problem. Jurisdictions have assessed penalties on property owners who have an excessive number of false alarms, but this may cause the property owner not to call when the alarm sounds or to shut down the system. For good reason, departments insist on a rapid call when there is an emergency. If the property owner investigates the alarm before calling the fire department, the call for assistance will be delayed.

Particularly difficult are automatic alarms sounding inside a tightly secured building. These situations often present the IC with a dilemma. Forcing entry can cause significant damage; however, failing to enter the building may place people or property at risk.

Fallacy	**Fact**
Alarms from automatic alarm systems are always false.	Automatic alarm systems are often activated when there is no fire. Different systems have different degrees of accuracy. Treating automatic alarms as false alarms can have disastrous outcomes.

Steel: © Sharpshot/Dreamstime.com

To assist fire department personnel in gaining entry after business hours, the building owner may be willing to provide a lockbox with emergency access keys or make similar arrangements. However, if a decision is made to force entry, it should be done in a manner that causes as little damage as possible.

Direction should be provided to the owner through code adoption, local ordinance, or official notification from the fire chief or fire marshal. Further, the fire department should work with the building owner to identify entry options in advance and include them in the preincident plan.

Chapter Summary

- If the building is equipped with an automatic fire suppression system, the primary tactic is to support the system and let it do its job.
- Preincident planning is essential for any building that is protected by an automatic fire suppression system.
- With few exceptions, a properly designed, maintained, and operated sprinkler system will prevent large loss of life and property.
- To avoid depriving the sprinkler system of an adequate water supply, fire department personnel should not take water from the same source to supply hose lines.
- The main objective when working in a building protected by an automatic fire protection system is to support the system and let the system do its job.
- When working in a building protected by a sprinkler system, back up the system with hose lines and ventilate.
- Hazards protected by deluge systems may pose additional dangers to fire fighters.
- A properly operating and maintained building standpipe system may be helpful in conducting offensive attacks, but it is not an automatic system; it must be manually operated.
- Five major types of standpipe systems exist: automatic wet and dry, semiautomatic dry, and manual wet and dry systems. Automatic wet standpipes are most common and most reliable.
- When working from a standpipe system, check for proper operation of the main control valve and fire pumps, and supply the fire department connection (FDC).
- Fire departments should provide equipment needed for standpipe operations along with standard operating procedures (SOPs) for standpipe use.
- Preplanning of buildings with non-water-based fire suppression is essential, as improper use of and/or exposure to some agents can be harmful to fire fighters.
- Unless there are victims inside, do not enter an area where a carbon dioxide or clean agent system has discharged if the fire is being controlled.
- If it becomes necessary to enter a room protected by a carbon dioxide or clean agent system to perform a rescue or for final extinguishment, members must wear full protective clothing and must don their self-contained breathing apparatus (SCBA) face pieces before entering. The fire department should not assume that any alarm is false, even if multiple false alarms have occurred from the same system.

Key Terms

backpressure Also known as elevation pressure, the pressure required to overcome the weight of water in a piping or hose system. Each vertical foot of water in a pipe, hose, or tank exerts a pressure of 0.434 psi (3 kPa) at the base.

Class A foam Foam for use on fires in Class A fuels such as vegetation, wood, cloth, paper, rubber, and some plastics.

deluge system A sprinkler system in which all sprinklers or applicators are open. When an initiating device, such as a smoke detector or heat detector, is activated, the deluge valve opens and water discharges from all of the open sprinklers simultaneously.

draft curtains Walls designed to limit horizontal spread of the fire that extend partially down (usually no more than 20% of the height of the compartment) from the underside of the roof.

Flames: © Dnz/Dreamstime.com; Steel texture: © Sharpshot/Dreamstime.com

dry pipe sprinkler system A system in which the pipes are normally filled with compressed air or nitrogen. When a sprinkler is activated, it releases the air from the system, which opens a valve so the pipes can fill with water.

foam house A fixed facility consisting of an enclosure housing a foam concentrate supply tank, a foam solution proportioning system, a pump, and sometimes an extra supply of foam concentrate that can be added to the proportioning system.

preaction sprinkler system A dry sprinkler system that uses a deluge valve instead of a dry pipe valve and requires activation of a secondary device before the pipes fill with water.

standpipe system A piping system with discharge outlets at various locations; in high-rise buildings an outlet will normally be located in the stairway on each floor level. Most are connected to a water source and the pressure is boosted by a fire pump.

wet pipe sprinkler system A sprinkler system in which the pipes are normally filled with water.

Challenging Questions and Activities

1. Using the information provided in this chapter and in the *Procedures, Preincident Planning, and Size-Up* chapter, develop a preincident plan for a sprinkler-protected property.

2. Using the information provided in this chapter and in the *Procedures, Preincident Planning, and Size-Up* chapter, develop a preincident plan for a building equipped with a standpipe system.

3. Compute the pump discharge pressure needed on the top floor of the tallest building in your response district, including areas covered by mutual/automatic aid agreements.

4. Compare standpipe equipment used by your jurisdiction to the suggested list of equipment in this chapter. Justify your present standpipe hose and nozzle combination or recommend a different hose size or nozzle type.

5. Using the information provided in this chapter and in the *Procedures, Preincident Planning, and Size-Up* chapter, develop a preincident plan for a facility equipped with a Class B foam system.

6. Using the information provided in this chapter and in *Procedures, Preincident Planning, and Size-Up* chapter, develop a preincident plan for a building protected by a non-water-based extinguishment system.

WRAP-UP

Endnotes

1. National Fire Protection Association. *NFPA 13R: Standard for the Installation of Sprinkler Systems in Low-Rise Residential Occupancies.* Quincy, MA: 2013.

2. National Fire Protection Association. *NFPA 13D: Standard for the Installation of Sprinkler Systems in One- and Two-Family Dwellings and Manufactured Homes.* Quincy, MA: 2013.

3. Hall, J. R., National Fire Protection Association. *NFPA's U.S. Experience with Sprinklers.* Boston MA, 3/2012, NFPA Fire Analysis and Research. Quincy, MA: NFPA; 2012.

4. National Fire Protection Association. *NFPA 13: Standard for Sprinkler Systems.* Quincy, MA: 2010.

5. Reproduced from: NFPA 13 E. Recommended Practices for Fire Department Operations in Properties Protected by Sprinkler and Standpipe Systems. Section 4.1.1 Copyright © 2010, National Fire Protection Association. This reprinted material is not the complete official position of the NFPA on the referenced subject, which is represented only by the standard in its entirety.

6. Reproduced from: NFPA 25. Standard for the Inspection, Testing, and Maintenance of Water-Based Fire Protection Systems. Section 15.3.1 and 15.3.2. Copyright © 2014, National Fire Protection Association. This reprinted material is not the complete official position of the NFPA on the referenced subject, which is represented only by the standard in its entirety.

7. National Fire Protection Association. *NFPA 14: Standard for the Installation of Standpipe and Hose Systems.* Quincy, MA: 2010

8. Klem, T. J. *One Meridian Plaza, Philadelphia, Three Fire Fighter Fatalities.* 2/23/91, Fire Investigation Report. Quincy, MA: NFPA; 1991.

9. Clark, W. E. *Firefighting Principles and Practices* (2nd ed.). Saddle Brook, NJ: PennWell; 1991.

Flames: Dr2/Dreamstime.com; Steel texture: © Sharpshot/Dreamstime.com

Flames: © Dx/Dreamstime.com; Steel texture © Sharpshot/Dreamstime.com; Chapter opener photo © Rubber Ball Selects/Alamy Images

CHAPTER

8

Offensive Operations

LEARNING OBJECTIVES

- Compare an offensive fire attack to a defensive fire attack, explaining the basics of each type of attack and identifying the rationale for each strategy.
- Describe trial-and-error methods of calculating rate of flow.
- Describe "softening the target" and how it is different from an indirect attack.
- Analyze rate of flow requirements using Royer/Nelson (V/100), National Fire Academy (A/3), and sprinkler calculations.
- Define and contrast ventilation-controlled and fuel-controlled fires.
- Describe "area of involvement" and how it applies to rate of flow calculations.
- Write a brief position paper outlining the advantages of using the Royer/Nelson (V/100) rate of flow formula.
- Explain why a fire attack meeting or exceeding the calculated rate of flow could fail to extinguish the fire.
- Explain the relationship between nozzle type, rate of flow, and nozzle reaction force.
- Describe extinguishment of ordinary combustibles by inhibiting pyrolysis.
- Define internal and external exposures.
- List factors to consider when evaluating external exposures.
- Describe the purpose of a backup line and how it can be used to protect fire fighters attacking the fire.

LEARNING OBJECTIVES (CONTINUED)

- Evaluate water supply requirements based on rate of flow and other factors.
- Examine the relationship of and proper use of ventilation during offensive extinguishment operations.
- Discuss apparatus management and the factors used to determine the number of apparatus needed at an offensive operation. Develop a list of advantages and disadvantages when using Class A foam during structural firefighting.
- Compute and compare the rate of flow for various areas using A/3 and V/100.
- Evaluate the available flow from standard preconnected hose lines and determine when the rate of flow for a structure should be preincident planned.
- Estimate the number and size of hose lines needed to apply a calculated rate of flow.
- Assess staffing requirements for an offensive attack based on rate of flow and life safety factors.
- Assess the probability of an imminent life-threatening situation.
- Using a fire scenario, assess the total water supply available and apparatus needs in terms of required fire flow.
- Given fire conditions and location, determine the ventilation possibilities and choose the best ventilation method(s).
- Evaluate the flow available from a standpipe system and standard fire department standpipe equipment based on a calculated rate of flow.
- Examine and evaluate various attack positions in a multistory building.
- Discuss factors involved in choosing an offensive strategy.

Case Study

Flames © Jag_cz/ShutterStock, Inc.; Steel texture: © Sharpshot/Dreamstime.com; Stacked photo background: © Vitaly Korovin/ShutterStock, Inc.; Paper: © silver-john/ShutterStock, Inc.; Photo: © Crystal Craig/Dreamstime LLC

View of side C during the initial stages of the fire as Engine 5 was entering from side A.

Courtesy of the Peabody Fire Department.

On December 23, 2011, at 12:55 PM, two engine companies, a truck company, and a deputy chief were dispatched to a report of fire in an apartment building. Fire companies on the initial alarm were staffed with a lieutenant, driver, and fire fighter. The deputy chief was first on the scene and assumed the role of incident commander (IC). He observed heavy smoke on the second floor, side A, of a three-story apartment building. He established command, declared a working fire, and entered the building, encountering fire in Apartment 3 on the second floor. The IC then instructed the first-arriving engine company to attack the fire in Apartment 3. The engine company advanced a 1¾-in. (45-mm) hose to the second floor while the truck company made their way to the third floor. After a delay, the lieutenant and fire fighter on the first-arriving engine company entered Apartment 3, which flashed. The lieutenant, with his helmet on fire, was pulled out, but the location of the fire fighter was unknown. The lieutenant returned to Apartment 3 with a thermal imaging camera but was unable to locate the fire fighter. A personal alert safety system (PASS) was sounding, but still fire fighters were unable to find the fire fighter. The officer of a later-arriving company finally located the fire fighter, who succumbed to a cardiac arrhythmia.

1. How does staffing affect safety during an offensive fire attack?
2. What is the value in "softening the target" before entering a fire area?
3. Why should an initial rapid intervention crew be established prior to making entry to the fire compartment?

Data from: NIOSH Report 2011-31 Career Fire Fighter Dies during Fire-Fighting Operations at a Multi-family Residential Structure Fire – Massachusetts

Introduction

An offensive fire attack is the preferred strategy whenever conditions and resources permit an interior attack. A defensive decision limits operations to the exterior, generally resulting in a larger property loss and limiting rescue options. The offensive versus defensive decision is based on staffing available to conduct an interior attack, water supply, ventilation, and most importantly a risk-versus-benefit analysis. Fire fighters should not enter a building that is in imminent danger of collapse or in which fire conditions do not permit safe entry. Likewise, resource capabilities in terms of staffing, apparatus, and water must be able to meet incident requirements for a safe and effective operation. Rate of flow, as described in this chapter, is a major factor in determining if resources are adequate. The objective of an offensive

fire attack is to apply enough water directly to the burning fuel to achieve extinguishment, thus providing a safer work environment for fire fighters, facilitating rescues, and reducing property damage. The *Fire Fighter Safety* chapter addresses fire fighter safety, and the *Life Safety* chapter describes rescue tactics. Offensive operations are the most effective way to save lives and property, but are potentially the most dangerous strategy. The offensive/defensive decision is critical and the basis of the entire incident action plan. The ability to extinguish the fire is critical to a safe and effective offensive strategy.

The essential question to answer when addressing extinguishment is how many gallons per minute (liters per minute) are required to extinguish a given fire with properly placed hose lines. Calculating the rate of flow allows the incident commander (IC) to match the number and size of fire lines to flow requirements. In most cases, applying water directly to the burning fuel completes the extinguishment process. However, large offensive fire operations require the IC to consider many variables to successfully save lives and extinguish the fire.

NOTE

An offensive attack is preferred when conditions and resources permit interior operations.

Texture: © Eky Studio /Shutterstock, Inc; Steel: © Sharpshot/Dreamstime.com

Calculating Rate of Flow

Notable authors of fire tactics textbooks have addressed rate of flow in different ways. In his book *Firefighting Principles and Practices*, William Clark uses a derivation of the fire compartment volume in cubic feet divided by 100 (V/100) to determine the rate of flow in gallons per minute.[1] In *Fire Command*, Alan Brunacini states, "When the IC is able to apply more water than the fire can match with heat, we win. Until the IC reaches this level, the fire will continue to burn and eventually win (if the IC cannot overpower it)."[2] Others in the fire service take a trial-and-error approach.

John Coleman notes that a hose stream's extinguishing capability will generally be determined in about 30 seconds. He further recommends a trial-and-error approach in stating, "If you are at the top of a stairway and have a line directing water at a well-involved second floor or attic and you don't darken down the fire within 30 seconds or so, get more water."[3]

NOTE

Rate-of-flow calculations as discussed in this text apply only to offensive attacks in ordinary combustible (Class A) materials.

Texture: © Eky Studio /Shutterstock, Inc; Steel: © Sharpshot/Dreamstime.com

This text takes an approach similar to Clark's methodology. However, trial-and-error methods can also be useful. With trial-and-error methods, fire fighters start with their favorite preconnected hose line, advance the hose line into the building, and, in many cases, make progress on the fire. Successes can reinforce bad habits, warding off any discussion of a more scientific approach. However, when trial-and-error fire fighters fail to make progress with their favorite line, they simply add more lines of the same kind. When they are forced out of a building after taking a severe beating during an unsuccessful effort, there is always an excuse. Seldom do they recognize their failure to flow enough water to overpower the fire. The continued use of initial attack lines when they are clearly overpowered by the fire is a sign of poor training and a lack of fire-ground discipline.

Many refer to this failure to recognize the need for larger fire streams as a "residential mentality." One method of avoiding the problem of always using the same size fire line is to write standard operating procedures (SOPs) requiring the use of 2½-in. (65-mm) attack lines when confronted with a fire in a large commercial building. This protocol provides a higher rate of flow but is less efficient than using a preplanned rate of flow. Some large commercial buildings are divided into small compartments where one or two 1¾-in. (45-mm) hose lines would be sufficient and easier to maneuver. As an example, office buildings or strip malls are often subdivided into small individual compartments. Other offices and strip malls have large open spaces where the flow from multiple 2½-in. (65-mm) hose lines would be needed for an advanced fire.

The success rate of the trial-and-error method is highly dependent on the flow rate of the standard preconnected hose line. The larger the flow from the preconnected line, the higher the success rate. With the advent of 1¾-in., 2-in., and 2½-in. (45-, 50-, and 65-mm) preconnected attack lines, success rates have improved dramatically from the days of 1-in. (25-mm) or booster line preference.

The trial-and-error method is probably the most commonly used method of determining flow rate. Unfortunately, many fire officers fail to recognize the need for larger flows when the situation arises.

NOTE

Placing a third small-diameter hose line in the same compartment is generally a tactical error.

Texture: © Eky Studio /Shutterstock, Inc; Steel: © Sharpshot/Dreamstime.com

All of the methods of calculating rate-of-flow requirements have some imprecision built into them, because it is not possible to take into account every variable that will be encountered at a fire scene. Simply put, how much water is needed to overwhelm the fire? Should operations begin with 2½-in. (65-mm) hose lines? Should the second hose line be a 2½-in. (65-mm) hose? Applying rate-of-flow principles should answer these questions.

Fallacy	**Fact**
Rate-of-flow calculations are used to determine the exact amount of water needed and then to apply a prescribed flow in gallons per minute.	Rate-of-flow calculations yield a rough approximation of the required flow. They assist the IC in deciding the number and size of fire lines needed for an offensive operation.

Steel: © Sharpshot/Dreamstime.com

Three rate-of-flow calculation methods are in general use, and all have certain advantages and disadvantages. They are described and compared here.

- Royer/Nelson: $\frac{\text{Volume in ft}^3}{100} = \text{GPM}$

  $\frac{\text{Volume in m}^3}{0.748} = \text{L/min}$

- National Fire Academy: $\frac{\text{Area in ft}^2}{3} = \text{GPM}$

  $\frac{\text{Area in m}^2}{0.074} = \text{L/min}$

- Sprinkler calculations: Fire load–specific calculations

NOTE

The purpose of rate-of-flow calculations is to determine the size and number of hose lines needed when a fire progresses beyond the normal room-and-contents residential fire when an interior, offensive attack is still practical.

Texture: © Eky Studio /Shutterstock, Inc; Steel: © Sharpshot/Dreamstime.com

Indirect Application Theory

Lloyd Layman's indirect attack theory[4] is based on the very efficient cooling effect of water being converted to steam. Indirect attacks and resultant heat absorption result in the ability to control fairly large fires with relatively little water. Fire tests conducted while Layman was in the Coast Guard are the basis of his theory. Theoretically, the indirect attack is valid. In practice, it places occupants' lives in jeopardy and is very ineffective under most structural fire conditions. This proves two important points about research:

1. A hypothesis can be theoretically sound, or even seem logical, yet be empirically incorrect.
2. Good test methods can lead to erroneous results when extended to conditions beyond the scope of the research.

When reading technical reports, it is important to keep in mind the conditions under which the experiment was conducted and the correlation to actual conditions experienced on the fire ground before implementing new tactics based on the research.

Softening the Target

Recent research conducted by Underwriters Laboratories indicates that "softening the target" by applying water into the fire compartment as quickly as possible, from the exterior or interior, can make conditions in the entire structure better.[5] After in-depth study, this research appears to be valid and is recommended in this text. However, it must be remembered that Underwriters Laboratories' vertical ventilation research was conducted using typical house fires where the fire had not entered concealed spaces to become a structure fire.

It seems that softening the target tactics could be applied to larger compartment fires and be a useful tactic for structure fires, but theoretical and empirical data are lacking at this time. One of the outcomes of nearly every research study is identifying areas for future research. Underwriters Laboratories' vertical ventilation research identifies future research needs as follows:

This project built on the previous horizontal ventilation study and a future study on positive pressure ventilation in houses like these would be very beneficial to have the three main types of ventilation studied under similar conditions. Additional vertical ventilation experiments should be conducted where the fire has transitioned to a structure fire. Experiments where the fire is in the attic space prior to fire department arrival and the corresponding impact of vertical ventilation may be assessed under different conditions. Future experiments should examine different suppression flow rates, utilizing various suppression media.

The need for additional research is clear. In addition, the fire service must share experiences in using the softening the target tactic, especially in larger compartments such as those found in commercial occupancies and in structure fires. At the time of this writing, Underwriters Laboratories is conducting research into positive-pressure ventilation and various ways to attack attic fires.

Many departments are applying the softening the target tactic. Undoubtedly, this tactic has many advantages, not the least of which is increased safety for fire fighters conducting an offensive fire attack. In most cases, a company officer will direct the initial attack. All fire department personnel should be trained in the proper application and limitations of this tactic. The research fires were all advanced fires in houses. The softening the target principle has wider application but does not apply to every fire. Some examples where softening the fire may not be applicable include, but are not limited to:

- A fire inaccessible from the outside at ground level (e.g., upper floor of a high-rise building)
- A compartmented area with no outside access point (e.g., center room in a large office building)
- A small fire that fails to gain intensity (e.g., food on the stove, electrical fire, clothes dryer fire)

The International Association of Fire Chiefs published a Firefighter Safety Call to Action in reaction to the Underwriters Laboratories research.[6] It says in part:

Water doesn't push fire or threaten trapped occupants:

- Water should be applied to a fire as soon as possible and from the safest location because research has proven it reduces thermal temperatures.
- Simply put, if you see fire, put water on it immediately. This greatly increases civilian and firefighter survivability as well as property conservation.

If written as an SOP this could be misinterpreted to say, "Water should be applied to a fire as soon as possible from the safest location." This would miss the critical point: "*if you see fire.*" Following this SOP could result in fire fighters opening doors or breaking windows and discharging water into an area within a building where no fire is visible from the outside or door opening. The Underwriters Laboratories research clearly applies to a fire that is rapidly growing toward flashover or one that is in temporary decay waiting for a breath of fresh air. In the absence of visible fire or heavy and/or pushing smoke conditions, the soften the target tactic does not apply.

Softening the Target Versus Indirect Application

Quite often, practitioners attempt to use a favorite tactic in every situation. Such predilections have often caused problems in the fire service. In the case of indirect application, fire departments have attempted to apply the principle to all fires, even situations that obviously did not resemble the tightly closed shipboard fires from which the theory evolved. The prescribed rate of flow for indirect applications, as described by Lloyd Layman, would be much less than any of the flow calculations in common use today.

Advocates of indirect application point to the reduced water usage and resultant reduction in property loss. In fact, the loss due to proper use of water in a direct attack is overstated, and failure to extinguish the fire will inevitably result in far more damage. It is important to note that indirect application disrupts the heat balance within the room, and the combination of high heat and humidity will greatly decrease the chance of survival for anyone occupying the compartment.

At first glance, softening the target, as advocated in Underwriters Laboratories' vertical ventilation research, seems to be similar to indirect application, but there are significant differences. In the case of indirect application, extinguishment is achieved by continued application of a high-pressure fine fog stream into a tightly enclosed space. During Underwriters Laboratories' experiments, water is applied for 15 seconds to soften the target. Live burns conducted at Governors Island, New York, resulted in a finding that the use of wide-angle fog or even rapid movement of a straight or solid stream operated through a window could have a negative ventilation effect. The correct way to soften the target is to apply a straight stream or smooth-bore stream at a steep angle through the window and off of the ceiling to allow water to rain down onto the burning contents and cool the hot gases. This method allows much of the window to remain a ventilation point for gases and steam, thus minimizing the negative impacts of heat and steam being forced along the flow path into the structure. When applying softening the target techniques, coordinated ventilation is recommended.

Pulsing is widely used to reduce fire growth and prevent flashover by directing short blasts of water toward the ceiling. Pulsing and softening the target are very similar in that they reduce heat so fire fighters can enter the fire compartment for final extinguishment.

The Royer/Nelson Formula

Keith Royer and Bill Nelson of Iowa State University developed fire flow calculations for structural firefighting based on test fires. They hypothesized that most fuels release 535 British thermal units (564 kilojoules) per pound when combined with 1 ft^3 (0.0283 m^3) of oxygen. Thus, the fuel load and fire type are not as important as the room volume and can be ignored because fires are **ventilation controlled**. This proposition is somewhat in conflict with standards developed by fire protection engineers in determining sprinkler flow calculations. Sprinkler flows are adjusted based on the size and type of fuel.

The **Royer/Nelson formula** is based on the premise that the best rate of application is one that results in control of the fire within 30 seconds of effective application (80% efficiency). Iowa State University conducted many test fires to evaluate fire behavior; the Royer/Nelson volume formula was a result of observations and measurements taken at actual test fires. Royer and Nelson calculated the rate of flow in gallons per minute as the volume of the fire area in cubic feet (V) divided by 100, resulting in the formula V/100, as illustrated in **Figure 8-1**. This is the formula adapted in Clark's book *Firefighting Principles and Practices*.

The major assumption of Royer and Nelson—that structure fires are primarily oxygen (ventilation) controlled—is true for many fires. This assertion was further authenticated in the recent Underwriters Laboratories research comparing flashover in legacy furnishing to contemporary furnishings. Under conditions of the Underwriters Laboratories experiments in small residential compartments the

Length × Width × Height = Volume

$$\frac{\text{Length} \times \text{Width} \times \text{Height}}{100} = \text{Rate of Flow (GPM)}$$

Figure 8-1 The Royer/Nelson rate-of-flow formula.

fire remained entilation limited even when the compartment was well ventilated.[7] However, as the volume of the compartment increases, the fire may be **fuel controlled**, thus negating the major assumption behind the Royer/Nelson theory. These large-area fires are the ones most likely to require larger hose lines and the application of rate-of-flow principles. **Oxygen- (ventilation-) controlled fires** require less water than a free-burning fire. Therefore, the Royer/Nelson formula may understate the needs in a well-ventilated large-area fire.

Gross underestimation is most likely to occur when fires reach a point beyond the capability of an interior attack and other factors such as the reach of fire streams, water supply, pump capacity, and staffing become more important. Therefore, the Royer/Nelson formula (V/100) will be valid for most fires where an interior attack is advisable.

Although Clark advocates the V/100 formula, he acknowledges that the progression to flashover is based primarily on the ratio of the surface area of the fuel to the size of the enclosure. If other factors such as fuel type and fuel load are held constant and the volume of the enclosure is increased, the time to flashover will also increase.

The process leading to flashover and backdraft is described in basic fire fighter training manuals, such as *Fundamentals of Fire Fighter Skills*.[8] Although

the conditions necessary for backdraft (an oxygen-deficient atmosphere) and resultant explosive effect are much different than flashover, the process leading to each event is similar. The major difference is the oxygen-starved condition necessary for backdraft and the oxygen-sufficient condition required for flashover. The progression to flashover begins with a fuel being heated until vapors being generated (via pyrolysis) ignite. The fire continues to burn and grow to a point of oxygen depletion. This results in a ventilation-controlled fire, followed by a vented or completely involved fire. If the fire is allowed to smolder at high heat levels rather than venting, the potential for backdraft exists. Too many variables occur in actual fires to accurately predict the progress of an ignition-phase fire to the fully developed phase. However, the effect of volume is fairly predictable. A large-fire area has more oxygen and therefore remains a fuel-controlled fire for a longer period. The heating of all combustibles is delayed as heat rises and reaches equilibrium in large-area fires. All of these factors lead to the conclusion that no rate-of-flow calculation is completely accurate. The suggestion by Underwriters Laboratories mentioned previously for additional research into rate-of-flow calculations is absolutely warranted given the fact that the last field-expedient flow research was conducted over 50 years ago using equipment and tactics no longer in use. However, to be useful, the new formula must be tested on large-volume fire areas and be easily applied. A formula requiring difficult and complex mathematical calculations would be less useful, and rate-of-flow calculations are not needed for small fire areas where standard attack lines can easily control the fire.

The National Fire Academy Formula

Rate-of-flow calculations are used to determine whether the fire is beyond the capacity of the standard attack and to determine how many and what size hose lines are needed for an interior attack. Most departments begin an interior attack using one

1¾-in. (45-mm) hose line with a required backup line at least as large as the initial attack hose line. The backup line can be used to augment flow if necessary. Therefore, the normal attack for most fire departments would be equal to the flow from two standard preconnected hose lines. Thus the flow delivered using SOPs will vary depending on the size of the backup line and the flow produced by the standard preconnected hose line. With this in mind, the more conservative and probably less accurate rate-of-flow calculation introduced by the U.S. National Fire Academy[9] cannot be fully discounted. Test fires, such as those used to develop the Royer/Nelson formula, were not used as a basis for the **National Fire Academy formula**. Therefore, the validity of the formula is yet to be proven.

The National Fire Academy developed the A/3 formula for use in their fire tactics courses **Figure 8-2**. Many experts believe that calculating the volume at the scene of a structure fire is not practical, thus justifying the simpler area calculations. An article by Royer and Nelson recognizes that the V/100 rate-of-flow formula is primarily a planning tool.[10]

The A/3 and V/100 formulas can yield very different numbers. Typically, the A/3 formula requires more water than the V/100 formula. The A/3 formula generally overstates the need, but as the ceiling height increases, the A/3 formula becomes less conservative.

Large-volume fires are the fires that require the application of a rate-of-flow formula. Any mathematical calculation on the fire ground is difficult, but experience can be gained through practice by estimating volume or area during training. These estimates can later be compared to the actual area and volume.

Given the difficulty in calculating the rate of flow at the fire scene, a strong recommendation is made to include the rate of flow as part of the preincident plan information when the flow exceeds the capabilities of the two standard initial attack hose lines used by the department. For example, many departments attach an automatic nozzle to four sections of 1¾-in. (45-mm) hose with a pump discharge pressure of 150 pounds per square inch (psi) (1034 kilopascals [kPa]) as their initial attack line. Depending on the hose and nozzle configuration, this 1¾-in. (45-mm) initial attack configuration yields a flow of approximately 125 GPM (473 L/min). The standard backup line is the same as the initial attack line. Therefore, this department would preincident plan any building with a compartment where the rate of flow exceeds 250 GPM (946 L/min). The rate of flow would then be listed for applicable compartments within the structure—in this case 25,000 ft³ (708 m³). If one 2-in. (51-mm) attack line with a 1-in. (25-mm) smooth-bore nozzle (approximate flow of 240 GPM [908 L/min]) were used as the initial attack hose line with a matching backup line, then a preincident plan would be suggested for any building with a compartment requiring more than a 480-GPM (1817-L/min) rate of flow, or 48,000 ft³ (1359 m³).

Length × Width = Area

$$\frac{\text{Length} \times \text{Width}}{3} = \text{Rate of Flow (GPM)}$$

Figure 8-2 The National Fire Academy rate-of-flow formula.

NOTE

Rate of flow should be included in preincident plans when the size of the largest compartment exceeds the rate of flow established by the department's SOPs for the initial attack, including provisions for a backup line.

Texture: © Ely Studio /Shutterstock, Inc; Steel: © Sharpshot/Dreamstime.com

Sprinkler Rate-of-Flow Calculations

Both the Royer/Nelson and National Fire Academy formulas ignore the fuel load and fuel type; this is a substantial weakness in these formulas. For this reason, **sprinkler system calculations** may prove useful in preincident planning. There are several sources of sprinkler calculations, including NFPA documents and Factory Mutual Data Sheets.

Fires go through a series of stages from ignition phase to flashover. These stages are characterized by being fuel dependent in the early stages and then oxygen dependent in later stages. A well-involved fire in a vented or large open area will be controlled by both fuel and oxygen. Therefore, the IC must consider the weaknesses of the rate-of-flow calculations being applied.

A fire in an area containing a small fire load will generally require less water than one in a similar area with a large fire load unless the fire is totally oxygen controlled in a tightly enclosed space. Once this space is vented, either by the fire or by fire fighters, the fire will react to the amount and type of available fuel. Remember, rate of flow is being used to determine the size and number of hose lines needed and whether the fire is beyond the capabilities of available resources. The IC decides when it is time to use larger, more powerful streams. The IC also uses rate-of-flow calculations to determine whether there are extenuating circumstances (i.e., reasons other than rate of flow) when operations are unsuccessful. If two 1¾-in. (45-mm) lines are not controlling a fire with a calculated rate of flow of 100 GPM (379 L/min), the problem is probably a failure to effectively apply water. Proper ventilation often will make the fire more visible and allow a fire fighter to get closer to the fire, thus improving the chances of successful direct application. However, ventilation will also increase fire intensity by increasing the air supply. By first softening the target, the fire intensity is reduced, allowing coordinated ventilation to control the flow path while effectively applying water onto the burning fuel.

Estimating the Size of the Largest Area

Rate-of-flow calculations are based on the area or volume of the compartment(s) on fire. There is some difference of opinion among practitioners of rate of flow as to whether to make allowances for compartmentation. Some advocate applying rate of flow to the entire area or volume on fire; others see each individual compartment as a separate fire area.

This text recommends calculating each room as a separate fire area. This has been a point of discussion over the years, but Royer and Nelson's research clearly indicates that each enclosure should be calculated separately.

In a discussion with Keith Royer, he emphasized that flows should be determined considering each room capable of holding heat as a separate fire area. Royer elaborated by stating that he was talking not just about areas enclosed by fire walls and fire doors, but about any enclosure, even if the door is missing. This is an extremely important distinction. Few large areas are undivided, and it is, in fact, easier to fight fires in a series of smaller compartments than to do battle in one large, undivided area. Fires completely involving a large, undivided area often require a defensive attack.

In looking at the plan view for the apartment building in [Figure 8-3], those who advocate calculating the entire area would apply the rate-of-flow formula to the entire space. Some would recommend adding each involved floor to the total. The more rational approach would calculate each individual space as a separate fire area. If each individual space is calculated separately, it becomes possible to move down the hallway, then into each apartment, extinguishing the fire in one room at a time. Instead of calculating the entire area, only the flow needed for the largest area (the open dining and living room in this case) is required.

Fallacy	Fact
The entire floor volume is calculated to determine the rate of flow.	Calculate the flow needed for the largest single area on fire. Each compartment (room) can be handled as a separate fire, and the fire can be extinguished in one room at a time.

Steel: © Sharpshot/Dreamstime.com

Figure 8-3 Calculating each fire compartment as a separate fire area.

Practical limitations apply to this logic, as it probably would not be possible to enter the floor if the entire floor were on fire. This would lead to a defensive attack where the rate of flow does not apply. If several rooms are on fire, it may be advisable to use multiple lines to expedite extinguishment by simultaneously attacking more than one room at a time.

Estimating the size of the largest involved area is extremely difficult on the fire scene. The IC does not always have an exact fire location or a detailed floor plan. Preincident planning the rate of flow where it exceeds the capabilities of two standard attack lines is highly recommended. When confronted with a fire exceeding the rate of flow for two standard preconnected hose lines with no preincident-planned rate-of-flow data available, the IC will be forced to estimate the size of the involved compartment to determine the rate-of-flow requirement or else revert to trial-and-error methods. Fire fighters making the initial attack may be able to give the IC an estimated compartment size. Also, for a large, undivided building, the IC can estimate the length, width, and height by comparing them to sections of fire hose or to the height of a pumper **Figure 8-4**. However, these methods are much more difficult and less reliable than consulting a preincident plan that includes rate-of-flow calculations. Departments that fail to preincident plan buildings requiring a large rate of flow place their ICs at a considerable disadvantage.

Either of these calculations (A/3 or V/100) is possible at the fire scene, and each provides only a rough approximation. Both calculations assume that the entire area is undivided. If the area is subdivided, the rate of flow will be less and possibly within the capabilities of 1¾-in. (45-mm) hose lines if the compartments are small. If the building is not subdivided, the large rate of flow calculated by either method would indicate that once a fire involves a considerable portion of the building, an interior attack would probably be unsuccessful even if fire fighters could safely enter. This undivided structure most certainly would *not* be a situation in which additional 1¾-in. (45-mm) lines should be used. If the building is undivided and involves a considerable area, a fire in this structure will require interior 2½-in. (65-mm) hose lines or interior master stream appliances.

Given that calculating rate of flow is not an exact science, most fire officers would estimate, round off, and shortcut the math. For example, a building that

Front of building approximately two 100 ft (30 m) of hose length = 200 ft (61 m)
Width of building approximately half the front = 100 ft (30 m)
Height approximately twice the height of a pumper = 25 ft (8 m)

Volume = 200 x 100 x 25 ft = 500,000 ft^3(14,640 m^3)

$V/100 = \dfrac{500,000}{100}$ = 5000 GPM (18,925 L/min)

Area = 200 x 100 = 20,000 ft^2 (1858 m^2)

$A/3 = \dfrac{20,000}{3}$ = 6666.67 or 6700 GPM (25,360 L/min)

Figure 8-4 Field-expedient area and volume estimations.

is 32 × 32 ft (9.8 × 9.8 m) would become 30 × 30 ft (9 × 9 m). Dividing the first 30 ft by 3 yields 10 × 30 ft, yielding 300 GPM (10 × 30 for the A/3 rate-of-flow calculation). The metric calculation yields 1136 L/min. This A/3 calculation results in a rate of flow exceeding the standard 1¾-in. (45-mm) pre-connected hose line. Multiple 1¾-in. (45-mm) hose lines may be sufficient. Advancing a 2½-in. (65-mm) hose line as a backup line would exceed the 300-GPM (1136-L/min) requirement for this fire.

By taking this concept of matching the hose line to the rate of flow one step further, it is possible to determine in advance the number and the size of hose lines needed to combat fires in dwellings, apartment buildings, and small businesses. This is a form of preincident planning—that is, preincident planning by general occupancy type. Examining a typical dwelling or apartment building will provide the fire officer with general rate-of-flow information that is very useful in initiating an offensive attack in these properties. For example, calculating the rate of flow for a 20- × 15-ft (6.1- × 4.6-m) living room with a 10-ft (3-m) ceiling yields:

$$A/3 = 20 \times 15 \text{ ft} = 300 \text{ ft}^2 \text{ divided by } 3$$
$$= 100 \text{ GPM (379 L/min)}$$

$$V/100 = \frac{20 \times 15 \times 10 \text{ ft}}{100} = 30 \text{ GPM (117 L/min)}$$

For this situation, 1½-in. (38-mm) or 1¾-in. (45-mm) lines are acceptable.

Calculating the rate of flow for a 20- × 25-ft (6.1- × 7.6-m) basement recreation area with a 10-ft (3-m) ceiling yields:

$$A/3 = 20 \times 25 \text{ ft} = 500 \text{ ft}^2 \text{ divided by } 3$$
$$= 166 \text{ GPM (628 L/min)}$$

$$V/100 = \frac{20 \times 25 \times 10 \text{ ft}}{100} = 50 \text{ GPM (189 L/min)}$$

A 1¾-in. (45-mm) hose line is acceptable only when using the V/100 formula. These smaller areas within residential properties tend to reaffirm the lower V/100 rate-of-flow calculations. When discussing rate of flow in terms of the advantages of 1¾-in. (45-mm) and larger hose lines, fire fighters who preferred using 1-in. (25-mm) booster hose ("red line") often challenged the need for 1¾-in. (45-mm) hose, stating that they successfully extinguished many fires with the "red line." Most fires occur in dwellings and many in very small rooms within a dwelling. A 10- × 12- × 8-ft, or 960-ft³ (3- × 3.7- × 2.4-m, or 26.6-m³), bedroom is common in a home. Dividing the volume (960 ft³ [26.6 m³]) by 100 (or in metric, 0.748) yields a rate of flow of 9.6 GPM (35.6 L/min). The "red line" easily produces this flow. Calculating the same room using the A/3 formula computes to 120 ft² divided by 3 (or in metric, 11 m² divided by 0.074), which yields 40 GPM (150 L/min). The maximum flow from the low-volume nozzles typically attached to the "red line" is 40 GPM (182 L/min). Thus, the successful encounters with the "red line" tend to verify the V/100 formula. However, anything less than a 1¾-in. (45-mm) line for a structure fire is unacceptable. The larger the flow from the initial hose line, the greater the chance of immediate success. However, larger 2½-in. (65-mm) hose lines are very difficult to maneuver inside the small rooms in most residential occupancies.

The preceding basement recreation area example represents a fairly large single area for a dwelling. Given that the National Fire Academy figures tend to overstate the case, and there should always be a backup line, it can be concluded with a fair degree of confidence that fires within one- and two-family dwellings are normally within the flow capabilities of a 1¾-in. (45-mm) or 2-in. (50-mm) hose line with backup line. Only extraordinarily large, open-layout dwellings would require rate-of-flow calculations and subsequent preincident planning.

Apartment buildings can also be included in this category, as they normally have smaller undivided areas, with two exceptions. Apartment buildings with common attics or basements can, and do, present large-area fires requiring additional water.

Incident Summary

On January 31, 2008, Cincinnati fire fighters were dispatched to a residential fire. Two engine companies arrived on scene simultaneously. One company began an offensive attack using a standard pre-connected 1¾-in. (45-mm) hose line. The second company initially knocked down fire threatening an exposure on the D side. The second company then made entry with a 1¾-in. (45-mm) hose line. One company attacked the fire in the foyer and hallway, while the other company extinguished the fire in two rooms on the first floor. They were able to extinguish the main body fire and prevent extension to the second floor with two 1¾-in. (45-mm) hose lines due to the limited-volume rooms.

Working fire in a two-story residential building with attic.

Courtesy of Denny Baker, Cincinnati Fire Department

Larger apartment buildings and complexes may also have common areas (e.g., recreation rooms, dining rooms) requiring a large rate of flow. Apartment complexes may be preplanned for reasons other than rate of flow. As an example, maps showing locations of buildings and streets may be needed to find specific addresses within an apartment complex made up of multiple buildings.

Estimating the Percentage of Area on Fire

The Royer/Nelson V/100 formula is based on fuel consumption being oxygen controlled by the volume of the enclosure and *does not* reduce the rate of flow by the percentage of volume of the enclosure involved in fire. The National Fire Academy rate-of-flow formula is based on the area of involvement rather than the volume of the enclosure; therefore, a reduction by the area of involvement is recommended. This difference has to do with an assumption by Royer/Nelson that the fire is oxygen controlled.

The National Fire Academy recommends a percentage-of-involvement modifier in applying the A/3 rate of flow. For example, if one fourth (25%) of the floor area is involved, divide the total flow by 4 (or multiply by 0.25). This modifier appears to be reasonable. It takes less water to extinguish 100 ft² (9 m²) of fire than to extinguish 400 ft² (37 m²) of fire.

It is important to point out that the total V/100 flow will not always be required even though a percentage of involvement is not calculated. During the early stages of a fire in a large area, the fire will be fuel controlled, and less than the total calculated flow will be needed. Most fires can be controlled with less than 1 GPM (4 L/min) during the ignition phase.

In most cases, the first company advancing to the fire area will provide the reconnaissance needed to determine the fire area in terms of percentage of involvement and compartmentation. However, heavy smoke conditions may make it difficult or impossible to accurately determine the percentage of involvement.

Advancing small-diameter preconnected hose lines (1¾ in. or 2 in. [45 mm or 50 mm]) is faster than advancing 2½-in. (65-mm) hose. Unless the size of the fire is obvious, the first company advancing to the fire area may decide to use the standard preconnect to size up the situation from the interior. It is important to get companies into all areas exposed to the fire as soon as possible to determine the life safety and extinguishment needs. Some departments specify the use of a 2½-in. (65-mm) hose line as the primary attack line for commercial buildings, which increases the chance of success when confronted with a large-volume fire. The use of 2½-in. (65-mm) hose lines is highly recommended, as preconnected lines reduce the time needed to deploy larger streams when a large rate of flow is required for extinguishment.

The differences between the V/100 and the A/3 formulas are less significant if the percentage involvement is considered. This text recommends the V/100 for preincident planning, but if A/3 calculation is used at the incident scene (because of a lack of preincident planning), a percentage involvement modifier should be used. Since the purpose of calculating rate of flow is to determine the size and number of hose lines needed, it is reasonable to apply either formula. However, V/100 is based on a substantial body of research and is preferred for that reason.

Comparing Rate-of-Flow Calculations

A three-story building that is occupied as an auto parts store will be used as an example. The office and sales area are on the first floor, as shown in **Figure 8-5**. The second floor is a storage area for metal automobile parts in cardboard boxes. The third floor is used to store automobile and truck tires, on end. The building is 40 × 60 ft (12.2 × 18.3 m) with 15-ft (4.6-m) ceilings on the first floor and 20-ft (6.1-m) ceilings on the second and third floors.

Royer/Nelson (V/100) rate-of-flow calculations would be as follows:

$$\text{First floor: } \frac{40 \times 60 \times 15 \text{ ft}}{100} = 360 \text{ GPM (1362 L/min)}$$

Figure 8-5 Auto parts store example.

This GPM is reduced by half for the first floor, as the first floor is subdivided into two 30- × 40-ft (9.1- × 12.2-m) areas, resulting in a flow of 180 GPM (681 L/min) for the first floor.

Second and third floors:
$$\frac{40 \times 60 \times 20 \text{ ft}}{100} = 480 \text{ GPM (1817 L/min)}$$

No reduction is made, as the second and third floors are not subdivided into smaller compartments.

The A/3 formula would calculate the same floors as follows:

First floor: $\dfrac{40 \times 30 \text{ ft}}{3} = 400$ GPM (1514 L/min)

Second and third floors: $\dfrac{40 \times 60 \text{ ft}}{3} = 800$ GPM (3028 L/min)

It is also possible to determine the required rate of flow by using sprinkler calculations. A number of variables would apply in making these calculations, such as the building type, the number of floors, the occupancy type, the commodity inside the structure, and the storage configuration of the commodity. These sprinkler requirements are addressed in NFPA 13, *Standard for the Installation of Sprinkler Systems*.[11]

Properties that are already equipped with a sprinkler system should have a plate on the riser with all of the required water flow requirements stamped into the plate. This information could be very useful for comparing the two methods outlined above with the engineered water requirements.

At first, sprinkler calculations may seem complicated, but with the proper training, they are fairly easy to apply. However, sprinkler calculations will require additional time and a degree of judgment regarding the commodity's classification. Sprinkler rate-of-flow calculations are most accurate, as they are based on actual fire experience and consider the important factors of area of involvement and fuel load.

NFPA 1142, *Standard on Water Supplies for Suburban and Rural Fire Fighting*,[12] identifies occupancies where extra water supplies may be needed. This list could be used as a reference for deciding when to consult the NFPA 13 commodity storage series of standards for extra hazard rates of flow.

Figure 8-6 and **Figure 8-7** compare the rate-of-flow calculations discussed in this chapter. Figure 8-6 uses the actual ceiling height of the auto parts store example. Figure 8-7 shows how A/3 and V/100 calculations are nearly the same as the ceiling height increases to 30 ft (9.1 m). Both figures calculate the sprinkler rate of flow for the rubber tire storage on the third floor using extra hazard sprinkler tables found in the NFPA 13. Sprinkler calculations require a degree of engineering judgment, but the much higher calculated flow shows how high-hazard fuel storage can affect the rate of flow.

In the 30-ft (9.1-m) ceiling height example, the calculated flow using A/3 is still slightly higher, but V/100 and A/3 calculations are much closer. Mathematically, the rates of flow would be equal for a 33⅓-ft (10.1-m) ceiling. Ceilings of this height would be common in warehouse or mercantile occupancies but extremely rare on the upper floors of a high-rise.

Sprinkler rate-of-flow calculations are not suitable for on-scene use unless the preincident plan identifies flow requirements. With a preincident plan, no field calculation is necessary.

Sprinkler calculations apply to rate of flow, but the sprinkler system is far more efficient than manual operations. Sprinklers are more effective in "getting the wet stuff directly on the red stuff," and they prewet the exposed fuel much as fire fighters with hose lines wet exposures. Engineers designing sprinklers allow a margin of safety.

A close look at the Royer/Nelson and National Fire Academy rate-of-flow calculation systems shows that the National Fire Academy system yields a higher rate of flow in most situations. Underestimating the flow will allow the fire to continue to make progress while fire fighters engage in a hazardous and futile battle. By contrast, if the need is overestimated, the fire is extinguished in less time with a reduction in the hazard to the fire fighters. Because all field-estimated rate-of-flow predictions

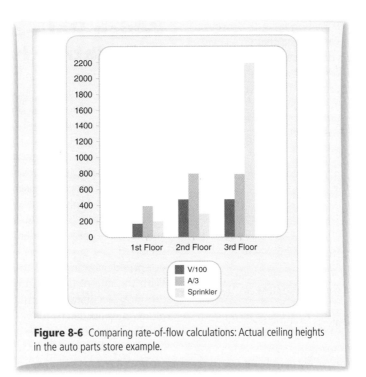

Figure 8-6 Comparing rate-of-flow calculations: Actual ceiling heights in the auto parts store example.

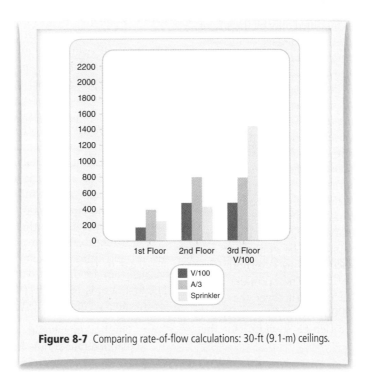

Figure 8-7 Comparing rate-of-flow calculations: 30-ft (9.1-m) ceilings.

are prone to error, overestimation is best. However, on closer examination it can be seen that gross overestimations could cause a delay in starting an aggressive interior attack or possibly lead to a defensive or nonattack decision because of a perceived lack of resources.

An article by C. Bruce Edwards in *Fire Engineering* points out that a rate of flow exceeding the minimum will extinguish the fire in less time and with less total water.[13] In calculating rate of flow at the incident scene, if in doubt, err on the side of higher rates of flow.

Remember that the National Fire Academy recommends a percentage-of-involvement modifier. If 25% of the auto parts store were involved in fire, the calculations would be as follows:

$$A/3 = \frac{40 \times 60 \text{ ft}}{3} = 800 \text{ GPM} \times 0.25 = 200 \text{ GPM (757 L/min)}$$

V/100 remains the same at:

$$\frac{40 \times 60 \times 20 \text{ ft}}{100} = 480 \text{ GPM (1817 L/min)}$$

The V/100 formula is modified only by the size of the compartment, not by the percentage of involvement, due to the assumption that fuels will produce a given heat quantity per volume of air regardless of fuel size and type.

The rate of flow for sprinkler systems is usually stated as a flow rate (e.g., 0.25 GPM/ft²) over a specific area, such as over the most hydraulically remote area (e.g., 3000 ft²). This estimate generally reduces the flow requirements in large undivided spaces and is based on an assumption that the sprinkler system will operate early in the fire to limit fire extension. This is a valid assumption for a sprinkler system but not for manual firefighting. Therefore, sprinkler flow densities should be applied to the entire compartment when used to determine the rate of flow using hose lines. The sprinkler flow rate example above (0.25 GPM/ft² over the most hydraulically remote 3000 ft²) would be calculated as 0.25 GPM × the area (length × width) for the entire compartment.

Which Rate-of-Flow Calculation Is Best?

As the preceding discussion shows, there is merit in each of the rate-of-flow calculation methods. There are also assumptions and limitations for each. The best approach would build on the relative strengths and weaknesses of the various methods. With this in mind, this text recommends that all buildings requiring a rate of flow exceeding the flow of two standard preconnected hose lines should be preincident planned applying the V/100 formula.

If the fuel load is heavy in terms of total volume and/or fuel type, sprinkler calculations would be preferred, the larger of the two calculations (V/100 or sprinkler) being used to determine rate of flow. With the rate of flow calculated in advance, the IC can more accurately determine the size and number of fire lines needed.

For purposes of discussion, let's assume that the standard preconnected hose line is four sections of 1¾-in. (45-mm) hose with an automatic nozzle and a predetermined pump discharge pressure of 150 psi (1034 kPa). Further, assume the flow rate for this standard preconnected line is 125 GPM (473 L/min). SOPs should require that a backup hose line at least as large as, or larger than, the initial hose line be placed in service whenever there is a working structure fire. Therefore, the preincident plan rate-of-flow threshold for this department would be 250 GPM (946 L/min). Applying V/100 in reverse order would indicate that the volume to be preincident planned in this case would be 25,000 ft³ (708 m³) or larger. As an example, a 25- × 50-ft (7.6- × 15.2-m) room with 10-ft (3-m) ceilings would equal 25,000 ft³ (708 m³). Very few homes will have a room this large.

Returning to the auto parts store example, the second and third floors are 48,000 ft³ (1360 m³), because they are undivided and have higher ceilings. The first floor is 18,000 ft³ (510 m³). Considerations for this area would include the following:

- This building would not require a preincident planned rate of flow for the first floor.

- V/100 would be used to calculate the ordinary hazard on the undivided second floor.
- Sprinkler calculations would be recommended for the extra hazard tire storage on the third floor.
- If there is no preincident plan for this building, after arrival on the scene, the IC may elect to use the A/3 or V/100 formula.

As discussed in the *Procedures, Preincident Planning, and Size-Up* chapter, rate of flow is but one of many reasons to preincident plan a building. When confronted with a fire in an area that is beyond the control of two standard preconnected hose lines and in which the fire flow has not been preplanned, the A/3 formula may be easier to apply. However, this places the IC at a distinct disadvantage, as the building size must be estimated, and calculations performed under stressful fire-ground conditions might not be accurate.

Whether rate of flow is calculated in advance or on the fire ground, the IC should modify the formula on the basis of trial and error. If the fire fighters on the fire floor report that they have enough lines, they probably do (trial and error). Of course, there are times when they fail to see hidden or extended fires, but this is not where rate-of-flow calculations are useful. Likewise, if fire fighters report that the calculated 500-GPM (1893-L/min) rate of flow is not getting the job done, the IC should consider the possibility that water is not being effectively applied to the burning materials. In such cases, a change of tactics or stream placement may be needed. If this part of the operation is not being supervised by a division or group leader, staffing a tactical-level management unit to supervise this critical function may improve extinguishment operations. This also could be a case where a lack of "truck work" is hampering the operation. Proper and coordinated venting of the area may allow attack teams to better apply water directly to the seat of the fire.

Royer and Nelson assume an 80% efficiency in getting water into the immediate fire area. If the nozzle team is attempting to cool the area by softening the target, thus making a close approach tolerable and allowing the engine company crew to get water directly onto the burning material, but these efforts are failing, then the 80% efficiency assumption is not valid. Royer and Nelson also assume the flow to be sufficient to knock down the fire in 30 seconds. This assumption supports John Coleman's assertion that progress should be evident within 30 seconds and that the attack teams should be making steady progress on the fire if the rate of flow is sufficient and being properly applied.

Once again, using the auto parts store as an example, assume that the responding department uses 125-GPM (473-L/min) initial attack lines (1¾-in. [45-mm] hose line with an automatic nozzle). Further, assume that the first attack hose line will be backed up with another attack hose line of the same size or larger. Looking at each floor, we can apply the rate of flow in terms of the number and size of fire lines needed to control a fire.

First Floor

The first-arriving engine company advancing a 1¾-in. (45-mm) hose line into the front or rear of the building would be well positioned to extinguish the fire with the single 1¾-in. (45-mm) hose line, as the rate of flow of 180 GPM (681 L/min) is nearly met.

The second- or third-arriving engine company would back up the initial attack, with another 1¾-in. (45-mm) hose line, providing enough water to easily overpower the fire (approximately 250 GPM [946 L/min] for a 180-GPM [681-L/min] fire).

Second Floor

The first-arriving engine company would advance a 1¾-in. (45-mm) hose line into the building and have little chance of success if the fire involved a substantial portion of the floor area. The recommended rate of flow of 480 GPM (1817 L/min) is far from being met with the 125-GPM (473-L/min) attack line.

If the first-arriving engine company reported anything other than extinguishment with the 1¾-in. (45-mm) hose line, the IC should order a 2½-in. (65-mm) hose line as the second line. Such a line can generate a flow of up to 350 GPM (1325 L/min). The 125 GPM (473 L/min) from the first hose line combined with a maximum 350 GPM (1325 L/min) from the 2½-in. (65-mm) hose line yields a total flow of 475 GPM (1798 L/min). This rate of flow nearly meets the 480-GPM (1817-L/min) requirements and has a good chance of success. If more water is needed, the IC could order an additional hose line or, in consultation with the first-arriving crew, increase the pressure in the 1¾-in. (45-mm) line. The IC could also opt to replace the automatic nozzle with a 7/8-in. (22-mm) solid stream and reduce the pressure while increasing the flow.

Third Floor

The initial 1¾-in. (45-mm) attack hose line would be effective only if the fire were contained to a relatively small area. For extinguishment on the third floor, 2½-in. (65-mm) lines or a master stream would be needed. Preincident plan information could prove crucial in this situation. Given the limited access to the third floor and the flow requirements for a fully involved fire, the department may decide that fires beyond a specific level of progression (e.g., beyond the capabilities of two 1¾-in. [45-mm] lines) would result in a quick and limited primary search and a move to a defensive attack.

When conditions permit, there is merit to initiating an attack on the fire using the standard preconnected hose lines, thus applying a trial-and-error method. Quite often, it is impossible to determine the area involved in fire or the size of the fire from the exterior. A crew advancing a 1¾-in. (45-mm) hose line may be in position to begin their initial attack while performing an interior size-up. However, when the fire is beyond the resources of the available units, a defensive attack is the only option. If the rate of flow is preincident planned, it is much easier to determine when the fire is beyond the department's capabilities to initiate a successful offensive attack.

> **NOTE**
>
> It is possible to have a small fire in a large building, but it is not possible to have a large fire inside a small building.

Texture: © Eky Studio /Shutterstock, Inc; Steel: © Sharpshot/Dreamstime.com

Selecting the Attack Hose Size

Many of the V/100 calculated flows will be less than the flow capacity of a 1-in. (25-mm) hose line, but given the variables involved in calculating rate of flow and the unknown factors in evaluating the fire from the exterior, booster hose is clearly inappropriate for structural firefighting. And, contrary to what many believe, the energy expended in stretching a booster line may be greater than that for a 1¾-in. (45-mm) line.

The use of 1¾-in. (45-mm) attack hose lines is recommended as a minimum. The flow rate for this size hose line will be adequate to extinguish most fires when the prescribed backup hose line follows the attack line. The work effort for 1½-in. (38-mm) and 1¾-in. (45-mm) hose is very similar, so why not go for the extra flow?

Preconnected hose length, pump discharge pressures, and hose diameter are highly dependent on the area being protected. If the department's jurisdiction consists of single-family dwellings, anything more than a 1¾-in. (45-mm) hose line is probably unnecessary in most offensive situations. If these dwellings sit well back from the street, a 250-ft (76-m) or longer hose line may be needed. When the average hose lay exceeds 250 ft (76 m), it would be wise to use 2-in. (50-mm) or larger preconnected lines to compensate for the additional friction loss. Some departments connect one to four sections of 2½-in. (65-mm) hose to the discharge and then connect sections of 1¾-in. (45-mm) hose to the 2½-in. (65-mm) hose to reduce friction loss in long hose lays while providing the mobility of a 1¾-in. (45-mm) attack.

A preconnected 2½-in. (65-mm) line with nozzle offers another tool in the arsenal and should be considered when the department responds to commercial buildings. Having a preconnected 2½-in. (65-mm) attack line can also play a role in reducing the "residential mindset." Local conditions would dictate the length of the preconnected 2½-in. (65-mm) line.

As the hose size increases to 2½-in. (65-mm), mobility decreases. Using a larger hose size is a trade-off of mobility for an increase in flow. Handling a 2½-in. (65-mm) hose line in a small area is often unnecessary and extremely difficult. If the fire requires entering small areas to achieve extinguishment, 2½-in. (65-mm) hose lines may be a liability. Conversely, for fires in large, open areas, the 2½-in. (65-mm) hose line is often the line of choice.

It is possible to use building features, such as fire doors, to hold the fire. However, this is not an ideal tactic in initiating an offensive fire attack. Either you advance on the fire or it advances on you. Remember that the V/100 rate of flow is based on extinguishment in 30 seconds. Some fires will take longer, but steady progress should be possible if the rate of flow meets or exceeds the requirement. If water is applied at less than the required rate of flow, extinguishment will be due to burning all available fuel or to a lack of oxygen. In most cases, either of these situations is less desirable than applying the amount of water necessary to reduce the fuel temperature.

Charts from manufacturers sometimes show flow rates for variable-stream nozzles based on pump discharge pressure and the length of hose attached (e.g., 150 GPM [568 L/min] at a 200-psi [1379-kPa] pump discharge pressure through 200 ft [61 m] of hose). Other manufacturers of variable-stream nozzles base flow rates on nozzle pressure. Smooth-bore nozzles are rated using nozzle pressure; therefore, comparisons are based on nozzle pressure as well. Nozzle pressure is the pump discharge pressure—or in the case of standpipes, the outlet pressure—minus friction loss in the hose and minus pressure loss due to elevation. Pressure loss due to elevation is 0.434 psi (3 kPa) per foot of elevation. In **Figure 8-8**, the friction loss for the hose is an estimate. The nozzle is operating

Pump discharge pressure		Friction loss pressure		Elevation loss		Nozzle pressure
PDP	–	FL	–	EL	=	NP
170 psi (1172 kPa)	–	61 psi (421 kPa)	–	9 psi (62 kPa)	=	100 psi (690 kPa)

Figure 8-8 Calculating pump discharge pressure.

on the third floor; therefore, if each floor is 10 ft (3 m) in height, the pressure loss due to elevation is 0.434 multiplied by 20 = 8.68 psi (60 kPa). Many departments simply use a rule-of-thumb calculation for elevation loss, such as 5 psi (34.5 kPa) per floor. As can be seen in Figure 8-8, the difference between nozzle pressure and pump discharge pressure can be significant.

Most of the loss in the Figure 8-8 example is due to friction loss in the 1¾-in. (45-mm) hose. Using 2½-in. (65-mm) hose would reduce the friction loss by approximately 75%, yielding a much higher flow rate. However, this higher flow rate would not be necessary in the small compartment shown in Figure 8-8. Calculating friction loss is a bit more challenging than elevation, because friction loss in hose varies by size, hose type, manufacturer, age, and condition. It is best to use calibrated pressure gauges to determine the actual friction loss in the hose carried on department apparatus. Using the manufacturer's friction loss information is a second alternative. Information provided on general friction loss tables, such as in Table 8-1 , is a less desirable method and could be significantly different than the actual friction loss.

Notice the difference in friction loss between 1½-in. (38-mm) hose and other hose sizes at 150 GPM (568 L/min). Many departments now use 2½-in. (65-mm) hose off standpipes to reduce the friction loss and improve the flow rate. Reducing the friction loss can dramatically improve flow through the nozzle when standpipe outlet pressures are low. However, 2½-in. (65-mm) hose can be difficult to maneuver through small rooms, such as those found in residential and some office occupancies.

Table 8-2 shows the flow rate for various types of nozzles based primarily on data provided by manufacturers. The flows for automatic nozzles were derived from several manufacturers' data and are provided as examples. Prior to 1998, combination nozzles were rated at 100 psi (689 kPa) of nozzle pressure. Today fire departments choose from a wide variety of combination nozzles. Nozzles are available that offer adjustable flow rates while others limit the options to either low or high pressure. Combination variable-stream and solid-stream nozzles, as well as variable-stream nozzles that are designed to operate at very low pressures, are also being used. Many older nozzle styles will shut down at low pressure.

Table 8-1	Approximate Friction Loss per 100 Feet (30 m) of Hose			
Flow in GPM (L/min)	1½-in. (38-mm) Hose, in psi (kPa)	1¾-in. (45-mm) Hose, in psi (kPa)	2-in. (50-mm) Hose, in psi (kPa)	2½-in. (65-mm) Hose, in psi (kPa)
40 (151)	4.5 (31)	3 (21)	1 (7)	Negligible
60 (227)	10 (69)	5 (34)	2.5 (17)	Negligible
100 (379)	25 (172)	12 (83)	6 (41)	3 (21)
125 (473)	37 (255)	21 (145)	10 (69)	4 (28)
150 (568)	54 (372)	26 (179)	13.5 (93)	6 (41)
175 (662)	Not Recommended	34 (234)	18 (124)	8 (55)
200 (757)	Not Recommended	45 (310)	24 (165)	10 (69)
250 (946)	Not Recommended	70 (483)	37.5 (259)	15 (103)
300 (1136)	Not Recommended	95 (655)	54 (372)	21 (145)
350 (1325)	Not Recommended	Not Recommended	78 (538)	28 (193)

© Ely Studio/Shutterstock, Inc.

Table 8-2 Nozzle Flows

Nozzle Type	Nozzle Pressure in psi (kPa)	Flow in Gallons per Minute (L/min)	Approximate Friction Loss for 150 ft (46 m) of 1¾-in. (45-mm) Hose, in psi (kPa)	Required Pump Discharge Pressure for 1¾-in. (45-mm) Hose, in psi	Approximate Friction Loss for 150 ft (46 m) of 2½-in. (50-mm) Hose, in psi	Required Pump Discharge Pressure for 2½-in. (50-mm) Hose, in psi
Standard automatic nozzle	30 (207)	Not rated				
	40 (276)	Not rated				
	50 (345)	50 (189)	6 (41)	56 (386)	Negligible	50 (189)
	75 (517)	93 (352)	18 (124)	93 (641)	4 (28)	79 (545)
	100 (689)	256 (969)	110 (758)	210 (1448)	23 (159)	123 (848)
Low-pressure automatic nozzle	30 (207)	110 (416)	23 (159)	53 (365)	5 (34)	35 (241)
	40 (276)	172 (651)	50 (345)	90 (621)	12 (83)	52 (359)
	50 (345)	193 (731)	63 (434)	113 (779)	14 (97)	64 (441)
	75 (517)	237 (897)	95 (655)	170 (1172)	20 (138)	95 (655)
	100 (689)	273 (1033)	122 (841)	222 (1531)	26 (179)	126 (869)
Smooth bore 7/8-in. (22-mm)	30 (207)	123 (466)	32 (221)	62 (427)	5 (34)	35 (241)
	40 (276)	142 (538)	37 (255)	67 (462)	9 (62)	49 (338)
	50 (345)	159 (602)	44 (303)	94 (648)	10 (69)	60 (417)
	75 (517)	195 (738)	65 (448)	140 (965)	16 (110)	91 (627)
	100 (689)	225 (852)	86 (593)	186 (1282)	18 (124)	118 (814)
Smooth bore 15/16-in. (24-mm)	30 (207)	142 (538)	37 (255)	67 (462)	9 (62)	39 (269)
	40 (276)	165 (625)	47 (324)	87 (600)	11 (76)	51 (352)
	50 (345)	184 (697)	59 (407)	109 (752)	14 (97)	64 (441)
	75 (517)	225 (852)	86 (593)	161 (1110)	18 (124)	93 (641)
	100 (689)	260 (984)	113 (779)	213 (1469)	23 (159)	123 (848)
Smooth bore 1¼-in. (32-mm)	20 (138)	206 (780)			15 (103)	35 (241)
	30 (207)	253 (958)			23 (159)	53 (365)
	40 (276)	292 (1105)			28 (193)	68 (469)
	50 (345)	326 (1234)			37 (255)	87 (600)
	75 (517)	400* (1514)			54 (372)	129 (889)
Vindicator® heavy attack (broken/ aspirated)	25 (172)	175 (662)	51 (352)	75 (517)	12 (83)	37 (255)
	40 (276)	210 (795)	75 (517)	115 (793)	16 (110)	56 (386)
	50 (345)	250 (946)	105 (724)	155 (1069)	23 (159)	73 (503)
	75 (517)	345 (1306)			42 (290)	117 (807)
	100 (689)	440* (1666)			66 (455)	166 (1145)

* Exceeds recommended maximum flow of 350 GPM (1325 L/min) for a handheld hose line

© Ely Studio/Shutterstock, Inc.

Incident Summary

A fire began in the fully sprinkler-protected Kmart warehouse in Falls Township, Pennsylvania, where aerosol cans containing petroleum-based liquid fell from a pallet after a forklift entered the aisle. A flammable liquid vapor released from the damaged aerosol cans ignited. The fire quickly spread to involve large areas of the warehouse. The aerosol cans became flaming missiles penetrating through conveyor openings in the fire wall. The fire overwhelmed the sprinkler system, making manual extinguishment necessary. The fire department had preincident planned the building and decided that the best tactic would be to confine the fire to a single quadrant.

Calculating the rate of flow shows that a large fire in any quadrant would be beyond the capabilities of the fire department; therefore, using the building's construction features to contain the fire was a good plan. The height of the structure was 30 ft (9.1 m). Quadrant A was the smallest compartment at 215,600 ft² (20,030 m²). If the fire flow were calculated for this building, it would be as follows:

Preincident plan of Kmart warehouse.

$$A/3 = \frac{215,600 \text{ ft}^2}{3} = 71,867 \text{ GPM } (272,017 \text{ L/min})$$

$$V/100 = \frac{6,468,000 \text{ ft}^3}{100} = 64,680 \text{ GPM } (244,814 \text{ L/min})$$

These calculations are for the smallest area. The fire actually began in the larger Quadrant B. Could any fire department apply 65,000 or 70,000 GPM (246,052 to 264,979 L/min) in an offensive mode? Could fire streams operated from entry points reach the minimum required distance of 440 ft (134 m) to the far wall of any compartment? Adding to the problem was an unprotected steel truss roof structure. Fire departments must carefully compare the risk to the benefit before committing to an interior operation that will ultimately prove futile. Even if the percentage modifier allowed in A/3 rate-of-flow calculations were used and only 1% of the area was heavily involved in fire, the value of property-saving operations would be questionable.

Data from: Richard Best, "$100 Million Fire in KMart Distribution Center," NFPA Fire Journal, March 1983, pages 36–42; rate-of-flow comparisons by authors.

In addition to variable-stream nozzles, there are smooth-bore and broken/aspirated nozzles.

In choosing a nozzle it is important to identify features that will improve fire-ground operations.

Problems associated with nozzle malfunctions increase with more complex nozzle designs. This is especially true if there are a wide variety of nozzles in use by the department. Do you really need to be able to adjust the flow at the nozzle? Is it important to change from a high- to a low-pressure setting?

One of the advantages of a preconnected hose line is establishing a standard pump discharge pressure, which eliminates the need to calculate hydraulics at the incident scene. However, it will be necessary to add or subtract elevation pressure to maintain a predetermined nozzle pressure. In Figure 8-8, a standard pump discharge pressure might be 150 psi (1034 kPa). With the hose line operating on the third floor, it would be necessary to add 9 or 10 psi (62 or 69 kPa) to the pump discharge pressure.

Data were also gathered from fire departments that tested nozzles. It is important to note that actual test results varied, with some nozzles of the same type and manufacture providing greater or lesser flows. For new nozzles, the flow rates were generally within 10% of the manufacturers' data. However, through actual field testing, departments discovered that improperly operating nozzles produced lower than expected flows, and hose of the same size yielded higher or lower friction loss. Both of these factors affected flow rates. Flow rates will vary according to nozzle pressure, pump discharge pressure, and length of the hose lay. Flow rates will also vary between manufacturers and even for different nozzles produced by the same manufacturer. Table 8-2 provides general comparisons for a variety of nozzles.

Probably the most important point to be made is that departments should flow test their nozzles on a regular basis to ensure they are operating properly and to confirm the flow that can be expected from the hose/nozzle configurations available on their apparatus. If any part of the apparatus, hose, or nozzle configuration is changed, flow testing should be conducted. Every department should perform flow tests on their hose and nozzle combinations using a calibrated flow meter. To do this, first place a flow meter in a hose line between a hydrant and the intake side of the pumper. The flow meter should not be connected directly to the hydrant. After making sure all drains are closed, and the tank to pump line is shut, advance the standard attack line. Place a pressure gauge in the hose line behind the nozzle and another 100 ft (30 m) back from the nozzle. Increase the pump discharge pressure to the pressure specified in the department's SOPs. The pressure gauges will measure the actual friction loss in the hose, and the flow meter will provide the actual flow for your hose and nozzle combination **Figure 8-9**. This is also a good time to try various hose layouts, including standpipe hose and nozzles. When testing standpipe hose and nozzle combinations, it is important to consider the possibility of extremely low pressures. Determine the rate of flow if the pressure at the standpipe discharge is only 40 psi (276 kPa) or 30 psi (207 kPa).

Most manufacturers do not recommend 1¾-in. (45-mm) hose layouts longer than 250 ft (76 m)

Figure 8-9 Flow/pressure test layout.

because of excessive friction loss in 1¾-in. (45-mm) hose at high flows. If longer hose lines are needed, larger-diameter hose should be used to lengthen the hose line to reduce the friction loss and increase the flow. If mobility is important, the last section of the hose line could be reduced to 1¾-in. (45-mm) hose. Try different nozzles with different lengths and sizes of hose to determine the best combinations.

Training is essential regardless of the hose/nozzle combination your department chooses. If the nozzle has flow or low-pressure selection devices, be sure fire fighters are completely familiar with how to operate these devices. If the nozzle has a flush mode, it is important that fire fighters be trained on proper operation of the nozzle. After flushing the nozzle, always check to make sure it is returned to the normal operating mode prior to being placed back in service. Most important, fire fighters must train in full turnout gear with all of the nozzles available to them under conditions that simulate those found on the fire ground, such as low visibility, lying on the floor, advancing the hose line through doorways, and directing the stream at different points in an enclosed area.

It is essential that fire fighters experience the nozzle reaction force during training. The fire ground is not the place to experiment. If the nozzle reaction force is too great, reduce the flow or replace the nozzle with one that is easier to control. Remember that handling a nozzle from a standing position with the hose in a straight line behind the nozzle is much different than controlling the nozzle from a crawling or kneeling position with the hose at an angle. When nozzle reaction is near the upper limit for handheld lines, a surge in the hose line could cause fire fighters to lose control. Always allow a margin of safety. Nozzle reaction force usually increases as the flow and nozzle pressure increase. However, some nozzles are designed to reduce nozzle reaction force or are equipped with devices, such as pistol grips, that assist in controlling the hose line. When purchasing nozzles, compare flow rates and nozzle reaction force in order to select a nozzle that produces the desired flow and can be safely handled under actual fire-ground conditions.

NOTE

Pump discharge pressure is generally set at 150 to 200 psi (1034 to 1379 kPa) for automatic nozzles. Pump discharge pressure can be directly read at the apparatus pump panel, which simplifies pump operations. Elevation loss and friction loss in the apparatus piping and hose reduce the pressure reaching the nozzle. The friction loss in a pumper and given length of hose varies, thus the flow will be different when operating from different apparatus and when using different hose.

Texture: © Eky Studio /Shutterstock, Inc; Steel: © Sharpshot/Dreamstime.com

As noted previously, a handheld hose line can generate a flow of up to 350 GPM (1325 L/min). It is possible to control a hose line flowing considerably more than 350 GPM (1325 L/min) by forming a hose loop. However, forming hose loops to counteract the nozzle reaction force is not practical inside most buildings.

Pressures available at standpipes can be significantly lower than those provided by a fire department pumper, and the pressure can be different depending on the building's elevation. Some high-rise buildings will have pressure- or flow-reducing valves on the standpipe (see the *Fire Protection Systems* chapter for details). These valves are designed to maintain predetermined pressures throughout the building. However, there have been occasions when these valves were improperly set, resulting in very low pressures at the standpipe outlet. At the Meridian Plaza fire in Philadelphia, Pennsylvania, improperly set pressure-reducing valves created a major problem (see the *High-Rise Buildings* chapter).

NFPA 14, *Standard for the Installation of Standpipe and Hose Systems*,[14] requires that a sign be posted, usually at the main control valve, stating the location of the two most hydraulically remote hose connections and the designed flow and pressure at these connections. It is important to note the standpipe pressure and flow on the building preplan. If pressure- or flow-reducing valves are used, note the pressure and flow at locations so equipped. Without pressure- or flow-reducing valves, the most hydraulically remote outlets should present the worst-case pressure and flow scenario, with other outlets in the building providing higher pressure and flow capability. Standpipe systems must be inspected and tested on a regular basis. During preincident planning, it is important to note the most recent residual test pressures.

Standpipe outlets with a residual pressure exceeding 100 psi (689 kPa) or a static pressure exceeding 175 psi (1207 kPa) are required to be equipped with pressure-reducing valves. If a pressure higher than this is needed for fire department operations, the pressure-reducing valve should be adjusted. For most situations, the maximum outlet pressure to a flowing hose line will be 100 psi (689 kPa). Some older standpipe systems are required to provide only 65 psi (448 kPa). There may be older standpipe systems with even lower pressures due to different code requirements at the time of installation and/or reduction in interior pipe diameter due to corrosion or other factors.

Some automatic nozzles are designed to operate at a 100-psi (689-kPa) nozzle pressure (nozzle pressure will be less than the pressure at the standpipe due to friction loss). A nozzle pressure below 100 psi (689 kPa) will result in a reduced—and possibly ineffective—fire flow in a nozzle designed to operate at a 100-psi (689-kPa) nozzle pressure. If the pressure is too low, some automatic nozzles will shut down. The low-pressure nozzles listed in Table 8-2 were designed to operate at lower pressures. Solid-stream nozzles will provide some flow at almost any pressure; however, at very low pressures, the stream will be ineffective, and at very high pressures, it will be difficult to safely handle the attack line. The flow increases in proportion to the nozzle pressure. Some automatic nozzles can be fitted with a smooth-bore tip, while others have a built-in solid-bore nozzle. When selecting a nozzle/hose combination for standpipe operations, consider the pressure at standpipe discharges and select standpipe hose and nozzles based on the lowest pressure. If there are large

pressure differences in standpipes or buildings that require a large rate of flow, it may be necessary to provide different nozzles and/or hose for these buildings. A pressure gauge with a control valve is included in the recommended standpipe equipment list in the *Fire Protection Systems* chapter. This device allows the fire fighter at the standpipe outlet to reduce high standpipe outlet pressures. Changing standpipe hose from 1¾-in. (45-mm) to 2-in. (50-mm) or 2½-in. (65-mm) will result in a greater nozzle pressure and a larger flow.

Aerial apparatus can be used as portable standpipes by placing the aerial apparatus in position at a window on the floor where a hose line is needed and attaching hose to fittings at the elevating platform or top of the aerial. This method of providing an improvised standpipe is especially effective when the aerial device has a prepiped waterway. This tactic does not place hose through the doorway from the stairs; thus it reduces smoke infiltration from the hallway into the stairway. Apparatus pumps will supply the pressure, thereby eliminating potential standpipe pressure problems. This tactic can be used to advance hose inside a building that is not standpipe equipped. When the aerial ladder or bucket is positioned at an ideal location close to the building, the placement of the portable standpipe outlet will normally be limited to the eighth floor, which is typically the maximum reach of a 110-ft (34-m) aerial ladder. One major disadvantage in using the aerial device for hose lines is that it is no longer available for removing occupants or fire fighters from the building.

Another option is using a rope to advance a fire line to an upper floor. This is a more time-consuming and complex tactic, as it typically requires several fire fighters to advance the hose line, and the hose may need to be supported at intermediate levels. Also, when using a rope to advance a hose line because the stairs are unsafe, the attack team members are at a disadvantage, because they do not have the stairway or aerial device available as a means of escape. A rope can be used as a means of escape; however, fire fighters may not have the equipment, training, or experience necessary to perform a safe and effective rope escape from an upper floor. An alternative is to use a rope to advance the hose line to the floor below the fire, then advance up the stairway to the fire floor, similar to a standpipe operation, if the stairway is safe to use.

When it is possible to conduct an offensive attack, the normal first action is to place the standard attack hose line in position, applying water to soften the target whenever possible then applying water directly on the burning material. Preincident planning or the on-scene size-up may indicate the initial use of 2½-in. (65-mm) hose.

Assuming that the first-arriving company will place a 1¾-in. (45-mm) or larger standard attack hose line in position, the trial-and-error method is generally applied. If this initial hose line does not allow the attack team to make immediate progress, more water is needed. The second crew may also lay the standard attack hose line as a backup line. Many times, department SOPs specify that the first-arriving engine company attack the fire and the second crew lay a backup line. Quite often, the IC is still in the process of developing an incident action plan as these preestablished procedures are implemented.

A mistake is made when the IC calls for another 1¾-in. (45-mm) standard attack hose line to combat a fire that is not being controlled by two 1¾-in. (45-mm) lines. When more than two 1¾-in. (45-mm) hose lines are needed to control the main body of fire, there is an obvious need for 2½-in. (65-mm) hose lines if an offensive attack is to continue. This text recommends the use of 2½-in. (65-mm) lines when the flow from two standard preconnected lines does not equal or exceed the calculated rate of flow. The reason for using 2½-in. (65-mm) hose during an offensive operation is to apply large quantities of water to the fire. The nozzle pressure and tip size will determine the actual flow from a 2½-in. (65-mm) hose stream. Fire departments should establish SOPs based on the nozzle types available and the flow requirement challenges in the community. The use of 2½-in.

(65-mm) lines inside a building is an arduous task and requires the hose crew to gain experience through training under simulated fire conditions. As the flow increases, the nozzle reaction force will also increase.

Selecting the Nozzle Type

A discussion of the best type of nozzle is sure to result in a lively dialogue. Obviously, those from the indirect application school prefer finely divided fog (vapor) streams and the variable-stream nozzle that is capable of providing this pattern. In reality, there is little need for fog streams during offensive structural firefighting. There are times when a fog stream may be appropriate, but these are mostly when the fire stream is being used for exposure protection or for Class B fires.

As Table 8-2 indicates, the variable-stream nozzle operated in the straight stream position is capable of delivering as much water as solid-stream nozzles typically attached to 1¾-in. (45-m) hose. From an empirical point of view, the solid core of water produced by the solid-stream nozzle would appear to deliver more water with better penetrating power than a straight stream or large droplet stream. A smooth-bore nozzle will deliver a fire stream even at very low pressures. Nozzle manufacturers have conducted technical tests measuring stream force at various distances that have disputed some of the claims made by smooth-bore advocates. Actual data from independent testing sources are sparse, so the controversy continues.

Stream force is an issue, as it affects the distance the stream will carry. In large undivided areas, stream reach can mean the difference between success and failure. Likewise, force allows the nozzle crew to access hidden fires behind dry wall and other lightweight building materials by penetrating the ceiling or wall with a powerful, compact stream.

In addition, there is sprinkler research that proves that large-orifice/large-drop sprinklers have better extinguishing qualities. The distribution pattern from these heads creates larger water droplets that are better at penetrating the fire plume and reaching the burning fuel. This research appears to favor the solid or straight stream nozzle.

It is possible to have the best of both worlds by attaching an automatic variable-stream nozzle to the standard preconnected hose lines but with a 1½-in. (38-mm) fitting above the shutoff. These nozzles can be changed to a solid-bore stream by removing the variable-stream tip. This arrangement allows the attack crew to quickly change the nozzle without shutting down the hose line at the pump. Some nozzles are designed with a built-in smooth-bore feature just above the shutoff. Be sure to measure the flow in each mode before deciding on a nozzle capable of a variable and solid stream. Fire fighters should be trained in changing nozzle tips and understand the potential difference in nozzle reaction force.

Selecting the Method of Attack

There are several types of fire attack. The most obvious types are offensive and defensive. This chapter explains different tactics used to control a fire using an offensive attack. The *Defensive Operations* chapter describes defensive tactics. Offensive fire attacks are normally described as a **direct attack** or an **indirect attack**. Generally speaking, an offensive attack involves applying water from an interior position or an exterior position in close proximity to the building.

Indirect Attack

The indirect attack has very limited application in structural firefighting. However, there are a few possibilities in which the indirect attack may be warranted. Because the steam produced by this attack will be dangerous to fire fighters and occupants, it is obvious that this is a poor choice in occupied areas, including areas occupied by fire fighters. Tightly enclosed areas aboard ships are the

best application of this attack, but shipboard fires are beyond the scope of this text. Nonetheless, picturing the tightly enclosed unoccupied hold of a ship provides some indication of when this attack may be appropriate.

In conducting an indirect attack, the idea is to keep ventilation to a minimum while introducing a wide-angle fog stream directed at the ceiling through the smallest possible opening, such as a partially closed door or window. This approach will result in maximum steam production disruption of heat layers and an extremely humid environment. For this reason, an indirect attack is usually considered inappropriate when there is any possibility that occupants remain in the building **Figure 8-10**. This tactic may prove useful in unoccupied basements, attics, and storage areas. Directing a piercing, cellar, or distributor nozzle into unvented areas has been used with success. Occasionally, a flat roof is covered by a peaked roof (rain roof) making it especially difficult to access the space between roofs. Limiting openings to a small hole made to insert a piercing or larger nozzle may be the solution to this difficult fire. At the time of this writing, Underwriters Laboratories is conducting research into controlling vented and unvented attic fires. Some of the research involves making a small opening in the ceiling to apply water into an unvented attic. Perhaps this research will uncover additional applications for an indirect attack. Although the indirect attack has extremely limited application, fire fighters should know and understand every tactical tool.

Direct Attack

The direct attack is preferred in the vast majority of situations where conditions permit an offensive attack. Underwriters Laboratories' horizontal and vertical ventilation research as well as live studies on Governors Island proved that softening the target and controlling ventilation result in a safer and more effective approach when confronted with a working residential fire where an offensive attack is warranted.

Figure 8-10 An indirect attack.

Applying water to the fire for 15 seconds in a small residential fire prior to making entry or just as entry is made softens the target, reducing heat. Larger rooms require water to be discharged for a slightly longer period. Controlling the door by opening it just enough to make entry, and minimizing the opening while advancing the hose line, limits air intake, retarding fire growth and allowing fire fighters to advance on the fire for a safer and more effective offensive attack. This tactic is unproven in large undivided areas such as those found in many commercial occupancies.

Proper ventilation timed with the fire attack can significantly improve conditions in the fire area. Proper ventilation is critical to prevent flashover and make interior conditions tenable for victims and fire fighters conducting an offensive attack under heavy fire conditions. In **Figure 8-11**, notice that the window near the fire is opened as the attack begins, which provides a ready flow path away from the fire fighters entering the room to attack the fire. A roof vent over the fire area would also provide a flow path away from fire fighters entering the door. However, to be effective, a roof vent would require opening the ceiling between the attic and fire area to allow smoke and heat to escape. This is a very effective vent but would place the attic in the flow path, possibly extending fire into the attic.

Ventilation is not always under the control of fire forces. During moderate weather conditions, windows and other openings are often left opened by occupants. In other cases, the heat produced by the fire causes a window to fail or burns through part of the structure creating an unplanned vent.

Figure 8-11 Softening the target and controlling airflow through the door followed by a direct attack and ventilation.

The softening the target tactic is highly recommended. This tactic is based on credible and exhaustive research by Underwriters Laboratories involving house fires and later live fires conducted at Governors Island. However, it must be noted that, to date, research was limited to rooms (compartments) inside residential buildings, including large open layouts with high ceilings. This research has not yet included large open-area commercial buildings.

There is a reasonable expectation that this tactic could be extended to larger compartments; however, the results may not be the same. Larger compartments have more air space and thus more available oxygen. Larger-volume fires are less likely to go into decay phase as they are less likely to become ventilation limited. Fire growth necessary to involve a large compartment could take considerably longer than rooms typically found in residential occupancies, depending on fuel load, volume of the compartment, and other factors. It is also possible to have a small fire in a large open area where the fire remains confined to a limited area prior to the arrival of fire forces.

Large commercial buildings are more likely to be protected by automatic sprinkler systems. If the system is properly designed and maintained, there is little chance of the fire spreading beyond the relatively small area.

For purposes of this discussion, either the area on fire is not protected by a sprinkler system or the sprinkler system fails to control the fire and the fire has sufficient air and fuel to develop into an uncontrolled large-area fire.

Large-volume commercial buildings tend to have a variety of fuels, some of which result in smaller, slower-burning fires, while others have fuel loads that greatly exceed those found in a residential property. The possibility of being able to soften the target from an exterior position or a position just inside the building is less likely in large commercial buildings. The size and complexity of many large commercial buildings may result in this tactic being less effective and more difficult to apply.

Consider the effects of softening the target in the buildings previously used to explain rate-of-flow calculations in this chapter. The apartment building in Figure 8-3 is a residential property with fairly small enclosures. It is a large building with small compartments (rooms). The height and configuration of the building may not be conducive to softening the target from the exterior in all areas and levels of the building. However, water can be applied from interior hallways into an apartment to soften the target." This same approach should be effective in a high-rise building except in large open areas, such as in the open-layout office used in the National Institute of Standards and Technology (NIST) high-rise staffing report[15] Figure 8-12 .

The supermarket shown in Figure 8-4 has limited access from the exterior. Typically, this mercantile occupancy would have front windows and doors, but no direct access on the other three sides of the building. The inside of the building is probably divided into several areas, including a rear loading dock and storage areas, pharmacy, and refrigerated storage. However, there will be a large open store area that accounts for the majority of the 500,000 ft³ (14,158 m³) in this example. Aisles make it difficult to apply water from a distance, and the fire could be a considerable distance from the front door. The wide-open store area provides sufficient air to support combustion for an extended period, and there could be many customers needing assistance in finding their way out of the building.

Thankfully, nearly all of these occupancies are sprinkler protected; unfortunately, sprinklers are sometimes poorly designed, poorly maintained, or simply shut down.

The auto parts store in Figure 8-5 presents an interesting scenario in terms of softening the target. The two first floor, 40- × 30-ft (12.2- × 9.1-m) compartments are not much larger than some of the biggest rooms in residential properties. There is a direct entry into each of these areas and the fire could go into decay. Unless there is a considerable quantity of flammable or combustible liquids in the area, softening the target could be an effective tactic on the first floor of the auto parts store.

The second floor provides a potential fire volume (48,000 ft³ [1359 m³]) that would exceed the volume of any of the fires used by Underwriters Laboratories in their residential experiments. The limited fuel supply (metal parts in cardboard boxes) would most likely produce a less intense and slower-growth fire than the modern furnishings used by Underwriters Laboratories in their research. Obstructions in the form of rack storage may hinder efforts to apply water to the fire prior to making entry or from a position just inside the door leading to this area. The efficacy of softening the target would be uncertain in this case.

The effectiveness of any offensive fire attack is questionable on the third floor of the auto parts store

Figure 8-12 Open-layout office in a high-rise (from the NIST high-rise report).

Reproduced from: NIST High-Rise Staffing Report- NIST Report on High-Rise Fireground Field Experiments, NIST Technical Study Note 1797, Jason Averill, Lori Moore-Merrill, 2013

once the fire gains significant headway. The size of the compartment (48,000 ft^3 [1359 m^3]) combined with the extra hazard fuel load has the potential to limit fire attack to a defensive mode once the fire extends to involve a major portion of this floor.

Rule of Engagement for Fire Fighter Survival

DO NOT risk your life for lives or property that cannot be saved.

Reproduced from: © 2011, IAFC Safety, Health and Survival Section. All rights reserved.

Texture: © Eky Studio /Shutterstock, Inc; Steel: © Shapshot/Dreamstime.com

No fire department using any tactic could change the outcome of the Kmart warehouse fire described in the earlier incident summary. The risk in making entry at the Kmart fire is extreme and the benefit is near zero. The only credible advice is to stay out of the building and collapse zone.

It is essential that fire departments take advantage of the information gained through valid research. However, it is not possible to conduct live fire research for every building type, especially large buildings. As practitioners, it is up to the fire service to extend the usefulness of research by sharing positive and negative fire-ground experiences.

Final extinguishment involves applying water directly on the burning material, thereby extinguishing the fire by reducing the temperature of the burning and superheated fuel. Class A fuels must be vaporized by heating the solid materials until combustible vapors are emitted; this process is known as **pyrolysis**. By applying water directly to the burning fuel, the temperature of the fuel is reduced, thus reducing or eliminating pyrolysis. Cooling ordinary combustibles to shut down the pyrolysis process is

Figure 8-13 Transitional attack.

much like shutting off the gas supply at a flammable gas fire.

When fire fighters encounter very high heat conditions, indicative of an impending flashover, directing a stream overhead (pulsing) is advised, but this should be done using short intermittent blasts with a straight or solid stream aimed at the ceiling in a pulsing fashion before actually entering the fire compartment. If done properly, pulsing will reduce the probability of flashover while not seriously disrupting the heat balance.

Figure 8-11 illustrates softening the target at the entry point. Another way to soften the target is to apply water to the fire from an exterior opening. This approach is sometimes referred to as a *transitional attack*. The transitional attack involves directing a stream from the exterior into a room that is involved in fire where the fire has self-vented Figure 8-13 . Just as with the softening the target attack in Figure 8-11, the fire stream is applied for a short time (usually 15 seconds or less) to knock down a heavy volume of fire, making entry possible. The fire stream should be a handheld solid or straight stream, not a fog stream or master stream. Live fire experiments at Governors Island demonstrated that a fog stream could have a negative impact on venting and affect the flow path. Rapidly moving a solid or straight stream from side to side or up and down had a similar negative effect.

From a safety standpoint, the officer should, before entering an area of extreme heat, conduct a risk-versus-benefit analysis. If there is a high potential for saving lives, the continued advancement may be justified. If the goal is keeping the fire out of the common hallway or stairs, it may be better to use the door to the apartment to contain the fire while protecting the stairs Figure 8-14 . Each situation is

Figure 8-14 Fire contained by building features.

different, and no single approach is correct for every incident.

There are times when the fire is shielded by nonburning materials, such as in a warehouse with rows of high-stacked commodities. If possible, the hose should be moved to a position that allows a direct attack. If this is not possible, then deflecting a solid or straight stream off of the ceiling and allowing the droplets to fall on the fuel is an alternative **Figure 8-15**. The objective is to get the water directly on the fuel. This tactic should be considered a derivation of a direct attack.

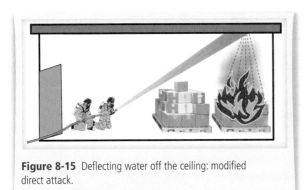

Figure 8-15 Deflecting water off the ceiling: modified direct attack.

Estimating the Number of Attack Lines

Previous examples showed that a single 1¾-in. (45-mm) hose line could probably extinguish most fires within a dwelling. It is good practice to lay a backup hose line to protect egress routes and bolster the attack. Even trial-and-error methods become more successful when provisions are made to double the size of the attack. The number of lines needed will be based on flow requirements, but there must always be a reserve. Simply stated, enough lines must be placed in service to apply the necessary rate of flow and protect egress routes for interior fire fighters, should conditions deteriorate.

In addition, the IC is responsible for predicting where the fire is going. Construction features may

hold a fire within a given area or allow it to travel horizontally and/or vertically throughout the structure. In the case of an apartment building with a common attic, it is necessary to have truck companies gain access to and check for extension in this hidden space while an engine company provides a hose line to limit fire extension. Just as it is necessary to perform search and rescue operations above the fire, it is also necessary to place hose lines above the fire.

Evaluating Exposures

Estimating the number of hose lines above the fire addresses the need to protect interior exposures. Hose lines are needed to prevent the fire from moving through natural pathways such as concealed spaces, stairs, chutes, and shafts. Fire can also extend vertically up the exterior of the building from windows or other openings on the fire floor(s). All of these potential routes of fire extension within the building are classified as **internal exposures** and need to be checked and protected with a hose line as required.

Fire can extend to adjacent buildings (**external exposures**) as well. Ventilation crews must be aware of this potential, as improper ventilation can spread the fire to exposed buildings—just as proper ventilation can modify a flow path, directing fire and smoke away from exposures.

Nearby buildings and building occupants may need to be protected once the fire breaks out of the containment of the building of origin. Evaluating the need to defend nearby buildings with additional hose streams involves many factors, including the following:

- Life hazard in the exposure
- Proximity to the fire building
- Wind direction
- Height of exposure compared to the level of fire involvement in the fire building
- Hazard presented by the exposed occupancy (e.g., explosive or chemical storage)

The threat to human life is always the first consideration. In the case of exposure buildings, it is important to keep fire from entering or otherwise threatening

Figure 8-16 Protecting occupants in an exposure using a master stream.

people in the nearby buildings. Often a protective stream can be placed between the fire building and exposure to limit the hazard **Figure 8-16**.

When evaluating exposures, be sure to consider the location of fire apparatus and equipment parked near the fire building **Figure 8-17**. Protecting exposures can involve hose lines on the exterior or interior of the exposed building. When working on the exterior, master streams are most often the streams of choice.

Estimating Backup Needs

Backup lines are needed to protect the crew on the initial attack line and to provide additional flow

when required. Given that a single 1¾-in. (45-mm) hose line is sufficient for most residential fires (in terms of rate of flow), the backup hose line is additional insurance. In larger-area fires, the backup hose line again provides insurance against rate-of-flow miscalculation and provides additional protection for retreating fire fighters. Backup lines should be at least as large as the initial attack line. If you started the attack with 2½-in. (65-mm) hose, the backup hose line should be a 2½-in. (65-mm) hose line. If the attack began with 1¾-in. (45-mm) hose, the backup hose line could be 1¾ in. (45-mm), 2 in. (50-mm), or 2½ in. (65-mm).

Although backup hose lines can be used to augment the initial attack, the defined purpose of a backup line is to provide flow in addition to what

Figure 8-17 External exposures.

is needed to extinguish the fire, thus providing a measure of protection for fire fighters combating the fire. If the backup line is used to augment the flow rate and the fire is not immediately extinguished, another backup line is needed.

Estimating the Number of Hose Lines Needed Above the Fire

It is a good practice to get a hose line on the floor immediately above the fire. Where there is a risk of extension to concealed spaces and attics, additional precautionary lines are needed at each of these areas. Backup lines may also be needed in other areas above the fire **Figure 8-18**.

Evaluating Other Hose Lines Needed

On occasion, hose lines are needed below the fire to protect egress routes and to protect external exposures. The total number of lines needed is as follows:

- Hose line(s) required to meet the rate of flow in the immediate fire area
- Backup hose line(s) for immediate fire area
- Hose line(s) to protect egress routes
- Hose line(s) to protect internal exposures (e.g., attic where vertical travel is likely, upper floors, concealed spaces)
- Other backup hose line(s) as needed
- Hose line(s) or master stream(s) to protect external exposures

Figure 8-18 Attack and additional hose streams.

Estimating Water Supply Needs

Prior to the widespread use of large-diameter hose, the officer assigned to the first-arriving unit in a hydranted area had to decide whether to rely on a marginal water supply by forward laying a single 2½-in. (65-mm) or 3-in. (76-mm) supply line or use a reverse lay with a direct hydrant connection,

which could result in placing the pumper a considerable distance from the fire building. Large-diameter hose eliminates this dilemma and has proven to be an effective tool in providing an adequate water supply at the fire scene **Figure 8-19**. In an area with hydrants, providing at least two water supplies for a working structure fire is considered good practice. In areas with a marginal or nonexistent water supply, estimating and providing an adequate water supply can be a substantial challenge. Fortunately, many of the fires requiring large fire flows occur in areas with reasonably good water supplies. When a significant water supply challenge exists in an area with a limited water supply, operational options should be outlined in SOPs and the preincident plan.

Attacking a fire with anything less than a water supply from a source of water that is capable of supplying flow requirements over an extended period of time is a gamble. Attack pumper tactics (operating with water from the apparatus tank) work well in many situations, especially in fighting a fire in a small dwelling. However, relying solely on water from the apparatus water tank can be dangerous. Many things can happen that will cause fire fighters to lose their water supply. When 1¾-in. (45-mm) lines are used, even a 1000-gallon (3785-L) water tank can supply water for less than 10 minutes. Often, this water

Figure 8-19 Large-diameter hose supply.
Courtesy of Captain Nick Morgan, St. Louis Fire Department

discharge time is sufficient. However, if fire fighters need to attack fire in multiple compartments or they are not able to rapidly control the fire for any reason, they may deplete their water supply before exhausting their SCBA air supply.

If an attack pumper concept is implemented, its use must be well understood by all members, and it should be limited to fires in relatively small structures. Where the water supply is limited or nonexistent, consider using two pumpers or other means, such as a water tender, to augment the water supply to the pumper.

NOTE

A 1000-gallon (3785-L) booster tank, compared to a 500-gallon (1893-L) tank, will add over 2 tons (1814 kg) to the weight of the pumper, which could be problematic, especially in hilly terrain.

Texture: © Eky Studio /Shutterstock, Inc; Steel: © Sharpshot/Dreamstime.com

---

Some large urban areas with excellent water supplies will have areas containing small ancillary buildings, such as a shed on a golf course located a considerable distance from a hydrant. In many cases, a long, complex relay is planned to deliver water to these properties. Rather than relying on relay operations, it might be better to use multiple attack pumpers to deliver the needed water to the scene.

It is important to remember that a large flow of water for a short period of time will require less water and be more effective than a lesser flow over a longer period. As an example, consider a 25- × 16-ft (7.6- × 4.9-m) room with 10-ft (3-m) ceilings in a single-family dwelling. The required rate of flow is as follows:

$$\frac{25 \times 16 \times 10 \text{ ft}}{100} = 40 \text{ GPM (151 L/min)}$$

If the apparatus booster tank contains 500 gallons (1893 liters), a 1-in. (25-mm) booster line could discharge 20 GPM (76 L/min) for 25 minutes, or longer than the air supply in a 30-minute SCBA air cylinder. Even a 45-minute SCBA could be depleted

as the crew would probably be "on-air" before actually discharging water. The entire 500 gallons (1893 liters) would likely be exhausted, and the fire would not be extinguished because the rate of flow was insufficient to suppress the fire.

If 150 GPM (568 L/min) were discharged, the fire would be extinguished in less than a minute, with less than 150 gallons (568 liters) of water. In the article by Edwards from *Fire Engineering*, his calculations would indicate extinguishment in about 2 seconds, requiring only 5 gallons (19 liters) of water at nearly four times the required flow. Seldom is water applied at maximum efficiency, but the fact remains that the higher-flowing 1¾-in. (45-mm) hose line would quickly extinguish the fire.

The need for a continuous supply of water having been established, standard guidelines should be followed. The use of a single, unpumped 2½-in. (65-mm) supply line is discouraged. From a water supply standpoint, this single line is much better than total reliance on the apparatus water tank, used with the attack pumper, and may be acceptable in residential fires. However, double 2½-in. (65-mm) or 3-in. (76-mm) lines are far superior to the single unpumped 2½-in. (65-mm) or 3-in. (76-mm) supply line. Furthermore, in using 2½-in. (65-mm) or 3-in. (76-mm) supply lines, both the volume and pressure can be improved by assigning a second pumper to relay the water from the hydrant. However, pumping into large-diameter supply hose (4 to 5 in. [102 to 127 mm]) during offensive operations is rarely necessary.

Fallacy	Fact
When the water supply is limited, ration the water so that it will last until a tanker shuttle, relay, or other reliable source of water can be supplied.	A rate of flow below the required flow will be ineffective and waste water. A rate of flow meeting or exceeding the flow requirement will be effective and use less water.

Steel: © Sharpshot/Dreamstime.com

More often, it would be better to position the extra relay pumper in a position to provide a separate water supply or assign it to staging. Only in situations in which flows in excess of 1000 GPM (3785 L/min) are anticipated from a single pumper or for extremely long relay lines (over 700 ft [213 m]) is it sometimes necessary to place another pumper in a large-diameter hose water supply layout. Further, once the pumper is placed in operation, the apparatus operator must remain at the pump panel. Thus, unnecessary use of apparatus also results in a reduction of available staffing.

Just as a backup hose line is needed inside structures, obtaining a second source of water is a good practice. This second source of water is essential at working fires, especially when operating off dead-end mains or limited flow hydrants. Water supply is not enhanced by securing a second source of water off the same dead-end main. Remember, for offensive operations, a minimum water supply layout must be capable of supplying the calculated rate of flow plus backup lines and other lines used to protect exposures. Having a second water supply provides a backup source if the initial pumper or hose layout is incapacitated. There have been many instances when a failure in the water supply resulted in a controlled fire getting out of control.

Water supply should be addressed in department SOPs. The initial water supply is crucial. Companies responding must be given clear instructions regarding the water supply. See the *Procedures, Preincident Planning, and Size-Up* chapter for additional information regarding preplanned water supplies.

Fallacy	Fact
Every pumper should be supplied with water.	There should be at least two separate sources of water whenever possible. In initiating an offensive attack, the total water supply should exceed the calculated rate of flow plus additional lines.

Steel: © Sharpshot/Dreamstime.com

Estimating Ventilation Needs

In the *Life Safety* chapter, venting is discussed as a life safety tactic. Venting is most often used to assist in extinguishment efforts by controlling the flow path and making conditions tenable for fire fighters and victims. Timing is crucial when the venting tactic is intended to support extinguishment. The ideal scenario calls for making the ventilation opening just as the attack team enters the fire area. Vertical and horizontal vent placement should be as close to the fire as safety permits and located to direct smoke and heat away from fire fighters making entry. The top three drawings in **Figure 8-20** illustrate a step-by-step progression in a properly timed roof vent. Improperly placed or poorly timed ventilation can hinder extinguishment by spreading the fire into unburned areas. The bottom drawing in the figure shows an improperly placed vent made after fire fighters enter the building to extinguish the fire. This vent location would cut off the pathway to the escape ladder on the roof and cause the fire to spread toward the attack team inside the building.

Horizontal and vertical venting should be timed and reversible (open the window rather than breaking it) whenever possible. Proper ventilation allows fire fighters to approach the fire and apply water directly to the burning materials. It also provides oxygen, which intensifies the burning rate, particularly if the fire is ventilation controlled. Therefore, when venting is being carried out to support extinguishment, attack crews with charged lines should be positioned before venting or before opening a door leading directly to the fire area. As entry is made, the door should be controlled to reduce the air flowing into the fire area.

Improper vent placement can channel the fire into concealed spaces or otherwise extend the fire. Remember, regardless of our best intentions, fire and the products of combustion are bound by the laws of physics. When a door leading from the hallway to the room on fire is opened, smoke and heat will vent into the hallway. If vertical ventilation is provided via

Figure 8-20 Venting for extinguishment.

the stairway, expect the fire to enter the stairs unless extinguishment efforts are immediately successful and an alternate flow path established.

If vent openings can be controlled, positive-pressure venting virtually ensures the vent timing, as opening the door to enter the fire compartment allows the pressure to enter the area and exit through an existing or created vent opening. As with natural ventilation, it is imperative that a hose line be ready to immediately attack the fire.

Fires in small compartments are easily ventilated using positive pressure. Research and field use proved the effectiveness of properly applied positive-pressure ventilation in residential and other small buildings. The effectiveness of positive-pressure venting in larger and more complex buildings was in question until NIST conducted extensive positive-pressure ventilation research in larger buildings. NIST studied the results of large-scale fires in an unused high school as well as high-rise buildings. The results of the school experiments proved that positive-pressure ventilation can be effective in very large and complex buildings.[16]

Of particular interest were live fire experiments in a 340,078-ft³ (9630-m³) school gymnasium. As expected, this large open compartment took longer to fill with smoke and in most cases did not become ventilation limited. Positive-pressure ventilation was very effective in reducing heat and clearing smoke during fires involving the moderate fuel loads expected in a gymnasium.

A larger fuel load was introduced to study how effective positive-pressure venting might be under a heavy fuel load. During this experiment, 160 pallets were added to the wooden bleachers and foam padding usually found in an auditorium. This fuel load resembles the fuel load found in many high-hazard storage facilities. This fire was allowed to fully develop. Positive pressure combined with vertical ventilation resulted in flashover, with the temperature reaching 1290°F (700°C) throughout the gym. This result would be invariably fatal to remaining occupants and fire fighters inside the building. However, it was noted that conditions inside the gym prior to positive-pressure ventilation and flashover were already lethal. The NIST experiments proved the usefulness of positive-pressure ventilation in large areas, but also revealed limitations in large building compartments containing large fuel loads. At the time of this publication, Underwriters Laboratories is conducting further research involving residential structures.

Positive-pressure ventilation, like all other ventilation, can also have negative effects. As an example, if the fire has entered large or vented concealed spaces, the vent openings will be uncontrolled and positive-pressure venting is likely to promote fire spread. Consider the effect of positive-pressure ventilation if there is an opening between the room on fire and the attic. The positive pressure could extend the flow path into the attic area and greatly accelerate the spread of fire into the attic. The application of positive-pressure ventilation in high-rise buildings is addressed in the *High-Rise Buildings* chapter.

Ventilation is an important tactic. The IC makes decisions as to what type of ventilation is best and coordinates hose team and vent crew activities. Fire officers should be completely familiar with the various ventilation methods.

Venting is normally considered an offensive tactic; however, the **trench cut** is sometimes referred to as a defensive tactic. Because the trench cut normally involves interior operations, it falls under the definition of offensive operations used in this text. According to another definition, defensive can mean establishing a fire break and stopping the fire spread at that point. In this sense, the trench cut is defensive, as the objective is to protect interior exposures. The trench cut provides a fire break separating the burning area from the uninvolved area. To be successful, it must be made well ahead of the fire and effectively separate the burned from the unburned. It takes considerable time and effort to cut a trench completely across a roof. It is imperative that this tactic be planned with sufficient lead time.

Trench cuts can be used in any large structure and are particularly suited to U-shaped apartment buildings and garden apartments. The largest trench cut operation on record was at Tinker Air Force Base in Oklahoma,[17] where several long trenches were cut in a roof covering 2,531,965 ft² (235,227 m²) for a fire operation spanning 3 days.

Calculating Staffing Needs

Staffing requirements are many and varied. Hose and ladder crews, rescue crews, rapid intervention crews (RICs), incident management system positions, and a tactical reserve are examples of assignments that must be adequately staffed at the scene of a working structure fire. In addition, the IC needs to consider the total extinguishment staffing requirements based on rate of flow, backup lines, placement of lines above the fire, and establishment of a secondary water supply. NFPA 1710, *Standard for the Organization and Deployment of Fire Suppression Operations, Emergency Medical Operations, and Special Operations to the Public by Career Fire Departments*,[18] defines tasks and minimum staffing for the initial response to a 2000-ft² (186-m²) two-story, single-family dwelling with no basement and no exposures. These tasks and staffing include the following:

1. Establish command (minimum one person).
2. Establish a minimum uninterrupted water supply of 400 GPM (1514 L/min) (minimum one person).
3. Establish an attack and backup line flowing at least 300 GPM (1136 L/min) (minimum four people).
4. Assign one support person per line (minimum two people).
5. Assign at least one search and rescue team (minimum two people).
6. Assign at least one vent/ladder team (minimum two people).
7. Establish an initial RIC (minimum two people).
8. If aerial is in use, assign a person operating the aerial (minimum one person).

On some occasions, the first-arriving unit(s) handles the fire, and the total response defined in

NFPA 1710 is not required. As an example, the first-arriving engine and truck companies conduct the initial attack on a mattress fire inside a residential property. The engine extinguishes the fire and provides an uninterrupted water supply as well as the two-out RIC. Truck company members conduct a quick search and rescue with one crew while the other crew vents the building.

The minimum staffing requirements outlined in NFPA 1710 were the object of considerable debate within the fire service, with many fire service and community leaders disputing the need for four-person minimum staffing or the total staffing required for a typical two-story single-family detached dwelling. In order to prove or disprove the NFPA 1710 staffing requirements, NIST constructed the 2000 ft^2 (186 m^2) two-story, single-family dwelling described in NFPA 1710 then conducted 60 full-scale experiments to determine the impact of crew size and total staffing.[19] A panel of fire service technical experts identified 22 tasks that should be completed. In these experiments, crew sizes varied from two to five fire fighters per company. The experiments provided scientific evidence that the smaller the crew size, the longer it takes to complete the tasks necessary to successfully combat the single-family dwelling fire identified in NFPA 1710. Especially important was the fact that, on average, it took the four-person crews 30% less time to complete the primary search than the two-person crews. The average time required to stretch the initial hose line was 87 seconds longer for the two-person crew compared to a four-person crew.

It is virtually impossible to conduct safe and effective operations at structure fires when remote fire companies are staffed with fewer than four fire fighters. When two or more fire department units respond from the same fire station, understaffed units may be able to combine staffing to form one adequately staffed company. When a single understaffed fire company located at a remote station encounters a working structure fire, they are forced to make difficult decisions regarding their personal safety versus their commitment to saving lives and property.

As an example, a single-engine company fire station (Engine 1) staffed by two fire fighters protects an outlying part of the jurisdiction. The second-due engine company (Engine 2) is 7 miles (11.3 km) from Engine 1. The average travel time between stations is 10 minutes. Continuing with this example, a house fire is reported a few blocks away from Engine 1's station. Engine 1 with two fire fighters arrives at the scene of a working house fire within 2 minutes. Engine 1 has several options, none of which are both safe and effective:

- One fire fighter advances a hose line inside the building while the other operates the pumps. (unsafe and ineffective)
- The pump operator sets up the pumper and both fire fighters make entry. (unsafe)
- Both fire fighters working together forward lay a supply line to the pumper then lay a 1¾-in. (45-mm) line to the front door, but do not make entry until Engine 2 arrives. (ineffective)

There are other possible tactical options, but none would be safe *and* effective. With only two fire fighters at the scene, either one or two fire fighters make entry with no IRIC, or extinguishment is substantially delayed, which could result in the total loss of the building and civilian fatalities if occupants are unable to self-evacuate.

A three-person crew would be better but still would result in either an unsafe and/or ineffective operation. Although NFPA 1500 permits crews to make entry without an IRIC when there is the potential to save lives, failing to staff an IRIC always compromises safety.

The 2000 ft^2 (186 m^2) two-story, single-family dwelling with no basement and no exposures is sometimes referred to as a "low-hazard occupancy." NFPA 1710 outlines the minimum safety requirements for this occupancy. Even in this typical two-story residential building, staffing needs could be greater than those listed in the standard.

The typical four-person staffing on a fire company consists of an officer, a driver, and two fire fighters. This minimum safe staffing also requires some compromises in terms of safety and effectiveness when responding from a remote station. Using the example above, but with four fire fighters on the first-arriving engine company, the IRIC by necessity will be involved in accomplishing other necessary tasks. In most cases, the pump operator is assigned as one member of the IRIC. In the event of a mayday, rescuing the interior fire fighters will require the pump operator to abandon the pump panel at the apparatus. In many cases, the fire fighter assigned to connect to a fire hydrant is assigned as the second member of the IRIC. This fire fighter could be located a considerable distance from the fire building or may not be able to complete the water supply assignment.

Working fires in a complex or larger building, a building with an extraordinary life hazard, buildings containing an extra hazard fire load, and areas with poor water supplies require additional staffing.

For a staffing example, consider a fire on the second floor of the auto parts store example shown in Figure 8-5. In this scenario, the building is believed to be unoccupied and has external exposures on three sides. Each fire company has an officer, a driver, and two fire fighters on duty. The standard preconnect is a 200-ft (61-m) 1¾-in. (45-mm) hose line with a 7/8-in. (22-mm) smooth-bore tip. The department's SOP establishes a 65-psi (448-kPa) nozzle pressure, resulting in a flow of 184 GPM (697 L/min).

Engine 1 (First-Arriving Engine Company)

Assume that the building has been preincident planned with an estimated rate of flow of 480 GPM (1817 L/min) based on V/100 calculations (sprinkler calculations are also 480 GPM [1817 L/min]

in this case). The first-arriving officer takes command and gives an initial report indicating a working fire on the second floor while working with another member of the crew in advancing a standard preconnected hose line to the second floor **Table 8-3**.

As per SOPs, a forward lay of 5-in. (127-mm) hose supplies the initial operation. The apparatus operator remains at the pumper, while the fire fighter assigned to the hydrant hooks up then advances toward the entry to assist in advancing hose into the building and controlling the door as part of the IRIC.

Engine 2 (Second-Arriving Engine Company)

In the absence of orders, SOPs for this department call for the second-arriving engine company to advance an additional hose line to the fire floor to increase the rate of flow or provide a backup hose line for the first-in attack crew. Otherwise, they are to advance a hose line to the floor above the fire. Realizing that there is a working fire in a preincident-planned building that requires a large rate of flow, the officer of Engine 2 plans to back up Engine 1 by advancing a 2½-in. (65-mm) hose line with a 1⅛-in. (29-mm) smooth-bore nozzle into the building.

Engine 2 does not secure a second source of water. The Engine 2 officer radios command to advise that a second source of water has not been secured. One fire fighter from Engine 2 assists in advancing the 2½-in. (65-mm) hose line to the entry then joins the fire fighter from Engine 1 to form a dedicated RIC. The remaining three crew members from Engine 2 advance a 2½-in. (65-mm) hose line off Engine 1's

Table 8-3	Staffing with One Hose Line in Operation				
Company	**Total Staffing**	**Staffing at Apparatus**	**Assignment**	**Staffing at Assignment**	**Flow**
Engine 1	4	2	2nd floor	2	184GPM (697 L/min)
Totals	**4**	**2**	**(N/A)**	**2**	**184 GPM (697 L/min)**

© Eky Studio/Shutterstock, Inc.

Table 8-4 Staffing with Two Hose Lines in Operation

Company	Total Staffing	Staffing at Apparatus	Assignment	Staffing at Assignment	Rapid Intervention Crew	Flow
Engine 1	4	1	2nd floor	2	1	184 GPM (697 L/min)
Engine 2	4	0	2nd floor	3	1	305 GPM (1155 L/min)
Totals	**8**	**1**	**(N/A)**	**5**	**2**	**489 GPM (1851 L/min)**

pumper **Table 8-4**. Engine 2 is parked out of the way of other arriving apparatus, but in a position to be ready for response to protect exterior exposures, should it become necessary.

There is a good chance that water applied from these two hose lines will be adequate to control the fire, as the flow being applied is slightly above the 480-GPM (1817-L/min) required rate of flow. Had the department opted for a 125-GPM (473-L/min) 1¾-in. (45-mm) nozzle, the flow rate would have been slightly below the needed rate of flow. This is where trial and error plays a role. If the fire is being extinguished with the present attack, the flow is sufficient. Otherwise, an increase in pump discharge pressure to 200 psi (1379 kPa) for the 1¾-in. (45-mm) hose line with automatic nozzle would increase the flow by approximately 50 GPM (189 L/min), thus meeting the required rate of flow when used with the 2½-in. (65-mm) line.

Reconnaissance is the best way to determine the need for more lines to control the fire. In any case, there is still a critical need to back up the hose lines operating on the fire floor, secure an additional water supply, and get a hose line to the third floor.

Truck 1 (First-Arriving Truck Company)

If Truck 1 arrives early, there may be a need for its crew to force entry into the building. Otherwise, the engine company crews will be responsible for this task. The truck company's crew will split up—two fire fighters providing truck duties such as horizontal ventilation for the engine companies working on the second floor, laddering the second floor, and controlling utilities. The other two members of the truck company will work above the fire, conducting the primary search and checking for fire extension on the third floor **Table 8-5**.

Table 8-5 Staffing with Two Engine Companies and a Truck Company

Company	Total Staffing	Staffing at Apparatus	Assignment	Staffing at Assignment	Rapid Intervention Crew	Flow
Engine 1	4	1	2nd floor	2	1	184 GPM (697 L/min)
Engine 2	4	0	2nd floor	3	1	305 GPM (1155 L/min)
Truck 1	4	0	2nd floor	2	0	0
Truck 1	(see above)	0	3rd floor	2	0	0
Totals	**12**	**1**	**(N/A)**	**9**	**2**	**489 GPM (1851 L/min)**

Chief Officer

The first-arriving chief officer establishes an exterior, stationary command post and confers with the company commander of Engine 1, who is serving as the IC (**Table 8-6**). Engine 1's officer advises the chief officer that slow progress is being made on the fire. After a careful size-up, the chief officer assumes the position of IC and develops the following incident action plan:

- Extinguish the fire.
- Conduct a primary search of all floors. (Companies currently on the second and third floors will be responsible for these floors.)
- Ladder the building to provide alternative egress for fire fighters.
- Fully staff the RIC with four fire fighters.
- Advance a backup hose line to the second floor.
- Advance a hose line to the third floor.
- Secure a second water supply.
- Stage an engine and truck company as a tactical reserve.
- Plan for a possible defensive operation.
- Set up rehab.
- Provide for medical treatment and transportation.

This would be a full first-alarm response for some departments, and in many cases mutual aid would be necessary to muster the needed staffing of 13 to begin operations. Most urban departments would send a larger response to a fire in a commercial building. Comparing the task being performed and the staffing at the scene, the requirements outlined in NFPA 1710 are being met for a low-hazard occupancy if the aerial ladder is not in use. A commercial occupancy such as the auto parts store would require additional staffing. In fact, the IC's incident action plan includes several additional tasks that have not been staffed and there is no tactical reserve. The two-person minimum is assigned as a RIC. Continuing operations will thus require additional resources.

Engine 3 (Third-Arriving Engine Company)

Engine 3 is instructed to supply its apparatus with a 5-in. (127-mm) supply line and to advance a 1¾-in. (45-mm) hose line with a 7/8-in. (22-mm) smoothbore tip to the third floor (**Table 8-7**).

Engine 4

Engine 4 will back up the two lines operating on the fire floor with a 2½-in. (65-mm) hose line laid from Engine 3's apparatus (**Table 8-8**).

At this stage, the members of Truck 1 should be free to complete a primary search and check for extension on the first floor.

Table 8-6 Staffing with Two Engine Companies, One Truck Company, and a Chief Officer

Company	Total Staffing	Staffing at Apparatus	Assignment	Staffing at Assignment	Rapid Intervention Crew	Flow
Engine 1	4	1	2nd floor	2	1	184 GPM (697 L/min)
Engine 2	4	0	2nd floor	3	1	305 GPM (1155 L/min)
Truck 1	4	0	2nd floor	2	0	0
Truck 1	(see above)	0	3rd floor	2	0	0
Chief officer	1	0	Command post	1	0	0
Totals	**13**	**1**	**(N/A)**	**10**	**2**	**489 GPM (1851 L/min)**

© Eky Studio/Shutterstock, Inc.

Table 8-7 Staffing with Three Engine Companies, One Truck Company, and a Chief Officer

Company	Total Staffing	Staffing at Apparatus	Assignment	Staffing at Assignment	Rapid Intervention Crew	Flow
Engine 1	4	1	2nd floor	2	1	184 GPM (697 L/min)
Engine 2	4	0	2nd floor	3	1	305 GPM (1155 L/min)
Truck 1	4	0	2nd floor	2	0	0
Truck 1	(see above)	0	3rd floor	2	0	0
Chief officer	1	0	Command post	1	0	0
Engine 3	4	1	3rd floor	3	0	184 GPM* (697 L/min)
Totals	17	2	(N/A)	13	2	489 GPM* (1851 L/min)

*The flow column reflects the rate of flow required to extinguish the main body of the fire. Engine 3's hose line is on the floor above; it is not added to the total flow calculation. The two water supplies (Engines 1 and 3) can easily handle the total potential flow of 673 GPM (2548 L/min) if the hose line on the third floor is operated.

Heavy Rescue, Truck 2, and Engine 5

Heavy Rescue 1 is assigned as the dedicated RIC with one member assigned as the accountability officer. The two members assigned to rapid intervention from Engines 1 and 2 can now rejoin their companies or remain with the RIC possibly surveying the building, providing additional egress for units working inside the building. Truck Company 3 and Engine Company 5 are staged two blocks from the scene.

Two medic units are assigned to rehab, treatment, and transport. Additional staff officers are assigned to fill the positions of safety officer, planning section chief, and Division 2 supervisor **Table 8-9** .

Table 8-8 Staffing with Four Engine Companies, One Truck Company, and a Chief Officer

Company	Total Staffing	Staffing at Apparatus	Assignment	Staffing at Assignment	Rapid Intervention Crew	Flow
Engine 1	4	1	2nd floor	2	1	184 GPM (697 L/min)
Engine 2	4	0	2nd floor	3	1	305 GPM (1155 L/min)
Truck 1	4	0	2nd floor	2	0	0
Truck 1	(see above)	0	3rd floor	2	0	0
Chief officer	1	0	Command post	1	0	0
Engine 3	4	1	3rd floor	3	0	184 GPM* (697 L/min)
Engine 4	4	0	2nd floor	4	0	305 GPM (1155 L/min)
Totals	21	2	(N/A)	17	2	794* GPM (3006 L/min)

*Although the backup hose line is not assigned to attack the fire, the additional flow is immediately available should the need arise.

Table 8-9 Staffing with a Full Complement

Company	Total Staffing	Staffing at Apparatus	Assignment	Staffing at Assignment	Rapid Intervention Crew	Flow
Engine 1	4	1	2nd floor	3	0	184 GPM (697 L/min)
Engine 2	4	0	2nd floor	4	0	305 GPM (1155 L/min)
Truck 1	4	0	2nd floor	2	0	0
Truck 1	(see above)	0	3rd floor	2	0	0
Chief officer	1	0	Command post	1	0	0
Engine 3	4	1	3rd floor	3	0	184 GPM* (697 L/min)
Engine 4	4	0	2nd floor	4	0	305 GPM (1155 L/min)
Rescue 1	4	0	RIC/accountability officer	0	4	0
Engine 5	4	4	Staged	0	0	0
Truck 2	4	4	Staged	0	0	0
Medic unit	2	0	Treatment/transport	0	0	0
Medic unit	2	0	Rehab	2	0	0
Staff	3	0	Staff	3	0	0
Totals	**40**	**10**	**(N/A)**	**24**	**4**	**794* (GPM (3006 L/min)**

*Although the backup hose line is not assigned to attack the fire, the additional flow is immediately available should the need arise.

While it may be possible to double up on some assignments shown on the various deployment charts, fire fighter safety must never be compromised. In this case, fully staffing the RIC was delayed. Many departments preassign RIC duties (e.g., second-arriving truck company). Doing so ensures a fully staffed RIC but sometimes results in a delay in establishing a RIC. In a large jurisdiction, this fire could be handled by on-duty forces. In smaller jurisdictions, fires such as this would require assistance from mutual aid companies. In some cases, mutual aid assistance would respond automatically to a reported structure fire.

Without an exposure or rescue problem, and with all companies taking appropriate action, this fire required what would be an extra alarm assignment for most departments. Seldom is this level of efficiency experienced under the stress and limited information at the incident scene. Adding a rescue

component to this fire problem could greatly increase the required staffing.

Note that three or four fire fighters are assigned to each hose line, with the exception of the first fire line, where only two fire fighters are available for interior assignments. Operating a hose line with two fire fighters is an absolute minimum and a very arduous task. Two fire fighters can safely operate a 1¾-in. (45-mm) hose stream during practice sessions. During an actual fire, it may be necessary to advance the hose line up stairways, around obstacles, and through multiple doorways, then operate the hose from a crawling or kneeling position, all of which greatly increase the effort needed to successfully place a hose line in position. Consider that Engines 2 and 4 in our scenario advanced 2½-in. (65-mm) lines to the second floor. Not only does assigning all available personnel on a single company to a hose line provide the staffing necessary to properly position and operate the attack line, it also maintains company unity and accountability.

On occasion, there may be a need to assign more than one company to a hose line even on departments that assign four people to a company, as was done in the previous auto parts store example. As an example, a company is assigned to advance a hose line to the fifth floor of a large warehouse without a standpipe. If the company is supplying its own water, one member must remain at the pumps, several fire fighters will be needed to advance the hose line through the entry door and up the stairway (many warehouse buildings have 20-ft [6.1-m] or higher floors), and the hose line must then be advanced through a door at the fifth floor level and possibly around storage racks and through aisles. In this scenario, multiple companies might be needed to advance the hose line and operate the nozzle that could be flowing as much as 350 GPM (1325 L/min).

Determining Apparatus Needs

The apparatus used to deliver staffing to the incident scene is generally more than sufficient to support an offensive operation. Typically, the greater challenge is to properly position on-scene apparatus to best utilize their capabilities. Apparatus that are not needed at the incident scene should be parked out of the way, preferably in positions that allow access to water supplies and are available to support a possible defensive attack. During large-scale incidents, apparatus are often assigned to a staging area. The IC should designate a staging officer to manage and coordinate all companies assigned to the staging area. A staged apparatus has the staffing necessary to function as a unit, such as an engine company, a truck company, or a medic unit. Apparatus without adequate staffing are classified as out of service or parked, rather than staged.

There is a temptation to use on-scene apparatus whether they are needed or not. In the auto parts

Fallacy	Fact
All nonoperating apparatus at the scene are staged.	Apparatus + staffing = staged. Apparatus − staffing = out of service or parked.

Steel: © Sharpshot/Dreamstime.com

store example, there is some justification for a third source of water, and a ladder truck could be used to accomplish additional venting or to ladder the building. However, more often than not, during offensive operations, there is a need for additional staffing, not additional apparatus. When companies are staffed with two or three fire personnel, there is an even greater number of unnecessary apparatus on scene. Placing unnecessary apparatus in the immediate fire zone tends to block access that may be needed should a change in tactics be required.

Fallacy	Fact
All apparatus at the scene should be operating.	Use only the apparatus necessary to meet tactical objectives.

Steel: © Sharpshot/Dreamstime.com

Class A Foam

The scope of this text is limited to structural firefighting. Class A foam has been used primarily in wildland and wildland–urban interface fire operations. Proponents of Class A foam claim that the product is superior to untreated water when used as a suppression agent or to protect exposures from radiant heat. Early tests and field experience indicate that these claims are valid.[20] However, several fire departments throughout the United States and Canada were surveyed and found that most fire departments were not using Class A foam to fight structure fires even when it was available on the apparatus.

Apparatus equipped with Class A foam capability may prove useful in areas without a reliable water supply, as buildings in these areas may require more water than is carried in apparatus water tanks. In these situations, the use of Class A foam could make a significant difference. Class A foam may also be beneficial during overhaul.

The second expected advantage of Class A foam is that there is less water damage because less water is used. A hose crew properly applying untreated water at or above the required rate of flow will normally use more water than a crew applying Class A foam; however, a few questions should be asked. Are the additional gallons of water saved enough to justify the expense of the foam? Finally, prewetting fuels and providing a foam layer on an exposure seem to be the most useful applications of Class A foam in an urban or suburban setting. More information is needed to better quantify these characteristics of Class A foam, but its use in these situations is promising.

Chapter Summary

- Resource capabilities in terms of staffing, apparatus, and water must meet incident requirements for a safe and effective operation.
- Rate of flow is a major factor in determining if resources are adequate.
- Methods for determining rate of flow include trial and error, the Royer/Nelson formula, the National Fire Academy formula, and sprinkler system calculations.
- Each rate-of-flow approach has strengths and weaknesses, and none is completely accurate, as variables such as fuel load, fuel type, ventilation, and size of the compartment can affect fire dynamics.
- Buildings requiring a rate of flow exceeding the flow of two standard preconnected hose lines should be preincident planned by applying the V/100 formula, with the exception of areas containing an extra hazard fuel load, which require use of sprinkler calculations.
- A 1¾-in. (44-mm) hose line is the minimum recommended size of hose line for an offensive attack at a structure fire.
- Using hose size larger than 1¾ in. (45mm) is a trade-off of mobility for an increase in flow.
- Nozzle and hose configurations should be flow tested regularly to confirm the flow and proper operation.
- Variable-stream nozzles provide fog or straight streams.
- The number and size of attack hose lines will be based on flow requirements and backup lines, plus lines needed to protect means of egress and internal and external exposures.
- Fire can extend within a building to involve other areas or attached structures (internal exposures) or to adjacent structures (external exposures).
- Backup lines should be at least as large as the initial attack line.
- Each hose line requires at least two fire fighters, but a full company per line is preferred.
- It is good practice to provide at least two water supplies for a working structure fire.
- When a significant water supply challenge exists, options should be outlined in standard operating procedures (SOPs) and the preincident plan.
- Use of an attack pumper tactic should be limited to fires in relatively small structures.
- A flow of water meeting or exceeding the rate-of-flow requirements applied for a short period of time will require less total water and be more effective than a lesser flow over a longer period.
- Proper ventilation aids in the extinguishment effort and provides for a safer working environment for fire fighters in the building.
- Staffing needs vary depending on tasks to be accomplished, but certain minimum staffing requirements can be predetermined.
- Safe and efficient positioning of on-scene apparatus is a key consideration.

Key Terms

direct attack Firefighting operations involving the application of extinguishing agents directly onto the burning fuel.

external exposures Buildings, vehicles, or other property threatened by fire that are external to the building, vehicle, or property where the fire originated.

fuel-controlled Refers to a fire in which the heat release rate and growth rate are controlled by the characteristics of the fuel,

Flames: © Dny/Dreamstime.com; Steel texture: © SharpshotDreamstime.com such as quantity, chemistry, and geometry, and in which adequate air for combustion is available.

indirect attack Firefighting operations involving the application of extinguishing agents to reduce the buildup of heat released from a fire without applying the agent directly onto the burning fuel; ventilation is kept to a minimum while a fog stream is directed at the ceiling. This approach is most useful in unoccupied, tightly enclosed spaces.

internal exposures Areas within the structure where the fire originates or within buildings directly connected to the building of origin that were not involved in the initial fire ignition.

National Fire Academy formula A rate-of-flow calculation that calculates the rate of flow as the area in square feet divided by 3 (A/3).

oxygen- (ventilation-) controlled fire A fire in which the heat release rate and/or growth of the fire are controlled by the amount of air available to the fire.

pyrolysis The chemical decomposition of a compound into one or more other substances by heat alone; the process of heating solid materials until combustible vapors are emitted.

Royer/Nelson formula A rate-of-flow calculation that calculates the rate of flow as the volume in the fire area in cubic feet divided by 100 (V/100). This formula is based on the assumption that structure fires are primarily oxygen controlled.

sprinkler system calculations Specific rate-of-flow calculations for sprinkler systems; based on the fuel load. These calculations can be found in various publications, including NFPA documents and Factory Mutual Data Sheets.

trench cut An opening in the roof that extends from bearing wall to bearing wall to prevent horizontal fire spread in a building.

ventilation controlled A fire within a compartment or building that is limited due to a lack of air (oxygen) even though sufficient vapor fuel is available to support continued burning.

Challenging Questions and Activities

1. The large ranch-style house pictured in **Figure 8-21** was built in 2010 and has an undivided 2500 ft² (232 m²) finished basement with ceilings 9 ft (2.7 m) high. There is a walk-out double patio door leading to the basement at grade level on the C side in addition to a first floor door. There are two doors at the front of the house. It is midafternoon on Monday, September 30, with clear and cool weather conditions. Fire hydrants are located every 500 ft (152 m), with 700 GPM (2650 L/min) flow available from each hydrant. The total flow capacity of the system is 100,000 GPM (378,541 L/min).

Figure 8-21 Large ranch-style house.
Courtesy of Bernard J. Klaene.

Flames: © Dtor/Dreamstime.com; Steel texture: © Sharpshot/Dreamstime.com

A. Calculate the rate of flow for the finished basement using A/3 and V/100.

B. Would the rate of flow from a single standard preconnected hose line in use by your department be adequate to extinguish a well-involved basement fire in this house?

C. Would the rate of flow from a standard preconnected hose line *and* a backup line in use by your department be adequate to extinguish a well-involved basement fire in this house?

D. Which exterior door would be best to use to gain access to the basement?

E. What is the probability that occupants remain in the house, and where would they most likely be located?

F. Would you enter this basement to combat a working fire?

G. Would you recommend a softening the target tactic for this fire? Explain your answer.

H. Develop an incident action plan for this fire, assign companies to implement the plan, and organize the operation using NIMS.

2. You are responding to a mixed-occupancy bar and residential fire. Assume the role of the first-arriving engine at the fire shown in **Figure 8-22**. It is 7:00 AM, Sunday, August 1. The weather is warm and very humid.

The first floor is occupied by a bar. Your walk-around reveals the building is approximately 40 × 40 ft (12.2 × 12.2 m) at the base. It appears that the second and third floors are occupied apartments with an access stairway and road on the D side. There is a full basement that is partially above grade with a rear entry door to the basement on side C. There is also a fire escape on side C. A parking lot with five parked cars is partially blocking access to

Figure 8-22 Mixed-occupancy building—bar and residential.

Courtesy of Bill Strite, Cincinnati, Ohio

the building and fire escape. The fire escape appears to be poorly maintained. There is a fenced-in parking lot on side B.

A. Based on a risk-versus-benefit analysis, would you make entry to attack the fire with the staffing on the first-arriving fire company in your department? If your staffing is less than four fire fighters per company, would this staffing level affect your offensive/defensive decision?

B. Will your standard preconnected hose line be adequate to extinguish a fire in any area of this building?

C. Based on the conditions shown in Figure 8-22, where would you place the first attack line and why?

D. Assuming you decide on an offensive attack, where would you make entry and how would you reach the fire area to attack the fire.

Flames: © Drx/Dreamstime.com; Steel texture: © Sharpshot/Dreamstime.com

E. Would it be possible to apply the softening the fire tactic?

F. Would door control be effective in this scenario?

G. Would you consider a vent, entry, isolate, search (VEIS) tactic? If so, where would you make entry?

H. You are in charge of the first-arriving truck company. Assuming the first-arriving engine company took the actions you have recommended so far, what tasks would your company perform (in order of priority)?

I. As the first-arriving chief officer, develop and implement an incident action plan, assuming the first engine and truck companies took the actions you have recommended so far.

J. Develop a NIMS organizational chart for this fire showing all on-scene companies.

3. Assume the role of the first-arriving battalion chief (IC) at the fire shown in [Figure 8-23].

Figure 8-23 Fire in a one-story commercial building.
Courtesy of Denny Baker, Cincinnati Fire Department

It is 1:00 PM, Monday, July 11. The weather is hot and humid with a light wind from the east. The building is on the north side of the street. The occupant is a costume manufacturer. The fire building is 70 × 90 ft (21.3 × 27.4 m). The three-story attached building on side D (28 × 40 ft [8.5 × 12.2 m]) is also owned by the costume company and used as office and sales space. A two-story single-family home is located 20 ft (6.1 m) to the rear. An 80- × 80-ft (24.4- × 24.4-m) one-story manufacturing business is located on side B.

All companies are staffed with an officer, a driver, and two fire fighters. The first-arriving engine and truck companies from your fire station are also on the scene.

A. Would you begin operations in the offensive or defensive mode?

B. What is your rationale for an offensive or defensive attack?

C. If you deploy an interior attack line, what size line and what nozzle would you order.

D. If you decided on a defensive attack, where would you place the first fire stream? Would you order that a 1¾-in. (45-mm), 2-in. (50-mm), 2½-in. (65-mm), or master stream be deployed? What would be the best nozzle?

E. What is the probability that this building and exposure are occupied at this time?

F. Would you ventilate the building? If yes, what kind of ventilation would be best and where would you vent?

G. What would you direct the truck company to do?

H. Develop and implement an incident action plan assuming the first engine company took the actions listed so far.

Flames: © Dre/Dreamstime.com; Steel texture: © Sharpshot/Dreamstime.com

I. Assign units to implement your incident action plan.

J. How many units and what kind would you maintain as a tactical reserve.

K. Develop a NIMS organizational chart for this fire showing all on-scene companies.

4. Assume the role of fire chief in a community of 50,000 people. There are a few stores, churches, restaurants, and schools. However, the community is primarily made up of single-family detached dwellings. No buildings are more than three stories in height. The city is in a financial emergency and needs to reduce the budget. The fire department is being asked to reduce staffing at the three single-engine fire stations from four to three fire fighters per station. A private ambulance service provides emergency medical services (EMS) for the community, with the fire department responding as first responders. You are the only paid administrative staff. Develop a response justifying four-person staffing for your department.

Endnotes

1. Clark, W. *Firefighting Principles and Practices* (2nd ed.). Saddle Brook, NJ: PennWell; 1991.

2. Brunacini, A. *Fire Command* (2nd ed.). Quincy, MA: National Fire Protection Association; 2002.

3. Coleman, J. *Incident Management for the Street Smart Fire Officer.* Saddle Brook, NJ: PennWell; 1997.

4. Lloyd Layman. *Attacking and Extinguishing Interior Fires.* Quincy, MA: National Fire Protection Association; 1955.

5. Kerber, S., UL Firefighter Safety Research Institute. *Study of the Effectiveness of Fire Service Vertical Ventilation and Suppression Tactics in Single Family Homes.* Northbrook, IL: Underwriters Laboratories; 2013.

6. International Association of Fire Chiefs, Safety, Health and Survival Section. Firefighter safety call to action. December 23, 2013. Retrieved from http://www.iafc.org/Media/articlePR.cfm?ItemNumber=7299.

7. Kerber, S., UL Firefighter Safety Research Institute. *Impact of Ventilation on Fire Behavior in Legacy and Contemporary Residential Construction.* Northbrook, IL: Underwriters Laboratories; 2010.

8. International Association of Fire Chiefs, National Fire Protection Association. *Fundamentals of Fire Fighter Skills* (3rd ed.). Burlington, MA: Jones and Bartlett Learning; 2014.

9. National Fire Academy courses. Emmitsburg, MD: National Fire Academy.

10. Royer, K., Nelson, F. *Water for Fire Fighting, "Rate of Flow" Formula, Extension Services Bulletin #18.* Ames, IA: Iowa State University; 1959.

11. National Fire Protection Association. *NFPA 13: Standard for the installation of Sprinkler Systems.* Quincy, MA: Author; 2013.

12. National Fire Protection Association. *NFPA 1142: Standard on Water Supplies for Suburban and Rural Fire Fighting.* Quincy, MA: Author; 2012.

13. Edwards, C. B. Critical fire rate. *Fire Engineering.* September 1992: 97–99.

14. National Fire Protection Association. *NFPA 14: Standard for the Installation of Standpipe and Hose Systems.* Quincy, MA: Author; 2010.

15. Averill, J. D., Moore-Merrell, L. *NIST Report on High-Rise Fireground Field Experiments.* NIST Technical Study Note 1797. Gaithersburg, MD: National Institute of Standards and Technology; 2013.

16. Kerber, S., Madrzykowski, D. *Evaluating Positive Pressure Ventilation in Large Structures: School Pressure and Fire Experiments*. NIST Technical Note 1498. Gaithersburg, MD: National Institute of Standards and Technology; 2008.

17. Goodbread, J. C. *Fire in Building 3001*. Quincy, MA: NFPA; July 1985 and August 1985.

18. National Fire Protection Association. *NFPA 1710: Standard for the Organization and Deployment of Fire Suppression Operations, Emergency Medical Operations, and Special Operations to the Public by Career Fire Departments*. Quincy, MA: Author; 2010.

19. Averill, J. D. *Report on Residential Fireground Field Experiments*. NIST Technical Note 1661. Gaithersburg, MD: National Institute of Standards and Technology; 2010.

20. Routley, J. G. *Compressed Air Foam for Structural Fire Fighting: A Field Test*. Emmitsburg, MD: U.S. Fire Administration; 1993.

Flames: © Dex/Dreamstime.com; Steel texture: © Sharpshot/Dreamstime.com

Flames: © Dre/Dreamstime.com; Steel texture: © Sharpshot/Dreamstime.com; Chapter opener photo: © Karela/Shutter Stock, Inc.

Defensive Operations

LEARNING OBJECTIVES

- Compare and contrast a defensive versus offensive fire attack, explaining the key differences.
- Explain why an offensive attack is preferred over a defensive attack.
- Enumerate conditions that would lead to a defensive attack.
- Describe how collapse zone dimensions are determined.
- Discuss the positive and negative effects of operating a hose stream into a window or roof opening.
- Describe how water should be applied when protecting an exposure from radiant heat.
- Compare and contrast the use of handheld hose streams versus master stream appliances during defensive operations.
- List two ways the water utility company may be able to increase the total water supply at the incident scene.
- Compare and contrast the use of fog versus solid streams during a defensive attack.
- Estimate staffing and apparatus needs when operating master streams.
- List common problems leading to conflagrations.
- Explain tactics used to control a conflagration.

LEARNING OBJECTIVES (CONTINUED)

- Discuss why conflagrations are likely to occur immediately after a natural disaster.
- List reasons for a nonattack strategy.
- Given a scenario, calculate the dimensions of the collapse zone.
- Evaluate safety and tactical problems associated with a transition from an offensive attack to a defensive attack.
- Prioritize exposures based on fire conditions, occupancy, and weather factors.
- Develop an incident action plan for a defensive fire and for a conflagration.
- Apply defensive tactics to a defensive fire attack and conflagration.
- Evaluate staffing, water supply, and apparatus needs for a large-scale defensive fire.
- Apply the National Incident Management System to a defensive fire scenario.
- Determine the probability of a conflagration for a specified response area.

Case Study

Defensive operation in a four-story building with exposures.

Courtesy of David J. Jones, Cincinanti, Ohio.

On January 8, 2010, fire fighters were called to a fire in an older section of the city containing many large commerical and residential buildings. Upon arrival, they determined that the fire was on the upper two floors of a four-story building of heavy-timber construction. An offensive attack was attempted, but fire conditions required the abandonment of the offensive attack. A defensive strategy was employed to combat the fire in the building of origin and protect three nearby exposure buildings. Additional alarms were sounded, bringing a total response of 85 fire fighters staffing 20 fire companies. After 3 hours, the fire was declared to be under control, but units remained on the scene for more than 13 hours.

1. How does the level of fire involvement affect the offensive/defensive decision?
2. When selecting a strategy, what role does water supply and staffing levels play?
3. What effect does construction type have on structural stability?

Flames: © Jag_cz/ShutterStock, Inc; Steel texture: © SharpshotDreamstime.com; Stacked photo background: © Vitaly Korovin/ShutterStock, Inc; Paper: © silver-john/ShutterStock, Inc; Photo: © Crystal Craig/Dreamstime LLC

Introduction

An **offensive** fire attack is the preferred strategy whenever conditions and resources permit an interior attack. A **defensive** decision limits operations to the exterior, generally resulting in a larger property loss and limiting rescue options. The offensive/defensive decision is based on staffing available to conduct an interior attack, water supply, ventilation, and most importantly a risk-versus-benefit analysis. A defensive operation is used whenever the risk-versus-benefit analysis determines that the risk to fire fighters' lives and safety outweighs any possible benefit that might be achieved through an offensive attack. For example, if a fire occurs in a building that is structurally unsound and the fire does not place any lives at risk, then clearly the operation should be handled defensively. However, situations are not always so clear-cut, and the fire officer must rely on solid fire-ground information, coupled with training and experience, to determine whether the operation should be offensive, defensive, or nonattack.

> **NOTE**
>
> The rate-of-flow formulas described in the discussion of offensive attacks in the *Offensive Operations* chapter do not apply to defensive operations.

Texture: © Eky Studio /ShutterStock, Inc; Steel: © SharpshotDreamstime.com

Classifying as a Defensive Attack

The first-arriving officer must evaluate the situation using the NFPA 1500 risk management process described in the *Fire Fighter Safety* chapter to determine whether to implement an offensive or defensive strategy. This is a critical decision upon which every other decision is based. The preferred fire attack mode is offensive, and the incident commander (IC) should apply an offensive strategy whenever it is safe to do so.

There are times when initial conditions justify an offensive strategy that ultimately results in a change to a defensive attack. For example, an offensive attack that was at first justified due to life safety may no longer be justified once the building is cleared of occupants. Deteriorating conditions may also require a change from an offensive to a defensive attack.

Each officer who assumes command must reconsider critical size-up factors to determine if the operation should continue in the present strategic mode. Generally, the longer an uncontrolled fire burns, the more hazardous the building becomes. Therefore, the risk-versus-benefit analysis should be assessed not only during the initial attack or during a change of command, but also at set times and whenever major tactical objectives are achieved.

Changing from an offensive to a defensive attack is a command challenge and potentially hazardous time on the fire ground for several reasons, including the following:

1. The defensive attack must be delayed until the building is completely cleared of fire fighters. It is imperative that a personnel accountability report be conducted prior to commencing the defensive attack. Offensive and defensive attacks must never be conducted in the same building at the same time; however, if warranted, a defensive attack may be conducted in a fire-separated building while an offensive attack is taking place in another building.
2. Fire companies must be reassigned to defensive operational tasks. This can be a major problem if there are a large number of companies operating at the scene when a decision is made to go defensive. Whenever possible, the IC should anticipate the possibility of a move to a defensive attack (plan B) and appoint a planning section chief to develop plan B, then

assign officers to relay instructions to companies as they evacuate the building. Firm command and control of the operation is essential during this critical transition.
3. Master stream appliances are primarily used in defensive attacks. Unlike the offensive attack, in which several fire fighters are needed to advance a handheld line, the master stream–oriented defensive attack typically requires only one fire fighter to direct the stream and another to operate the apparatus pumps. Once a water supply has been established, unassigned personnel must be managed to avoid dangerous freelancing.

Typically, defensive attacks are conducted in the following situations:

- Structural integrity concerns, fire conditions, or other hazards prohibit entry.
- Resource needs outweigh resource capabilities.
- A risk-versus-benefit analysis indicates that the risk is too great compared to what can be saved.

When the IC initiates a defensive fire attack, the objective is to save property that has not already been destroyed and/or to protect the environment. Often, this means sacrificing the building of origin in favor of saving surrounding external exposures. At other times, a defensive attack can knock down the heavy volume of fire that prevents entry, limiting or stopping the fire spread, thus saving the remainder of the building. The overhaul portion of a defensive attack, where part of the building is still standing, can be extremely dangerous.

Defensive fires appear to be a greater challenge and more spectacular, but in reality they are easier to handle and pose fewer risks to fire fighters if the proper precautions are taken. The fire officer must keep in mind the factors that lead to a defensive decision. Fire fighters should never be needlessly placed in a dangerous environment. When a fire is of such magnitude that the building's structural support system is threatened, the building should be evacuated and a defensive operation initiated.

IC Rule of Engagement for Fire Fighter Safety

Do not risk fire fighters' lives for lives or property that cannot be saved. Seriously consider a defensive strategy.

Reproduced from: © 2011, IAFC Safety, Health and Survival Section. All rights reserved.

Texture: © Eky Studio /Shutterstock, Inc; Steel: © Sharpshot/Dreamstime.com

Establishing a Collapse Zone

When a fire seriously threatens the structural integrity of a building, fire fighters should move to the outside of the collapse zone to avoid life-threatening emergency retreats. Figure 9-1 shows fire fighters in a defensive mode working close to a building showing obvious signs of collapse.

Rule of Engagement for Fire Fighter Survival

You are required to abandon your position and retreat before deteriorating conditions can harm you.

Reproduced from: © 2011, IAFC Safety, Health and Survival Section. All rights reserved.

Texture: © Eky Studio /Shutterstock, Inc; Steel: © Sharpshot/Dreamstime.com

Many theories exist regarding collapse zone distances. Some theorists believe that a building will fall within one third of its height. Generally, this and other theories are based on sound principles but with questionable assumptions being made about how the collapse will occur. Buildings sometimes collapse within themselves; other times, the walls fall away from the supporting structure as a unit. This type of structural failure results in the collapse zone being equal to the height of the building plus an allowance for debris to scatter Figure 9-2 .

Any collapse zone that extends less than the building's height plus an allowance for debris scatter is a calculated risk. The IC must ask whether the expected benefit is worth the risk.

NOTE

A good rule-of-thumb method of estimating the collapse zone is to use a distance equal to 1½ times the height of the building.

Texture: © Eky Studio /Shutterstock, Inc; Steel: © Sharpshot/Dreamstime.com

Figure 9-1 Fire fighters working within the collapse zone.

Courtesy of Mr. David J. Jones, Cincinnati, Ohio.

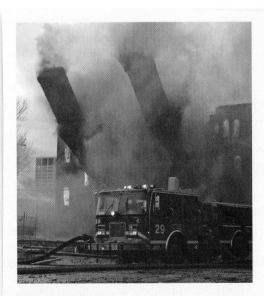

Figure 9-2 Wall collapse.

© Brian Kersey/AP Photos.

The ability to apply water from a distance greater than the height of the building may lead to a nonattack strategy for tall buildings that are in danger of collapse. Fire departments should measure the effective water application distance for master streams carried on their apparatus. Collapse zones can be preplanned. The actual height of the building can easily be determined during preincident planning. Rough approximations must be made at the time of an incident. For older multistory residential and office buildings, using a factor of 12 ft (3.7 m) per story is reasonably accurate. Modern buildings normally have lower ceilings and are often less than 12 ft (3.7 m) per story, whereas warehouse buildings or special occupancies could have much higher ceilings. Using the 12 ft (3.7 m) per story estimate, a five-story building would be 60 ft (18.3 m) in height, requiring a collapse zone of approximately 90 ft (1.5 × 60 ft) (1.5 × 18.3 m = 27.5 m). The question becomes, can available master streams effectively apply water from a distance of 90 ft (27.5 m)? Wind conditions must also be considered when estimating the effective distance for a fire stream. Solid-stream nozzles, as compared to straight-stream and fog nozzles, generally maintain better stream continuity for long distances, particularly when the stream is operated in opposition to the wind or there is a crosswind.

Another limiting factor for apparatus and fire stream placement is the width of the street. If a 90-ft (27.5-m) collapse zone is required and streets are only 40 ft (12.2 m) wide, personnel, apparatus, and equipment placed in the street would be within the collapse zone. Before committing people and apparatus to a potential collapse zone, the IC must seriously consider the expected benefits. If a decision is made to position personnel and equipment within the potential collapse zone, evaluate the building to determine the safest possible location. For example, positioning companies at the corners of the building is usually safer than a frontal attack. Again, anyone who is seriously considering being an IC is strongly encouraged to read texts such as *Brannigan's Building Construction for the Fire Service* to become familiar with the building's structural components.[1]

Evaluating Exposures

Exposures are generally divided into internal and external exposures. Internal exposures deal with fire extending from one area of a building to another within a structure, as described in the discussion of offensive fire attacks in the *Offensive Operations* chapter. Protecting the interior exposure is only marginally effective in a defensive mode. However, if the building is tightly constructed with closed doors separating areas, there is a possibility that the fire will be contained to the area already involved in fire.

Many times, in a defensive operation the building of origin is recognized as a total loss; therefore, emphasis is placed on protecting external exposures. The IC should evaluate external exposures in terms of life safety, extinguishment, and property conservation.

The distance between exposed structures and the volume and location of the fire have much to do with prioritizing exposure protection. Radiant heat increases as the size of the flame front increases. Knocking down the main body of fire can reduce the flame front, thus reducing radiant heat energy. Therefore, the most effective way to protect exposures is to extinguish the main volume of fire. When inadequate resources, collapse probability, or fire volume makes a direct attack on the fire ineffective, applying water directly on the exposed structures is the best tactic.

The energy levels for radiant heat are inversely proportional to the square of the distance between the heat source and the exposure. The closer the buildings are located to each other, the more radiant heat will be transferred from the fire building to an adjacent structure. Exposure buildings that are higher than the fire building are also at a greater risk.

The fire officer must determine the best method of protecting potential exposures. Available staffing, as well as matching apparatus resources to incident requirements, will greatly affect the IC's options. Wetting the exposure is generally the most effective way to prevent ignition. Radiant heat will travel through transparent materials such as water. Therefore, directing a water stream between buildings is less effective than applying water directly to the exposed building.

As was previously mentioned, Class A foam appears to present possibilities in terms of protecting exposures. The compressed air variety, in particular, tends to cling to exposures and do a better job of prewetting surfaces.

Evaluating a Direct Attack

Hose streams applied directly into the burning structure is an exposure protection tactic that has merit, but the fire intensity and depth may be such that it is impossible to effectively extinguish the fire or knock down the flame front. If a direct attack fails to control the fire or the effectiveness of a direct attack is questionable for any reason, first cover nearby exposures and then direct as much water as possible onto the main body of fire from a safe distance.

NOTE

The IC must consider the possibility of fires starting inside exposed buildings due to radiant heat igniting combustibles that are near windows facing the fire building. Radiant heat passes through intact panes of glass.

Texture: © Eky Studio /Shutterstock, Inc; Steel: © Sharpshot/Dreamstime.com

Sometimes it is possible to place a master stream in a position where it can be operated either directly on the fire or on an exposure without moving the apparatus or master stream appliance **Figure 9-3**.

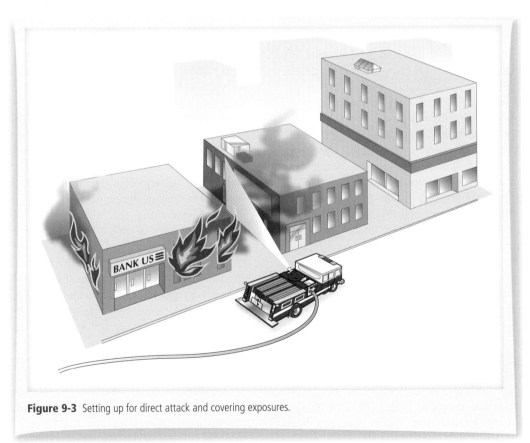

Figure 9-3 Setting up for direct attack and covering exposures.

Estimating the Number and Type of Master Streams Needed

There are times when master streams are used to augment an offensive attack. However, master streams are used primarily during defensive attacks. Master streams are the tools of choice for defensive operations. A master stream can apply more water from a greater distance and requires fewer personnel to operate as compared to a handheld hose line.

Master streams can be operated from ground level, the top of an apparatus, or an elevated position. However, it is important to remember that any tactic that hinders the upward and outward movement of heat and smoke is usually detrimental to the operation. Elevated master streams applying wide-angle fog streams are particularly problematic in terms of interrupting the upward and outward flow path of the heated products of combustion. Sometimes, enough of the building has been destroyed that it is impossible to change the flow path. If the IC considers the building a total loss, directing streams into vent openings may be necessary to protect nearby exposures.

It is extremely difficult for a person to see where a stream is hitting from behind the nozzle. A fire fighter or officer should step back or to the side to get a better view of the stream, giving directions to the fire fighter who is operating the nozzle from this improved vantage point.

Once master streams are in position, staffing needs diminish. Personnel inside the fire zone should be kept to a minimum. This is a good time to start rehabilitation. Handheld lines have little or no value during defensive attacks on large structures and often invite unsafe tactics. Anything less than a 2½-in. (65-mm) line would be ineffective. A common problem is that fire fighters who are not operating pumps or master streams are standing around at a big fire, and they may occupy themselves using handheld lines that serve no purpose. It is the IC's responsibility to manage this situation by placing fire fighters without a specific assignment in a safe location. If handheld lines were used during an offensive attack that has been abandoned, it is recommended that these lines be disconnected from the apparatus during defensive operations to ensure that they are not used inappropriately. However, there are occasions when the fire is being fought defensively in one building and offensively in another. In this case, the use of smaller-diameter hose may be acceptable on the interior of an exposure.

> **NOTE**
>
> When handheld lines are needed as part of a defensive attack, reduce fatigue by using hose loops whenever possible.

Texture: © Eky Studio /Shutterstock, Inc; Steel: © Sharpshot/Dreamstime.com

Estimating Water Supply Needs

Providing water at defensive operations may prove more challenging than for most offensive attacks. Many master stream appliances are capable of flowing 1000 GPM (3785 L/min) or more, thus requiring the total pump capacity of an apparatus. When several of these master stream appliances are operating, the water system can be exhausted even in areas with high-volume and reliable water supplies. Supply lines should be large-diameter hose, and there may be a need to relay or shuttle water.

When evaluating the water supply, consider the flow capacity of individual hydrants as well as the total flow capacity of the water system. Some water supply systems are made up of several separate systems that can be connected (cross-tied) when necessary. System connections can be accomplished by pumping from one system to the other. Typically, the water utility company will open street valves to cross-tie the systems. Cross-tie locations should be identified and preplanned. The water utility company may also be able to bring additional pumps on line, increasing the volume or pressure at the fire scene. Consider including water utility representatives in the multiple alarm call list.

Selecting the Stream Position

The first consideration in placing exterior streams is safety. The second is the ability to apply water to exposures and to the interior of the building. Quite often, these two considerations are at odds. The best direct application positions are often in the collapse zone.

Unstaffed ground monitors are sometimes the answer to this dilemma. However, the effectiveness of these streams is often limited, which results in sending fire fighters back into the collapse zone to redirect the streams. Elevated streams have a natural advantage when the fire is in the upper stories of the building but require the apparatus to be parked fairly close to the building. Likewise, the apparatus-mounted master stream device is easier to place in service but places the pump operator and apparatus in the collapse zone when directing an attack on tall buildings. Remember, there is no justification to risk injury to fire fighters during a defensive operation, and there is seldom justification to risk damage to expensive apparatus and equipment.

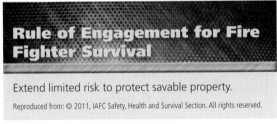

Rule of Engagement for Fire Fighter Survival

Extend limited risk to protect savable property.

Reproduced from: © 2011, IAFC Safety, Health and Survival Section. All rights reserved.

Texture: © Eky Studio /Shutterstock, Inc; Steel: © Sharpshot/Dreamstime.com

Selecting the Nozzle Type

Is the solid stream (smooth bore), straight stream, or fog stream best for defensive operations? They all have their place, and it is important for the fire officer to know the advantages and disadvantages of all available tools. The solid stream has the greatest reach and penetrating ability and is best suited to situations when the attack is on the main body of the fire in a large structure.

The fog pattern can be gently applied to an exposure, covering a wider area without breaking windows. Straight streams from a variable-stream nozzle are not generally considered equal to solid-stream (smooth-bore) nozzles, but most are capable of penetrating the main body of fire. Using a variable-stream nozzle allows the stream to be used for an attack on the main body of fire, and it can also be repositioned to cover exposures if the apparatus or master stream is properly positioned.

Calculating Staffing Needs

Unlike the offensive attack, in which several fire fighters are needed to advance a handheld line, the master stream–oriented defensive attack typically requires only one fire fighter to direct the stream and one to operate the pumper, once a water supply has been established. The officer should stand to the side to direct the stream and provide direction to the fire fighter operating the stream. Unassigned personnel may be rotated through rehabilitation or assigned to staging, making companies available for other responses or to augment the attack.

Determining Apparatus Needs

During the offensive attack, there may be an excess of apparatus, whereas the defensive attack is apparatus intensive. Offensive attacks are people intensive. Where two or more fire fighters are required to advance a 2½-in. (64-mm) line flowing 250 GPM (946 L/min) in an offensive attack, one fire fighter can direct a 1000-GPM (3785-L/min) or greater master stream.

A master stream that is being supplied from a distant hydrant may require one or more pumpers to supply the water. Further, a large master stream appliance may require one or more pumpers to supply it. However, 1250-GPM (4732-L/min) or 1500-GPM (5678-L/min) pumpers connected to a reliable water supply with large-diameter hose may be able to supply two master streams.

Elevated streams can also require additional pumpers, particularly if the aerial apparatus are not equipped with prepiped waterways and pumps.

When water supplies are very limited, tanker shuttles or drafting from distant water sources will require a substantial commitment of personnel, apparatus, and time.

Conflagrations and Group Fires

Conflagrations occur less frequently now than in the past, but they remain a possibility. Fire departments should recognize the extraordinary challenge presented by a conflagration and determine the probability of a conflagration in their response area. Special tactics are needed to resolve a fire incident involving a large geographic area.

Early conflagrations, such as the burning of Rome, are not well documented. More is known about later conflagrations that destroyed large tracts within urban centers. London had conflagrations in 798, 982, 1212, and 1666. Constantinople is the undisputed conflagration champion, with conflagrations recorded in 1729, 1745, 1750, 1756, 1782, 1791, 1798, 1816, 1870, 1908, 1911, 1915, and 1918. In 1607, the first permanent American colony was founded at Jamestown, Virginia, which was destroyed by fire in January of 1608. Plymouth, Massachusetts, was founded in 1620 and was substantially destroyed by fire 3 years later. Boston experienced nine conflagrations before the Revolutionary War. During the Revolutionary War, like all other wars, fire was used as a weapon, destroying many villages in colonial America. Conflagrations continued with great frequency in the New World until the early 1900s. More recently, terrorists used fire as a weapon of mass destruction at both World Trade Center attacks.

Some of the more significant conflagrations to have occurred in North America in the past two centuries are noted in **Table 9-1**. The list is not all inclusive. Many group fires and some conflagrations are not investigated by national organizations and, therefore, go unnoticed outside the region where they occur.

The term conflagration is often misused. Many people refer to any large fire as a conflagration. It is not unusual to hear a news anchor refer to the conflagration in a downtown tenement building. There are several definitions for conflagration. In an urban setting the term conflagration is used to describe a fire that spreads from building to building over a considerable distance, beyond a natural or artificial barrier (e.g., beyond a city block).

Group fires are similar to conflagrations; however, unlike a conflagration, a group fire is confined within a complex or among adjacent buildings. Some definitions describe a group fire as not extending beyond a complex or confined to the city block of origin. Most group fires have the potential to become conflagrations, but either they are confined early or no exposures are readily available to allow fire spread. Group fires are generally of smaller scale than a conflagration; however, there are exceptions. For instance, the main body of fire at the Santana Row Fire in San Jose, California, was contained within the complex, which would be defined as a group fire even though it did cover a large geographic area. However, this fire does not completely fit the definition of a group fire, as flying brands traveled a considerable distance, igniting buildings well beyond the complex where the fire originated. From a strategic point of view, conflagrations and group fires present similar problems and have similar solutions.

Wildland–urban interface fires spreading from wildland into an urban area, and destroying large numbers of buildings, are properly defined as conflagrations. The most notable wildland–urban interface fire occurred in Peshtigo, Wisconsin, the same day as the Great Chicago Conflagration—October 7, 1871. The Peshtigo fire killed 1200 people (though some estimate as many as 2000 fatalities) and destroyed 17 towns. A wildland fire that does not destroy a large number of buildings should not be labeled a conflagration.

The focus of this text is structural firefighting; thus tactics used to combat wildland–urban interface fires are beyond the scope of this book. However, Table 9-1 lists several conflagrations that began as wildfires, because once a wildland fire enters a populated area, many of the tactics described in this text would apply. The discussion here is limited to the tactics used when confronted with a large-area fire in a built-up area.

When conflagrations are studied, common contributing factors are evident, including:

1. Closely built structures, especially combustible structures
2. Wood shingle roofs

Table 9-1 Conflagrations of Significance

Year	Place	Number of Buildings Involved
1835	New York City, NY	700 buildings
1845	Pittsburgh, PA	1000 buildings
1851	San Francisco, CA	2500 buildings
1866	Portland, ME	1500 buildings
1871	Chicago, IL	17,430 buildings (250 fatalities)
1871	Peshtigo, WI	17 towns (approx. 1200 to 2000 fatalities)
1901	Jacksonville, FL	1700 buildings
1904	Baltimore, MD	2500 buildings
1906	San Francisco, CA	28,000 buildings (earthquake)
1908	Chelsea, MA	3500 buildings
1914	Salem, MA	1600 buildings
1916	Paris, TX	1440 buildings
1917	Atlanta, GA	1938 buildings
1918	Minnesota Forest	4000 buildings (559 fatalities)
1922	Ontario Forest	18 townships (44 fatalities)
1922	New Bern, NC	1000 buildings
More recent conflagrations and group fires include:		
1961	Bel Air, CA	630 buildings
1973	Chelsea, MA	300 buildings
1979	Houston, TX	25 buildings
1981	Lynn, MA	28 buildings
1982	Anaheim, CA	51 buildings
1983	Dallas, TX	6 buildings (125 apartments)
1991	Oakland/Berkley, CA	2886 buildings (25 fatalities)
2002	San Jose, CA	10 large buildings
2012	New York, NY	130 buildings

© Eky Studio/Shutterstock, Inc.

3. Poor water supplies or fire suppression weaknesses (automatic and manual)
4. Dilapidated structures, especially abandoned buildings in large numbers
5. Large-scale combustible construction or demolition projects
6. Residential and/or commercial developments near wildlands
7. Built-up areas near high-hazard locations, where a transportation or industrial fire/explosion could quickly involve large numbers of buildings

Incident Summary

On Monday, August 19, 2002, a conflagration that began in the Santana Row construction site of San Jose, California, spread via burning embers to the Huff/Moorpark area, which is a half mile away.

The building of origin that was under construction within the Santana Row complex was a very large six-story structure covering 6 acres (24, 281 m²). This fire created burning embers as large as 2 in. × 4 in. (51 mm × 102 mm). The high wind carried embers to an apartment complex and townhouses in the Huff/Moorpark area where they ignited wood shake roofs. Some of the roofs had been replaced by composite roofing materials. The damage was largely confined to 10 large apartment and townhouse buildings with wood shake roofs. Buildings with composite roofs were, for the most part, undamaged.

Data from: John Lee Cook, Jr., USFA-TR-153, 1/2004, Santana Row Development Project Fire, San Jose, California

The fire at the Santana Row construction site, from which burning embers traveled and created a conflagration half a mile away.

© Justin Sullivan/AP Photos.

Response areas with one or more of these seven conditions have an increased probability of a conflagration, especially when experiencing dry and windy weather conditions. Most of the urban conflagrations that occur have two or more of these factors combined with windy weather conditions. However, of the conditions listed, the wood shingle roofing has historically posed the greatest problem and is most often cited as a conflagration factor. Codes allowing untreated wood shingle roofs ignore a real and present danger.

An understanding of how large fires spread is essential in order to effectively apply the tactics involved in controlling a conflagration or group fire. A large flame front created by multiple burning structures creates extremely high radiant heat that rapidly spreads the fire. As the fire gets larger, the flame front expands exponentially, greatly accelerating the rate of fire spread. In addition, wood shingle roofs or other burning materials (flying brands) move upward via convection currents and are then carried by the wind to other areas.

The flying brand hazard is addressed by sending brand patrols into areas downwind, particularly areas containing materials that are likely to ignite.

Flames: © Jag_cz/ShutterStock, Inc.; Paper: © silver-john/ShutterStock, Inc.

The 2002 Santana Row fire in San Jose, California, serves as an excellent example of the dangers associated with flying brands and wood shingle roofs. At the Santana Row fire, flying brands ignited wood shingle roofs a half mile away.

Understanding the methods of heat transfer is the first step in developing a conflagration strategy. Once the flame front widens, radiant heat is the primary means of fire extension from building to building and from groups of buildings to other groups of buildings. If the fire is extending along a wide radiant flame front, the primary tactics are narrowing the flame front and setting up primary and secondary lines of defense while protecting exposures. In setting up lines of defense, take advantage of natural fire breaks, such as wide streets.

As always, life safety is the top priority followed by extinguishment. In large group fires and conflagrations, being proactive and evacuating people in the fire's path long before they are actually threatened is the key to success. Remember that the evacuation area must be beyond the secondary line of defense, and the police department is typically the best agency to manage the actual evacuation. Isolating the evacuation area from well-intentioned "would be" victims and traffic control are extremely important; thus a wide fire perimeter should be maintained. Establishing a law branch, under police supervision, is generally a good organizational tactic at large-scale fires.

A fully staffed planning section is of great value when commanding a fire that is spread over a wide geographic area; however, many times the initial line of defense is overrun by the fire before planning can be properly staffed.

Figure 9-4 shows three lines of defense for a Houston, Texas, apartment complex fire in 1979. The first two lines of defense had to be abandoned before a successful stop was made at the third line of defense. A battle-line strategy established without allowing enough time to position apparatus and equipment results in the IC expending scarce resources in a losing battle.

The Breezy Point (Queens, New York) conflagration of 2012 occurred during a hurricane. Streets were flooded and access to the fire area was very difficult. Still the IC managed to set up fire streams to narrow the fire front using a flanking maneuver, applying a three-pronged attack that limited fire spread.[2] Given the flooded conditions, adverse weather, and massive fire, the fact that no lives were lost during this massive conflagration testifies to a well-executed strategy **Figure 9-5** .

> **NOTE**
>
> When setting up primary and secondary lines of defense for a conflagration, place apparatus so they can be rapidly redeployed if the line of defense must be abandoned.

Texture: © Eky Studio /Shutterstock, Inc; Steel: © SharpshotDreamstime.com

Halting or preventing a conflagration can seriously challenge any water supply system. The use of water must be prioritized with individual units, realizing the importance of water conservation. In most cases, water being discharged into a flame front is of little value. The top priorities are maintaining the fire break at the lines of defense and protecting exposures. Once the decision has been made to write off an area, do not waste water trying to save what will surely be lost. Large cities are generally served by several water service areas, and the planning section chief should identify auxiliary water sources well in advance. For instance, when water mains were destroyed in the Marina District fire area during the 1989 San Francisco earthquake, the San Francisco Bay became an auxiliary water source as fireboats relayed water to the shore.

Collapsed structures, accumulations of combustible debris, utility problems (electrical shorts, gas leaks, etc.), and heating and lighting with open flames are only a few of the many reasons that areas struck by a natural disaster are prone to fires and conflagrations. Cincinnati, Ohio, suffered a conflagration in 1937 during a major flood. Petroleum tanks floated off their bases, spreading flammable and combustible liquids on top of the water. The petroleum products were ignited and spread fire to many buildings over a wide area. Fire hydrants and access roads were under

Figure 9-4 The Houston, Texas, conflagration in 1979.

Figure 9-5 Aftermath of the 2012 Breezy Point, Queens, conflagration.

Courtesy of Andrea Booher/FEMA.

water, hampering fire control efforts. Many buildings burned to the water line.

The six tactical elements of a successful conflagration strategy are:

1. Evacuate and rescue people in imminent danger.
2. Evacuate and rescue people in the endangered area beyond the secondary line of defense.
3. Set up a line of defense in an area with natural or artificial fire breaks.
4. Establish a secondary line of defense.
5. Narrow the flame front.
6. Provide flying brand patrols beyond the line of defense.

Incident Summary

A conflagration that destroyed three city blocks in Perkasie, Pennsylvania, was believed to have been started by juveniles setting fires in grass next to buildings. The fire entered a lumber shed and appeared at first to be a small fire. First attempts at extinguishment seemed to be successful, but hidden fires reignited and wind-swept the flames through the lumber shed. The fire continued to burn out of control and eventually either destroyed or damaged seven buildings and two fire trucks.

Data from: Jay Bradish, "Conflagration Destroys Three City Blocks," Fire Command, January 1989, pages 22, 23 and 26.

Buildings damaged or destroyed in the Perkasie, Pennsylvania, fire.

Flames: © Jag_cz/ShutterStock, Inc.; Paper: © silverjohn/ShutterStock, Inc.

These tactics are interrelated and must be accomplished simultaneously or in rapid succession. As an example, evacuating people beyond the secondary line of defense requires that the location of the secondary line of defense be known. Actually protecting the secondary line of defense protects the evacuation.

Classifying as a Nonattack

It is difficult to stand by and watch a building being consumed by fire, but this is sometimes the only safe approach. Nonattack postures are underused because ICs often fail to recognize a total loss.

Often, in nonattack operations, environmental concerns are paramount. In these situations, the IC may assign resources to protect rivers or streams from contaminated runoff water or evacuate downwind neighborhoods. As an example, a Sherwin Williams warehouse fire in Dayton, Ohio, threatened the community's water supply.[3] A nonattack posture avoided a major environmental catastrophe. If a safe offensive attack is not possible and there is little or nothing to be gained by initiating a defensive attack, choose the nonattack option.

Flames: Dx/Dreamstime.com; Steel texture: © SharpsheUDreamstime.com

Chapter Summary

- Defensive attacks are used when the building's structural integrity or other hazards prohibit entry, when resource needs outweigh capabilities, or when the risk to fire fighters is too great in terms of what can be saved.
- Solid fire-ground information, training, and experience are used to determine whether to use an offensive, defensive, or nonattack strategy.
- Defensive attacks seek to save intact property and/or to protect surrounding exposures.
- The collapse zone is the height of the building plus an allowance for debris scatter, usually estimated as 1½ times the height of the building.
- A master stream can apply more water from a greater distance and requires fewer personnel to operate as compared to a handheld hose line.
- Water supply and apparatus positioning are key considerations during a defensive attack.

- A defensive attack is apparatus intensive but generally requires fewer fire fighters than an offensive attack.
- Apparatus at large-scale fires must be positioned for rapid redeployment.
- A successful conflagration strategy includes evacuating civilians, creating fire breaks for a primary and secondary line of defense, narrowing the flame front, and patrolling for fire brands.

Key Terms

defensive A type of fire attack in which exterior fire suppression operations are directed at protecting exposures.

offensive A type of fire attack in which fire fighters advance into the fire building with hose lines or other extinguishing agents to overpower the fire.

Challenging Questions and Activities

1. Assume the role of the first-arriving battalion chief (IC) at the fire shown in **Figure 9-6**. It is 5:30 PM on Monday, September 15. The weather is hot and humid with a light wind from the east. The building is on the north side of the street. The building is occupied by a costume manufacturer. The fire building is 70 × 90 ft (21.3 × 27.4 m). The three-story attached building on side D (28 × 40 ft [8.5 × 12.2 m]) is also owned by the costume company and used as office and sales space. A two-story single-family home is located 20 ft (6.1 m) to the rear. An 80- × 80-ft (24.4- × 24.4-m) one-story manufacturing business is located on side B. All companies are staffed with an officer, a driver, and two fire fighters. The first-arriving engine and truck

Figure 9-6 Fire in a one-story commercial building.

Courtesy of Denny Baker, Cincinnati Fire Department.

companies (Engine 1 and Truck 1) are on scene when you arrive.

- Engine 1 has secured a 5-in. (125-mm) water supply. The pump operator remains at the apparatus while the remaining three company members are preparing to advance a 2½-in. (65-mm) hose line through the door next to the loading dock on side A. You approve!
- Truck 1 has divided into two crews.
  - Truck Crew 1A is forcing entry and entering the building with search rope and thermal imaging camera to perform search and rescue. Truck 1A is working with Engine 1 as they attack the fire.
  - Truck Crew 1B has entered the office/sales building on side D to evacuate the building and check for fire extension.

- Engine 2 has secured a second source of water via a 5-in. (125-mm) hose line. The pump operator remains at the apparatus while three company members advance a 2½-in. (65-mm) hose line inside the building alongside Engine 1.
- Engine 3 sets up a portable master stream appliance inside the fire building; water is supplied by Engine 1's pumper. Engine 1's captain is assigned as the Division 1 supervisor.
- Engine 4 lays a 2 ½-in. (65-mm) backup line supplied by Engine 2's pumper.
- You assign Truck 2 to evaluate the roof structure and vent the roof near Side C if roof operations can be safely accomplished.
- Engine 4 is assigned to lay a 1¾-in. (45-mm) interior exposure line into the first floor of the building on side D. They are also instructed to check the second and third floors of the building for fire extension. Truck Crew 1B is directed to report to the Engine 4 officer.
- Truck 3 is assigned as the rapid intervention crew (RIC).
- Heavy Rescue 1 is assigned to Division 1 to assist in the primary search.
- Medic 1 arrives on scene and is ordered to stand by, ready to treat and/or transport as needed. You also order an additional medic unit.
- Engine 5 is assigned to secure their own water supply and operate an apparatus-mounted master stream covering the side B exposure.

Flames: Dro/Dreamstime.com; Steel texture: © SharpshotDreamstime.com

- Engine 6 is instructed to secure their own water supply and set up a master stream on side C to protect the rear exposure.
- Truck 4 is assigned as a second RIC.
- Engines 7 and 8, along with Truck 5, are staged two blocks away as a tactical reserve. Engine 4's officer reports an "all clear" for the office/sales building on side D. The officer also reports that there is smoke in the building, but no visible fire in this exposure. Engine 4 requests reassignment, but you direct Engine 4 and Truck Crew 1B to remain in the exposure and lay an additional line to the second floor.
  a. Evaluate the present operation in terms of safety and effectiveness.
     i. Are all necessary tactics being accomplished?
     ii. If not, what else needs to be done? (Add units as necessary.)
     iii. Was there an initial rapid intervention crew (IRIC)? If not, should there have been?
     iv. Is the water supply adequate for the present operation?
     v. Should an alternative defensive action plan be developed?
  b. Develop an incident action plan and NIMS organizational chart for the operation in progress, adding additional units as you deem necessary.

2. Use the same building, fire, and deployment described in Question 1. Fire conditions are worsening as you receive the 20-minute notification from dispatch. Truck Crew 1A sounds a mayday, stating,

"Mayday, mayday, mayday! We are inside the rear of the building and cannot find our way out. We are running low on air."

As the IC, you acknowledge the mayday and ask Division 1 if the location of Truck Crew 1A is known. In reply, the officer states that their exact location is unknown, but they were last assigned to search the C–D quadrant. You decide to begin a transition to a defensive attack but first want to locate and assist Truck Crew 1A to safety.

A. Explain the deployment and actions that should be taken by the RIC teams.

B. As the IC, explain your assignments and reassignments of units other than the RIC during the mayday operation.

C. Restructure your incident action plan and NIMS organizational chart to reflect changes in the operation.

3. Assume you are successful in rescuing the two missing members of Truck 1 and decide to move to a defensive attack. Fire conditions have changed, as shown in **Figure 9-7** .

A. List the necessary safety measures to be taken as you transition to a defensive attack.

B. Develop an incident action plan for the defensive attack.

C. Explain your reassignments, location of apparatus, and use of handheld and master streams. Also describe the water supply necessary to support the defensive operation.

D. What should be done with the numerous hand lines that were used during the offensive attack?

Flames: Dx/Dreamstime.com; Steel texture: © Sharpshot/Dreamstime.com

Figure 9-7 Defensive fire conditions in a one-story commercial building.

Courtesy of Mr. David J. Jones, Cincinnati, Ohio.

E. Once the defensive attack is in place, additional personnel will be available (e.g., engine company personnel not operating pumps or fire streams). Describe how you would handle excess personnel.

F. Develop a NIMS organizational chart for the defensive operation.

4. The fire continues. It is now 8:30 PM (first call occurred at 5:30 PM). Four fire fighters have taken it upon themselves to set up handheld lines in the street using lines previously deployed to the interior . Figure 9-8 .

A. Do you feel the building is in danger of collapse?

B. Are the fire fighters shown in Figure 9-8 within the collapse zone?

C. Are the hose streams shown having a positive effect on the outcome?

Figure 9-8 Three hours into the operation at one-story commercial building.

Courtesy of Mr. David J. Jones, Cincinnati, Ohio.

Endnotes

1. Corbett, G. P., Brannigan, F. L. *Brannigan's Building Construction for the Fire Service* (5th ed.). Burlington, MA: Jones and Bartlett Learning; 2015.

2. Pfeifer, J. W. Conflagration Hurricane Sandy—NY, NY, Breezy Point, Queens. *Fire Engineering*. May 2013: 61–69.

3. Isner, M. S. $49 million loss in Sherwin-Williams warehouse fire. *NFPA Fire Journal*. March/April 1988: 65–73, 93.

Flames: Dxd/Dreamstime.com; Steel texture: © SharpshotDreamstime.com

Flames: © Dra/Dreamstime.com; Steel texture: © Sharpshot/Dreamstime.com; Chapter opener photo: © Jones and Bartlett Learning. Photographed by Glen E. Ellman.

Property Conservation

LEARNING OBJECTIVES

- List the three tactical priorities, in priority order, and explain how property conservation can be accomplished simultaneously with the life safety and extinguishment priorities.
- Define primary and secondary damage.
- Explain how proper ventilation and forcible entry actually reduce property damage while protecting occupants and fire fighters.
- Given a pressure and discharge orifice size, estimate the flow from a single and multiple sprinkler heads.
- List and explain six tactics used to reduce water damage.
- Explain why removing property from a building is not generally a good property conservation tactic.
- Calculate the weight of water from a nozzle discharging 350 GPM (1326-L/min) over a 10-minute period.
- Describe the importance of ventilation in property conservation.
- Discuss the importance of overhaul.
- Explain how thermal imaging cameras can be used to reduce overhaul damage and the precautions necessary when using thermal imaging cameras to find hidden fires.
- Enumerate safety issues related to overhaul and fire investigation.
- Develop a fire scenario and apply property conservation tactics for a fire controlled by the sprinkler system or an accidental discharge from the sprinkler system.

A fire fighter uses a thermal imaging camera while another uses a pike pole, and a third awaits with a fire hose.
© Jones and Bartlett Publishers. Courtesy of MIEMSS.

An electrical fire in a four-story apartment building was very difficult to overhaul. The fire started in an electrical outlet on the third floor and extended into the wall. The fire department disrupted the power supply to the apartment, opened up the wall, and extinguished the fire. However, the fire had extended into the apartment above and the attic. Even after the wall and ceiling were opened up, some hidden fire was missed. Eventually, all of the fire was found and completely extinguished.

Wires should be traced through the wall space, as heating could occur anywhere along the circuit. Thermal imaging cameras could also be a valuable tool during overhaul operations.

1. What part does a thorough overhaul play in reducing property damage?
2. When trying to locate concealed fire, what are the advantages and limitations in using a thermal imaging camera?

Data from: Frank C. Montanga, Chasing down electrical fires. Fire Engineering, July 1999, pp. 83–86

Introduction

Proficient fire officers realize the importance of property conservation and recognize that property conservation tactics can substantially reduce property loss. Although property conservation is the lowest of the three operational priorities, being listed as the third priority does not mean that it must wait until all life safety and extinguishment tactics are completed. Extinguishment is the ultimate property conservation tactic. Once staffing permits, the incident commander (IC) should simultaneously attend to all three operational priorities during offensive operations.

Fallacy	Fact
Property conservation must be delayed until all life safety and extinguishment activities are complete.	Property conservation is prioritized behind life safety and extinguishment. Be certain that resources are assigned to accomplish these priorities first. When staffing is adequate, property conservation can and should be conducted simultaneously.

Steel: © Sharpshot/Dreamstime.com

NOTE

The three tactical priorities are (1) life safety, (2) extinguishment, and (3) property conservation.

Texture: © Eky Studio /Shutterstock, Inc; Steel: © Sharpshot/Dreamstime.com

Good public relations results from professionalism shown in property conservation efforts. Few citizens (or elected officials) recognize a good extinguishment effort, generally because expertly conducted offensive operations are hidden from view.

Property conservation tasks are obvious, however, even to laypeople. Efforts directed at saving valuable property show the citizens that you care about them and their property. Remember, fire departments are expected to save both people and property. In a residential fire, the occupants' personal property—property that is often irreplaceable—is being protected. The occupants will appreciate close attention to salvaging what can be saved.

NOTE

The golden rule of property conservation is "Do unto the property of others as you would have them do unto yours."

Texture: © Eky Studio /Shutterstock, Inc; Steel: © Sharpshot/Dreamstime.com

Fire departments often receive unfair criticism when they conduct forcible entry, ventilation, and overhaul operations. The untrained observer does not understand why it is necessary to cut a hole in the roof or tear walls apart to check for extension. To prevent unfair criticism, the department must educate the media, the public, and politicians about the purpose and importance of these tactics. Most important, the IC should explain the overhaul process to the owner and occupants. Although most secondary damage is essential, training should emphasize keeping the damage to a minimum while stopping the forward progress of the fire and preventing a rekindle.

Property conservation is prioritized behind life safety and extinguishment, but property conservation activities need not be delayed until the fire is extinguished or until the primary and secondary searches are complete. Priority activities are assigned first, and property conservation is delayed when staffing is not sufficient to attend to all of the life safety and extinguishment tasks. However, proficient ICs include staffing for property conservation in their request for assistance or reassign forces to property conservation activities early in the battle.

The purpose of this chapter is to explain the primary components of property conservation operations. However, this is not an exhaustive reference on how to complete the various tasks involved in property conservation. Other sources of information, such as *Fundamentals of Fire Fighter Skills*, are dedicated to the actual tasks involved in protecting property.[1]

In commercial and industrial occupancies, employees may be extremely helpful in locating valves, drains, and mechanical controls. Large industrial complexes sometimes have trained fire brigades, many of which are dedicated to property conservation duties.

Classifying as an Offensive Attack

Property conservation is generally limited to offensive attacks. If fire crews have been withdrawn from the building because of a lack of resources or because of deteriorating fire and building conditions, property conservation is no longer an issue. It makes no sense to send fire fighters into a building to throw salvage covers or otherwise protect property when the entire building is at serious risk of being destroyed by fire. Remember, fire fighters should *never* be placed at risk attempting to save what is already lost, or ultimately will be lost.

Rule of Engagement for Fire Fighter Survival

Size up your tactical area of operation.

Reproduced from: © 2011, IAFC Safety, Health and Survival Section. All rights reserved.

Texture: © Eky Studio /Shutterstock, Inc; Steel: © Sharpshot/Dreamstime.com

Estimating Indirect Damage

A balance must be struck between doing necessary damage to halt the loss being caused by the fire and minimizing the damage caused by fire-ground activities. Categorizing damage helps in understanding this relationship.

Primary damage is caused by the products of combustion. **Secondary damage** results from fireground activities or the operation of a fire protection system. Primary damage will consume the entire property if fire fighters fail to force entry or ventilate. On the other hand, if the building is washed off its foundation, nothing has been saved. In a fire, both primary and secondary damage will occur. Using caution and good judgment, the IC must do everything necessary to limit all types of property damage. However, fire fighters should never be placed in grave risk to save property.

Rule of Engagement for Fire Fighter Survival

Extend limited risk to protect savable property.

Reproduced from: © 2011, IAFC Safety, Health and Survival Section. All rights reserved.

Texture: © Eky Studio /Shutterstock, Inc; Steel: © Sharpshot/Dreamstime.com

Fire protection systems often cause secondary damage. In the *Fire Protection Systems* chapter, the importance of letting the system control the fire and not shutting down the system prematurely was emphasized. Smoke and steam reduce visibility, making it difficult to determine the proper time to close valves and shut down systems. Therefore, when the IC decides to shut down a system, a fire fighter should remain at the control valve so that the system can be charged again if needed.

Sprinkler systems have an excellent record of extinguishing or holding fires in check with minimal water damage. On occasion, system piping or sprinkler heads are damaged, causing an unwanted or accidental water flow. Although this situation is not a fire, the IC needs to be prepared to mitigate it promptly.

Flows from sprinkler heads can vary depending on the size of the orifice opening, the type of sprinkler head, and the water pressure in the system. As **Figure 10-1** shows water flows of up to 80 GPM (303 L/min) per head are possible.

Whether the head was accidentally damaged or opened automatically because of a fire, the flow of water will continue until action is taken to stop it. A 30-GPM (114-L/min) head flowing during a 10-minute response time will discharge 300 gallons (1136 L) of water. Once the fire is under control, it is necessary to stop the flow of water, either by shutting the system down at the riser, by closing division valves, or by using sprinkler stops.

Other fire-suppression agents can also harm equipment and stock. The discharge of a dry chemical system, for example, will require considerable cleanup and can damage certain types of equipment, such as computers, telephone equipment, and other sensitive electronics. Generally, the cleanup should be the responsibility of the property owner, not the fire department.

Evaluating Water Damage

A fallacy exists regarding secondary damage and rate of flow. Many believe that larger-than-needed hose streams cause unnecessary water damage. In reality, the opposite is true. Larger-than-needed hose streams, if properly applied, will result in less water damage, as the fire is quickly extinguished and the time that water is flowing is reduced. Conversely, choosing a fire line that is too small to quickly extinguish the fire will result in both increased water and fire damage, as water will be applied for an extended time without effectively extinguishing the fire.

Life safety and extinguishment activities concentrate on the fire floor and floors above, while efforts to control water damage are concentrated on the fire floor and floors below. Water damage is reduced by doing the following:

1. *Promptly extinguishing the fire, and avoiding wash downs.* In addition to reducing water damage, prompt extinguishment activities also reduce fire and smoke damage.

2. *Stopping the flow of water from sprinkler systems.* Promptly shutting down hose lines is one method of stopping the flow of water into the building. In the case of sprinkler

Figure 10-1 Water discharge rates of typical nominal 1/2-in. (12.7-mm) and 17/32-in. (13.5-mm) orifice automatic sprinklers.

systems, or other fixed fire suppression systems, it is important to locate control valves. Using sprinkler stops or controlling the water at the main or divisional control valve is the primary way to reduce the flow from an automatic sprinkler system. Be careful never to shut a system down too soon, as the result will be additional water and fire damage.

3. *Containing runoff.* Containing the water before it has a chance to run off or migrate to another location is another property conservation tactic. Catchalls that are made with salvage covers, or containers (e.g., drums, buckets), are placed under a ceiling leak below the fire

floor. Also, pumps and water vacuums can be very effective in property conservation operations. Preventing further migration of water will reduce damage.

4. *Channeling the water into drains or chutes or otherwise directing it out of the building.* Fire fighters have devised unique ways of channeling water out of the building. Makeshift chutes are sometimes fashioned using ladders, pike poles, and salvage covers. These chutes are then directed out a window, down a stairway, or to a drain.

   Fire debris will tend to clog floor drains; it is important to keep these drains clear. If floors do not have drains, a common way to provide one is by removing a toilet, creating a drain through the 4-in. (102-mm) opening in the floor. If this is done carefully, the only damage to the toilet will be to the seal and floor bolts. As with any drain, however, it is critical to ensure that it does not become clogged with debris. Removing water from the floor is not only important to property conservation; it is also a safety issue. A large buildup of water on a floor can cause structural failure. The IC must determine whether it is safe to conduct interior operations once a large quantity of water has been applied. This consideration is particularly important after long defensive attacks when it is time to move into the building for final extinguishment and overhaul. It is also important to ensure that any standing water in areas such as basements is not hiding hazards, such as openings, changes in elevation, or steps. If it is necessary to send fire personnel into a basement or other confined space that contains standing water, it is crucial to ensure that the electric power has been shut down to avoid any possibility of electrocution.

The nature of the commodities that are stored in a building can affect the weight loading on the floor. If absorbent materials are being stored in the fire area, the weight of the water that they will absorb will remain for a considerable time and these absorbent materials may expand, placing pressure against outside walls, which can also affect the structural integrity of the building. Water weighs 8.33 pounds per gallon (lb/gal) (1 kg/L). When multiple master stream appliances are flowing water into a structure, the weight of the water can rapidly increase the collapse potential. Depending on conditions, the water may flow out of the building or be contained. **Table 10-1** assumes that materials stored in the building or fire debris is containing the water being discharged from large master stream appliances.

It is difficult to estimate the amount of water remaining inside a building. Water streaming out of a building may represent only a fraction of the total amount of water being applied.

5. *Covering valuable property.* It is important for the IC, as well as all other members on the scene, to be aware of the need for property conservation. By placing salvage covers on exposed property, an alert fire fighter can often prevent water damage to valuable property. When using salvage covers, crews should group together furniture and other items to protect them from flowing water. This will allow several items to be protected by a single cover. Whenever possible, materials being protected should be raised several inches off the floor using available objects. In

© Ely Studio/Shutterstock, Inc.

Table 10-1 Weight of Water from Master Streams

Number of Master Streams at 1000 GPM (3785 L/min) Each	Weight Added Each Minute at 8.33 lb/gal (1 kg/L)	Total Weight at 30 Minutes	Total Weight at 60 Minutes
1	8330 lb (3778 kg)	249,900 lb (113,353 kg)	499,800 lb (226,705 kg)
3	24,990 lb (11,335 kg)	749,700 lb (340,058 kg)	1,499,400 lb (680,116 kg)
10	83,300 lb (37,784 kg)	2,499,000 lb (1,133,527 kg)	4,998,000 lb (2,267,055 kg)

an industrial setting, wooden or plastic pallets are often available. In residential properties, place salvageable items on top of a table, bed, or couch, for example.

6. *Moving or removing valuable property.* Water flows downward through a building, following the path of least resistance. Stairs, elevator shafts, and drains provide paths of least resistance. At times, the water is blocked by debris, or the volume simply overwhelms the natural flow out of the building. Even when water is flowing through the ceiling, the path of least resistance will be followed. This is generally around openings in the ceiling, such as light fixtures. If a ceiling is holding water runoff for a period of time, the weight of the water is likely to cause the ceiling to collapse. This ceiling collapse can injure fire fighters assigned to the area and increase damage to the contents below. Using a pike pole to drain the ceiling, being careful to avoid electrical equipment, can alleviate this problem.

As soon as possible, property should be moved away from natural flow paths, grouped together, and covered with salvage covers.

On occasion, property is completely removed from the building by fire crews to save it from water damage (Figure 10-2). This is a very labor-intensive way to conduct property conservation operations and has very limited application. If materials are moved outside, they must then be protected from inclement weather, vandalism, or theft.

Figure 10-2 Moving valuable materials.
© The Messenger-Inquirer, John Dunham/AP Photos

Evaluating Smoke Damage

Smoke can cause considerable property damage. Quite often, the smoke damage greatly exceeds water damage. Smoke generally follows an upward path but can cause damage below the fire as well. In previous chapters, ventilation was discussed in terms of life safety and extinguishment. However, ventilation is also an important property conservation tactic. Blowers and fans are generally used to remove smoke during property conservation operations. Of course, ventilation tactics used to remove smoke for life safety and extinguishment purposes also help to prevent property damage.

Calculating Staffing Needs

Once critical life safety and extinguishment positions are staffed and a tactical reserve has been established, remaining personnel can be assigned to property conservation. The earlier property conservation is addressed, the more successful it will be. Property conservation that is delayed too long will be of little value, as the exposed materials will already be damaged by water and/or smoke.

Property conservation staffing needs vary greatly. For most offensive operations, at least one crew should be assigned to start property conservation activities on the floor below the fire. Several crews may need to be assigned to property conservation duties, depending on the amount of damage, the size of the area to be protected, and the type of property that needs to be protected.

Evaluating Property Conservation Needs

Property conservation activities should not be delayed until other priorities have been completed. This idea needs to be emphasized. It is very easy to become overwhelmed with fire attack activities and neglect property conservation until the fire is completely extinguished. Implementing a planning section greatly assists the IC in defining necessary property conservation activity. Staffing and equipment are assigned to life safety and extinguishment first, but if property conservation tasks still remain to be assigned, then there is a need to call for additional resources.

Crews generally view property conservation as an undesirable assignment, preferring to be involved in life safety and extinguishment activities. Companies generally do not freelance into property conservation tasks. The ability to simultaneously conduct life safety, extinguishment, and property conservation tasks is a sure sign that the IC is effectively managing the incident.

Evaluating Overhaul Needs

Overhaul is extremely important; some tactical priority models list it as a separate priority. Overhaul is the completion of the extinguishment priority. Overhaul can result in what might appear to be

additional damage to the property, but this damage is warranted to prevent further primary damage. The use of a thermal imaging camera can substantially reduce damage caused by opening walls and ceilings to locate hotspots **Figure 10-3**. However, it is important that members using the thermal imaging camera be well trained in using the device and interpreting readings. If a thermal imaging camera is used to examine walls and ceilings, and any doubt remains as to the presence of hidden fire, the wall and/or ceiling must be opened.

Figure 10-3 Checking for fire extension using a thermal imaging camera.

The purpose of overhaul is to ensure that the fire is completely extinguished and that the building is safe for personnel such as investigators or the property owner to reenter. To permit people to reenter a building only to have fire rekindle is a serious failure on the part of the IC. Overhaul is an activity that is appealing to few fire fighters, which sometimes leads to a decision to simply wet down everything in sight. This "wash down" is not effective in getting to hidden fires and often results in unnecessary property damage.

There is a tendency for fire scene discipline to deteriorate during the overhaul phase. Fire fighters are often injured as they rush to complete this distasteful chore so they can return to the fire station or home. As the smoke clears, many fire fighters take the opportunity to remove their breathing apparatus and other items of protective clothing. Studies discussed in the *Fire Fighter Safety* chapter indicate that toxic gas levels are present during overhaul operations. If the IC does not maintain discipline, overhaul can become an uncontrolled, dangerous activity.

To keep property damage to a minimum and to ensure that the fire is completely extinguished, overhaul must be accomplished in a well-planned, systematic manner. A good overhaul technique is to assign a fully protected crew inside as a fire watch while the building is being ventilated. This will allow other fire fighters on the scene to rest, thereby reducing the potential for injuries. If it is practical, keep crews outside at least 15 minutes. After the building is well ventilated, the rested troops can return to a much safer environment to complete the overhaul.

Full protective clothing, including breathing apparatus, must be worn during overhaul activities **Figure 10-4**. Opening walls and ceilings will expose smoky, toxic, smoldering debris. In addition, ceilings and walls may contain asbestos or other materials that are capable of harming the unprotected fire fighter. Members operating inside the structure, including fire investigators, cannot be permitted to remove their face pieces.

Many departments use carbon monoxide meters or multiple gas meters to determine when it is safe to remove the breathing apparatus face piece. Carbon monoxide is usually the most prevalent fire gas, and low readings after a fire indicate some measure of smoke and toxic gas removal. However, there are many other inhalation hazards that are not read by carbon monoxide or multiple gas detectors. The best practice is to remain "on air" throughout the overhaul phase of the operation.

A Word About Fire Investigation

Improper overhaul can destroy any hope of properly conducting a fire scene investigation. A determination of what areas are to be left undisturbed should be made while fire suppression forces are resting between fire control and overhaul. The IC should attempt to make a preliminary determination as to what areas will be of interest to the investigators. When a fire investigator is on the scene during overhaul operations, this person is an invaluable resource to the IC. However, proper and complete overhaul should always be done to ensure that the fire is completely extinguished.

Figure 10-4 Overhaul in full personal protective equipment.

© Glen E. Ellman

Chapter Summary

- Property conservation is the third priority but can be performed simultaneously with life safety and extinguishment when staffing permits.
- Property conservation is limited to offensive attacks.
- Ventilation and forcible entry generally cause less property damage than delaying extinguishment.
- Covering or moving valuable property can help prevent water damage.
- Smoke damage generally occurs above the fire but can damage areas below the fire as well.
- Proper ventilation is a life safety, extinguishment, and property conservation tactic.
- Staffing and equipment are assigned to life safety and extinguishment first, but if property conservation tasks still remain to be assigned, call for additional resources.
- To ensure complete extinguishment and minimize property damage, overhaul must be accomplished in a well-planned, systematic manner.

Key Terms

primary damage Damage caused by the products of combustion.

secondary damage Damage caused by fire-ground activities or the operation of a fire suppression system.

Challenging Questions and Activities

1. A fire has occurred in the store shown in **Figure 10-5**. It appears that the sprinkler system has the fire under control but is still operating.

   A. Assume the role of IC and assign companies to the various tasks necessary to limit damage and prevent a rekindle.

   B. Describe how you would reduce property damage to other merchandise not already damaged by the fire.

   C. Explain the use of a thermal imaging camera to find hotspots and how it would be used in overhaul at this fire.

Figure 10-5 Overhaul and salvage at a mercantile occupancy.

Courtesy of Bernard J. Klaene.

Endnote

1. International Association of Fire Chiefs, National Fire Protection Association. *Fundamentals of Fire Fighter Skills* (3rd ed.). Burlington, MA: Jones & Bartlett Learning; 2014.

Flames: © Dtx/Dreamstime.com; Steel texture: © Sharpshot/Dreamstime.com

Flames: © Dra/Dreamstime.com; Steel texture: © Sharpshot/Dreamstime.com; Chapter opener photo: © kazoka/ShutterStock, Inc.

The Role of Occupancy

LEARNING OBJECTIVES

- Classify occupancy risks using an occupancy factor matrix.
- Given the dimensions, number of floors, and occupant density, determine the maximum number of people who could be in a building.
- Define assembly occupancy and provide examples of different types of assembly occupancies.
- Compare and contrast the risk to fire fighters when fighting a fire in an assembly occupancy compared to a residential occupancy.
- Define educational occupancy and compare the life hazards in elementary schools, high schools, and colleges.
- Define health care occupancy and explain how evacuations in health care occupancies are different from evacuations in most other occupancies.
- Compare and contrast occupants in hospitals, nursing homes, and limited care facilities, and describe how the occupant characteristics in each affect life safety during a structure fire.
- Define residential board and care occupancy and compare these facilities to nursing homes.
- Define detention and correctional occupancy and explain the special challenges associated with combating a fire in a large correctional facility.
- Define residential occupancy and compare various types of residential buildings in terms of life safety.

LEARNING OBJECTIVES (CONTINUED)

- Evaluate and discuss civilian fire deaths and fire fighter line-of-duty death rates in residential occupancies.
- Define mercantile occupancy and compare older-style shopping centers to enclosed malls and lifestyle centers in terms of life safety and extinguishment.
- Examine life safety and extinguishment problems related to "big-box" stores.
- Explain how search and rescue procedures in a large commercial structure differ from search and rescue in a residential building.
- Define business occupancy.
- Compare and contrast business and residential occupancies in terms of the risk to fire fighters during fire-ground operations.
- Define storage occupancy and evaluate the effect of fuel load on manual firefighting.
- Describe the problems associated with changing the commodities stored in a sprinkler-protected storage occupancy.

- Compare and contrast storage and residential occupancies in terms of the risk to fire fighters during fire-ground operations.
- Define industrial occupancy and the effect of hazardous materials related to life safety and extinguishment.
- Compare and contrast industrial and residential occupancies in terms of the risk to fire fighters during fire-ground operations.
- Define multiple, mixed, and separated occupancies, and explain the difference between a mixed and separated occupancy.
- Explain the increased hazard to occupants in multistory buildings where the first floor is occupied by stores and shops, with apartments above.
- Describe the fire hazards associated with buildings under construction, renovation, or demolition.

A Pittsburgh church fire on March 13, 2004, results in 29 fire fighter injuries and two fire fighter line-of-duty deaths.

©REUTERS/Jason Cohn/Landov

An electrical fire was reported in the basement of a historic church in Pittsburgh, Pennsylvania, on March 13, 2004. Operations began as an interior attack with several hose lines. The fire spread horizontally through the concealed space between the basement ceiling and first floor. The fire then spread vertically via concealed wall spaces to the structural members, framing, and interior furnishings. Deteriorating structural conditions and a backdraft caused the incident commander to withdraw and begin a defensive attack, which eventually brought the fire under control. During overhaul, the bell tower collapsed, trapping fire fighters working on the interior. It took several hours to extricate the trapped fire fighters. The collapse resulted in two fire fighter on-duty deaths and 23 injuries. The earlier backdraft injured an additional six fire fighters.

1. What role does occupancy type play in determining the occupied status of a building?
2. What effect does occupancy type have on the required rate of flow for extinguishment?
3. How does occupancy type affect fire fighter safety?

Data from: Fire Fighter Fatality Investigative Report 2014-17, CDC/NIOSH, Morgantown, WV.

Introduction

As a part of the size-up process, the incident commander (IC) must consider the building's occupancy type. Once the occupancy type is known, the IC can apply a strategy applicable to the occupancy and can begin determining the life safety, extinguishment, and property conservation tactics required to bring the incident to a successful conclusion. This information will assist the IC in determining the danger to occupants and risk to fire fighters. **Figure 11-1** lists occupancy types discussed in this chapter. Residential, industrial, and assembly occupancies all require different strategies. Time of day and day of week are also critical factors. This chapter reviews the process of evaluating occupancies and formulating safe and effective fire-ground strategies.

Some occupancies are dangerous places not only for the occupants but also for fire fighters. As an example, fire fighters working at a store or office occupancy fire are 5½ times more likely to be killed than when working at a residential fire (Figure 11-1).

Classifying the Occupancy Type

The function or use of a building has much to do with life safety. Codes and standards are written with a focus on the special hazards presented by the occupancy. Some of the major factors related to occupancy type are as follows:

- Mobility of the occupants
- Age of the occupants
- Leadership (Will there be an organized evacuation?)
- Awareness (Are occupants sleeping? Mentally impaired? etc.)
- Occupant density and total number of occupants
- Familiarity (Do they know the building layout?)
- Time factors (When are people present in the structure?)

__Occupant density__ is a measure of the number of people in a given area. Codes usually state occupant density as the number of square feet (ft^2) per person. In some assembly occupancies, the maximum

Flames: © Jag_cz/ShutterStock, Inc; Steel texture: Stacked photo background © Sharpshot/Dreamstime.com; © Vitaly Korovin/ShutterStock, Inc; Paper: © silver-john/ ShutterStock, Inc.

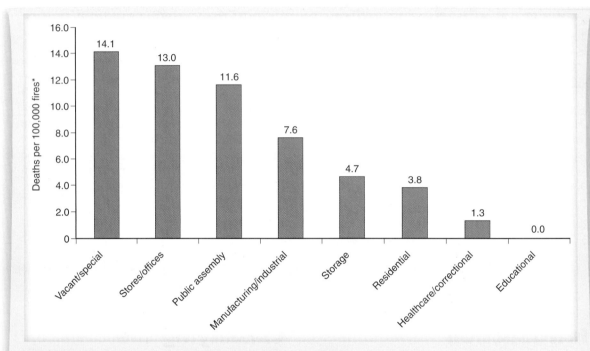

Figure 11-1 Fire fighter fire-ground fatalities by occupancy type, 2007 to 2011.

Modified from: Fahy, Rita F. Firefighter Fatalities in the United States-2013. National Fire Protection Association.

density is 7 ft² (0.65 m²) per person (occupant density factor). In other words, 100 people could occupy a 700-ft² (65-m²) area. Some business occupancies have a maximum density of 100 ft² (9.3 m²) per person, meaning the same 700-ft² (65-m²) area would be limited to 7 people. Most codes require the maximum number of people allowed in a given area to be posted in assembly occupancies and sometimes in other occupancies. If the area is not posted, a rough approximation of the number of people who could safely occupy an area can be determined by dividing the area in ft² by the occupant density factor. Occupancy factors affecting the ability to evacuate without assistance will be rated as part of the discussion of each major occupancy type. A scale of 1 to 10 will be used to indicate whether each factor is positive (10), neutral (5), or negative (1) for the specific occupancy type. As an example, an awareness rating of

10 would mean that all occupants within this occupancy would be expected to be fully awake and mentally unimpaired. An awareness rating of 1 would be given to an occupancy where occupants are totally unaware, such as people who are under general anesthesia. The numeric ratings are general in nature, and buildings of the same occupancy classification may have different ratings. As an example, some ambulatory care facilities anesthetize patients, which would result in a very negative awareness rating, while other ambulatory care facilities do not anesthetize or use medications that reduce patient awareness.

The following occupancy types will be addressed in this chapter:

- Assembly occupancies
  - Churches
  - Eating and drinking establishments
  - Sports arenas

- – Convention centers
- – Theaters
- Educational occupancies
  - – Elementary schools
  - – Middle, junior high, or high schools
  - – Colleges/universities[*]
- Health care occupancies
  - – Hospitals
  - – Nursing homes
  - – Limited care facilities
  - – Ambulatory care facilities
- Residential board and care occupancies
- Detention and correctional occupancies
- Residential occupancies
  - – One- and two-family dwellings
  - – Apartment buildings
  - – Dormitories
  - – Hotels/motels
- Mercantile occupancies
  - – Shopping centers
  - – Enclosed shopping malls
  - – Lifestyle centers
  - – Big-box stores
  - – Multilevel department stores
- Business occupancies
- Storage occupancies
- Industrial occupancies
- Multiple occupancies
- Buildings under construction or demolition
- Renovated buildings

[*]Colleges are technically categorized as business occupancies when classrooms accommodate less than 50 students and as assembly occupancies when the room accommodates 50 or more students.

Occupancy and suboccupancy types will be discussed in terms of the operational priority list: life safety, extinguishment, and property conservation.

Codes are promulgated to avoid the repetition of past mistakes. However, strengthening codes is not enough. Each of the major loss-of-life fires described in this chapter contains lessons for the fire officer. In many of these cases, existing codes were ignored, resulting in a large loss of life. The various occupancies are used to discuss common problems and solutions but also to offer history lessons. The NFPA's

Fire Investigations Department publishes fire investigations and research reports. The U.S. Fire Administration (USFA), National Institute for Occupational Safety and Health (NIOSH), and National Institute of Standards and Technology (NIST) produce technical reports that are available for recent large loss-of-life fires. Periodicals such as the *NFPA Journal* and *Fire Engineering* have offered technical reports to the fire service for over 100 years.

> **NOTE**
>
> The philosopher and poet George Santayana said, "Those who cannot remember the past are condemned to repeat it." As practitioners of fire strategy and tactics, it is imperative that we learn from the past.

Texture: © Eky Studio /Shutterstock, Inc; Steel: © Sharpshot/Dreamstime.com

Assembly Occupancies

NFPA 101, *Life Safety Code*,[1] 3.3.190.2, defines an assembly occupancy as an occupancy (1) used for the gathering of 50 or more persons for deliberation, worship, entertainment, eating, drinking, amusement, awaiting transportation, or similar uses; or (2) used as a special amusement building, regardless of occupant load.

The discussion here is limited to larger places of assembly and tactical difficulties related to the occupancy. Included under the places of **assembly occupancy** category are the following:

- Churches
- Eating and drinking establishments
- Sports arenas
- Convention centers
- Theaters

An occupancy factor matrix for assembly occupancies is shown in **Table 11-1**. Note that occupant familiarity and occupant density tend to be negative factors, but mobility and awareness tend to be positive factors for most assembly occupancies.

Table 11-1 Occupancy Factor Matrix: Assembly Occupancies

Assembly Occupancies	Mobility of Occupants	Age of Occupants	Leadership	Awareness	Occupant Density	Familiarity of Occupants
Churches	8	5	9	10	2	5
Eating and drinking establishments	8	7	5	4	2	3
Sports arenas	8	7	6	7	2	2
Convention centers	8	6	5	10	2	1
Theaters	8	6	5	10	2	2

1 = extremely negative factor; 10 = very positive factor.

Places of assembly are not occupied at all times. However, when they are occupied, large numbers of people gather in a relatively small area. In some instances, a population density of 7 ft² (0.65 m²) per person can be allowed; stated another way, nearly 1500 people can occupy a 100- × 100-ft (approximately 930-m²) room, and even more people are allowed where there is fixed seating. This compares to 333 people in the same area of a grade-level mercantile occupancy, 100 in a business occupancy, or 50 in a hotel. The primary challenge in managing an incident in an occupied assembly building is removing people to the outside or to a place of safe refuge.

The fuel load will vary by the assembly classification. The concern is not only for the total fuel load; flame spread rate is also critical. Decorative materials with high flame spread rates were significant contributing factors in the large loss of life at the Cocoanut Grove, Beverly Hills, and Station Nightclub eating and drinking establishment fires (discussed later). Fire control efforts must place a high priority on keeping the fire out of the exits. Aside from the main tactical consideration of removing victims during times of peak occupancy, responders must remain mindful that these structures are unattended for long periods when they are not in use, allowing fires to gain considerable headway before discovery. The probability of total involvement on arrival in these situations

dramatically affects fire fighter safety. Fires in places of assembly are fairly rare, accounting for approximately 3% of all structure fires,[2] but the potential for a large loss-of-life fire in these buildings requires preincident planning and special tactical considerations.

Churches

Churches often use open flames as part of religious services and tend to be overcrowded on religious holidays. Churches are also left unattended for long periods, and many church fires occur when the building is unoccupied. However, churches are occupied at times other than during religious services. Meetings, educational programs, fund-raisers, and social events take place within the church or attached rooms. Some churches provide a wide array of social services, such as child care, elderly programs, and shelter for the homeless. These services require supplies that add to the fuel load. Collecting clothing and food for poor or homeless people in the community is a common practice. These supplies—food and clothing—are often stored in places that are not separated from the main church and can pose a severe fire threat.

In the case of old Gothic-style churches, fires quickly involve large concealed spaces such as roofs and bell towers **Figure 11-2**. Bell towers typically contain large, heavy bells that are higher than the

Figure 11-2 Gothic-style church.

Courtesy of Bernard J. Klaene.

church and present a potential collapse hazard. Further, many churches are attached to other buildings, such as rectories or schools, which can create an exposure problem.

Older Gothic-style churches typically have large timbers supporting the roof structure, whereas modern churches often have truss roof structures Figure 11-3 . Recent fire fighter fatalities have occurred due to truss roof collapses with fire fighters working on the interior (see the *Fire Fighter Safety* chapter for a detailed analysis). The roof structure should be noted during preincident planning.

Figure 11-3 Modern church.

Courtesy of Bernard J. Klaene.

The large open spaces in places of worship can result in a large-volume fire that is extremely difficult to extinguish once the fire gains considerable headway. In addition to these traditional church structures, religious services are sometimes held in buildings not originally designed as a church.

Statistically, fires and loss of civilian lives are rare in religious properties. On average only 1760 of 517,130 structure fires (0.3%) per year occurred in religious properties from 1999 to 2002.[2] However, the potential for a large loss of life is a concern. Most fires in these occupancies take place when services are not being held, resulting in an extended alarm time that allows the fire to gain considerable headway prior to the fire department's response. Many times fires that occur when a church is unoccupied are not reported until fire breaks through the roof.

The fire loss in places of worship is 1.1% of the total loss, with only 0.3% of the fires occurring in these properties.[2] These occupancies tend to have priceless artifacts, stained glass windows, and other high-value property. Unfortunately, large fire volume and structural damage/collapse may render property conservation efforts ineffective. Large church fires usually dictate a defensive operation and often result in full or partial building collapse.

Eating and Drinking Establishments

Overcrowding is a common problem in eating and drinking establishments. If a fire occurs while the occupancy is overcrowded, exits may quickly become blocked. Many tragic fires have occurred in these occupancies during the past 100 years; most notable are the ones listed in Table 11-2 . The largest loss of life in an eating or drinking establishment in the United States was in the Cocoanut Grove fire in Boston, Massachusetts, where nearly 500 people were killed, many of them service men on their way to fight in World War II.

An analysis of eating and drinking establishment fires reveals the killing potential of fires in assembly occupancies. Occupants will often be mobile, but they may be impaired because of alcohol consumption. Fire probability is fairly high, owing to smoking, cooking, and highly combustible decorations

Table 11-2 Notable Fires in Assembly Occupancies		
Fire	**Year**	**Deaths**
Iroquois Theater, Chicago, IL	1903	602
Rhoads Opera House, Boyerton, PA	1908	170
Rhythm Club, Natchez, MS	1940	207
Cocoanut Grove, Boston, MA	1942	492
Ringling Brothers and Barnum and Bailey Circus, Hartford, CT	1944	168
Indiana State Fairground, Indianapolis, IN	1963	75
Beverly Hills Supper Club, Southgate, KY	1977	165
Happy Land Social Club, New York, NY	1990	87
The Station Nightclub, West Warwick, RI	2003	100

© Ery Studio/Shutterstock, Inc.

and furnishings. Fire fighters must quickly confine the fire while protecting the exits.

Most patrons will attempt to exit the same way they entered. Although alternative exits are available, they are often underutilized and overcrowding occurs at the main entrance. Fire fighters should direct people to alternative exits while keeping the crowd at the main entrance moving. It is essential to identify alternative exits, including windows and other openings that could be used to provide emergency egress, during preincident planning for eating and drinking establishments. Preincident plans should also consider the possibility of using forcible-entry tools to create alternative exits or to widen existing exits during an emergency evacuation.

Several large loss-of-life fires have occurred outside of the United States, including a January 28, 2013, fire in Santa Maria, Brazil, that killed 238. Like The Station nightclub fire, pyrotechnics used by entertainers were the cause of ignition.

Many times occupants escaping the building stop as soon as they reach safety, thereby unintentionally blocking the exit for others. Someone should be stationed at exit discharges to keep people moving away from the building. This is best done by fire fighters, but police, security personnel, or property management can be assigned this task when fire department staffing does not permit assigning fire fighters. However, consideration must be given to fire and smoke conditions at the exit discharges. If heavy smoke or fire is present at the exit discharge, only fire fighters will be able to facilitate movement at the exit.

Panic has long been blamed for many deaths at nightclub fires. In reality, panic is a scapegoat. Often the real problem is a lack of proper fire protection features. Extensive human behavior research was done after the Beverly Hills Supper Club fire **Figure 11-4** .[3] The findings indicated that people actually assisted

Figure 11-4 Beverly Hills Supper Club, Southgate, Kentucky, 1977.

© AP Photos

Incident Summary

At The Station nightclub in West Warwick, Rhode Island, on February 27, 2003, pyrotechnics used as part of a stage act ignited a highly flammable polyurethane foam insulation material used on the walls and ceiling around the stage, resulting in a rapidly advancing fire in a crowded nightclub. At first occupants failed to realize the need to evacuate, but 30 seconds after ignition the band stopped playing and occupants began to evacuate. Smoke quickly filled the building and guests began stacking up at the main entrance approximately 100 seconds after ignition. Some occupants escaped via windows at the front of the building. Flame was visible at the front of the building 312 seconds after ignition. The foam covering and crowded conditions provided little time for evacuation, resulting in 100 civilian fire deaths.

Data from: William Grosshandler, Nelson Bryner, Daniel Madrzykowski, Kenneth Kuntz, NIST NCSTAR 2: Vol. I, Report of the Technical Investigation of The Station Nightclub Fire, National Institute of Standards and Technology, U.S. Department of Commerce, 2005.

The floor plan of The Station nightclub showing exit locations. Ignition points are also shown.

Source: Grosshandler, W., Bryner, N., Madrzykowski, D., Kuntz, K. *NIST NCSTAR 2: Vol. I, Report of the Technical Investigation of the Station Nightclub Fire*. National Institute of Standards and Technology, U.S. Department of Commerce; 2005.

one another as long as options were available. Only in desperation, when no chance of escape was available, did panic occur. As long as there is hope for survival, panic should not be a significant problem.

People are known to respect role models under emergency conditions; children will follow their parents, and workers will look to the manager. At Beverly Hills Supper Club, patrons listened to their server while ignoring the bus person and others. Fire fighters most certainly represent an authoritative role model. People should be expected to follow any reasonable directions given by fire officials during an emergency. The challenge is not persuading the occupants to follow orders; the challenge is formulating clear, concise, and correct instructions.

As mentioned previously, extinguishment, the second tactical priority, often is the most effective means of accomplishing the top tactical priority of life safety. Extinguishing or confining the fire to areas away from exit pathways will do more to save lives than individual rescues when there are large numbers of victims and the fire is controllable. The first-arriving company at a major fire in an occupied eating or drinking establishment is faced with a critical life safety decision: Should efforts be directed toward removing endangered occupants or should the operation begin with an offensive fire attack? If the fire can be controlled quickly and many occupants are in danger, extinguishment is generally the best choice.

Sports Arenas

In 1985, 56 people died and 300 were injured in a Bradford, England, stadium fire.[4] The 77-year-old grandstand was an outside facility that resembled wooden grandstands of the past. Problems would be greatly magnified in an inside facility holding nearly 100,000 people, such as the huge, domed stadiums of today.

Fortunately, newer facilities limit the amount of combustible material; nonetheless, there is still plenty to burn, largely in the form of plastics. Stadium owners may point to their facility's modern construction, stating that "it can't happen here." However, that type of thinking is a recipe for disaster. These occupancies

pose an extreme life hazard threat, and a close examination of evacuation tactics is a must. The logistics associated with evacuating 10,000 or 100,000 people are very complex. When conducting preincident planning at these facilities, the fire officer should keep in mind that fire is not the only type of incident that may require a rapid evacuation. Today, any place that attracts large crowds is a potential target for terrorists.

A fire and explosion occurred in Indianapolis, Indiana, in 1963, killing 74 people at the Indiana State Fairgrounds Coliseum. A situation of this nature should be handled like a nightclub fire: protecting egress while directing and facilitating evacuation. Of course, a stadium or coliseum fire presents a significant life safety hazard. On the positive side, open-air arenas allow smoke and toxic gases to dissipate. Even coliseums and covered stadiums have large open-air spaces that will allow more time for evacuation. Newer facilities, such as the sports arena shown in **Figure 11-5**, are equipped with sprinklers, which greatly improve

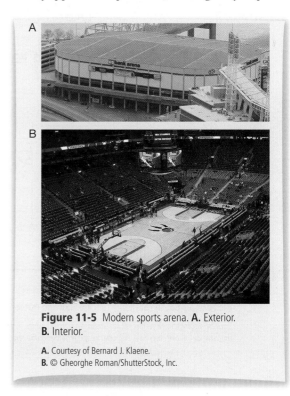

Figure 11-5 Modern sports arena. **A.** Exterior. **B.** Interior.

A. Courtesy of Bernard J. Klaene.
B. © Gheorghe Roman/ShutterStock, Inc.

life safety. Sports arenas are multi-purpose facilities used for a variety of non-sports functions.

Convention Centers

The 1967 McCormick Place convention center fire in Chicago, Illinois, caused $52 million in damages, the largest property loss in U.S. history at that time. Since the McCormick Place fire, automatic fire sprinklers have become commonplace in convention center construction. Sprinklers greatly diminish the potential loss of life and property. Large loss-of-life fires are unheard of in fully sprinkler-protected buildings when the sprinkler system is installed and maintained in compliance with nationally recognized codes and standards.

However, sprinkler systems are not the total solution; conditions such as blocked sprinkler heads, closed valves, or a fuel load exceeding the design capacity can render the system ineffective. Also, event organizers may be reluctant to cancel events when the sprinkler system is not operating. Fire companies that respond to convention centers should be aware of the event schedule and conduct on-site inspections of the property whenever large events are planned. It is also a good idea for fire companies to participate in the code enforcement process for buildings located in their response districts.

Improvements in fire protection features allow for larger and more complex convention center buildings. The new McCormick Place convention center is a good example. Finding and accessing the fire location is a key element of a preincident plan in any large, complex structure. Under normal conditions, it is a challenge to navigate throughout these large multilevel structures even with a building diagram in hand and interior signage pointing the way. Adding to the confusion is the facility's ability to change the interior layout using movable dividing walls. Locating a fire when visibility is limited and with people rushing to evacuate the buildings is extremely difficult.

The main entrance is not always the best route to the fire area. Contingency plans are a must. Part of the contingency plan should address staging later-

arriving first-alarm fire companies at a predetermined location so they can be deployed based on information gained by the first-arriving units. The preincident plan should delineate the best access locations for each major building area. In many cases, security personnel will meet the first-arriving unit at a designated location to provide incident information and to direct fire fighters to the alarm location. However, fire conditions or injuries to security personnel require contingency planning.

Large quantities of combustibles may be brought into convention centers, and hazardous demonstrations may be performed. Because of the combination of large fuel loads and large numbers of people, suppression efforts can be very difficult. As previously mentioned, it is important to ensure that the building's fuel load does not exceed the designed capacity of the sprinkler system. If a fire does overwhelm the sprinkler system and involves a large area, the rate-of-flow requirement may be beyond the fire department's resources and capabilities. In this case, extinguishment may not be a viable option. Fire departments must be prepared to combat a large fire combined with an extraordinary rescue operation in these occupancies. Tactical considerations for convention centers, such as protecting egress routes, are similar to those for nightclubs and similar public assembly properties.

Theaters

Movie theaters built during the first half of the 20th century were built to accommodate live shows or to present both live shows and movies. The construction of these facilities was similar to that of the Iroquois Theater in Chicago, where a 1903 fire resulted in 603 fatalities (mostly children). Some older theaters have now been subdivided into several smaller units.

Modern movie theaters are housed in very large buildings. Generally, these are one story, but there is a movement toward mega-theaters that have multiple stories. These buildings are usually subdivided into several smaller theaters **Figure 11-6**.

The current trend in movie theaters toward smaller theaters within very large buildings should

Figure 11-6 Modern movie theater. **A.** Exterior. **B.** Interior.

A. Courtesy of Bernard J. Klaene. **B.** Courtesy of Ed Comeau, writer-tech.com

the building complex. Furthermore, modern movie theaters are often equipped with sprinkler systems.

While movie theaters are getting smaller, venues for live entertainment are not. The fire during a live stage show at the Iroquois Theater taught building designers some important lessons, but it was not until recently that theaters were required to be constructed with sprinkler systems. If an older theater has not been equipped with a fire sprinkler system and its exit design updated, it can be just as dangerous as the Iroquois Theater was in 1903. Today, many live performances, formerly staged in theaters, are presented in large arenas, in stadiums, or outdoors using temporary stages and seating.

Educational Occupancies

NFPA 101, *Life Safety Code*,[1] defines an **educational occupancy** as follows:

> ***3.3.190.6—Educational Occupancy.*** *An occupancy used for educational purposes through the twelfth grade by six or more persons for 4 or more hours per day or more than 12 hours per week.*

College classrooms are not technically considered educational occupancies. However, there are similarities between college and other classrooms in terms of life safety, extinguishment, and property conservation. Therefore, the discussion of college classrooms is included here. For purposes of this comparison, three types of educational occupancies will be discussed:

- Elementary schools
- Middle, junior high, or high schools
- Colleges and universities

An occupancy factor matrix for educational occupancies is shown in **Table 11-3**. Mobility factors tend to be very good in school settings. In the case of elementary and high schools, most of the occupants would be mobile. At the college level there will be an older population, but this should

make the fire fighter's job less difficult. However, the interior layout is sometimes very complex. Tactics are much the same as for the eating and drinking establishments, except that there are generally fewer people in an individual movie theater unit. The modern concept of having numerous but smaller theaters located in a single large building also increases the probability that extinguishment efforts will be effective. In addition to the multiple, smaller theaters allowing for a reduced rate of flow, the tendency for everyone to exit via the main entrance is also reduced in these buildings, as alternative exits should be well marked and easily identified by patrons. Many newer theaters have plainly marked exits at several locations within the individual theater and throughout

Table 11-3 Occupancy Factor Matrix: Educational Occupancies

Educational Occupancies	Mobility of Occupants	Age of Occupants	Leadership	Awareness	Occupant Density	Familiarity of Occupants
Elementary schools	10	6	10	10	4	10
Middle, junior high, or high schools	10	10	7	10	4	10
Colleges/universities	8	8	6	10	5	5

1 = extremely negative factor; 10 = very positive factor.

not pose a problem. However, younger elementary school students may not be capable of taking independent action, thus the lower age ratings for elementary school occupancies in the educational matrix. A high degree of leadership is provided by classroom teachers, especially at the elementary school level. At the high school and college levels, students move from area to area and from building to building; therefore, there are times when they are not under the direct control of a teacher.

Students at all levels should be aware of fire conditions and be able to hear the evacuation alarm. Fire drills held on a regular basis in elementary and high schools increase awareness and better establish teacher-to-student leadership. Occupant density maximums are 20 ft² (1.9 m²) per person in elementary and high school classrooms.[1] There are fewer people in the same relative area in elementary and high school settings as compared to an assembly occupancy, but there are more people in these settings than in many other occupancies.

Elementary Schools

Most tragic fires involving educational occupancies occur in primary or elementary schools, mainly because younger children are less likely to take appropriate action on their own. All other things being equal, the younger the student, the greater the danger.

Some notable school fires are listed in Table 11-4 . Figure 11-7 shows the scene at Our Lady of Angels School in Chicago in 1958.

Only recently have schools been protected with automatic sprinkler systems. Very few older school buildings are sprinkler protected. Placing large numbers of children within a small, nonsprinkler-protected

Table 11-4 Notable Fires in Educational Occupancies

Fire	Year	Deaths
Lakeview Grammar School, Collinwood, OH	1908	175
Cleveland School, Beulah, SC	1923	77
Consolidated School, New London, TX	1937	294
Our Lady of Angels, Chicago, IL	1958	95
Star Elementary School, Spencer, OK	1982	7

Figure 11-7 Our Lady of Angels fire, Chicago, Illinois, 1958.
© AP Photos

electronic locking systems, have improved in recent years due to multiple "active shooter" incidents. It is important that the fire department review new security measures to determine what effect security has on evacuation and fire department access to the building and the fire department's role during an active shooter incident.

Knowing how long it takes to totally evacuate the school is important, but not nearly as important as knowing whether students are under control

area creates the potential for a tragedy. In addition to not being sprinkler protected, older schools tend to be three or more stories high **Figure 11-8**.

The saving factors in elementary schools are that children become disciplined through frequent fire drills, and teachers are natural role models for the children to follow in an emergency. In the event of a fire in a school, it is necessary to coordinate operations with the school principal to determine the evacuation status and the location of students, teachers, and staff members who are missing. If rescue efforts are necessary, fire fighters can be directed to the most probable location for trapped students, teachers, and staff on the basis of this information.

Fire department personnel should observe school fire drills whenever possible. Observing a school fire drill provides invaluable training for fire fighters, who should know the prescribed location for students and the school contact person. Being present during a fire drill also provides an opportunity to evaluate accountability and evacuation procedures.

Elementary and high schools maintain visitor logs. Overall security measures, such as the installation of

Figure 11-8 Buildings used as schools.
A. Multistory, older high school. **B.** One-story school. **C.** Modern school with sprinkler protection.

Courtesy of Bernard J. Klaene.

and whether the accountability system is reliable. Determining whether the school evacuation plan is consistent with fire department operations is also important. Sometimes students are placed in areas that could obstruct or slow fire department access. Other times the school contact person is located in an area that is not in the line of travel for arriving fire units. Minor changes in the school evacuation plan can significantly improve fire department response. Many jurisdictions conduct unannounced fire drills that are timed and otherwise evaluated by the fire department.

Some newer one-story schools have doors from each classroom leading directly to the outside, greatly reducing the life hazard. In older schools, heavy reliance is placed on the use of hallways and stairways. As with any building, the stairs in multi-story school buildings must always be controlled by the fire department during evacuations. Evacuating trapped victims through the interior stairs is typically the most effective and safest option.

When absolutely necessary, aerial ladders, elevated platforms, and ground ladders can be used to evacuate a multistory school. Ladders allow pupils to form a steady evacuation stream with teachers assisting from the inside. However, fire department ladders should be used only when fire fighters cannot gain control of the interior stairs. Establishing an occupant accountability system that ensures complete evacuation and accounting for all students, teachers, and staff is essential. It is also important to make a sound evaluation as to whether the immediate priority is rescue or fire control.

As mentioned previously, many school buildings pose special firefighting challenges. Fire extension should be expected in school buildings of ordinary construction. Where mobile buildings are used for temporary classrooms, complete destruction may occur in a short time. Many schools add to the fire control problem by storing school supplies, clothing, furniture, and other materials within the school, sometimes under stairwells and in corridors. Such storage practices greatly increase the fuel load and the threat to life safety.

Middle, Junior High, and High Schools

Fire drills are required in all kindergarten through grade 12 schools. As mentioned previously, the student accountability process may not be as efficient at the high school level.

Fire fighters are inclined to think schools are occupied only during school hours. When determining the probability of a school being occupied, it is important to remember that teachers, maintenance, and administrative personnel may occupy the school outside of normal classroom hours. There are also many sporting and social events that occur at schools outside the normal school day. At high schools and colleges, there will usually be more of these events than at the elementary school level. The fire department should obtain a list of scheduled events and include event information with the pre-incident plan. A basketball court or auditorium will likely be classified as an assembly occupancy within the educational occupancy. High schools may also have evening classes or permit citizens to use computers, gym equipment, and other school resources outside normal school hours.

Colleges and Universities

Colleges and universities usually have much larger campuses as compared to elementary or high schools, with many more evening and weekend activities. The adult population is better able to take independent action to evacuate a building; however, students move from building to building at different time intervals and are not under close supervision. Further, typically there is no occupant accountability system, and universities may be situated on a very large and complex campus, requiring a map to find specific buildings. Most campuses will have one or more large classroom facilities that hold 200 or more students. These classrooms resemble an assembly occupancy and are classified as such. Any college classroom with 50 or more students is considered an assembly occupancy.

Fires in classrooms, except for the very large college classrooms or large laboratory and shop areas, will generally be within the flow capabilities of standard preconnected hose lines. Rate of flow should be precalculated for any large undivided areas within the school property.

Schools contain library collections, computers, and other high-value contents. Therefore, property conservation tactics should be carefully thought out and included in the preincident plan.

Health Care Occupancies

NFPA 101, *Life Safety Code*,[1] defines a **health care occupancy** as follows:

> *3.3.190.7—Health Care Occupancy. An occupancy used to provide medical or other treatment or care simultaneously to four or more patients on an inpatient basis, where such patients are mostly incapable of self-preservation due to age, physical or mental disability, or because of security measures not under the occupants' control.*

Four types of health care facilities will be discussed:

- Hospitals
- Nursing homes
- Limited care facilities
- Ambulatory care facilities

An occupancy factor matrix for health care occupancies is shown in **Table 11-5**. Mobility is a major concern in hospitals and nursing homes. Some patients are considered ambulatory in these facilities, but ambulatory and mobile are not the same. An ambulatory person is able to move but may move very slowly and require a walker or cane. Many patients in hospitals and nursing homes are nonambulatory, meaning they are unable to self-evacuate and will need assistance if it is necessary to move to a place of safe refuge. The nursing staff generally provides this assistance, but the staff available may not be sufficient if large numbers of people need assistance. Occupants in limited care facilities are more mobile than in a nursing home, but many move slowly or need walkers, canes, wheelchairs, or other types of assistance.

Ambulatory care facilities treat minor illnesses and injuries. However, some of these facilities perform surgical or testing procedures that require the patient to be anesthetized or restrained. If this is the case, patients will be immobile and have limited or no cognitive ability. Part of the preincident planning process should include notation of ambulatory care facilities where patients are anesthetized or restrained.

Table 11-5	Occupancy Factor Matrix: Health Care Occupancies					
Health Care Occupancies	**Mobility of Occupants**	**Age of Occupants**	**Leadership**	**Awareness**	**Occupant Density**	**Familiarity of Occupants**
Hospitals	2	4	10	3	7	2
Nursing homes	2	1	8	3	7	7
Limited care facilities	4	2	7	5	8	9
Ambulatory care facilities	6	5	6	5	7	2

1 = extremely negative factor; 10 = very positive factor.

© Ely Studio/Shutterstock, Inc.

Residents of nursing homes and limited care facilities are typically much older than the general population. The nursing staff provides leadership and assistance during an emergency. In a hospital or nursing home, patients will look to their nurses for direction in an emergency. Limited care facilities will generally have staff personnel who can provide leadership, but patients are more independent. Patients at ambulatory care facilities may not recognize the staff, and conscious patients will be more likely to take independent action in an emergency. If patients are restrained in any way, a notation should be made on the preincident plan showing the location of the restraining devices and, if necessary, instructions on releasing the restraints.

The nursing home and assisted living facility is home to the occupant, thus most patients will be more familiar with the layout of the building. However, residents may not remember details of the building when under stress, and some patients may not be able to follow directions. Patients and visitors in hospitals and ambulatory care facilities may be unfamiliar with the building.

Large numbers of people located in a single building always represent life safety concerns. Physically or mentally challenged people present additional difficulties for rescuers. Fire fighters should expect facility personnel to assist with evacuations; this should be reinforced through training. In the hospital or nursing home, nurses and attendants will be needed to care for patients.

These occupancies often require fire forces to rely on people who are not trained to work under fire conditions. Health care facilities use places of safe refuge within the facility during evacuations. Hospital and nursing home personnel move patients through fire doors to safe areas. Seldom are hospital or nursing home patients evacuated to the outside. Codes require doors and hallways to be wide enough to allow the patients to be moved while remaining in bed. Moving people on beds with life support equipment is difficult, to say the least; therefore, the defend-in-place strategy is the preferred option.

In addition, hospital and nursing home corridors are often congested with computer carts, laundry bins, and food service equipment that partially block the pathway to a place of safe refuge.

Hospitals

Building features assist in accomplishing the life safety mission at hospitals, with building areas being separated by substantial fire walls and fire doors. Making sure that fire separations are not compromised is an essential part of a hospital fire protection plan. Fire doors are often left open in hospital facilities, allowing smoke and fire to enter common hallways and other patient areas, as was the case in a hospital fire in Petersburg, Virginia, which killed six patients Figure 11-9 .

The fire department may be needed to assist in moving patients from the fire area to another designated defend-in-place location. The IC must make an early determination as to progress being made in moving patients to an area of safe refuge and whether the place of safe refuge is defendable. Evacuation to a lower floor or to the outside is a mammoth undertaking and is rarely necessary. The nursing staff should be able to account for patients, but in most cases, they will not be able to account

Figure 11-9 Hospital in Petersburg, Virginia.
Courtesy of the U.S. Fire Administration, Major Fires Investigation Series

for visitors or doctors. Therefore, a primary search must be conducted.

There are large open areas within hospitals and some of these areas, such as laundries, storage areas, and refuse areas, have a heavy fuel load. Further, the presence of oxygen can increase flame intensity and fire spread. However, patient rooms in hospitals are usually within the flow capacity of a standard preconnected hose line or fire department standpipe hose line. It is essential to consider the possible smoke spread to the area of safe refuge when a fire door is opened to advance a hose line.

Hospitals will often have very expensive equipment. Some of this equipment is extremely sensitive to water and smoke damage. Given the high life safety problem, property conservation could be delayed. However, the IC should call for enough assistance to assign personnel to property conservation activities as soon as possible.

Nursing Homes

Most nursing home residents are unable or only partially able to assist themselves. In some cases, special security provisions such as restraints or exit doors that require a code number to open hinder self-rescue efforts. This special population presents many evacuation and rescue problems, requiring a significant commitment of fire department resources whenever a working fire is encountered in a nursing home.

During the past few decades, as the population aged, the need for nursing facilities increased. A larger percentage of the population moved into these occupancies, and large loss-of-life fires became common. Most areas of the United States now have codes requiring sprinklers in nursing homes, a trend that has greatly reduced the number of large loss-of-life fires in health care facilities. Multifatality fires still occur in nursing homes but not where sprinklers have been properly installed and maintained. The two most recent multifatality fires were in nonsprinkler-protected nursing homes Table 11-6 .

Table 11-6	Notable Fires in Hospitals and Nursing Homes	
Fire	**Year**	**Deaths**
Hospital, Cleveland, OH	1929	125
Hospital, Effingham, IL	1949	74
Psychiatric facility, Davenport, IA	1950	41
Nursing home, Hillsboro, MO	1952	20
Nursing home, Warrenton, MO	1957	72
Hospital, Hartford, CT	1961	16
Nursing home, Fitchville, OH	1963	63
Nursing home, Marietta, OH	1970	31
Nursing home, Buechel, KY	1971	10
Nursing home, Honesdale, PA	1971	15
Nursing home, Cincinnati, OH	1972	10
Nursing home, Denmark, WI	1972	10
Nursing home, Springfield, IL	1972	11
Nursing home, Philadelphia, PA	1973	11
Nursing home, Wayne, PA	1973	15
Hospital, Osceola, MO	1974	8
Nursing home, Chicago, IL	1976	23
Nursing home, Little Rock, AR	1984	2
Hospital, Riverside, CA	1986	5
Health care center, Memphis, TN	1988	6
Nursing home, Norfolk, VA	1989	12
Nursing home, Dardanelle, AR	1990	3
Hospital, Brooklyn, NY	1993	3
Hospital, Petersburg, VA	1994	6
Health center, Hartford, CT	2003	16
Nursing home, Nashville, TN	2003	14

When a properly installed and operating sprinkler system is in place, fires are generally limited to one room, and the primary tactic is to move patients out of smoky areas to places of safe refuge while supporting the sprinkler system. If the nursing home is not sprinkler protected or the sprinkler system is not operating properly, use fire doors to contain the fire whenever possible. Even though self-closing fire-rated doors separate rooms from hallways, obstructions may prevent the door from closing, which allows fire and smoke to spread to the hallway and other rooms. Moving patients out of harm's way is the primary tactic, but extinguishment is essential to defend the place of safe refuge. If the fire is not extinguished, smoke and possibly fire will eventually enter the area of safe refuge, requiring further evacuation and rescue.

Some notable fires in health care facilities (nursing homes and hospitals) are listed in Table 11-6. Note that from 1999 to 2002, more than 90% of the civilian fire fatalities in health care occupancies occurred in nursing homes.

Nursing homes and hospitals make a good case for total sprinkler protection, as is evident in Table 11-6. Prior to 1970, few nursing homes were equipped with total automatic sprinkler protection. The few multifatality fires that have occurred since were in nonsprinkler-protected buildings. Likewise, the majority of hospitals were not sprinkler protected in the early 1970s. At last count, 64% of hospitals are now protected by automatic sprinklers. Large loss-of-life fires were more devastating and more frequent when fewer facilities were equipped with automatic sprinklers.

Limited Care Facilities

Limited care facilities, sometimes called assisted living facilities, are similar to nursing homes, but the occupants do not require continuous nursing care Figure 11-10 . Some limited care facilities provide services that are similar to comprehensive nursing home care, while others establish independent living criteria for occupants. Often a nursing complex will include both assisted living and full nursing care. Elderly residents generally move from their homes into the assisted living facility with a degree of self-reliance. Over time, these residents require additional

Figure 11-10 Limited care facility.

Courtesy of Bernard J. Klaene.

nursing services, and many eventually occupy a full nursing facility. Most codes now require automatic sprinkler protection for all new nursing homes, but they may not require existing homes to be retrofitted with sprinklers or sprinkler protection in limited care facilities.

Part of the fire protection plan for a nursing home requires a minimum number of on-duty staff to assist patients in an emergency. Staffing will generally be less at a limited care facility and there may not be places of safe refuge where a defend-in-place strategy can be implemented. Many limited care facilities resemble apartment buildings; therefore, a primary search and quick extinguishment are the principal tactics. Occupants of these facilities, who are mostly elderly, would have great difficulty descending a ladder; therefore, attempting to rescue residents using fire department ladders would be an option of near last resort.

Many apartment buildings rent exclusively to elderly occupants. These buildings will be classified as a residential occupancy and have less stringent fire protection code requirements. However, the firefighting challenges will be very similar to those in a limited care facility.

Ambulatory Care Facilities

As noted previously, these facilities provide a variety of services. Some are group practice physicians who provide emergency care for minor illnesses and injuries. Others perform outpatient surgery or testing that requires full anesthesia. The emergency care facility where patients are conscious and unrestrained requires the same tactics as an office building. If patients are anesthetized, they will have to be physically removed from the building and provided medical care. If restrained, they will need to be released from the restraining equipment (e.g., MRI tunnel). These facilities should be preincident planned with notations regarding the types of service provided and a special notation if patients are anesthetized or restrained.

Residential Board and Care Occupancies

NFPA 101, *Life Safety Code*,[1] defines a **residential board and care occupancy** as follows:

3.3.190.12—Residential Board and Care Occupancy. An occupancy used for lodging and boarding of four or more residents, not related by blood or marriage to the owners or operators, for the purpose of providing personal care services.

Table 11-7 shows an occupancy factor matrix for residential board and care occupancies.

Awareness is given a rating of three due to the diminished mental and/or physical capacity of many of the residents and resultant problems in evacuating the building. There may be one or more staff members on duty depending on the facility size and the mental/physical status of patients being housed. In most cases, these staff members are less qualified that hospital and nursing home personnel. Thus leadership is a positive factor, but not to the extent found in schools and hospitals.

During the 1970s, mentally challenged patients who were formerly housed in large, fire-resistive structures were moved to board and care facilities. This created a problem similar to the problems experienced in nursing homes in the past. However, the board and care problem has been more difficult to regulate, as fewer patients are kept in a single structure. Staff requirements are less in board and care facilities, and many resemble a residential occupancy. Facilities classified as board and care occupancies in NFPA 101 are identified by a variety of designations such as group homes, rest homes, shelters, halfway houses, and other names.

The board and care fire problem is substantial, and tactics must be planned in advance. It is essential that the fire department identify and inspect board and care facilities, developing preincident plans for occupancies located within their jurisdiction. Tactics should take into account the limited ability of residents to escape on their own. Residents will be categorized as independent, but will usually be mentally or physically disadvantaged. Many of

Table 11-7 Occupancy Factor Matrix: Residential Board and Care Occupancies						
	Mobility of Occupants	Age of Occupants	Leadership	Awareness	Occupant Density	Familiarity of Occupants
Residential board and care occupancies	5	5	7	3	7	7

1 = extremely negative factor; 10 = very positive factor.

© Eky Studio/Shutterstock, Inc.

Table 11-8	Notable Fires in Residential Board and Care Occupancies	
Fire	**Year**	**Deaths**
Pleasantville, NJ	1973	10
Connellsville, PA	1979	10
Farmington, MO	1979	25
Washington DC	1979	10
Pioneer, OH	1979	14
Brady Beach, NJ	1980	24
Detroit, MI	1980	5
Keansburg, NJ	1981	31
Pittsburgh, PA	1982	5
Eau Claire, WI	1983	6
Gladstone, MI	1983	5
Worcester, MA	1983	7
Gwinnett County, GA	1983	8
Cincinnati, OH	1983	7
Lexington, KY	1984	5
Southfield, MI (hospice facility)	1985	8
Johnson City, TN	1989	16
Colorado Springs, CO	1991	10
St. Isidore, Quebec, Canada	1991	5
Detroit, MI	1992	10
Mobile, AL	1994	6
Broward County, FL	1994	6
Laurinburg, NC	1996	8
Mississauga, Ontario, Canada	1996	8
Sainte-Geneviève, Quebec, Canada	1997	7
Harveys Lake, PA	1997	10
Arlington, WA	1998	8
Maryville, TN	2004	5
Anderson, MO	2006	11
Mt. Carmel, CA	2011	5

these buildings are residential buildings; therefore, the standard preconnected hose line should be sufficient to extinguish most fires that occur. Converted two- or three-story residential buildings typically have only one means of egress from upper floors via an unenclosed stairway.

The NFPA has issued several fire investigation reports for fires in board and care facilities. Some of these facilities are occupied by mentally challenged or drug-impaired residents; others are scaled-down facilities for elderly patients who do not require full nursing care.

Some notable board and care fires are listed in Table 11-8 . Widespread use of residential board and care facilities is a relatively new phenomenon; therefore, the multifatality fires shown in Table 11-8 are fairly recent.

The number of multifatality fires in these facilities seems to be trending downward, as shown in Table 11-9 .[5] This improvement is due in part to more stringent codes and better code enforcement, but also to increased awareness among fire departments that these occupancies exist in their communities.

Table 11-9	Structure Fires in Residential Board and Care Facilities by Year[5]		
Year	**Total Fires**	**Civilian Deaths**	**Civilian Injuries**
2003	2020	3	74
2004	2100	5	44
2005	2060	10	55
2006	2250	29	124
2007	1940	2	51
2008	1990	6	55
2009	1730	2	35
2010	1710	10	39

Detention and Correctional Occupancies

NFPA 101, *Life Safety Code*,[1] defines a **detention and correctional occupancy** as follows:

> *3.3.190.5—Detention and Correctional Occupancy.* An occupancy used to house one or more persons under varied degrees of restraint or security where such occupants are mostly incapable of self-preservation because of security measures not under the occupants' control.

An occupancy factor matrix for detention and correctional occupancies is shown in **Table 11-10**. Residents of nursing homes and hospitals are characterized as having physical and/or mental challenges that hamper self-evacuation, whereas prisoners in detention facilities are generally physically restrained. During a prison fire, inmates are moved to other areas within the confines of the detention property.

The prison staff provides leadership for inmates. Close cooperation between prison staff and the fire department is needed for successful operations in a jail or prison environment. Some detention and correctional occupancies are no more than cells within a police station designed to temporarily hold prisoners. Other correctional facilities are nearly self-sufficient minicities with laundries, storage areas, large-scale food preparation areas, and many other facilities included within the prison complex.

During a detention facility fire, maintaining security is extremely important; however, dealing with tight security measures while implementing fire department tactics is sometimes exasperating for the responders. Close cooperation between the IC and prison administrators is an absolute necessity.

Prisoners present a physical threat to the fire fighter, so close coordination with prison officials is needed to enable the fire fighters to operate safely and suppress the fire. Modern prisons are usually constructed with a fair degree of fire safety in mind, but older, less fire-safe, structures are still being used as prisons. Generally, older prisons are not sprinkler protected and have fewer fire protection features. These less protected prisons and jails will complicate operations by permitting the free circulation of smoke and toxic gases as well as allowing rapid fire extension.

Modern and old prisons alike tend to be overcrowded, holding more prisoners than they were designed to house. Any form of overcrowding increases the life hazard.

Detention and correctional structures in Latin America and elsewhere sometimes ignore fire and building codes that are generally enforced in the United States. As a result, at least 9 large loss-of-life fires with 10 or more fatalities occurred during the 10-year period from 2003 through 2012. Most notable was a fire in Comayagua, Honduras, on February 14, 2012, that resulted in 361 inmate deaths.[6]

Table 11-10	Occupancy Factor Matrix: Detention and Correctional Occupancies					
	Mobility of Occupants	Age of Occupants	Leadership	Awareness	Occupant Density	Familiarity of Occupants
Detention and correctional occupancies	1	6	9	6	7	6

1 = extremely negative factor; 10 = very positive factor/

Table 11-11 Notable Fires in Detention and Correctional Occupancies

Fire	Year	Deaths
Raymond, MS	1913	35
Columbus, OH	1930	320
Lancaster, SC	1972	11
Sanford, FL	1975	11
Williamsport, PA	1975	3
Marion, NC	1976	9
Point Pleasant, WV	1976	5
Danbury, CT	1977	5
St. John, New Brunswick, Canada	1977	18
Columbia, TN	1977	42
Leavenworth, KS	1979	7
Biloxi, MS	1982	29
Bakersville, NC	2002	8
Jersey City, NJ	2002	7

Several notable fires in detention facilities are listed in Table 11-11.

Preincident planning is an absolute necessity for a detention or correctional facility. Securing a water supply and advancing hose lines may require special precautions. The details of fire department interaction with security are worked out during preincident planning. Questions that must be answered in advance include the following:

- How will fire fighters be provided access to secured areas? This generally involves a security person meeting the fire department at a predesignated location and escorting them while inside the prison. Be sure to detail how security will provide access when smoke or fire conditions require fire fighters to advance ahead of security personnel into areas requiring self-contained breathing apparatus and full protective clothing.

- Are there areas within the prison complex that do not house prisoners, and can they be accessed without assistance from security? Work areas such as laundries may be occupied by prisoners during certain times of the day but may be unoccupied at other times.
- What are the emergency procedures in the event of system malfunctions, an automatic lockdown, or a power failure?
- What are the evacuation routes, and where are the locations of safe refuge for prisoners?

The goal is to conduct a safe and effective operation while avoiding direct contact with the prison population. There may be a need for special rapid intervention crews (RICs) and accountability and rehabilitation procedures during a detention and correctional occupancy fire.

Residential Occupancies

NFPA 101, *Life Safety Code,*[1] defines a **residential occupancy** as follows:

> *3.3.190.13—Residential Occupancy. An occupancy that provides sleeping accommodations for purposes other than health care or detention and correctional.*

Residential occupancies include:

- One- and two-family dwellings
- Apartment buildings
- Dormitories
- Hotels/motels

An occupancy factor matrix for residential occupancies is shown in Table 11-12. Note that mobility and age are near average (5), except in the dormitory, where occupants tend to be young adults who are very mobile. There is little leadership in most hotels and motels except when provided by a parent traveling with children. In one- and two-family dwellings, the head of household should be in a

Table 11-12 Occupancy Factor Matrix: Residential Occupancies

Residential Occupancies	Mobility of Occupants	Age of Occupants	Leadership	Awareness	Occupant Density	Familiarity of Occupants
One- and two-family dwellings	5	5	7	4	8	10
Apartment buildings	5	5	5	4	6	9
Dormitories	7	7	5	3	4	8
Hotels and motels	6	5	2	3	4	1

1 = extremely negative factor; 10 = very positive factor.

strong leadership position, particularly if the family has practiced a home escape plan. Awareness is rated slightly negative in all residential occupancies because occupants are sleeping part of the time that they occupy the building. With the exception of dormitories, hotels, and motels, there are very few people in a fairly large area; thus exit capacity is not generally considered a problem, although the number of exits from upper floors of a home are generally very limited. People living in a residence or dorm should be familiar with their surroundings. Hotels and motels are transient in nature; therefore, most occupants will not know the location of alternative means of egress.

Most fires, most fatal fires, and most property loss occur in residential fires. In 2012, 83% (2380) of civilian fire deaths and 76% of structure fires occurred in the home.[7] Most of the deaths in residential fires occur one or two at a time. However, the annual Catastrophic Multiple-Death Fires Report for 2013[8] includes twelve fires occurring in one- and two-family dwellings where five or more fatalities resulted. Unfortunately, 28 of the 67 deaths in these multiple-death home fires were children under the age of 6 years. This is not an unusual circumstance; every year several multiple-death fires occur in single-family homes. A list

of notable residential fires for the past 100 years, including those in homes, would have over a thousand entries.

The residential occupancy classification includes more than just one- and two-family dwellings; it also includes apartments, hotels, motels, and dormitories. Large loss-of-life fires have occurred frequently in apartment buildings. **Table 11-13** lists apartment building fires that resulted in five or more civilian fatalities from 2001 through 2012.

Table 11-14 lists some significant residential fires, excluding one- and two-family homes and apartment buildings, over the past 100 years.

In residential occupancies, as with most occupancy types, there is a direct correlation between life safety and the time of day. In residential occupancies, the hours when residents are sleeping are the most dangerous. Wide acceptance of smoke detectors has greatly diminished the number of deaths in residential properties. Many high-rise buildings are now constructed for use by senior citizens and family living. Expect to find many people throughout the structure in need of assistance, including the possibility of the **convergence cluster phenomenon**.[9]

Convergence cluster behavior was studied in depth after the MGM Grand Hotel and Casino fire in Las Vegas. Many guests attempted to use stairs

Table 11-13 Notable Fires in Apartment Buildings, 2001 to 2012

State	Year	Deaths
Pennsylvania	2001	8
Pennsylvania	2001	6
Alabama	2001	6
New York	2001	5
Alabama	2001	5
Illinois	2001	5
Maryland	2001	5
New York	2001	5
Tennessee	2001	5
New York	2002	7
New York	2003	5
Ohio	2003	5
Ohio	2004	10
Ohio	2004	7
Texas	2004	5
New York	2004	5
New York	2005	5
Pennsylvania	2006	5
West Virginia	2007	9
California	2007	5
New York	2008	5
Ohio	2008	5
Mississippi	2009	9
Illinois	2010	7
New York	2010	5
Illinois	2011	6
Washington	2011	5
Mississippi	2011	5
Texas	2012	5
Missouri	2012	5

Table 11-14 Notable Fires in Residential Occupancies Other Than One- and Two-Family Dwellings or Apartment Buildings

Fire	Year	Deaths
Gulf Motel, Houston, TX	1943	54
La Salle Hotel, Chicago, IL	1946	61
Winecoff Hotel, Atlanta, GA	1946	119
MGM Grand Hotel and Casino, Las Vegas, NV	1980	85
Hilton Hotel and Casino, Las Vegas, NV	1981	8
Dupont Plaza Hotel and Casino, San Juan, Puerto Rico	1986	96
Motel, Hagerstown, MD	1990	4
Ramada Hotel, Evansville, IN	1992	11
Paxton Hotel, Chicago, IL	1993	20
Fraternity house, Chapel Hill, NC	1996	5
Off-campus student housing, Columbus, OH	2003	5
Assisted living facility, Nashville, TN	2004	5
Residential hotel, Reno, NV	2006	12
Homeless shelter, Paris, TX	2009	5
Boarding house, Cleveland, OH	2009	4
Motel, Hoover, AL	2010	4
Bed and breakfast, New Ulm, MN	2011	6

to exit the building but encountered smoke as they descended. They then sought refuge in a room other than their own. This resulted in several people being in a single room, rather than the expected one to five guests per room. It would be logical to assume that guests had evacuated the floor successfully after finding several rooms empty, but this assumption could have serious, deadly consequences. The convergence cluster phenomenon should reinforce the importance of conducting a complete, systematic search.

One- and Two-Family Dwellings

One- and two-family dwellings are an important and large part of the fire problem. During the 5-year period of 2005 through 2009, U.S. fire departments responded to an estimated average of 373,900 home structure fires per year. These fires caused an annual average of 2650 civilian deaths, 12,890 civilian fire injuries, and $7.1 billion in direct property damage. Home fires accounted for 73% of all reported structure fires, 92% of civilian structure fire deaths, 86% of the civilian structure fire injuries, and 68% of direct structure fire property loss.[10]

Fire fighters generally feel confident attacking a fire in a one- or two-family dwelling "bread and butter" fire. However, more fire fighter line-of-duty deaths occur in these properties than any other occupancy, which is primarily due to the large number of fires that occur in these properties. The point that must be emphasized is that no fire should ever be considered routine. Proper precautions must be taken at every structure fire.

One- and two-family dwellings are primarily occupied at night and on weekends, but people can be home and asleep at any time. Whenever it is safe to conduct an offensive attack, a primary and secondary search should be conducted, unless there is a *credible* and *verifiable* report that everyone escaped. Reports that everyone is out of the building should be carefully scrutinized for reliability. When in doubt, conduct a primary and secondary search. When the fire occurs late at night, the primary search should normally begin in bedrooms.

Except for some very large open-layout buildings or mansions, fires in one- and two-family dwellings should be within the capacity of one or two preconnected hose lines. As is most often the case, extinguishment is usually the best way to achieve the life safety objective.

The actual monetary value of the building's salvageable contents will vary, but it is important to remember that in many cases, personal items cannot be replaced and are of great value to the occupants. Property conservation efforts must take into consideration that personal mementos are invaluable and cannot be replaced.

Apartment Buildings

In terms of life safety, extinguishment, and property conservation, apartment buildings are much like one- and two-family dwellings. However, a larger number of people are concentrated in a smaller area and the probability of someone being able to provide a *credible* and *verifiable* report that everyone has successfully escaped is unlikely. The life safety problem is directly proportional to the number of units in a single building. Larger buildings with more apartment units require a larger staffing commitment to complete the primary and secondary searches.

With the large number of people potentially exposed to the fire and the lack of a reliable evacuation status report, extinguishment becomes even more important. Most fires in apartment buildings will require flows that are within the capacity of one or two standard preconnected hose lines. However, some apartment buildings have large open storage or common spaces that may require additional flow.

As is the case in other residential properties, the occupants' personal items will be of great value.

Exposure problems are more likely in large apartment complexes; a preincident plan should, at a minimum, include a map showing the building layout with an address for each unit within the complex, as well as areas requiring a rate of flow exceeding the flow capacity of two standard preconnected hose lines. Preincident plans should also note access problems. Some apartment buildings have back-to-back units with fire separations between units or a center hallway separating units. Apparatus access may be limited to the front side of the building, with no road access to the rear units. A serious fire in a rear unit may not be visible from the front of the building, and access to rear apartments may be limited.

If a fire is being attacked in a defensive mode at an apartment complex where buildings are closely spaced, the primary concern is that other buildings will quickly become involved in fire, as was the case in the Houston, Texas, apartment complex conflagration discussed in the *Defensive Attack* chapter.[11] It is important to get ahead of the fire, for both extinguishment and evacuation purposes, as it is fairly

Incident Summary

In a four-story apartment complex fire in Bremerton, Washington, four residents were killed in a fast-spreading early morning fire. The fire originated in a third floor apartment that was unoccupied at the time and quickly spread from the apartment door that was left open when the fire was detected. The fire moved into the attic space and then throughout the complex. There were four fire separation walls in the attic space, but openings in these walls allowed the fire to spread beyond them. This case shows how an attic fire can create a severe challenge for the fire department in attempting to contain the fire and protect the occupants.

Data from: Edward R. Comeau, Apartment Complex Fire, Bremerton, Washington, Fire Investigation Report. Quincy, MA: NFPA.

Flames: © Jag_cz/ShutterStock, Inc.; Paper: © silver-john/

easy to alert and evacuate residents who are not in immediate danger. However, if the fire spreads and threatens the occupants, the rescue effort will be much more complex.

High-rise residential properties would also fall into the category of apartment buildings. The special perils and tactics associated with high-rise buildings are discussed in the *High-Rise Buildings* chapter.

Dormitories

College and university dormitory housing is supervised. University policies and jurisdictional codes require a specified level of safety in dormitories. However, there are other types of housing for college students. Students rent housing in apartments or homes that have no affiliation with the college or university. Fraternity and sorority housing may or may not be controlled by the college. College dormitories tend to be larger facilities than off-campus housing; some are high-rise buildings. The number of people in both on- and off-campus structures is fairly high compared to most residential property. Dormitories tend to provide small living spaces for each student, with common public areas. Many fraternity, sorority, and private houses rented by students are old residential properties that have been modified to hold more tenants than the original design intended.

The occupant load is increased substantially when guests are invited to social events. A 2003 fire in private housing for students near Ohio State University claimed the lives of five students. The three-story building had 12 residential units plus a large number of visiting students. In another deadly fire, three students perished in an off-campus housing unit near Miami University in Oxford, Ohio, in 2006.

Off-campus statistical information is difficult to obtain, because many fire reports categorized these fires as apartment occupancies and many smaller fires go unreported. Approximately 75% of college students live in off-campus housing, compared to 25% living in dormitories. A dormitory fire at Seton Hall University and a fraternity house fire at North Carolina University prompted officials to require sprinkler protection. The American Fire Sprinkler Association, Campus Firewatch, NFPA, National Fire Sprinkler Association, USFA, and other organizations and fire safety advocates are working to require sprinkler protection in dormitory, sorority, and fraternity housing, but many on- and off-campus housing units remain unprotected.

Life safety is a critical issue in dormitory and off-campus housing. Alcohol consumption can result in a lack of awareness and reduces mobility. Many victims of fatal campus housing fires have high blood alcohol levels. When a fraternity or sorority house (or other private student housing unit) invites other students to social events, alcohol consumption is often excessive and many of the invited students spend the night at the off-campus housing unit. The fire officer must consider the possibility that more students than expected may occupy an off-campus house and occupants may be in areas other than normal sleeping areas. Many of the multifatality fires in off-campus housing occur after a party. Further, arson is a leading cause of fire in student housing; therefore, a heavily involved fire with multiple victims is not uncommon.

Most rooms in dormitory or off-campus housing will be within the flow capacity of standpipe or preconnected hose streams. Common areas used for study and recreation will be larger, but most will still be within the capabilities of one or two standard hose lines. Irreplaceable personal heirlooms are not as common in dormitories and off-campus student housing as compared to some other residential occupancies; however, as with other residential properties, the importance of personal items should be considered during incident operations. Students may have course assignments or research on a computer that took considerable time and effort to compile.

Hotels and Motels

Like dormitories and off-campus housing, hotels and motels have a much greater occupant density than the typical home or apartment. Similar to the occupants of hotels and motels, guests in the dormitories and off-campus housing will not be familiar with building and exit facilities. Small motels may have doors that lead directly to the outside at grade level or doors leading to a walkway with direct access to ground level. Larger hotel and motel room doors typically lead to an interior hallway with well-marked exits. Motels and hotels are required to post evacuation information inside each unit. Unfortunately many people fail to read the evacuation instructions or familiarize themselves with exits when they are assigned a room. Occupants in a rush to evacuate will seldom read the instructions in their room, and exit signage may be obscured by smoke at the time of the fire.

Many times fire apparatus access will be limited to only one or two sides of a hotel or motel; therefore, fire department aerial devices will not be able to reach occupants on the nonaccessible sides of the building. Further, fires in units on the opposite side of back-to-back units or units with an interior hallway may not be visible upon arrival. Expect to find occupants still in their rooms, possibly with occupants from several rooms converged in a single unit, disoriented in hallways, in elevator lobbies, on exterior balconies, and in stairways. If the roof is accessible, occupants may try to seek refuge there as well.

When encountering heavy fire and smoke conditions, several fire companies will be needed to conduct search and rescue operations in a large hotel or motel property. The desk clerk may have a list of people occupying the hotel, but the hotel staff will not know which occupants are in the facility or be aware of people who share rooms without notifying hotel management. Most large hotels will have assembly, food, and drinking establishment areas within the building. Because of the potential for multiple rescues and a large rate of flow, it might be necessary to assign multiple companies to these areas. Most of the large loss-of-life fires in Table 11-14 are in high-rise hotels. High-rise tactics are addressed in the *High-Rise Buildings* chapter.

Mercantile Occupancies

NFPA 101, *Life Safety Code*,[1] defines a **mercantile occupancy** as follows:

> **3.3.190.9—Mercantile Occupancy.** *An occupancy used for the display and sale of merchandise.*

An occupancy factor matrix for mercantile occupancies is shown in **Table 11-15**. Shoppers and employees should be fully alert, and most will be able to evacuate with little assistance. However, there may be people who need assistance in an emergency, particularly in multistory mercantile buildings. Leadership is not provided in most stores, and during busy shopping times, the occupant density can be very high (as much as one person per 30 ft^2 [2.8 m^2][1]).

Table 11-15 Occupancy Factor Matrix: Mercantile Occupancies

	Mobility of Occupants	Age of Occupants	Leadership	Awareness	Occupant Density	Familiarity of Occupants
Mercantile occupancies	7	5	3	9	3	2

1 = extremely negative factor; 10 = very positive factor.

Time factors are critical in sizing up a mercantile occupancy. During times of peak occupancy, the life hazard can be high and content losses can be significant in these buildings. Property loss is typically the largest problem facing fire forces in mercantile occupancies, but don't be misled; the danger to life may be substantial. Consider the life safety problem in a crowded department store during the Christmas season.

Mercantile occupancies include shopping centers and malls, individual stores, and shops. Individual stores and shops can range from very small shops or convenience stores to mammoth "big-box" stores. Small convenience stores tend to be more congested, with stock in the aisles making it difficult to navigate even under normal conditions. Each type of large store, whether a big-box retailer, multistory department store, shopping center, lifestyle center, or enclosed mall, presents special challenges.

Any building with large undivided spaces requires special search and rescue techniques. The right- and left-hand searches used in residential occupancies are not effective in a store that could be hundreds of feet in each direction, with aisles, cross aisles, and various obstructions. These large areas require team searches, which are difficult to coordinate and are labor- and time-intensive. Team search techniques using ropes and thermal imaging cameras will be needed when vision is obscured. However, it takes longer for smoke to fill these large-volume compartments, allowing many evacuations to be completed in a relatively clear atmosphere. If smoke is filling the store, proper venting can restore visibility and may allow occupants to evacuate with little assistance.

Large undivided spaces will require a substantial rate of flow. Extinguishment may not be possible within the store of origin once the fire gains sufficient headway. Rate of flow should be computed during preincident planning for each store and storage area exceeding the flow capacity of two standard preconnected hose lines. Consider using 2½-in. (65-mm) hose lines or master streams as initial attack lines inside large stores.

Shopping Centers

Smaller shopping centers are usually configured as a line of stores (strip malls) with direct front and rear access to each store. Larger shopping centers characteristically will have anchor stores on each end. Anchor stores are typically big-box stores or multistory department stores. Having direct access to the outside makes evacuation less confusing as compared to evacuating stores located inside enclosed malls. Most occupants of stores in a shopping center will be able to find the main entrance with little difficulty. The rear of the store is often used as a storage area, thus rear exits will require occupants to go through unfamiliar surroundings to reach the outside. Rear doors tend to be heavily secured during nonbusiness hours.

Many shopping centers lack sprinkler protection, although sprinkler systems are becoming more commonplace. As stated previously, the rate of flow could be very large, and it may be beyond the fire department's ability to extinguish a fire in a large store. Having stores attached on both sides creates an internal exposure hazard. Exposure hose lines should be deployed to stores on each side of the store of origin as soon as critical life safety and initial extinguishment assignments are made or if a defensive attack becomes necessary. Fire can penetrate anywhere along the common fire wall, but the roof line and any place with utility penetrations are most likely extension paths. Older strip malls may have a common attic, which could allow the fire to spread horizontally to all attached stores. To deal with this scenario, hose lines must be quickly extended into these false spaces on each side of the fire. Some shopping centers have double fire walls between buildings with no penetrations. Extension is less likely with this higher degree of compartmentation.

Enclosed Shopping Malls

The enclosed shopping mall took shopping convenience one step further by enclosing the total shopping area with entrances to various shops from an interior walkway, thus providing access to shops in a climate-controlled environment. Shopping malls tend to be larger than shopping centers and may have several anchor stores located at the ends and off multiple wings. Shopping malls are usually two or more levels. The layout of these structures is much more confusing than in shopping centers. It is virtually impossible to find a specific smaller store without a map and/or guidance from building management. Preincident plans should include exterior access areas for fire apparatus, and floor layout drawings showing the location of each store should be carried on the apparatus of the first-arriving units. Primary and alternative access locations for each general area inside the mall should also be noted. If preincident plans are computerized, it is possible to search electronically for specific stores and identify the best exterior access points.

Incident Summary

A fire in Altoona, Pennsylvania, demonstrates the problems associated with shopping malls that are not sprinkler protected. The fire began in an unprotected store and rapidly spread to the mall walkway area and other stores. This shopping mall fire also demonstrates another common fire problem in strip and enclosed malls: common concealed spaces. The fire began in a utility closet then spread to the ceiling space, involving a large concealed area before breaking out into the mall area. The result was the total destruction of 15 stores and damage to 37 other stores within the mall in the weeks before Christmas.

Data from: Michael S. Isner, Shopping Mall Fire, Township of Logan, Pennsylvania, December 16, 1994, Fire Investigation Report. Quincy, MA: NFPA.

Areas of damage from the mall fire in Altoona, Pennsylvania.

Reprinted with permission copyright © 2007, National Fire Protection Association, Quincy, MA. All rights reserved.

Flames: © Jag_cz/ShutterStock, Inc.; Paper: © silver-john/ShutterStock, Inc.

Stores within a shopping mall will be separated by partition walls of noncombustible construction but will be open to the interior walkway area. During business hours, large door openings are designed to provide easy customer access to the store from the mall walkway. At night, metal grates are lowered and secured. These metal grates provide a barrier to criminals but will not stop the forward progress of a fire. Most modern shopping malls are sprinkler protected, greatly reducing the potential fire problem. Without a properly maintained and operational sprinkler system, these properties would present an unmanageable life hazard for occupants and fire fighters.

The size of enclosed shopping malls has increased dramatically. Some of the largest malls cover several blocks and are virtually small cities inside a single building. These larger malls contain many occupancy types, including places of assembly. One such mall in Edmonton, Alberta, Canada, has hundreds of stores, an amusement park, a hotel, and a wave pool inside the structure.

Along with sprinkler protection, smoke control systems are also an integral part of the fire safety system within a mall. If not controlled early, fire in one store will likely spread smoke and heat into the walkway area and other stores. Smoke control systems and large open walkways should allow enough time for a successful evacuation.

Lifestyle Centers

The latest trend in shopping is toward lifestyle centers. The lifestyle center is laid out as blocks of stores, much like a city street. Large anchor stores are usually surrounded by smaller shops at grade level, and each individual store can be accessed directly from the outside. The same basic tactics can be used as in the shopping center, but lifestyle centers will have fewer stores in each cluster and nearly all will be protected with automatic sprinklers. Finding individual shops and stores will normally be more difficult in the lifestyle mall than in a shopping center. Like the enclosed mall, a drawing showing the location of each store is needed as part of the preincident plan.

There is generally a regular turnover of tenants in shopping centers, enclosed malls, and lifestyle centers. It is important to communicate with building management on a regular basis and update the preincident plan whenever a store's occupant changes.

"Big-Box" Stores

Many big-box stores are anchor stores at shopping centers, but there are also many stand-alone stores or big-box stores with a few small stores nearby. The typical layout for a big-box store is an open-layout store to the front with an attached storage area to the rear. The big-box store should be sprinkler protected. As long as the sprinkler system is operating properly and store management has not increased the fuel load beyond the capabilities that the sprinkler system was designed to protect, the sprinkler system operational guidelines provided in the *Fire Protection Systems* chapter will be sufficient. Many of these stores will have standpipe drops from the sprinkler system, but it may be better to lay hose directly from the apparatus to avoid robbing water from the sprinkler system.

The sprinkler system is the only positive factor in these buildings. Variances in construction methods, exit facilities, and size are permitted owing to the sprinkler protection. This is a valid concept provided the sprinkler system is properly maintained and is fully operational. If the sprinkler system is out of service or unable to control the fire, the big-box store is a very dangerous place for fire fighters and occupants.

Even with a properly operating sprinkler system, the height and configuration of storage can require substantial intervention to achieve final extinguishment. As an example, consider an indoor lumber yard where sheets of plywood are stacked with no in-rack sprinkler protection, such that the upper stacks of plywood shield the lower racks from overhead sprinkler water. The storage racks are typically metal; therefore, a serious fire in a lower rack could result in a collapse of the entire rack. Further, these stores may contain merchandise of an extra-hazardous nature (e.g., chemicals, compressed gas cylinders, aerosols, racked clothing), which are extremely difficult to extinguish and create a serious life hazard for occupants and fire fighters.

Multilevel Department Stores

As noted previously, multilevel department stores can be part of a shopping center, enclosed mall, or lifestyle center. They can also be stand-alone buildings. Life safety and extinguishment are more difficult in these occupancies because of the multiple stories. Evacuation from upper floors requires the use of stairs or escalators. In many stores there is little separation from floor to floor; thus fire on a lower level is likely to extend to upper floors. Due to the large fuel load, extinguishment will be difficult. Property conservation is also more problematic in a multilevel structure. Property conservation is a major issue in all mercantile occupancies where high-dollar inventories are common and many items are extremely sensitive to smoke and water damage.

Business Occupancies

NFPA 101, *Life Safety Code*,[1] defines a **business occupancy** as follows:

> *3.3.190.3—Business Occupancy. An occupancy used for the transaction of business other than mercantile.*

An occupancy factor matrix for business occupancies is shown in Table 11-16 . The occupants of a business occupancy should be awake and aware

while occupying the building. Occupant mobility is generally not a major problem, but there may be physically or mentally challenged people in the building who require special assistance. Further, many business occupancies are located within high-rise buildings that will require the use of stairways for evacuation as well as places of safe refuge for occupants with special needs.

In business settings, the people occupying the premises are typically between the ages of 18 and 65 years. Leadership depends on the specific occupancy, but many business occupancies rehearse evacuation plans, particularly in high-rise buildings. Occupant density can be as high as one person per 100 ft^2 (9.3 m^2),[1] which is considerably less crowded than most places of assembly but more densely populated than many other occupancies. Employees in a one- or two-story office building with exits leading directly to the outside will probably be familiar with egress. People working in high-rise buildings usually take elevators to their work areas and may be unfamiliar with stairway locations. The large open areas and complex layouts place fire fighters at greater risk.

Business occupancies have a relatively good history in terms of life safety for occupants. However, the trend toward more and larger high-rise structures increases the probability of a large loss-of-life fire in a business occupancy. For many years, fire professionals predicted that a future high-rise fire would result in the loss of a thousand or more lives. Unfortunately, this prediction became a reality on September 11, 2001; however, the circumstances of the terrorist attack on that day were much different

Table 11-16	Occupancy Factor Matrix: Business Occupancies					
	Mobility of Occupants	**Age of Occupants**	**Leadership**	**Awareness**	**Occupant Density**	**Familiarity of Occupants**
Business occupancies	7	7	7	9	5	6
1 = extremely negative factor; 10 = very positive factor.						

© Eky Studio/Shutterstock, Inc.

from the fire the experts envisioned. The 2001 attack on the World Trade Center was the deadliest fire in U.S. history, but not the first terrorist attack on a high-rise building, nor even the first attack on the World Trade Center. The first World Trade Center attack, occurring in 1993,[12] and the Alfred P. Murrah Federal Building in Oklahoma City[13] were also the result of terrorist activity, and both included fire and explosions as part of the scenario. Historically, fire has been used as a weapon, and most experts agree that high-occupancy buildings will continue as targets of choice for terrorists. The World Trade Center had approximately 50,000 occupants at the time of the first bombing in 1993. The Murrah Building is believed to have been occupied by 350 people at the time of that explosion. At nine stories, the Murrah Building was small in comparison to many high-rise office buildings. High-rise buildings create special problems and are discussed in detail in the *High-Rise Buildings* chapter.

At night, when most business buildings are unoccupied, fires can progress for a long period if the building is not protected by an alarm and/or suppression system. Most new construction in business occupancies provide built-in automatic fire suppression and alarm systems.

In office occupancies, separation between individual offices is typically poor or nonexistent. Many times, movable partitions (cubicles) are used to separate work stations, allowing heat and smoke to spread quickly throughout a large, undivided area. Furthermore, this layout makes search and rescue operations very complex, since the configuration can resemble a maze. Wall separations are sometimes not full fire barriers, as some walls separating offices do not extend beyond the suspended ceiling. The area above the suspended ceiling provides an easy path for fire extension.

Fire control efforts vary widely in business occupancies, depending on the fuel load and the size of the undivided area. Fire control will not always be possible or could be substantially delayed owing to large flow requirements. Business occupancies tend to have a light-to-moderate fuel load. As for other wide-open areas, the rate of flow requirements will depend on the volume of the largest undivided compartment. If the sprinkler system does not control a fire in a large open-layout office, heavy-volume solid or straight-stream appliances may be needed for suppression.

Property conservation in business occupancies must emphasize the importance of protecting sensitive electronic equipment and business records.

Storage Occupancies

NFPA 101, *Life Safety Code*,[1] defines a **storage occupancy** as follows:

> *3.3.190.15—Storage Occupancy. An occupancy used primarily for the storage or sheltering of goods, merchandise, products, or vehicles.*

The factors affecting occupant evacuation in Table 11-17 are mainly positive for a storage occupancy. Most employees in these occupancies are expected to perform physical labor; thus most will be mobile. Generally speaking, the very young and elderly are at greatest risk during a structure fire.

Table 11-17 Occupancy Factor Matrix: Storage Occupancies						
	Mobility of Occupants	Age of Occupants	Leadership	Awareness	Occupant Density	Familiarity of Occupants
Storage occupancies	9	8	6	9	9	8

1 = extremely negative factor; 10 = very positive factor.

Age should not be an evacuation factor in a storage occupancy with a working-age population. Some storage facilities have an emergency action plan that prescribes internal leadership, while others will lack leadership in the initial stages of an emergency. Employees should be awake and alert, as well as familiar with the building layout and exit facilities.

Occupant density is generally low in storage facilities, especially the automated warehouse. Many modern warehouses are automated, with conveyors or robotics extending into remote and difficult-to-access locations. The increase in online sales is expected to continue with a subsequent increase in the number and size of automated warehouses.

The risk of a fast-moving fire or explosion may be high within the storage occupancy, which could have an immense fuel load. Yet the fire safety record in these occupancies has been good, primarily because of steps taken by responsible owners in protecting their workforce and property from fire.

Storage facilities vary in size from small self-storage modules to mammoth facilities like the Kmart warehouse in Falls Creek, Pennsylvania, or the general merchandise warehouse in New Orleans, Louisiana. Most one-story modern warehouses are of noncombustible construction with metal truss roof structures. Like the big-box stores described under *Mercantile Occupancies*, these storage buildings can have tremendously high fuel loads and the roof structures make them prone to early roof collapse. On the positive side, most of the large modern warehouses are sprinkler protected. Conversely many older warehouses will be of fire-resistive, heavy-timber, or ordinary construction, though they are often multistory structures with limited egress and confusing layouts. Many of these older warehouse buildings lack sprinkler protection.

Fighting a fire in a large commercial building is more challenging tactically than attacking a fire inside a small residential building. The size and complexity of the building, coupled with the possibility of an extremely high fuel load, make these buildings more hazardous for fire fighters working inside. The large volume of air inside each compartment will provide

oxygen for an extended time, allowing the fire to grow rather than become ventilation controlled.

Escape time will be much longer from a building that may be 1,000,000 ft² (92,900 m², approximately 9 hectares) or greater as compared to a residential structure, where room dimensions seldom exceed 1000 ft² (92.9 m²).

A fire in a vacant warehouse in Worcester, Massachusetts, resulted in six fire fighter fatalities. The interior of this multistory warehouse was very confusing. Fire fighters are more likely to be killed in these large commercial structures than in residential fires, but the danger increases exponentially in large vacant buildings where the fire fighter fatality rate is much greater than in the typical home fire.

In warehouses with an emergency action plan, including provisions for employee accountability, the primary search and rescue will be much easier because the search team can focus on a targeted area, provided the emergency action plan has been successfully implemented.

If the property is sprinkler protected, supporting and augmenting the sprinkler system will be the primary extinguishment tactic. However, facility management will sometimes increase the fuel load inside a warehouse beyond the design capacity of the sprinkler system. Additional water supplied by fire department pumpers could make the difference between a successful operation and a total loss of the building when the fuel load exceeds the design capacity of the system.

When a fire overwhelms the sprinkler system and reaches an advanced stage, extinguishment will be difficult. Every effort should be made to increase the effectiveness of the sprinkler system before attempting to control a major fire with an offensive attack using handheld lines.

A common practice in warehouse facilities is to place large stacks of idle pallets in a staging area. This creates a tremendous challenge to the sprinkler system. Only an extra-hazard sprinkler system is capable of handling a fire in large quantities of wooden pallets. When a new occupant leases a storage facility, the new owner/occupant may store a

Flames: © Jag_cz/ShutterStock, Inc.; Paper: © silver-john/ShutterStock, Inc.

Incident Summary

In 1996, the New Orleans Fire Department was successful in containing a fire that was beyond the capabilities of the sprinkler system, even though 30 overhead and 17 in-rack sprinkler heads operated. After manually bringing the fire under control, all valves controlling the systems were closed to eliminate additional water damage within the heavily damaged 930,020-ft² (86,400-m²) undivided warehouse. Subsequently, the power was restored to the warehouse area, resulting in a second fire caused by damaged wiring. The second fire resulted in a total loss of the warehouse and contents valued at $280 million.[14]

different commodity, possibly exceeding the design capacity of the sprinkler system. As an example, a warehouse sprinkler system may be designed to protect metal parts stored in cardboard boxes. If a new owner or occupant were to store rubber tires, flammable liquids, or plastics in this same area, the system, which was designed for the lower-hazard commodity, would be inadequate to extinguish the fire. The result would likely be a total loss.

The Kmart fire in Falls Creek, Pennsylvania, is a good example of the loss potential in a storage occupancy fire.[15] The Kmart warehouse, covering nearly 1.3 million ft² (approximately 12 hectares), was fully sprinkler protected and compartmentalized, yet it was completely destroyed in a $100-million fire **Figure 11-11**. Aerosol cans of carburetor cleaner, which exceeded the design capacity of the sprinkler system, were first ignited at this incident. The burning aerosol cans rocketed from the compartment of origin to other compartments, eventually spreading the fire to all four building quadrants. Had the fire chief failed to see the danger to his personnel, this fire could well have resulted in several fire fighter fatalities. By developing a preincident plan for the structure, the chief recognized the collapse potential

and the futility of manual firefighting operations when confronted with an advanced fire.

The rate of flow must be calculated during pre-incident planning. Standard V/100 calculations (discussed in the *Offensive Operations* chapter) may

Figure 11-11 Kmart warehouse fire, Falls Creek, Pennsylvania, June 1982.
© AP Photos

not be sufficient for storage occupancies. Sprinkler system calculations should be used when dealing with extra-hazard storage. It is important to remember that sprinkler calculations assume the fire will be contained to a limited area due to quick operation of the system, with water discharging directly on to the fire. These assumptions are not valid when manually attacking the fire with handheld hose lines, because sprinklers are much more efficient in applying water directly on the fire. Therefore, the calculated rate of flow for hose lines may result in a flow that is beyond the available water supply and fire department staffing and pumping capacities. With that said, when calculating rate of flow and developing preincident plans, remember that in many cases the fire will be limited to a small area within a very large building.

Many firms that suffer a large fire go out of business. Even with insurance, the loss of production and subsequent loss of customers are more than many businesses can overcome. Competent businesspeople apply risk management principles to protect their profitability. One of the most common risk management measures involves installation of suppression systems or other built-in fire protection.

Most warehouses contain large quantities of valuable commodities. Some of these commodities are very susceptible to water and smoke damage. Storage in racks or on pallets reduces water damage from run-off. Storage of materials that absorb water, especially if the materials are stored on the floor, will greatly increase the water damage and add significant weight to the floor.

Industrial Occupancies

NFPA 101, *Life Safety Code*,[1] defines an **industrial occupancy** as follows:

> *3.3.190.8—Industrial Occupancy. An occupancy in which products are manufactured or in which processing, assembling, mixing, packaging, finishing, decorating, or repair operations are conducted.*

The occupant profile in an industrial setting is much like that of the storage occupancy, with mobile, alert workers who are familiar with their surroundings (Table 11-18). The occupant density is usually higher in an industrial occupancy facility as compared to a storage occupancy. Industrial plants are required to have emergency evacuation plans; therefore, leadership is usually better in the industrial occupancy than in many other occupancies. Emergency action plans often include an accountability process that can assist the IC in facilitating an orderly evacuation and accounting for plant personnel.

Recognizing hazards and equipment limitations is paramount during an industrial fire. In the large industrial building or industrial complex, preincident planning is critical to fire fighter safety and efficient operations. However, preincident planning is not enough. Cooperation between plant personnel and fire fighters is essential. In-plant employees are plant experts; they know valve locations, how to navigate the plant, and how to shut down processes. The fire fighter is the fire suppression expert. Successful industrial fire suppression activities involve plant staff assisting fire department personnel. Anything less

Table 11-18	Occupancy Factor Matrix: Industrial Occupancies					
	Mobility of Occupants	**Age of Occupants**	**Leadership**	**Awareness**	**Occupant Density**	**Familiarity of Occupants**
Industrial occupancies	9	8	8	9	7	8

© Ely Studio/Shutterstock, Inc.

than a cooperative effort invites total destruction and places fire fighters at unnecessary risk. The number of hazards and the potential for harm dictate that special tactics be developed for each large manufacturing and storage property within a jurisdiction.

Industrial occupancy fires are notable because they tend to cause a large loss of life and/or large financial loss. Some notable large loss-of-life fires in industrial occupancies are listed in **Table 11-19**. Several of the multifatality fires in industrial properties involved hazardous materials, such as ammunition, fireworks, explosives, chemicals, and flammables. Table 11-19 lists industrial fires that killed at least six people. There were many additional oil refinery and tank fires that claimed three to five lives that are not listed in the table.

In the industrial setting, the main tactical activity often involves controlling a manufacturing process or otherwise stabilizing the incident to protect life and property. Depending on the type of burning fuel, the use of water may be counterproductive. Using knowledgeable facility personnel as advisors to the IC during an incident is imperative, as they are familiar with the building, systems, and processes and can advise the IC on a course of action to take (or not take) to safely mitigate the problem.

Fire control efforts must be customized to the hazard. Preincident planning, which includes identifying and outlining the responsibilities of plant personnel, is key to successfully controlling fires in industrial occupancies.

In addition to the normal salvage activities, property conservation includes stabilizing manufacturing and operating processes. Many industrial facilities have interlocks that sense a drop in pressure in the water supply and/or fire protection system. These interlocks sometimes interrupt production or dump materials being processed. Flowing water from a fire hydrant located on the property may unintentionally interrupt a process, which could result in significant property loss.

Damage or destruction of high-value storage and manufacturing equipment at industrial properties results in many large-dollar losses. A million-dollar fire is commonplace in an industrial occupancy. The

Table 11-19 Notable Fires in Industrial Occupancies

Fire	Year	Deaths
R. B. Grover Shoe Factory, Brockton, MA	1905	50
Triangle Shirtwaist, New York City, NY	1911	145
Eddystone Ammunition Corporation, Eddystone, PA	1917	133
Aetna Chemical Company, Oakdale, PA	1918	193
Semet-Solvay, TNT Manufacturing, Split Rock, NY	1918	50
Hercules Powder Company, Kenvil, NJ	1940	52
Elwood Ordnance Plant, Joliet, IL	1942	54
East Ohio Gas, Cleveland, OH	1944	130
Munitions depot, Port Chicago, CA	1944	300
Chemical plant, Pasadena, TX	1989	23
Chicken processing plant, Hamlet, NC	1991	25
Fireworks manufacturing, Osseo, MI	1998	7
Power plant, Dearborn, MI ($1 billion property loss)	1999	6
Automobile insulation manufacturing, Corbin, KY	2003	7
Oil refinery, Texas City, TX	2005	15
Sugar refinery, Wentworth, GA	2008	14
Hydrocarbon refinery, Anacortes, WA	2010	7
Power generating plant, Middletown, CT	2010	6
Grain elevator, Atchison, KS	2011	6
Ammonium nitrate fertilizer plant, West, TX	2013	15*

* Including 9 fire fighters and an emergency medical technician.

© Ely Studio/Shutterstock, Inc.

$100-million-plus fires at Kmart in Pennsylvania,[16] Tinker Air Force Base in Oklahoma,[16] Central Storage and Warehouse in Wisconsin,[17] and West Fertilizer Company in Texas[18] are examples of large-loss fires eclipsed by the billion-dollar fire at a Texas oil refinery in 2005.

Multiple and Mixed Occupancy Buildings

NFPA 101, *Life Safety Code*,[1] defines a **multiple occupancy** as follows:

> *3.3.190.11. Multiple Occupancy. A building or structure in which two or more classes of occupancy exist.*

Multiple occupancies are further categorized as *mixed* or *separated*, as follows:

> *3.3.190.10—Mixed Occupancy. A multiple occupancy where the occupancies are intermingled.*

> *3.3.190.14—Separated Occupancy. A multiple occupancy where the occupancies are separated by fire resistance–rated assemblies.*

Separated multiple occupancy buildings can be handled as two buildings due to fire-resistive separations. **Mixed multiple occupancy** buildings are more problematic because fire can easily travel from one occupancy to another. Having an unoccupied area in the same building where other areas are occupied can result in a severe life safety threat. As an example, many buildings have a series of small stores or shops on the first floor with residential units above. Fuel loads in the shops will typically be greater than the fuel loads found in residential occupancies and often include hazardous processes. Further, the shops are usually unoccupied outside normal business hours, allowing a fire to go undetected in an area below sleeping residents. By the time residents realize there is a fire, egress routes may be untenable and the fire may have already extended into the residential sections of the building.

Buildings Under Construction, Renovation, or Demolition

Buildings under construction, renovation, or demolition are not classified as specific occupancies, but these buildings are vulnerable to fire and warrant special attention. Buildings under construction or undergoing extensive renovation may lack fire protection equipment and structural features designed to impede fire until late in the construction process. Conversely, in buildings being demolished, these same fire protection features and equipment are often the first to go, including disabling the sprinkler system and removing interior walls.

Buildings under construction will not have fire barriers or exterior coverings that retard fire growth, and large quantities of building materials may be stored within the interior framework of a partially constructed building. Like the building under construction, the building being renovated may have large stacks of building materials stored inside the building. When a building is being razed, some interior materials are removed for salvage, then work begins to demolish the outer shell with large quantities of debris accumulating around and within the structure.

Buildings under construction, renovation, or demolition are typically unoccupied when work crews leave for the day; however, these structures may be illegally occupied by youths or homeless individuals. In either case, after-hour fires in these structures usually gain considerable headway before being reported, allowing the fire time to spread throughout the structure and involve external exposures. Free-burning fires involving wooden framing, combustible building materials, or debris can develop a large fireball with subsequent high levels of radiant heat. These buildings can also produce flying brands that can ignite fires far from the building of origin, as occurred at the Santana Row fire in San Jose, California (see the incident summary in the *Defensive Operations* chapter). As discussed in the *Defensive Operations* chapter, the radiant heat ignites nearby structures while flying brands can travel a considerable distance beyond the building of origin. The open status of buildings under construction, renovation, or demolition also makes these structures vulnerable to flying brands from nearby fires.

A building that is only partially constructed or one that is partially razed is more likely to collapse than a finished structure. Many times scaffolding is placed around buildings under construction, tempting fire fighters to use this scaffolding to gain rapid access to upper floors of the building. However, fire fighters should avoid using scaffolding, because water from extinguishment efforts can undermine the base of the scaffolding and the fire can destroy the scaffolding connections to the building. Either scenario could result in a total collapse of the scaffolding system, injuring or killing fire fighters who were working from the scaffolding at the time of collapse, while leaving fire fighters who used the scaffolding to access upper floors stranded with no means of escape.

Renovated Buildings

Once the renovation process is completed, the occupancy classification may change. Many multistory, downtown buildings that were once offices, stores, schools, or factories have been converted to other occupancy types. Many old school buildings are now apartments, condominiums, and restaurants; old department stores lofts and buildings that were built as factories are now offices. These are but a few examples of the kinds of conversions taking place in many urban areas.

In many cases, converted buildings are safer, because building and fire codes require upgrades when occupancy classifications are changed or when a significant portion of the building is renovated. However, these converted occupancies can create challenges. For instance, interior walls are often removed to create large open spaces to accommodate the new building use. In addition, false ceilings are added to reduce the ceiling height for aesthetic and energy conservation reasons.

The present trend of moving back into the central city core is expected to continue. And, with preservationists using their political clout to protect older buildings from being razed, much of the new residential and commercial space will be located in renovated structures. Many of the potential problems in renovating these older buildings can be avoided by enforcing building and rehab code requirements. Unfortunately, local and state government officials, in their zeal to revitalize the urban core while at the same time appeasing special-interest groups, may bow to pressure to approve code waivers. For fire fighters, this means combating fires in buildings that contain many false spaces that can hide fire and provide channels for fire extension, confusing floor plans, and large open areas that will complicate search and rescue efforts, possibly requiring a large rate of flow.

General Considerations for Special Occupancy Fires

Special occupancy fires require fire departments not only to preplan for specific properties but also to develop plans of action for general use within particular types of buildings and occupancies. Many of the fire fighter fatalities that have been recorded over the years were in residential properties, including single-family detached dwellings. However, the size and complexity of business and industrial buildings place the fire fighter at additional risk when combating fires in these structures. Extra precautions and more frequent accountability procedures may be necessary in combating assembly, mercantile, storage, business, and industrial fires.

It is fair to say that fire fighters are at much greater risk of dying when confronted with commercial fires than when working a residential structure fire. Obviously, the threat to a fire fighter's life is substantial at the residential fire, but the fire officer must recognize the even greater potential danger when working at a commercial fire.

Using operational priorities as a basis, fire suppression forces can wage a safe and effective operation at commercial and other special occupancies. Due to the relatively rare incidence of these special occupancy fires, few ICs gain confidence through actual firefighting experience; therefore, it is imperative that they learn through training and from the experiences of others.

Most of the large loss-of-life fires in commercial and other special occupancies occurred in the fairly distant past. The number of civilian fire deaths and line-of-duty deaths for fire fighters has shown much improvement in the recent past. Two important points should be made about these positive statistics:

1. The fire threat to human life is real and present. Maintaining a properly staffed, well-trained fire force is absolutely necessary.
2. Improvements in codes and standards and the widespread use of smoke detectors, fire sprinkler systems, and public education programs are directly related to life safety improvements.

Estimating the Number of Potential Victims

Determining the total number of occupants is closely related to the type of occupancy. The high-rise office building could have thousands of occupants, compared to one or two occupants in many single-family detached dwellings. Occupant load is also time sensitive. The office building will most likely be fully occupied during normal working hours, while the residential building will most likely be occupied at night.

The staffing requirements for a fire department response to a single-family detached dwelling is usually inadequate for larger building fires because of the additional search and rescue requirements combined with a need for larger fire flows. The more complex the structure, the greater the need for personnel. However, some commercial buildings will be more substantially built and are more likely to be protected by fire suppression systems. These fire-protective features will lower the risk to fire fighters and occupants, provided that the fire safety features designed into the building are maintained and operating properly.

The first step in determining evacuation needs is an evaluation of the time of day, building size, and occupancy type. Classifying these variables will provide a rough estimate of how many people could be in the building; however, it is important to remember that this evaluation is, at best, only an estimate.

WRAP-UP

Chapter Summary

- Occupancy type, time of day, and day of week are critical factors in determining fire-ground strategy.
- Major factors related to occupancy type are number, mobility, and age of the occupants; leadership in guiding evacuation; occupant awareness; occupant familiarity with the structure; and time of day, week, and year.
- Assembly occupancies include churches, eating/drinking establishments, stadiums, convention centers, and theaters where 50 or more persons gather.
- Mobility factors are usually above average in school settings, but younger elementary school students may not be capable of taking independent action to safely evacuate.
- Leadership and preparedness are generally positive factors in educational occupancies.
- Operations in detention/corrections occupancies are complicated by the fact that prisoners cannot self-evacuate and must be kept secure during firefighting operations.
- Preincident planning is an absolute necessity for a detention or correctional facility.
- Residential occupancies include single- and multiple-family dwellings, apartment buildings, dormitories, and hotels/motels.
- Most fires, fatalities, and property loss occur in residential fires.
- In many occupancies, including residential, there is a direct correlation between life safety and the time of day.
- More fire fighter on-duty deaths occur in residential properties than in any other occupancy due to the large number of fires that occur in these properties.
- Most fires in one- and two-family dwellings should be within the capabilities of one or two preconnected hose lines.
- Property loss is typically the largest problem facing fire forces in mercantile occupancies; however, depending on time factors, danger to life safety may be substantial.
- Preincident plans should include exterior access areas for fire apparatus and floor lay-out drawings showing the location of each store in shopping centers, malls, and lifestyle centers.
- High turnover of tenants in shopping centers and malls makes it important to update pre-plans on a regular basis.
- Large fuel loads in mercantile occupancies make extinguishment difficult.
- The risk of a fast-moving fire or explosion is high in a storage occupancy, especially if the fuel load exceeds the design capacity of the sprinkler system.
- Most large, modern warehouses are of light-weight construction but sprinkler protected; older warehouses are often of fire-resistive, heavy-timber, or ordinary construction, but often lack sprinkler protection.
- Emergency evacuation and preincident plans are required in most industrial facilities.
- Mixed and multiple occupancies combine the fuel loads and life safety hazards of two or more occupancy types.
- Structural collapse and/or collapse of construction scaffolding is highly likely in buildings under construction.
- Occupant load is related to building size, type of occupancy, time of day, and day of week.

Key Terms

assembly occupancy An occupancy used for the gathering of 50 or more persons for deliberation, worship, entertainment, eating, drinking, amusement, awaiting transportation, or similar uses; or used as a special amusement building, regardless of occupant load.

business occupancy An occupancy used for account and record keeping or the transaction of business other than mercantile.

convergence cluster phenomenon A reaction to fire conditions in which groups take shelter together to provide mutual support.

detention and correctional occupancy An occupancy used to house four or more persons under varied degrees of restraint or security where such occupants are mostly incapable of self-preservation because of security measures not under the occupants' control.

educational occupancy An occupancy used for educational purposes through the 12th grade by six or more persons for 4 or more hours per day or more than 12 hours per week.

health care occupancy An occupancy used for purposes of medical or other treatment or care of four or more persons, where such occupants are mostly incapable of self-preservation due to age, physical or mental disability, or security measures not under the occupants' control.

industrial occupancy An occupancy in which products are manufactured or in which processing, assembling, mixing, packaging, finishing, decorating, or repair operations are conducted.

mercantile occupancy An occupancy used for the display and sale of merchandise.

mixed multiple occupancy A multiple occupancy where the occupancies are intermingled.

multiple occupancy A building or structure in which two or more classes of occupancy exist.

occupant density A measure of the number of people in a given area; usually stated as the number of square feet per person.

residential board and care occupancy A building or portion thereof that is used for lodging and boarding of four or more residents, not related by blood or marriage to the owners or operators, for the purpose of providing personal care services.

residential occupancy An occupancy that provides sleeping accommodations for purposes other than health care or detention and correctional.

separated multiple occupancy A multiple occupancy where the occupancies are separated by fire resistance–rated assemblies.

storage occupancy An occupancy used primarily for the storage or sheltering of goods, merchandise, products, vehicles, or animals.

Flames: Dvn/Dreamstime.com; Steel texture: © Sharpshot/Dreamstime.com

Challenging Questions and Activities

1. Classify specific occupancies in your jurisdiction using an occupancy factor matrix similar to the general occupancy factor matrices used in this chapter.

2. Compute the maximum number of occupants in a 100- × 100-ft (30.5- × 30.5-m), 10-story office building assuming 100 ft² (9.3 m²) per person.

3. For this activity, use the plan-view drawing showing the exit doors and ignition points from The Station nightclub incident summary. Change the ignition sequence to a slower-moving fire (one not involving foam plastic materials). When you arrive as the officer in charge of the first-in engine company, there are heavy smoke and fire conditions with people jammed at the main entrance blocking people attempting to escape.

    A. What is the best course of action for this first-arriving unit assuming the company is staffed with an officer, apparatus operator, and two fire fighters? Explain the rationale for you decision.

    B. If you decide to lay a hose line, what would be the best place to enter the building to attack or control the fire?

    C. If you decide to perform manual rescues, explain how you could rescue the most people in the least amount of time.

    D. Assume the role of a battalion chief arriving to take command. Develop an incident action plan and deploy units to implement the plan. Organize units deployed and in reserve using a National Incident Management System (NIMS) chart.

4. Review an evacuation plan for a local elementary or high school in your jurisdiction.

    A. Does the plan clearly spell out the location of the school contact person for a fire emergency? Who are the primary and alternate contact persons?

    B. Are students assembled in areas that interfere with fire department access or operations?

    C. If the school has an "active shooter" plan, how does it affect fire response?

5. Obtain a list of residential board and care occupancies in your response area. What are the occupant characteristics in these facilities in terms of their ability to escape without assistance?

6. It is 2:00 AM and you are responding as the officer of the first-due engine company to a 30-unit apartment building. Upon arrival, you can see fire in the main stairway, and the fire appears to have originated in the basement storage area. Some residents are already out of the building; others are at windows or on balconies calling for help. What is your best first action? Explain.

7. An insurance company's claims adjustment unit is housed in a 250- × 250-ft (76- × 76-m) building of noncombustible construction. Office space is divided using 4-ft (1.2-m) high cubicles with aisle and cross-aisle access to individual work areas **Figure 11-12** . Describe a search method that could be used under conditions of low visibility in this building.

Figure 11-12 Insurance claims office subdivided using movable partitions (cubicles).

© Jude Lazaro/ShutterStock, Inc.

Endnotes

1. National Fire Protection Association. NFPA 101, *Life Safety Code®*. Copyright © 2012, National Fire Protection Association. This reprinted material is not the complete official position of the NFPA on the referenced subject, which is represented only by the standard in its entirety.

2. Ahrens, M. *U.S. Fires in Selected Occupancies 1999 to 2002*. Quincy, MA: NFPA Fire Analysis and Research Division; March 2006.

3. Best, R. L. *Reconstruction of a Tragedy: The Beverly Hills Supper Club Fire, Southgate, Kentucky, May 28*. Quincy, MA: NFPA; 1977.

4. Isner, M., Foley, S. *Bradford Soccer Stadium*. Fire Investigation Report. Quincy, MA; NFPA.

5. Evarts, B. *Structure Fires in Residential Board and Care Facilities*. Quincy, MA: NFPA; May 2012.

6. Moncada, J. A. Lessons of Comayagua. *NFPA Journal*. September/October 2012: 50–59.

7. Karter, M. J., Jr. Fire loss in the United States during 2012. *NFPA Journal*. September/October 2013.

8. Badger, S. G. Catastrophic multiple-death fires in 2013. *NFPA Journal*. September/October 2014.

9. Bryan, J. A phenomenon of human behavior seen in selected high-rise building fires: Convergence clusters. *NFPA Fire Journal*. November 1985: 27–30, 86–90.

10. Ahrens, M. *Home Structure Fires*. Quincy, MA: NFPA Fire Analysis and Research Division; May 2011.

11. NFPA Fire Investigations Department. *Wood Shingle Roofs Fuel Conflagration*. NFPA Fire Command. Boston, MA: Author; February 1980.

12. U.S. Fire Administration. *World Trade Center Bombing: Report and Analysis*. Emmitsburg, MD: Author; undated.

13. Comeau, E., Foley, S. Oklahoma City, April 19, 1995. *NFPA Journal*. July/August 1995.

14. Comeau, E., Puchovsky, M. *Warehouse Fire, New Orleans, Louisiana*. Fire Investigative Report. Quincy, MA: NFPA; March 1996.

15. Best, R. $100 million fire in K-Mart distribution center. *NFPA Fire Journal*. March 1983: 36–42, 80.

16. Goodbread, J. Fire in Building 3001. *NFPA Fire Command*. July 1985: 34–37.

17. Isner, M. $100 million fire destroys warehouses. *NFPA Fire Journal*. November/December 1991: 37–41.

18. Rueters. Ammonium nitrate stores exploded at Texas plant: State agency. May 7, 2013. Retrieved from http://www.reuters.com/article/2013/05/07/us-usa-explosion-texas-idUSBRE9460GP20130507

Flames. © Dix/Dreamstime.com; Steel texture. © SharpShot/Dreamstime.com; Chapter opener photo. © ALAN GRETH/AP Images

High-Rise Buildings

LEARNING OBJECTIVES

- Describe the magnitude of the high-rise fire problem in terms of number of fires, number of fire fatalities, and property loss.
- Define a high-rise building from a fire department perspective.
- Explain elevator recall and the responsibilities of lobby control.
- Explain why using an elevator during a fire emergency is a calculated risk and list the rules for elevator safety.
- Explain the duties of ground support and calculate the number of fire fighters needed to staff ground support for a fire on the 30th floor where the elevators are unsafe to use.
- Identify negative and positive aspects of using helicopters at a high-rise building fire.
- Compute the approximate pump discharge pressure needed to supply a hose stream operating on the 20th floor.
- Explain when the command post should be located in the lobby, in a fire command room, outside the building, or at a remote location.
- List four occupancy types that are commonly found in high-rise buildings.
- Compare old-style tower construction to modern planar-style high-rise buildings.
- Define and explain the advantages of a smoke-proof tower.
- Discuss methods that can be used to maximize the limited stairway capacity in a high-rise building and how fire department operations can hinder evacuation.

LEARNING OBJECTIVES (CONTINUED)

- Describe an emergency voice/alarm communications system (EVACS) and identify the advantages and disadvantages as compared with other notification systems.
- Explain how wind, stack effect, and heat affect smoke movement in a high-rise building.
- Enumerate measures taken to counteract the effects of a wind-driven fire.
- Discuss the importance of extinguishment to life safety in a high-rise fire.
- Explain what is meant by a "wrap-around" fire.
- Describe the hazards involved in ventilating upper floors by removing window glass and how to protect fire fighters and civilians on the street below.
- List pathways for floor-to-floor fire extension in a high-rise building.
- Define "lead time" and extrapolate the estimated lead time for a fire on the 40th floor of a high-rise building when elevators are unavailable.

- List practical forms of non-radio communication that can be used at a high-rise building fire.
- Discuss when interior and exterior staging would be used at a high-rise fire.
- List the duties of the interior staging officer and lobby control.
- Compare the designed "accidental aircraft" impact at the World Trade Center to the actual impact of the aircraft on September 11, 2001.
- Describe the conditions leading to structural collapse at the World Trade Center.
- Describe how a "convergence cluster" could affect search and rescue operations at a high-rise building.
- Using a simulated high-rise fire scenario, size up the incident, develop an incident action plan, assign units to carry out the plan, and develop a National Incident Management System organization.

Case Study

Cook County Administration Building.

Courtesy of NIST, Special Publication

On October 17, 2003, at 5:00 PM a fire on the 12th floor of the 37-story Cook County Administration Building in Chicago, Illinois resulted in six civilian fatalities. The fire was believed to have originated in a closet within a 2629-ft² (244-m²) suite of offices on the 12th floor. Actual fire damage was contained to the office suite due to fire-resistive building features, but smoke and fire migrated to the entire 12th floor. The fire self-vented through eight exterior windows, creating the potential for floor-to-floor fire extension via the exterior.

Building occupants were notified to evacuate the building via the stairways. However, some occupants used the elevators to escape while others used the east and west stairways to evacuate. The fire department advanced a hose line into the 12th floor via the east stairway. When the door to the 12th floor was opened to advance the hose line into the fire area, smoke levels in the east stairway increased dramatically. As far as can be determined, there were 13 occupants above the 12th floor in the east stairway when the fire department began attacking the fire. Doors from the hallways on each floor were locked on the stairway side, trapping these occupants in the stairway above the fire. Seven of the occupants in the stairway found a door that did not completely latch on the 27th floor and were able to leave the east stairway and escape. Six others perished in the stairway.

Fire fighters were faced with an intense fire that they were unable to extinguish from their hallway position. Elevated master streams were used to knock down the fire from the exterior. Interior hose streams were then redeployed to achieve final extinguishment.

1. What evacuation options are available to occupants beyond the reach of aerial devices?
2. How is extinguishment related to rescue on upper floors of a high-rise building?
3. What role does compartmentation play in saving lives and properties?

Flames: © Jag_cz/ShutterStock, Inc.; Steel texture: © Sharpshot/Dreamstime.com; Stacked photo background: © Vitaly Korovin/ShutterStock, Inc.; Paper: © silver-john/ShutterStock, Inc.; Photo: © Jones & Bartlett Learning. Photographed by Glen E. Ellman.

Introduction

The basic tactics and strategic objectives used in high-rise firefighting are essentially the same as those that apply to any other structure fire, but with special considerations because of the height of the building. The differences are great enough to deserve this separate chapter.

The most recent high-rise fire report from the NFPA[1] indicates that there are approximately 15,700 fires, 53 civilian deaths, and $235 million in direct property damage per year in U.S. high-rise fires. The World Trade Center bombings in 1993 and attack in 2001 are not typical high-rise fire situations. But this is the nature of the high-rise problem. It is not unusual for there to be intervals of several years between major loss-of-life or large property-loss fires in high-rise buildings in the United States; yet the potential always exists. The overall numbers of fires, civilian fire deaths, and property loss seem to be trending downward. This is probably due to more high-rise buildings being sprinkler protected. In the mid-1970s, most jurisdictions required new high-rise buildings to be sprinkler protected, but very few required existing buildings to be retrofitted. The World Trade Center was fully sprinkler protected, but the system was severely damaged on impact and the jet fuel provided a fire load beyond the design capacity of the system.[2] There is absolutely no doubt that a sprinkler-protected building is a safer building for fire fighters and occupants; however, fire departments should not place total reliance on the sprinkler system. Another factor contributing to fewer fire deaths is the common use of pressurized stairways beginning in the 1970s.

High-rise fires represent an extraordinary challenge to fire departments and are some of the most challenging incidents a fire department encounters. High-rise buildings can hold thousands of people well above the reach of fire department aerial devices, and the chance of rescuing victims from the exterior is near zero once the fire is above the operational reach of ladders or elevated platforms.

NFPA 1710, *Standard for the Organization and Deployment of Fire Suppression Operations, Emergency Medical Operations, and Special Operations to the Public by Career Fire Departments*, 5.2.4.2.3, states "Fire departments that respond to high, medium or low hazard occupancies that present hazards greater than those found in the low-hazard occupancy described in 5.2.4.2.2 shall deploy additional resources on the initial alarm."[3] The National Institute of Standards and Technology (NIST) issued its Report on Residential Fireground Field Experiments[4] in 2010 by conducting experiments in a two-story house to verify the staffing requirements specified in NFPA 1710 for a low-hazard occupancy. The results of that study have serious safety implications and are discussed in the *Fire Fighter Safety* chapter. One of the results of the residential study was a recommendation to determine the staffing required for a successful outcome in a high-hazard building. In 2012, NIST conducted field experiments in a 13-story high-rise with a working fire on the 10th floor in an effort to determine the effect of company-level staffing on successful outcomes during a high-rise building fire. In the NIST Report on High-Rise Fireground Field Experiments,[5] crew sizes ranging from 3 to 6 were compared. Deployment based on the number of companies responding on the initial and subsequent alarms was also evaluated. Other variables included sprinkler protection (sprinkler protected versus non-sprinkler protected) and availability of elevators for

fire department use (elevators available for use versus unsafe elevators). Each field experiment was conducted multiple times to verify accuracy. This important report will be referenced throughout this chapter.

Case histories provide an excellent learning tool. By studying past fires, it is possible to gain experience that may not be gained in any other way. Outside of New York, Los Angeles, Chicago, and a few other very large cities, few chief officers experience enough working high-rise fires to gain confidence in managing such incidents. Communities with one or two high-rise buildings may never experience a serious fire, yet the fire department is expected to be prepared if a high-rise fire occurs. For this reason, several high-rise fires are discussed at the end of this chapter. Studying these and other high-rise fire case studies is strongly recommended.

Realistic training sessions applying high-rise standard operating procedures (SOPs), tactics, and tasks are essential to maintain the skills necessary to successfully combat high-rise fires. Many fire companies from the Washington, DC area departments were involved in the NIST High-Rise Fireground Field Experiments. The study notes how these departments used the experiment scenarios to improve fire-ground skills and cooperation between the many departments likely to work together at a major high-rise fire in the region.

Training can be as simple as testing a single function such as the ability to apply the National Incident Management System (NIMS) or a tabletop exercise using scenarios such as the RGB scenario presented in the "Challenging Questions and Activities" at the end of this chapter. Full field exercises where standpipe equipment is actually moved to the fire floor and hose lines are connected to the standpipe and advanced to a simulated fire area, combined with search and rescue activities, are very difficult to develop and manage. Maintaining adequate staffing for real fire responses and taking precautions not to cause damage to the structure used for the scenario can be a significant challenge. The NIST High-Rise Fireground Field Experiments explain many ingenious ways to provide realistic training with a low probability of injury or property damage.

Developing and Revising High-Rise Standard Operating Procedures

Preincident planning and code enforcement can reduce the scope of the high-rise problem. However, special tactics will be needed to control fire forces working in different areas within these large structures while providing the necessary logistical support. Using NIMS and developing high-rise SOPs can do much to ensure successful operations.

A difference of opinion exists as to the definition of a high-rise building. Most codes define the high-rise building in terms of height and/or stories. Fire departments tend to think of a high-rise building as being beyond the reach of the aerial fire equipment available to them. Since the focus of this text is on fire-ground tactics, the fire department definition is most appropriate. However, do not forget the obvious: an 8-story building will not present the same challenges as an 80-story high-rise building. Logistics and access problems increase with height. The more floors that are located above the fire, the more people are likely to need fire department assistance and the more fuel there is to burn.

High-rise buildings were once found exclusively in larger cities, but today they are commonly found in small and midsized communities as well. In most cases, high-rise buildings in these smaller communities are newer, lower in height, and protected by automatic sprinkler systems. Even if your department does not respond to a high-rise building at present, if urban sprawl continues as expected, it probably will in the future. If you have a mutual aid contract with a jurisdiction that contains high-rise buildings, you will likely be called upon to help in the event of a working high-rise fire. Several larger cities that once contained just a few notable high-rise structures have experienced an incredible growth in high-rise construction. Given the special problems that these buildings present, each department that could reasonably be expected to respond to a high-rise fire should have a high-rise SOP and train accordingly.

Fire Fighter Safety

The risk to fire fighters and occupants increases in proportion to the height of the building and the height of the fire above grade level. Once fire fighters are operating above the reach of aerial devices, the only viable means of egress is the interior stairs; the extra protection afforded by laddering the building is not possible.

Good tactics and fire fighter safety cannot be separated. The tactics that are explained throughout this chapter improve fire fighter safety. The safety considerations that are discussed in the *Fire Fighter Safety* chapter apply to high-rise firefighting, but additional measures need to be taken during a high-rise operation.

Occupants should not use elevators to escape a fire except under special circumstances and then only under fire department supervision. Modern high-rise building elevators have fire department controls. When the alarm system is activated, elevators return to the ground floor and remain there "locked out" for fire department use. An elevator key is required to unlock the elevators. Older buildings may have fire department controls but may not have the feature that automatically returns elevators to the ground level.

A responsibility of **lobby control** is to control, operate, and account for all elevators. Some fire service professionals say that elevators should never be used under fire conditions or suspected fire conditions until their safety can be verified from the fire floor. This may not be practical where fire companies respond to alarms in high-rises several times each day. Requiring fire fighters to ascend 20 flights of steps to check an odor of smoke is not a productive use of resources. A more reasonable approach is to develop procedures and conduct training to reinforce the safe use of elevators.

Fire Fighters' Use of Elevators

A critical variable in high-rise fire operations is the availability of reliable elevators. If fire fighters can safely use the elevators, fire-ground logistics are dramatically improved. When the fire is located many floors above ground level, there is a strong inclination to use the elevators. However, elevators often stall or act erratically under fire conditions. Fire fighters who are trapped in a stalled elevator become part of the problem, as other fire fighters are then needed to rescue the rescuers. Therefore, the department SOPs should address the safe use of elevators, including circumstances when it is unsafe for fire fighters to use them. These procedures should include alternative measures for getting needed equipment to the fire floor when elevators cannot be safely used.

Fire department service controls and other elevator safety features will vary by age and elevator manufacturer. Building preplans should include information and instructions for using the elevator system.

One of the dangers in using elevators is that the doors may open on the fire floor, exposing fire fighters on the elevator to smoke and heat before they are in position with hose lines to attack the fire. Once elevator doors open on the fire floor, sensing devices (e.g., electric eyes) may prevent the doors from closing, thus trapping the fire fighters. This is a potentially deadly mistake that should be avoided at all costs. If fire fighters are caught in this situation, they may be able to use an override switch to force the doors to close. If not, they may be forced to exit the elevator in an attempt to escape the fire floor via the stairway. Similarly, elevators can stall in the shaft at or above the fire floor, also trapping fire fighters in the elevator car. Fire fighters should never, under any circumstances, use an elevator if there is a chance the elevator will travel to or above the fire floor. This is another possibly fatal mistake.

Elevators are equipped with redundant safety systems to prevent them from falling. In many cases, an elevator car has to be nearly destroyed before it

will fall. The elevator shaft is a fire-protected enclosure, but a fire of sufficient intensity can invade the shaft, and smoke and toxic gases will most certainly enter elevators that are stalled above the fire floor. This could prove deadly for fire fighters, who should be fully equipped with personal protective clothing, including self-contained breathing apparatus (SCBA). Civilians without protective gear have even less chance of survival when trapped in an elevator on or above the fire floor. Maintenance and security people should never be taken into the elevator until it has been verified that the elevator is completely safe.

The first rule of elevator safety is to avoid the use of elevators unless they will substantially improve operations. Fires on lower floors do not warrant the use of an elevator unless someone on the fire floor can verify its safe use. The first-arriving companies should use the stairways for fires on lower levels. Some departments prohibit the use of elevators for fires on the seventh floor or below. Department SOPs and preincident plans must address this issue specifically. Once the safety of an elevator has been established, fire fighters can use it under the close supervision of lobby control. Two other factors to be considered are fire separations between the elevator bank and fire location and whether the elevator goes to or above the fire floor.

High-zone/low-zone (split-bank) elevators are not uncommon, and some split-bank elevators are divided into several zones. For example, a three-zone bank of elevators would have one bank serving lower floors only, another bank serving only the middle floors, and a third bank providing access to upper floors only. This is important information that should be included in the preincident plan. For example, for a confirmed fire on the 10th floor, it is fairly safe to use elevators that terminate at the 8th floor **Figure 12-1** .

When there is a fire separation between the elevators and the fire area, it may be possible to safely use the elevators in another building zone and then travel horizontally to the fire. This approach is much easier than ascending the stairs. Large high-rise

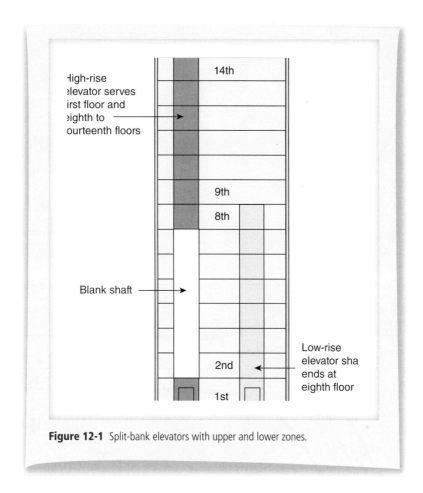

High-rise elevator serves first floor and eighth to fourteenth floors

Blank shaft

Low-rise elevator sha ends at eighth floor

14th

9th

8th

2nd

1st

Figure 12-1 Split-bank elevators with upper and lower zones.

buildings, especially hospitals, often have building zones. Fire walls and fire doors create a horizontal barrier to smoke and heat from the fire. Hospital patients are usually moved horizontally to another zone, rather than taken down to the ground floor or evacuated from the building. If another building zone is safe for the patients, it should certainly be safe for fire fighters.

Other occupancies also use horizontal separation as a fire protection strategy. The NFPA report on the 1981 Las Vegas, Nevada, Hilton fire[6] includes a plan view of a floor divided into three horizontal zones.

These zones provide a possible area of safe refuge for guests and a staging area for fire fighters. This hotel is actually three separate buildings built at three different times **Figure 12-2** .

When separate low-zone/high-zone elevators are unavailable or the building is not adequately zoned horizontally and the fire is reported to be on upper floors of the building, extreme caution should be used if the incident commander (IC) decides to use elevators to transport fire fighters.

The following list describes several rules that should be observed in taking the calculated risk of

Figure 12-2 Drawing of Las Vegas Hilton floor plan.

using an elevator when responding to a reported fire situation, including alarms from an alarm system. These rules for elevator safety should be considered when writing department high-rise SOPs and rigidly enforced during fire alarm response.

1. *Do not use an elevator of questionable safety for a fire on a lower level.* Elevators should not be used for fires on lower floors in the building or if there is any doubt about the safety of the elevator.

2. *Never take an elevator directly to the fire floor or above.* This is the cardinal rule of elevator safety and must be rigidly enforced. Department SOPs should state that the elevator should be taken two floors or more below the fire, and then fire fighters should walk up the stairways to the fire floor. This rule also applies to split-bank elevators that do not travel to the fire floor. If split-bank elevators are available, use an elevator that does not travel to or above the fire floor. Again, exit the elevator at least two floors below the fire and use the stairway to reach the fire floor.

3. *Place the elevators under independent (fire department) control.* Keys should be made available to the fire department so that elevators can be placed under independent control. Newer elevators cannot be operated without a key once the fire alarm has sounded. This independent control greatly increases fire fighter and civilian safety, as the elevator will not be responding to calls from occupants.

4. *Control all elevator cars in multiple hoistways.* Doing so not only provides the fire department access to upper floors, but also prevents erratic response to other calls within the building. Controlling elevators is a responsibility of lobby control.

5. *Never overcrowd elevators.* This principle is doubly important for fire fighters, not only because it is unsafe to exceed the weight limit on an elevator, but also because fire fighters may need space to don SCBA or to use tools to force their way out of the elevator.

6. *Wear personal protective clothing, including SCBA, and bring forcible entry tools.* Air supply should be conserved by not donning the SCBA face piece until approaching the fire area or when conditions indicate there is a need to be on air. *Never forget that using an elevator is taking a calculated risk.* Even when every reasonable precaution is taken, what lies at the top of the ascent is unknown. Forcible entry tools must be available to escape the elevator if necessary. Understanding elevator door operations can be extremely valuable if personnel become trapped in an elevator that stalls. In trying to escape a stalled elevator, always activate the emergency stop switch. Opening an emergency escape door or ceiling panel should also stop the elevator, as the interlocks are designed to prevent accidental movement when people are trying to escape.

7. *Send equipment rather than fire fighters on elevators.* Many times elevators that are considered unsafe for fire fighters can be used to transport tools and equipment to the interior staging area (usually two or more floors below the fire floor). Fire fighters can then safely ascend the stairs without the burden of heavy tools and equipment. It is not always possible to send an unstaffed elevator to a desired floor. When the fire alarm is activated, it may be necessary to mount the elevator and hold a floor button or elevator key before the elevator will move. In this case, equipment cannot be sent above without a fire fighter. Freight elevators are usually less safe than passenger elevators, but they may be suitable for sending equipment without placing fire fighters in the elevator car.

When ascending in an elevator, it is good practice to stop periodically, possibly every five floors, to check for smoke. This is accomplished by opening the top elevator escape panel, when the elevator is so equipped, and using a flashlight to see whether smoke is present in the shaft. This is also a test of the elevator's ability to stop on demand. It is further recommended that fire fighters stop the elevator three or more floors below the fire floor, step out into the hallway, and check the general floor arrangement to get a feel for the building layout. By examining an uninvolved floor, fire fighters can become familiar with landmarks and can quickly identify the location of secondary exits, floor configurations, and potential ventilation locations. The first floor, mezzanine floors, and second floors are generally not typical floor layouts; therefore, it is better to check an upper floor that will likely be similar to the fire floor. Many high-rise office buildings have floor layouts that resemble a maze, making it very easy for a fire fighter to become disoriented and lost. Therefore, fire fighters should constantly ask themselves, "Where am I in relation to the stairways?" Crews working away from a hose line should use guide ropes.

Rule of Engagement for Fire Fighter Survival

Size up your tactical area of operation.

Reproduced from: © 2011, IAFC Safety, Health and Survival Section. All rights reserved.

Texture: © Eky Studio /Shutterstock, Inc; Steel: © Sharpshot/Dreamstime.com

The fire officer must always conduct a careful risk-versus-benefit analysis before placing fire fighters in an elevator. Furthermore, there is seldom justification for placing civilians in elevators before the fire is completely extinguished and the smoke is ventilated from the building. SOPs regarding the use of elevators should be kept current, and all fire fighters should be thoroughly familiar with them. These procedures should consider the height of buildings in your jurisdiction, automatic fire suppression equipment, type of occupancy, and the safety considerations outlined in this text.

Code officials in some countries are now including elevators in the civilian evacuation plan. In the United States, the use of elevators by the occupants of high-rise buildings is being examined by NIST and others. At the World Trade Center attack on September 11, 2001, many building occupants successfully escaped using the elevators, although elevators were not part of the evacuation plan. It is estimated that 3000 people safely evacuated the South Tower using the elevators in the 16 minutes between attacks. Of particular concern is the evacuation of handicapped occupants. To allow elevators to be used for evacuation, it is recommended that fire-resistive construction be improved in elevator shafts and lobbies.

Ground Support

Moving equipment via a stairway relay function was formerly named "stairway support" but is now properly termed "ground support." The easiest and quickest way to move cylinders, nozzles, hose, first aid supplies, forcible entry tools, and other materials to the interior staging area in a high-rise building is using the elevator, but what if the elevator is not safe to use? As previously discussed, many times elevators are out of service or unsafe to use. Moving supplies and staff up 10, 20, 30, or more stories is an arduous task. If it is not properly managed, no one will reach the fire floor with the physical stamina necessary to fight the fire. Imagine returning to the apparatus for a fresh air tank from the top floor of the 1454-ft (443-m), 110-story Willis Tower (formerly the Sears Tower) in Chicago. When elevators cannot be used, gaining access to the upper floors of a building will take significant time. Getting fire fighters and equipment to an interior staging area will reduce the time required to support a high-rise fire operation.

Ground support is a high-rise support unit used to move supplies to the interior staging area when using elevators is deemed to be unsafe. Fire fighters assigned to ground support ascend two stories with air cylinders and other equipment, where the next fire fighter picks up the equipment and relays it two additional floors. After moving the equipment up two stories, fire fighters descend two stories empty handed, providing a rest period. During extended operations involving many companies in rescue and suppression activities, it may be necessary to place fire fighters on every floor or possibly recycle air cylinders to grade level for refill.

NOTE

Interior staging is set up in a safe area two or more floors below the fire floor when it is impractical for fire fighters to go outside at ground level to change SCBA cylinders and for rest and recuperation (rehab).

Texture: © Eky Studio /Shutterstock, Inc; Steel: © Sharpshot/Dreamstime.com

Moving equipment up through the building when the elevators are out of service is a mammoth undertaking; ground support provides a reasonable but slower alternative to using elevators. During the NIST High-Rise Fireground Field Experiments, it was found that moving the required equipment to the eighth floor staging area via ground support took 10 minutes longer than moving the same equipment using elevators. Obviously, setting up ground support and moving equipment to a higher floor would require more time and staffing. However, once each person assigned to ground support is in place, they would move up and down only two stories.

Depending on conditions in the stairway, it may be possible to allow members who are assigned to ground support to work without their SCBA and turnout gear. This will preserve their energy and allow them to transport more equipment in less time. For example, experience indicates that wearing rubber boots while ascending the stairs is particularly fatiguing. However, it is essential to always err on the side of caution. If there is any chance that members assigned to ground support will encounter smoke or other hazardous conditions, they should be required to wear appropriate protective equipment. Even when personal protective equipment is not being used, it should be immediately available. Ground support should be one of the first assignments given when the fire is on the upper floors in a building without elevator service. Ground support should be explained in the department's high-rise SOP.

Life Safety

As search and rescue teams proceed with a systematic search, they must provide status reports to their supervisor and mark areas that have been searched. A simple marking system involves placing a chalk mark "X" on doors to rooms that have been searched and/or indicating that the whole floor has been searched by marking hallway doors or walls opposite the elevator. Sidewalk chalk is preferred as it is easier to handle while wearing gloves. Door hangers or other marking devices are commercially available for this purpose as well. A method of indicating areas searched should be part of the department's high-rise SOP.

Many high-rise buildings lock the doors from the stairs to the hallway, further complicating search and rescue efforts. This was a major factor in the loss of life at the Cook County Administration Building fire in Chicago, Illinois, where occupants encountered fire on the 12th floor while attempting to escape. Fleeing occupants then attempted to reenter floors above the fire where doors leading to the hallway were locked on the stairway side. Forcible entry tools should be carried by rescue and extinguishment teams to force entry when necessary. When selecting the tools to carry, remember that power saws might not operate because of heavy smoke conditions. Further, using gasoline-powered equipment inside of a building can create other hazards if the area is not adequately ventilated.

Once doors have been opened, it is important to prevent them from relocking behind fire fighters entering the floor. Some departments use a piece of rubber with two holes that cover the doorknob and lock to prevent the door from relocking. For most door types, this method works very well. During preincident planning, examine the doors to determine which method would work best for the style(s) of doors in that building. Whatever method you decide to use, be careful not to leave doors propped open, as doing so will most certainly allow smoke to move from the stairs into the hallway or from the fire floor into the stairs. Open doors to hallways also reduce the effectiveness of stairway positive-pressure ventilation and can provide air to a ventilation-limited fire.

The primary search should also include a search of all elevators. Elevators should be brought down to the ground level and checked by lobby control. If elevators are stalled or otherwise located above the ground floor, they must be checked for victims. It is essential that all elevators be accounted for and checked for occupants.

Rescuing and Evacuating Occupants

Helicopter Rescues

There have been occasions when helicopters were successfully used to rescue occupants during high-rise fires. **Figure 12-3** shows a helicopter rescue operation in progress at the MGM Grand fire in Las Vegas. However, helicopter rescues are extremely dangerous and in most cases unnecessary. Few instances warrant the use of a helicopter in removing occupants from a roof, and many roofs make poor helicopter landing zones. A few fire departments have developed programs that use helicopters to place fire fighters on roofs, sometimes by having them rappel from the helicopter to the roof. Fire fighters placed on the roof can calm the occupants, control or limit the fire threat, and also descend to areas that are inaccessible from below.

These programs have merit but require constant training on the part of the rappel team and the helicopter crews. At last count, New York City had 32 buildings over 600 ft (183 m) in height. New York and other very large cities can no doubt justify the training and associated expenses related to such a program. However, the need or justification for such a program is doubtful for a department with only a few well-protected high-rise buildings within their response area. It is unlikely that members of such a department would have the expertise or equipment needed to safely operate from helicopters.

Helicopters can sometimes be used for reconnaissance. An aerial view can provide information that is not available from the interior or the exterior at ground level. At the World Trade Center on September 11, 2001, police in helicopters could see signs of an impending structural collapse. Unfortunately, they were unable to communicate this message to the fire department.

Helicopter operations above a burning building can actually create additional risks owing to the thermal updraft. Helicopters flying near a burning building can create high winds that negatively affect operations. Remember, just as with any tactic, before deciding to use helicopters for reconnaissance, for assisting fire fighters in gaining access, or for rescuing occupants, always conduct a risk-versus-benefit analysis. Is the risk to occupants, fire fighters, and the helicopter crew warranted? Are other less risky options available?

Figure 12-3 MGM Grand helicopter rescues.
© Master Sergeant Richard Diaz/AP Photos

IC Rule of Engagement for Fire Fighter Safety

Conduct an initial risk assessment and implement a safe action plan.

Reproduced from: © 2011, IAFC Safety, Health and Survival Section. All rights reserved.

Texture: © Eky Studio /Shutterstock, Inc; Steel: © Sharpshot/Dreamstime.com

Partial or Sequential Evacuation

A decision for partial or sequential evacuation can be made in advance of a fire and incorporated into the alarm system. Alternately, the IC may make this decision at the time of the fire. When the fire is small or smoke has entered the stairway, the IC may decide that it would be best to leave occupants in their rooms rather than attempting a complete evacuation. A defend-in-place tactic places a great burden on the IC, as a successful outcome depends on the fire being promptly extinguished. However, many times, occupants are placed in greater danger by attempting to evacuate a building through smoke, using resources that could have been assigned to extinguishment duties.

If the fire is not quickly controlled, the people who remain in the building will be in great danger because of the non-evacuation decision. Much of this decision-making process has to do with fixed fire protection features that are designed into the building. For example, there is less risk associated with the non-evacuation decision if the building is sprinkler protected.

Many departments permit the use of alarm systems that advise only the occupants in affected areas to evacuate the building. Rather than sounding a general alarm throughout the building, the evacuation alarm may sound only on the fire floor and on one floor above and one floor below. Occupants on the other floors may receive a prealarm notification but be advised to remain on the floor until further notification.

This evacuation tactic reduces stairway traffic but also leaves occupants inside the building. Use of a partial or sequential evacuation system requires a sophisticated fire alarm system and/or manual control. The IC should have preincident plan information about the operation of the system readily available at the command post.

Relying on building management or security to notify occupants of the need to evacuate is generally a mistake. Maintenance or security people have a propensity to investigate alarms first, thus delaying notification of the fire department. This has been the case in many deadly high-rise fires, such as the First Interstate Bank Building in Los Angeles, California, and One Meridian Plaza in Philadelphia, Pennsylvania. When a fire is discovered, many times in-house employees are not properly trained in how to initiate a safe, orderly evacuation. The answer lies in requiring direct and immediate fire department notification, regardless of the type of internal alarm system, fire protection, or method of evacuation.

Any time a decision is made to leave occupants inside a burning building, the IC is taking a calculated risk with the lives of the remaining occupants. Preincident plan information related to construction and fire protection systems along with accurate, timely, and continuous status reports will provide the IC with the information necessary to make the correct decision.

Emergency Voice/Alarm Communications System (EVACS)

Many high-rise buildings are equipped with an emergency voice/alarm communications system (EVACS). The EVACS uses recorded messages to notify occupants of a fire and provides specific directions for reaching places of safe refuge within the building. The EVACS sometimes directs people above the fire to higher floors within the structure, avoiding the need to pass the fire floor. People on the fire floor and the floor below are directed to areas two or three floors below the fire or to evacuate to the outside. When occupants are sent to other floors within the building, the occupants on the receiving floors are also notified of the evacuation. The EVACS notifies people on elevators that the elevator is responding to the ground floor, where they are to exit. By directing occupants away from the immediate fire area soon after the fire is detected, the EVACS helps to solve many high-rise evacuation problems.

Because a fire scenario presents many uncontrolled variables, caution is in order when using any automated system such as EVACS. Stairways that would be perfectly safe if the fire were to occur in one floor area might not be safe in another circumstance. All evacuation variables should be thought out in advance, and a manual backup system must be in place. The EVACS will generally have manual overrides that allow

Flames: © jag_cz/ShutterStock, Inc.; Paper: © silverjohn/ShutterStock, Inc.

Incident Summary

A New York City high-rise fire occurred in a 10-story apartment building. The fire started in an apartment on the 10th floor, and three fire fighters were killed during firefighting operations. Another fire occurred in an apartment in a 51-story high-rise building in Manhattan. Four civilians were killed in this fire. All of the fatalities were remote from the fire and occurred when the victims were overcome while trying to exit from upper floors.

Source: Duval, R., Solomon, R. Unpublished NFPA Fire Investigation Report. Quincy, MA: NFPA.

fire fighters or building management to direct the evacuation. Manual directions could include which occupants are to evacuate and the stairway(s) to use. Most of these systems can notify occupants on specific floors or all of the building's occupants. Sequential evacuation may be best to reduce stairway congestion. Stairways should be identified by marking both sides of the stairway door using an alphabetic or other easily understood designation.

In most cases, fire companies will be using one stairway to attack the fire. Evacuating occupants should be directed to other stairways. Assigning fire fighters to clear the attack stairwell from top to bottom assures that occupants will use the safest stairways for egress and will not hinder fire operations.

At the Cook County Administration Building fire in Chicago, Illinois, security unintentionally notified all building occupants to evacuate; the evacuation message was intended for the fire floor only. In this case, the fire department was criticized for not using the internal communications system to notify occupants to avoid the stairway where fire operations were taking place in favor of the stairway on the opposite side of the building. However, occupants often disregard verbal instructions, which was the case at the Cook County Administration Building fire, where many occupants ignored instructions to use the stairs and instead used the elevators. Another example can be found in occupants' response to the World Trade Center attack of 2001. Occupants had been trained to sequentially evacuate the building; yet, when aircraft struck the buildings, many

occupants self-evacuated and ignored the EVACS as well as their training. In this instance, the decision to ignore the system proved to be the best decision, as a sequential evacuation would have resulted in more people being in the buildings when they collapsed.

Reliance on the EVACS should be limited to compartmentalized buildings with full sprinkler system coverage. If the sprinkler system fails to control the fire and fire forces are unable to quickly extinguish or confine the fire, all occupants above the fire floor should be evacuated. Further, the IC and lobby control officer should have a thorough knowledge of the EVACS system.

Extinguishment

What makes high-rise firefighting different? In the lower portions of the building, the main difference is the exposure of many floors above the fire to vertical fire spread. Above the eighth floor, most exterior defensive fire control tools are no longer effective. Elevated streams can sometimes reach the fire floor, but the angle of deflection inside the window diminishes with each floor above the maximum height of the appliance. To improve fire stream penetration, it is necessary to move the nozzle farther from the building, thus reducing the angle of deflection. In most cases, the width of the street is the limiting factor in determining how far away the appliance can be placed from the building. If the width of the street is not a limiting factor, the effective reach of the nozzle will determine the maximum distance the nozzle can be placed from the building. At the Cook County Administration Building fire an elevated master stream was used to control a fire on the 12th floor; however, a delay occurred while the apparatus was repositioned.[7] In operating above the eighth floor, it is unlikely that ground-based appliances will deliver effective stream penetration to the fire area 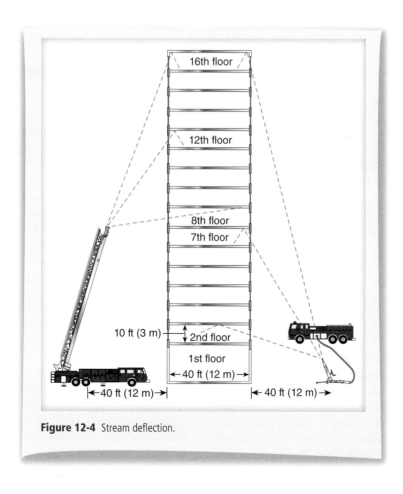 **Figure 12-4** . Occupants are also beyond the reach of platforms, aerials, and ground ladders; therefore, attempting an exterior rescue would be extremely dangerous for both fire fighters and occupants. In fact, the building's stairways are the best means of egress, even when floors are within the reach of aerials. For these reasons, it is extremely important that fire forces prevent fire extension into the stairways, because stairways are the safest and most effective egress routes.

Standpipe operations and equipment are discussed in the *Fire Protection Systems* chapter, including pressure- or flow-reducing valves. Generally, the pressure at the standpipe discharge will be lower than the pump discharge pressure used to supply hose streams connected directly to a pumper. Departments compensate for the lower pressure by using smooth-bore or low-pressure nozzles. Some departments have purchased hose that has less friction loss or use 2-in. (50-mm) or 2½-in. (65-mm) hose for standpipe operations. Any method of reducing the friction loss in the hose will increase the nozzle pressure.

Taller building standpipes could be equipped with pressure- or flow-reducing valves. If a building is not equipped with pressure-reducing valves, the pressure on the lower floors of the building will be

Figure 12-4 Stream deflection.

higher than the pressure on the top floors. It is also possible to increase the pressure in the standpipe system by pumping into the fire department connection. Pumping into the fire department connection may be an effective means of boosting the pressure and increasing the water supply. Very high buildings have complex water supply systems where pumping into the fire department connection may not increase pressure at the standpipe outlet. In these buildings, fire department pumpers alone cannot supply sufficient pressure for firefighting on upper floors, and the building's internal fire pumps must provide primary pressure. Although the ceiling height of individual floors will vary in high-rise buildings, a fair estimate is allowing 20 ft (6 m) for the first floor and 10 ft (3 m) for each additional floor.

The loss of pressure due to gravity (elevation) is 0.434 pounds per square inch (psi) per foot (3 kilopascals [kPa] per 0.305 m). The pressure loss for a hose line operating at the top floor of a 40-story building (39 floors above grade) with 10-ft (3-m) floors would be calculated as follows:

First floor	=	20 ft (6 m)
38 floors × 10 ft per floor	=	380 ft (116 m)
Total elevation	=	400 ft (122 m)

400 ft × 0.434 psi/ft = 174 psi pressure loss
(122 m × 3 kPa/0.305 m = 1200 kPa loss)

If the floor height is changed to 12 ft, then the pressure loss would be calculated as follows:

First floor	=	20 ft (6 m)
38 floors × 12 ft/floor	=	456 ft (139 m)
Total elevation	=	476 ft (145 m)

476 ft × 0.434 psi/ft = 207 psi pressure loss
(145 m × 3 kPa/0.305 m = 1426 kPa loss)

In addition to the pressure loss due to elevation, friction loss in the standpipe system and the hose line, as well as nozzle pressure, must be added. Friction loss will vary depending on the type of hose used, and nozzle pressure will vary depending on the type of nozzle. It is obvious that standard fire

department pumpers alone could not support hose streams operating on the upper floors of an ultra-high-rise building.

Pumping into the fire department standpipe connections can assist internal fire pumps and is a good practice. If provisions are not made to provide an external backup supply and the internal pumps fail, the only protection will be the fire-resistive nature of the structure (fire enclosures). In high-rise building fires, when the fire is above the reach of aerial ladders and the building is not protected by an automatic sprinkler system, the fire can be expected to burn out all floors above the fire unless the fire can be brought under control by an offensive attack. Contingency plans for supplying the standpipe and/or reservoirs in case of breakdown, developed in cooperation with building management, could prevent a disaster. These contingency plans should be included in the preincident plan.

Command Post Location

Fire department SOPs often list the lobby as the preferred command post location. Many times this is true, but on occasion the lobby is a poor choice for a command post. When fire or the products of combustion threaten the lobby, the command post should be located elsewhere. In the chapter titled *Organizing, Coordinating, and Commanding Emergency Incidents*, it was mentioned that the larger the incident, the farther away the command post should be. The idea is to isolate the IC from disruption so that the command process can be carried out efficiently. When large numbers of people are attempting to exit through the lobby and fire fighters are gathering there in an effort to ascend toward the fire, the lobby is a poor location for the IC. The September 11 World Trade Center fire provides an example of the lobby being a poor choice for the command post. At this incident, the buildings collapsed. However, even if the structures had not collapsed, the heavy traffic on the first floor of each building would have made the lobby a poor choice for the command post.

Most newer high-rise buildings are equipped with a command center, which is usually near the lobby. For most incidents, this is the best command post location, as these centers often provide good communications and the needed work space for command activities.

Once the command post has been established, its location should be communicated to all responding companies. Some department SOPs use a street name, such as Main Street Command, to notify companies of an exterior command post location. A command post located inside a building may be identified using the building's name, such as Rockefeller Command.

Developing Building-Specific High-Rise Preincident Plans

All high-rise buildings should be preincident planned. Knowing the location of interior command rooms is just one more of the many issues that should be addressed in the high-rise preincident plan. A good high-rise preincident plan will address all of the issues covered in a standard building preincident plan, including occupancy type. The chapter titled *The Role of Occupancy* describes the importance of occupancy types, and the chapter titled *Procedures, Preincident Planning, and Size-Up* lists occupancy type as one of the major subjects to be addressed in a preplan. Knowing the building's use is essential to preplanning. Most high-rise buildings are businesses, hotels, apartment buildings, or health care occupancies. Some high-rise buildings are mixed occupancies with assembly or mercantile occupancies on the first floor. In addition to the standard preincident planning factors, there are many special considerations in a high-rise, including the following:

- Elevator operations and elevator key location
- Access/egress issues such as stairway doors that automatically lock
- Standpipe operations, especially if field-adjustable pressure-reducing valves are present

- Floor layouts for each floor that is different or special (e.g., When tenants occupy several contiguous floors, there may be internal open stairs between floors.)
- Ventilation, such as the use of the heating, ventilation, and air-conditioning (HVAC) system, and window venting considerations, including the presence of windows that can be opened, tempered glass, or special windows (e.g., hurricane resistant)
- Procedures or operations that are unique to the building

High-rise buildings are generally made of fire-resistive construction, with the older high-rise buildings often being superior to the new in many construction features Figure 12-5 . Older tower buildings have better compartmentation, more fire-resistive components, and generally better exit facilities.

Many newer buildings have pressurized stairways, and they are often fully sprinkler protected. However, there was a period when high-rise buildings were constructed using the modern, lightweight construction methods but were not protected with automatic sprinklers. These new-style, nonsprinkler-protected high-rise buildings are likely to be most problematic. Many high-rise buildings are not protected by sprinkler systems. NFPA statistics quoted in the NIST High-Rise report indicate that 41% of high-rise office buildings, 25% of high-rise healthcare facilities, 45% of high-rise hotels, and 54% of high-rise apartment buildings are not protected by sprinkler systems.

Since the mid-1970s, codes used in the United States, with one notable exception, required high-rise buildings to be sprinkler protected. In New York City, until a change was made in March 1999, it was possible to construct a residential high-rise building without a sprinkler system. Two tragic fires led to a change in the building codes, and now all new high-rise buildings built in New York are required to have a sprinkler system installed. Also, within the past few years, many major cities such as New York and Chicago have implemented high-rise sprinkler retrofit programs for some existing high-rise buildings. For instance, the City of Chicago adopted an

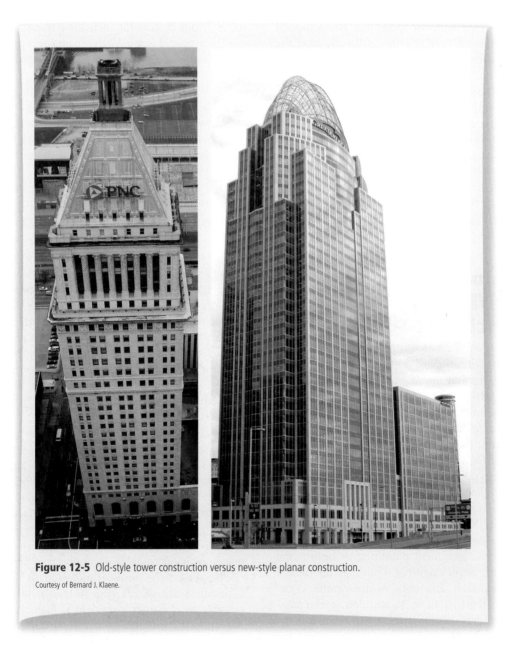

Figure 12-5 Old-style tower construction versus new-style planar construction.

Courtesy of Bernard J. Klaene.

ordinance that required high-rise buildings above 80 ft (24 m) to be retrofitted with sprinklers. However, if residential properties and certain other high-rise buildings that are classified as historical pass a locally developed life safety evaluation, they may opt out of the retrofit requirement.

In many new-style high-rise buildings, where core construction methods are used, fire attack and evacuation tactics are further complicated. In these buildings, stairways are usually located in the center of the building **Figure 12-6** . Therefore, if occupants are trapped or if the fire occurs near the core

Figure 12-6 Central core layout.

or between the occupants and the core, evacuation and fire attack positions are limited. Preincident plan drawings will prove to be extremely valuable to the IC in developing evacuation and fire attack plans in these buildings. Some core construction methods move the core away from the center of the building, which is sometimes referred to as side core construction **Figure 12-7**.

Interior building configurations can be radically different in buildings that appear to be the same from the exterior. As an example, in Figure 12-6, there is an open floor layout with individual offices separated by cubicles and standpipes in the stairway. In Figure 12-7, individual offices are separated by full floor-to-ceiling walls with the standpipe located in the core area. These seemingly subtle differences can have a major impact on firefighting and rescue operations.

The stairway at the periphery of the side core layout (Figure 12-7) is a **smoke-proof tower**.

Figure 12-7 Side core layout.

Smoke-proof towers are built as a separate structure, thus reducing the possibility of smoke entering the stairs. Also, the stairs could be pressurized to reduce smoke infiltration. Unless there is a compelling reason to use another stairway, the smoke-proof tower or the pressurized stairway should be the stairway of choice for occupant evacuation. This should be so noted in the preincident plan.

Analyzing the Situation Through Size-Up

All of the factors discussed in the *Procedures, Preincident Planning, and Size-Up* chapter regarding size-up apply to high-rise fires. However, size-up at the scene of a high-rise fire will often be particularly difficult, as visual information may be nonexistent or misleading. Merely finding the fire can be a perplexing chore. A large structure can contain a substantial fire without displaying any external signs. If flame and smoke are visible, it is both a good sign and a bad sign. The good news is that the fire has self-vented, and the IC has some indication of the location and intensity of the fire. The bad news is that the fire has probably gained considerable headway before the arrival of fire fighters and has the potential to be a wind-driven fire.

Observing a high-rise fire from the exterior could result in confusion in directing interior assignments. High-rise buildings often do not have a floor numbered 13, or the mezzanine may not be counted as a floor. There may also be half-floor equipment rooms within the building. A heavy volume of fire that originated at a central core may appear at a single window, while a fire that originates near the periphery may cause several windows to break. Visible evidence gained from the exterior, many floors below the fire, may be unreliable. Occupants may report smoke or odors, but these reports typically come from occupants who are many floors above the fire. Information may be available from building occupants regarding evacuation status, but this information is seldom completely accurate in a large building. Even internal alarms are sometimes

misleading, as smoke detectors sense smoke above the fire floor or in other areas where the HVAC is depositing smoke. For these reasons, even with fire and smoke showing outside the building, finding the seat of the fire may require assigning several fire companies to different areas of the building.

Matching the occupancy classification to time factors generally gives the IC a good idea of the life hazard potential. There is a tremendous life safety difference in a high-rise office building during working hours compared to the same building during nonbusiness hours. There may be people in the building during nonbusiness hours, but the occupant load during normal working hours will be much greater. How many people would you expect to find in an office building during normal working hours? As a rule of thumb, the average office building[8] has one person for every 100 ft^2 (9.3 m^2). The preincident plan would most certainly aid in determining the life safety problem.

The validity of dispatch information varies depending on the alarm type. A report of a smoke detector alarm on the 30th floor could be for a fire several floors below or for a localized problem on the 30th floor. A report of an odor of smoke from a building occupant may be no more reliable than the smoke alarm in determining the location and nature of the problem. However, a report of visible smoke or flames on the 30th floor in Suite 3012 provides the IC a wealth of information.

It is not unusual for the dispatch center to receive follow-up calls from the public during a high-rise fire. Many of these calls repeat and verify information already received. In many fires, building occupants called dispatch after the initial alarm to report people trapped, request evacuation instructions or to report fire conditions in specific areas of the building. This information is critical and should be immediately relayed to the IC. The IC should notify the operations chief and/or divisions, groups, or companies working in these areas or deploy companies to evaluate the situation. Communications is often a problem; in some past cases, the IC had vital information but was unable to relay it to units working within the building.

In high-rise buildings, structural stability will typically be very good. The First Interstate Bank Building and Meridian Plaza fires were both in new-style high-rise buildings, and both sustained large volumes of fire over several floors for an extended period of time. While there was structural damage, especially at the Meridian Plaza fire, neither building collapsed. Further, the old-style high-rise building can be expected to endure even more fire exposure than the modern, lightweight construction. The Empire State Building serves as a good example of the structural stability of the older, tower-type high-rise buildings. On July 28, 1945, a bomber crashed into the Empire State Building. The aircraft's burning fuel cascaded down and through the building. The building, which sustained both the crash impact and the resulting fire, still stands as one of America's famous landmarks.

Fire resistance in older high-rise buildings was achieved through massive structural members encased in concrete. The airplane that struck the World Trade Center was much larger, carried more fuel, and was traveling at a higher rate of speed upon impact compared to the bomber that crashed into the Empire State building. Therefore, it is difficult to compare the two incidents. However, one of the major factors leading to collapse at the World Trade Center was aircraft debris removing sprayed-on fire-resistive coatings from steel structural members. This sprayed-on insulation was not used in the construction of the older tower-type high-rise buildings, although it is sometimes used when older buildings are renovated. There is a high likelihood that the concrete covered structural members would have survived the barrage of aircraft debris.

Evaluating resource needs is extremely difficult in a high-rise building. However, the staffing and resources necessary to combat a large fire are significant, as you will see in the analysis of high-rise case studies later in this chapter.

The high-rise decision tree shown in **Figure 12-8** considers a few major factors that could lead to the call for additional alarms at the scene of a high-rise fire. This decision tree by no means provides a complete size-up of the potential challenges that a high-rise fire presents. However, it does offer a thought process to assist in determining resource requirements. An examination of the decision tree demonstrates that a large fire, located beyond the reach of ground equipment and with elevators unavailable, will require extensive resources.

Some department SOPs require that the first-arriving company request an additional alarm for any working fire in a high-rise building. There is justification for this precaution, given the life hazard, extinguishment, and property conservation potential. In most instances, additional personnel will be needed to augment the initial attack and to provide logistical support. The actual staffing requirements will vary depending on the circumstances, as demonstrated by the scenarios in this chapter.

Smoke Movement

Weather conditions, together with fire intensity, can have a significant effect on smoke movement within a high-rise structure. While the IC should consider these factors, a degree of unpredictability should be expected regarding the effects of weather and fire intensity on smoke movement.

Heat of the Fire

The intensity and size of the fire will determine the extent to which combustion gases are heated and how high they will rise inside the building. In lower structures there is generally enough heat energy to cause the heated fire gases to rise to the highest level in the structure. In high-rise buildings, the smoke and toxic gases will tend to rise until they reach temperature equilibrium, at which point they will stratify. It is not unusual to have heavy smoke on a midlevel floor and smoke-free floors above. This stratification can endanger occupants who enter a smoke-free stairway and then discover smoke several stories below. Many times, doors leading back into a floor area are locked, forcing the fleeing occupant to seek refuge in the stairway or proceed through the smoke.

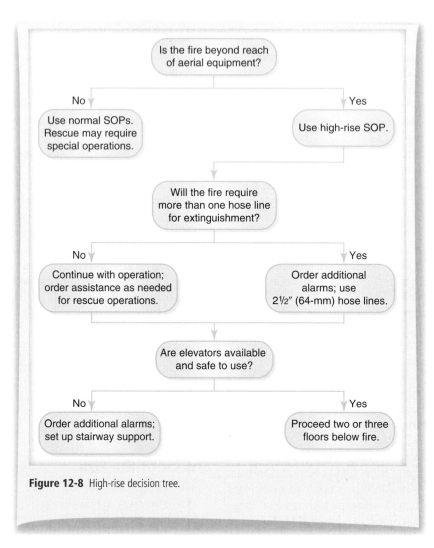

Figure 12-8 High-rise decision tree.

Stack Effect

The airtightness of the structure has much to do with **stack effect**—the vertical airflow caused by temperature differences within and outside the building. The unpredictable behavior of smoke within a high-rise is due, in large part, to stack effect. Many fire departments have been surprised when they tried to ventilate a high-rise building using a technique that was previously effective, in the same or a similar building, only to get an entirely different outcome. In some buildings the stack effect is so great that it interferes with the proper operation of the HVAC system. The colder it is outside and the warmer it is inside, the greater the positive stack effect (upward movement). Conversely, the stack effect can be negative (downward) on a warm day within an air-conditioned building. The heat of the fire and stack effect are interdependent. The chances of smoke stratification are less on a cold day than on a warm day. There are formulas for calculating stack effect, but they have little application for field use.

Incident Summary

In several fires, the smoke migrated to locations that were remote from the fire. At high-rise fires studied by the NFPA, victims have been killed in locations that were a considerable distance from where the fire originally started. In a hotel fire in Pattaya, Thailand, the fire started on the second floor, yet much of the smoke damage and a number of the victims were found on the top floors. Similar situations occurred at the MGM Grand fire, the Las Vegas Hilton fire, and a high-rise fire in 1998 at 124 West 60th Street in New York City that killed four people.

Data from: Duval, R., Solomon, R. Unpublished NFPA Fire Investigation Report. Quincy, MA: NFPA.

Smoke from the fire that swept through the 36-story President Hotel in downtown Bangkok on February 23, 1997.

© Sakchai Lalit/AP Photos.

Figure 12-9 Neutral pressure plane.

Wind

There is a point within a high-rise structure of sufficient height called the neutral pressure plane (NPP). Below the NPP, air is moving into the building; at the NPP, forces are neutral (air is not moving in or out); above the NPP, air is moving out of the building **Figure 12-9**.

The NPP is affected by heat from the fire and the stack effect. Wind also plays a major role **Figure 12-10**.[9]

Ground-level winds are not always a good indication of wind direction and speed at higher elevations. Downtown areas of large cities containing large numbers of high-rise buildings are like giant canyons. Wind entering the high-rise canyon is redirected and becomes very turbulent. These wind gusts also prevail high above the ground but possibly in another direction and/or at a higher velocity.

Wind passing over a roof opening has a pulling or venturi effect. In addition, the wind will push smoke and fire back into the building on the side facing the wind and help to vent smoke out of the building on the opposite side of the building **Figure 12-11**.

In reality, predicting the wind factor on the fire ground is more an art than a science. Wind direction and velocity can change dramatically and without warning, even when atmospheric conditions are not changing significantly.

Figure 12-10 Effect of wind on the neutral pressure plane.

Wind-Driven Fires

Much research has been done regarding wind-driven fires in high-rise buildings. Several possible tactics have been introduced as a result of this research:

- Controlling windows and doors
- Applying positive-pressure ventilation
- Deploying wind control devices
- Using exterior fire streams

Of these, controlling windows and doors is the most practical.

Controlling Windows and Doors

Even under high-velocity wind conditions, it is possible to protect stairs and hallways by maintaining barriers between pathways to hallways and stairways. The usual pathway to the stairway is through a window opening, and subsequently through one or more doorways leading to the stairs. Any tactic that interferes with this pathway can be effective in retarding fire extension and minimizing the effects of high wind.

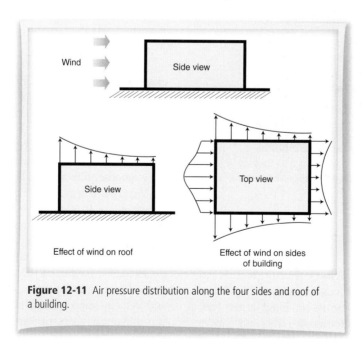

Figure 12-11 Air pressure distribution along the four sides and roof of a building.

Given the unpredictability of the wind and internal conditions such as stack effect, breaking windows to vent the fire is highly discouraged. Nonetheless, the heat of the fire often causes windows to fail, allowing wind to enter. In a residential building or health care occupancy there will likely be a fire door leading to a relatively small room or apartment. This door provides a fire barrier that can retard fire growth and delay extension into other areas. Fire fighters must be cognizant of the fact that a virtual "blow torch" fire could erupt when a door is opened to the fire area under high-wind conditions. A properly staffed charged hose line must be in place before opening doors leading to an active fire area and the door must be controlled. In the case of residential and health care occupancies, there will usually be at least one additional fire door in the hallway before reaching the stairway. Multiple doors provide barriers to fire extension into the stairway. In these buildings it may be possible to seek temporary safe refuge in another room if the fire breaks out of containment, although the stairway is a much better place of refuge if it can be safely reached.

Large undivided office spaces or common areas in other occupancies may have a single fire door between the fire area and the stairway. These areas are particularly hazardous under wind-driven fire conditions, as the wind could well push the fire directly into the stairway when the door to the fire floor is opened and there may not be an enclosed room in which to seek safe refuge. Controlling the door leading to the fire area in a high-rise fire is critical.

Applying Positive-Pressure Ventilation

Positive-pressure ventilation (PPV) was introduced to the fire service in the 1970s and is an effective tool when properly implemented. Like all ventilation tactics, improper venting can have a detrimental effect. Departments that protect high-rise buildings had many questions regarding the effectiveness of PPV in high-rise buildings, particularly on upper floors of taller buildings. A study completed by NIST in 2007[10] compared various PPV equipment and techniques, providing answers to many questions regarding PPV

tactics at high-rise fires. Effective PPV can clear stairways during a high-rise fire in two ways:

- Push smoke out of the building either through windows or vertical vent openings
- Pressurize the stairway to keep smoke from entering the stairway

Both methods rely on creating a positive pressure in the stairway. These methods can be very effective; however, there are several challenges. Stairways that terminate in a large open space such as a first floor lobby are more difficult to pressurize due to the large open air space. Open areas on the first floor are subject to depressurization when doors are opened. Sometimes larger, more powerful blowers or blowers in tandem can overcome this problem. It is far more efficient to pressurize a stairway leading directly to the outside than it is to pressurize a stairway terminating in a large lobby area.

When pressurizing a stairway, open doors leading from hallways to the stairway can result in a loss of stairway pressure. During the NIST PPV experiments, pressure loss was inconsequential when doors were opened just enough to advance a fire hose from the stairway into the hallway. However, when doors were opened wide, as they may be by escaping occupants or during firefighting operations, the drop in stairway pressure was significant. Placing additional blowers operating into the stairway on upper floors below the fire floor can increase stairway pressure and thus increase PPV effectiveness.

High-wind conditions can have an adverse effect on PPV methods, particularly if the wind is blowing into the building through an opening such as a window. This leads to a discussion of other wind-driven fire tactics.

Deploying Wind Control Devices

Realizing the probability of a wind-driven fire and the fact that wind-driven fire conditions are exacerbated during a high-rise fire, experiments were conducted to evaluate devices designed to minimize the effect of high winds entering a window opening.[11] At present, wind control devices are basically a weighted mat that is placed over a window to keep the wind from entering the building through an existing opening. These devices are deployed from an area above and in line with the fire floor. NIST evaluations determined wind control equipment to be effective in partially blocking the wind entering a window.

Currently, these devices seem to have very limited application and are not in widespread use. Questions need to be answered regarding how to best deploy the wind control device. For example, which window is best to use when placing the device in service, and how should a sealed window be opened? If the wind is blowing directly at one side of the building, then obviously placing a wind control device on that side of the building would be correct. However, winds high above grade level tend to be turbulent and may be striking the building at an angle, making it more difficult to cover window(s) below.

Using Exterior Fire Streams

Experience in using master streams operated from the exterior of a building in a defensive mode indicates that this is a less than ideal way to combat a fire and should be used only when the fire area cannot be safely entered to conduct an offensive attack. However, pulsing and softening the target tactics are used to knock down heavy volumes of fire before entering the immediate fire area. With this in mind, there are situations in which a limited exterior attack may be warranted. On lower floors of a high-rise building, solid or straight streams operated from an aerial device can be used for this purpose. On floors beyond the limitations of aerial devices, a nozzle with extension pipe could be operated into the fire floor from the floor below. At present, one nozzle manufacturer is marketing a special nozzle designed to be operated from the floor below the fire floor into a window on the fire floor. This nozzle was found to reduce heat in the compartment of origin during the NIST wind-driven fire research.

Questions arise as to how to position nozzles to apply an effective stream on the floor above. In many cases, high-rise buildings have windows that are not meant to be opened or removed. Breaking windows is a questionable tactic. Determining the best window to use from the floor below the fire is difficult, as is the accurate application of the fire stream. Actual field experience using exterior streams operated from the floor below the fire is very limited.

Developing and Implementing an Incident Action Plan

Incident priorities (life safety, extinguishment, and property conservation) remain the same, regardless of the type of structure. Each type of building or occupancy presents a special set of hazards to the occupants and the property. When people and property occupy buildings that are 110 stories high, the risk to life safety and property is obvious.

The primary rescue tactic in a high-rise building is a well-placed offensive attack. In a high-rise, more than in any other building, quickly confining and extinguishing the fire are critical, as ventilation and evacuation options are limited. Once the fire is extinguished, the toxic products of combustion are no longer being produced and the operation becomes more manageable.

How much water flow will be needed to extinguish a well-involved fire in a large high-rise building? Compartmentation plays a major role in fire flow calculations. Some office occupancies are divided by substantial walls; others have glass panels or portable dividers separating work areas. Further, there are often large, undivided spaces within high-rise buildings. These undivided areas create a high hazard requiring large flows. In the case of the MGM Grand Hotel fire, which is described later in this chapter, the required rate of flow was beyond a reasonable flow capacity, and the size of the area made it impossible for fire streams to reach the opposite interior walls from the access points.

There have been cases in which master stream appliances were brought into upper stories of a high-rise building and operated on the interior, but few standpipes could flow a sufficient volume of water to support master streams. To compensate,

5-in. (125-mm) hose was used in the building at One Meridian Plaza. However, even this tactic was unsuccessful in controlling or suppressing the fire.

Using master streams within a building would be an unusual tactic, but this could be the only realistic way to control a large-area fire on the upper levels of a high-rise building. Methods have been developed for advancing large-diameter hose up interior stairways for use as a portable standpipe, but the time and staffing required to advance the hose to an upper story present significant challenges and become unrealistic at higher elevations. Master stream appliances could be supplied by fire department pumpers located at street level, provided that the elevation pressure, friction loss, and required nozzle pressure do not exceed safe operating pressures. If master streams are operated on or into the interior, the structural integrity of the building must be closely monitored. Water from master streams could add several hundred thousand pounds of weight within the structure in a relatively short time. It is also easy to imagine the damage that internal master streams would do to property within the structure.

Until ventilation is accomplished, the floor area of a fire-resistive building can be very dangerous for fire fighters operating hose lines. Like all other energy and matter, the fire will follow the path of least resistance. If the floor is not properly ventilated, fire, smoke, and heat will follow pathways through hallways or concealed spaces. This can be extremely dangerous in a closed high-rise building, as the fire can "wrap around," moving through these spaces and getting between fire fighters and their exit route. The floor should be ventilated as soon as possible; at the same time, fire fighters should be aware of the special hazards associated with breaking windows on the upper floors of a high-rise building. Backup lines should always be in place to protect exit routes.

Company Operations

In attacking a fire in a high-rise building, as in any structure fire, engine company and truck company operations must be coordinated. Truck company members will be needed to force entry to floors and rooms. On reaching the fire floor, a thermal imaging camera should be used to check heat levels in the ceiling area above a suspended ceiling. If there is any doubt regarding fire entry into this false space, remove a ceiling tile under the protection of a hose line to control fire that might intensify as air is introduced to this space. Fire fighters should continue to monitor this space to ensure that the fire is not getting behind the fire attack crew. Suspended ceilings are common concealed spaces and a likely place for a "wrap-around" fire. This area sometimes contains cables and other equipment that can drop down, creating an entanglement hazard.

The attack line is used as the lifeline to safety. Members should stay within range of this protective line, not only for fire suppression purposes but also as a means of finding the stairway in heavy smoke conditions. It is also good practice to place a fire fighter at the stairway opening to the floor. This fire fighter will be needed to help extend the hose and control the door and can direct fire fighters to the exit if necessary. Department SOPs should require the use of lifelines (ropes) in situations where fire fighters are working away from hose lines.

Use of Stairways

The occupants of an office building are generally mobile and able to escape on their own, provided that stairways are available for their use. When fire fighters are trying to advance up stairways while occupants are attempting to evacuate, the result is gridlock. What can the fire department or occupants do to alleviate the problem? Many high-rise building managers and fire officials recognize that it is not always the best policy to have all of the building's occupants in the stairways at the same time. The people on the fire floor and the floor immediately above are in the greatest danger and should be among the first groups evacuated. When the stairway is filled with people from other floors, those in greatest danger cannot safely evacuate the building.

Mass evacuations also complicate fire operations, as fire fighters are forced to "swim upstream" in a stairway full of evacuees moving downward. Furthermore, when there is a mass evacuation, occupants above the fire must pass by the fire floor to reach safety. Fire fighters tend to exacerbate this problem by blocking doors open as they progress with fire lines, thereby venting the products of combustion into the stairway. These were contributing factors to the loss of life at the Cook County Administration Building fire in Chicago, Illinois. There is no easy or, for that matter, best way to deal with this dilemma. Several ideas are presented here that may or may not work in buildings protected by your department.

First, search efforts must be systematic and include a complete primary search of the fire floor and floors above the fire. In searching above the fire, it is important to check all stairways. One way to accomplish this is to enter the floor using one stairway and exit the floor using another stairway. However, the evacuation stairway must not be contaminated with smoke from the fire floor or other floors where smoke has infiltrated the hallways. It is best to assign a crew to check and clear the attack stairway of building occupants. Some important questions that need to be answered include:

- Are people in the stairway?
- Are any stairways filled with smoke? Smoke may stratify; therefore, do not assume that the entire stairway is clear simply because it is clear on an upper floor.
- Are doors from the stairway to each floor level locked?

Forcible entry is often needed, so the search and rescue team must carry proper tools. Forcing doors not only will result in damage but also will physically exhaust the team. Master keys, supplied by building management, can be invaluable in gaining access to rooms. It is necessary to check individual rooms to ensure that the occupants on any floor that is endangered by the fire or smoke have escaped.

As previously mentioned, another method of facilitating both evacuation and extinguishment is to designate separate stairways for occupant and fire department use. Fire companies must have control of the evacuation and good communications with occupants to successfully use this tactic. An internal public address system or EVACS can be used to direct occupants to evacuation stairways and away from the fire operations stairway. Even though occupants are directed to use stairways other than the fire operations stairs, the fire operation stairway must also be checked, as people do not always follow instructions. As was noted earlier, smoke-proof towers or pressurized stairs may be the best option for occupants. Controlling the stairs and dedicating one stairway for firefighting also reduces the possibility of opposing lines on the fire floor, as it is easier to control the entire attack if it is being made from one entry point. A stairway with standpipe outlets is the preferred fire operations stairway.

Ventilation

Ventilation and control of the HVAC system should be done by using reversible methods. Breaking windows high above grade level creates a serious hazard below in addition to the possible wind-driven effect mentioned previously. Plate glass windows tend to break into large, irregular pieces of glass, known as shards. These shards of glass create a very dangerous situation as they drop to the ground below. Tempered glass breaks into very small pellets and is less hazardous to people below. Even if windows are pulled into the building or are made of tempered glass, opening a window is much preferred to breaking a window. If a window is opened and the effect is negative, it can be closed. Unfortunately, most high-rise buildings have sealed windows, which cannot be opened. Sometimes fire codes require that a percentage of the windows be operable or made of tempered glass so that manual ventilation is possible. Codes that require these vent windows generally require that windows designated for ventilation be marked as such. The location of tempered glass or operable windows should be noted on the preincident plan.

The operation of the HVAC system is usually reversible. If operating the HVAC system does not have the desired effect or spreads the smoke, shut it down.

Interior Exposures

All floors above the fire are exposures. Companies need to check above the fire floor for extension and be equipped to fight the fire. In modern buildings with curtain walls, fire can extend upward, inside the building, near the exterior wall. Stairways and elevator shafts are vertical openings through which fire can spread. Additionally, pipe chases and utilities penetrate floors, creating openings through which fire can spread.

Fires can extend up the exterior of the building via auto-exposure by "lapping" from floor to floor. While upward extension is the main concern, the fire can spread downward through melting expansion joints or burning materials dropping below via the HVAC or by other means. Therefore, areas below the fire must also be checked. The search and rescue team can check floors above the fire while fire fighters who are assigned to property conservation can check floors below the fire.

Property Conservation

As previously stated, the first priorities are life safety and extinguishment. However, property conservation should also be considered early in the incident. High-rise office buildings typically contain valuable contents, such as computers and other office equipment and important files and documents. High-rise residential buildings will contain personal items that cannot be replaced. Often, protecting these valuables is overlooked, owing to the labor-intensive nature of rescue and extinguishment challenges in a high-rise building fire.

When considering property conservation, the height of the building comes into play in an opposite way. In life safety and extinguishment, people and property on and above the fire floor are normally considered to be most at risk, because fire, smoke, and heat will first affect these floors. However, the greatest property conservation exposure is often downward as water flows through curtain walls, electrical fixtures, and other openings, damaging valuable property beneath the fire. In addition to moving and covering valuables, fire fighters should channel the flow of water down stairs or through drains. Like life safety and extinguishment, property conservation operations must be planned and well executed to reduce the loss.

Lead Time

Experienced ICs understand that when an assignment is given, it takes time to see results. Ordering a portable master stream appliance to be deployed on the 10th floor could take a considerable amount of time. The level where the fire occurs has the greatest impact on lead time, especially when elevators are not available. **Table 12-1** displays the results of a training program conducted at a 48-story building. The results clearly demonstrate an increased lead time related to the height of the building. For purposes of this study, fire recruits and instructors were in full turnout gear, but not on air. Each five-person company (four recruits and an instructor) was assigned 150 ft (46 m) of 1¾-in. (45-mm) hose, nozzle, extra SCBA cylinders, and hand tools. The average time to walk up one floor for the first 9 floors was 20.8 seconds or a total of 3 minutes 7 seconds to reach the 10th floor level. The last eight floors took 59.0 seconds per floor. On average it took 28 minutes and 52 seconds to reach the top floor of this 48-story building. This does not include time needed to connect to the standpipe, place the SCBA in service, and advance on the fire. This building is less than half the height of some high-rise

Table 12-1 Climbing Stairs to the Top of 48-Story Building	
Floors	**Average Time Per Floor in Seconds**
1–10	20.8
11–20	27.8
21–30	33.6
31–40	45.9
41–48	59.0

buildings and the participants were in better than average physical condition. Studying high-rise fire reports and conducting tests reveals that fire fighters are sometimes unable to reach the fire floor or need to stop and rest during the ascent.

The NIST Report on High-Rise Fireground Field Experiments compared the time required to place the first attack line in position when elevators were available versus time needed to perform the same task using the stairway to access the 10th floor. In the NIST experiments, it took 2 to 4 additional minutes to reach the fire floor via the stairs. Any delay in extinguishment or rescuing occupants could have serious consequences. A fire on a higher floor would delay operations to a much greater degree, as indicated in Table 12-1.

Achieving a successful outcome when confronted with a working high-rise fire requires that an adequate firefighting force be rapidly assembled. During the NIST High-Rise Fireground Field Experiments, 66 to 120 fire fighters were assembled to carry out the 38 identified tasks. These experiments did not include property conservation or overhaul. It was found that the total number of fire fighters responding was not as important as the response time for resources allocated to complete critical tasks. Amassing the necessary personnel early in the operation resulted in reduced times to carry out the life safety and extinguishment objectives. The simulated fire on the 10th floor of an open-layout office required two 2½-in. (65-mm) lines for extinguishment. A 2½-in. (65-mm) line was also advanced into the 11th floor. Search and rescue operations were conducted on the 10th and 11th floors, with one victim located on each floor. Other tasks, such as ventilation and evacuation of the remainder of the building, were also conducted in priority order.

Crew sizes were three, four, five, and six with high and low responses. During the exercise, the crew size remained the same regardless of how many units responded. As an example, a low response of three engine companies and three truck companies each with a crew of 3 would total 18 fire fighters on the first alarm plus ambulances and battalion chiefs. Each alarm duplicated the first-alarm fire company response with changes in other responses such as

higher ranking chief officers. A high response was four engine companies and four truck companies per alarm. Eventually three-person staffing provided adequate on-scene fire fighters. However, many critical tasks were delayed as more distant companies responded to the scene. Using the worst-case scenario, a nonsprinkler-protected building with elevators unavailable took the low response three-person crew nearly 14 minutes longer to rescue the victim on the fire floor than the low response four-person crew.

The fire dynamics simulator modeled fire growth, and the fractional effective dose scale was used to estimate the impact toxic gases would have on victims. The toxic effects on the victim exposed to fire and toxic gases for 14 additional minutes were significant. Higher levels of staffing (five- and six-person per company) further improved times to complete critical tasks. This study indeed proved that the crew size makes a dramatic difference in saving occupant lives.

If the IC waits to give assignments until the need is obvious, he or she has waited too long. To be successful, the IC must be able to anticipate needs and assign resources in advance.

Establishing a Wide Fire Zone

Immediately adjacent to the exterior of a high-rise building fire is an unsafe place, particularly if glass is falling. A perimeter should be established to keep civilians out of danger and to provide a safe working area for fire fighters outside the building. Fire fighters who must work within the perimeter should be kept within vehicles or under protective structures. The importance of establishing zones changes with each incident. If there is danger from falling glass, a 200-ft (61-m) perimeter should be enforced around the building. As mentioned previously, plate glass falls to the street in large shards and at times will float a considerable distance from the base of the fire building. Opening windows by pulling the glass inward greatly increases safety at high-rise building fires. This technique has been tried with limited success using tape, suction cups, and other methods. When there is a working fire in a high-rise building, fire fighters may be forced to break windows on the upper floors or the fire may self-vent without warning. Therefore, the IC should expect glass to fall at any time and direct that a safe perimeter be established early in the incident.

Applying NIMS to a High-Rise Fire

NIMS should be used at every structure fire, and a high-rise building fire is certainly no exception. Proper use of NIMS is a test of the department's training, preincident planning, and discipline.

A look at the logistical requirements and multiple staging areas makes it apparent that the span of control can be quickly exceeded at a major high-rise fire. In the case of a high-rise fire, the IC is well advised to hand off the operations and/or planning sections early in the incident so that other priorities can be addressed. Record keeping is essential. If the IC has an assigned aide, this person usually acts as the initial planning officer and records situation reports and resource status and assignments. Units operating on the fire floor and staging area can be managed by a forward rescue/suppression branch director or operations section chief who documents which companies are in staging, which are in rehab, and which are working on the fire floor.

Communications

Communication and accountability are simplified by using NIMS. Companies working on the fire floor may change; however, the geographic designation remains the same. For example, if Truck 1 and Engine 2 are assigned to the 15th floor for search and rescue, they would be identified as Division 15. If Truck 1 and Engine 2 are later reassigned to rehab and replaced by Truck 3 and Engine 4, these replacement companies become Division 15. Remember that "division" is the NIMS term used to identify a geographic location or assignment.

Suppose a fire is on the tenth floor of a 20-story building. A rescue and evacuation group is doing the primary search on floors 12 to 20. Fire attack units are on the tenth and eleventh floors. **Figure 12-12** is a NIMS chart and communications network for this operation. Communications for this fire would be as follows:

- The rescue and evacuation group would be communicating with companies in their group as to the status of various floors.
- Division 10 (fire floor) would be communicating both with the operations section or IC (if operations is not staffed) and with companies assigned to Division 10.
- Division 11 would be doing much the same as Division 10.
- Companies assigned to ground support would be communicating with each other and with the logistics section.
- Operations, if assigned, would be communicating with Divisions 10 and 11; otherwise they would report to a branch manager or the IC.
- The staging officer would be communicating with the operations section or IC, which would then order the necessary staffing, equipment, and supplies for the staging area.
- Safety as well as the operations and logistics sections would be communicating with the IC.
- If staffed, the command staff positions of liaison and information, as well as the planning and administration/finance sections would most likely be at the command post location, communicating face to face with the IC.

Communications is the most frequently cited problem during major emergency operations. Communications can be very complicated within a high-rise building. If separate radio frequencies or hardwired communications are not provided at the incident scene, a breakdown in communications is likely to occur.

IC Rule of Engagement for Fire Fighter Safety

Maintain frequent two-way communications and keep interior crews informed of changing conditions.

Reproduced from: © 2011, IAFC Safety, Health and Survival Section. All rights reserved.

Texture: © Eky Studio/Shutterstock, Inc; Steel: © Sharpshot/Dreamstime.com

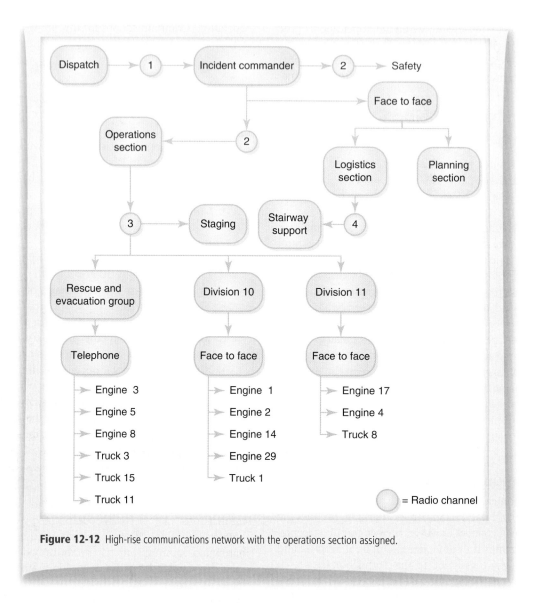

Figure 12-12 High-rise communications network with the operations section assigned.

Radio discipline *must* be maintained, and alternative methods should be arranged (e.g., messenger, hardwired telephones). Note that the rescue and evacuation group in Figure 12-12 is communicating via telephone. Hardwired telephones are abundant in most high-rise buildings; consider using them as a means of emergency communications. In Figure 12-12, for example, the search and rescue group supervisor could be located in a room off the lobby, on a lower floor, or in another building where two or more telephones are readily available. All units working in this group would be given the contact numbers for the group supervisor, which would be used to call in status reports and receive new assignments. Alternate communications must also be available because the fire could damage the telephone system. During a large-scale search and rescue operation, there is a need for considerable two-way communications within the search and rescue group. Complicating the communications problem at a high-rise building is the fact that using radios is difficult within many structures. In fact, it is often impossible to receive radio transmissions from inside a structural steel building.

Tactical Worksheets

Fire departments have devised a wide assortment of tactical worksheets to be used in documenting and tracking activities on the fire scene, including some that are specifically designed for high-rise fires. Most worksheets contain checklists to remind the IC of important functions.

A good tactical worksheet will include sections to record the fire location and the location of companies at the incident scene. During high-rise fires, companies will be working at different levels; therefore, the tactical worksheet should provide a generic sketch of a high-rise building that can be used to record activities on each floor.

Occupants on the fire floor and above are normally in the greatest danger, but it is necessary to search below the fire as well. A search and rescue group normally works above the fire to conduct a primary search and to assist in evacuation. This group is generally very active, moving from floor to floor. Therefore, it is very important that the group leader keep the IC (or operations chief) current on the group's location in the building and the status of the primary search.

The high-rise worksheet in **Figure 12-13** lists various high-rise positions described in this chapter and provides space to enter companies working on each level.

The value of an aide is clear. The IC needs to know the location of all companies operating in the building but does not always have time to keep the worksheet up to date. An aide, acting in a planning capacity, can assist in tracking and recording this information.

Base (Exterior Staging for High-Rise Fires)

The term "base" is used by wildland fire fighters to identify a location housing reserve equipment and personnel (base camp). This term is not normally used for structural firefighting but has utility in managing high-rise fires. The base for high-rise fires is a location where support equipment and personnel are kept on the *exterior* of the building. Depending on interior conditions and the location of the fire, an exterior staging area may be established near the fire perimeter. The reason for this distinction is that the staging area is normally moved inside the structure during a high-rise fire. Unless there is a possibility of moving to an exterior operation or the fire is involving other structures or is located on a lower floor, it would be unusual to amass a large force on the exterior at a high-rise fire when the fire is on an upper floor. However, some department SOPs maintain a predetermined reserve force in base as a tactical reserve.

Staging (Interior)

In a high-rise situation, most of the reserve force is moved through the lobby and then to the interior staging area. This area is normally two or more floors below the fire. A rehabilitation area may be set up on the same floor as the staging area or on another floor. The concept is the same as in exterior staging: to provide a readily available reserve force. To avoid the confusion of having a different name for the exterior staging area, the interior staging area could be identified as interior staging or as the resource area. Department SOPs should address this issue, using either the suggested terms of "base" and "staging" or other terminology that is consistent throughout the region.

The supervisor in charge of the staging area or an assistant can act as the accountability officer for crews that are working out of the staging area. Tracking crew rotations can be a challenging task. If an extended operation becomes necessary, three fire fighters will be needed for each position on a hose line. For example, a four-person fire company operating a 2½-in. (65-mm) line on the fire floor will exhaust its air supply in approximately 15 to 30 minutes. To continually operate the line on the fire floor, this crew will need to be relieved by an available crew in staging.

Upper Search and Rescue Group				Engine Cos.	Truck Cos.	Others

Upper Search and Rescue Group

Floors: From _____ to _____

Supervisor _____

	Engine Cos.	Truck Cos.	Others

Floor # _____

Division # _____

Supervisor _____

Floor above fire
Floor # _____

	Engine Cos.	Truck Cos.	Others

Division # _____

Supervisor _____

Fire floor
Floor # _____

	Engine Cos.	Truck Cos.	Others

Division # _____

Supervisor _____

Floor # _____

	Engine Cos.	Truck Cos.	Others

Division # _____

Supervisor _____

Staging
Floor # _____

	Engine Cos.	Truck Cos.	Others

Division # _____

Supervisor _____

Floor # _____

	Engine Cos.	Truck Cos.	Others

Division # _____

Supervisor _____

Lower Search and Rescue/Property Conservation Group

Floors: From _____ to _____

Supervisor _____

Engine Cos.	Truck Cos.	Others

Floor # _____

	Engine Cos.	Truck Cos.	Others

Division # _____

Supervisor _____

Lobby control

Supervisor _____

Company _____

Stairway support

Supervisor _____

Companies _____

Figure 12-13 High-rise tactical worksheet.

Rule of Engagement for Fire Fighter Survival

Maintain continuous awareness of your air supply, situation, location, and fire conditions.

Reproduced from: © 2011, IAFC Safety, Health and Survival Section. All rights reserved.

Texture: © Eky Studio /Shutterstock, Inc; Steel: © Sharpshot/Dreamstime.com

As the crew leaves the fire floor, it will move to rehab, and another crew will move to the ready position **Figure 12-14**.

Figure 12-14 Fire companies rotating from rehab.

The duties of the staging area manager are as follows:

- Report to the operations section, if it is staffed. Otherwise, interior staging would communicate directly with the IC. The IC could elect to have the interior staging officer communicate directly with the logistics section.
- Maintain records of the companies that are in staging and rehab.
- Maintain a minimum reserve of engine and truck company personnel as established by the IC.
- Request additional resources to maintain the established reserve force.
- Maintain an adequate supply of air cylinders and other equipment as needed.
- Supply first aid equipment and medical services for units that are involved in rescue and suppression.

The concept of an interior staging area should not be reserved for high-rise fires. Any time an extended interior attack is in progress on an upper floor of a building where fire fighters expend more than one tank of air, interior staging and rehabilitation may be a good idea. As an example, a fire on the top floor of a six-story building would not by definition be a high-rise fire. However, establishing a rehab/staging area on the fourth floor could result in a more efficient operation.

Lobby Control

The primary accountability officer may be part of the lobby control crew or co-located in the lobby. Lobby control is established regardless of whether the elevators will be used. The duties of lobby control include the following:

- Controlling, operating, and accounting for all elevators
- Locating and controlling all interior stairs
- Directing incoming companies to the proper elevator or stairway
- Consulting with the building engineer
- Controlling or shutting down the HVAC system after consulting with the IC

Controlling the elevators and stairways is the only way to effectively gain access to the upper floors of a high-rise building. Many department SOPs assign these initial lobby control tasks to a member of the first-arriving truck company. Another important lobby control duty is to gain control of the HVAC system. If operated properly, the HVAC system can prevent heat, smoke, and toxic gases from reaching occupants. Conversely, if it is operated incorrectly, it may spread the products of combustion well beyond the immediate fire area and into occupied areas.

In some buildings, it is virtually impossible to know what will happen when the HVAC system is operated because of the many unknown variables, such as heat produced by the fire, stack effect, and wind. Furthermore, few building engineers are completely knowledgeable about the operation of the HVAC system. For these reasons, many fire departments require an emergency shutdown switch near the lobby.

High-Rise Case Histories

The One Meridian Plaza Fire

The One Meridian Plaza Fire in Philadelphia, Pennsylvania, occurred at 8:40 PM on Saturday evening, February 23, 1991 **Figure 12-15** .[12,13] Because the fire occurred outside normal working hours, the threat to occupants was minimal. However, this fire claimed the lives of three fire fighters.

Fire fighters provided a considerable occupant load in the fire building, as the operation resulted in 316 fire fighters being called to the scene. The building was in total darkness and without elevator service.

Perhaps the One Meridian Plaza fire would not have gained the same headway had the building been occupied, because the fire probably would have been detected earlier. The reports do not provide the total number of occupants in the building during normal working hours, but the worst-case scenario was alarming. Each floor had 17,000 ft² (1579 m²) of usable space. According to the NFPA *Life Safety Code* rule of limiting occupancy to no more than one person per

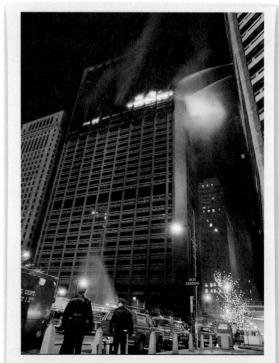

Figure 12-15 One Meridian Plaza, Philadelphia, Pennsylvania.

© George Widman/AP Images

every 100 ft² (9 m²), if the building had been fully occupied, there could have been 170 people per floor.

The fire began on the 22nd floor and dropped down to the 21st floor. The top (28th) floor is a mechanical floor, meaning that the occupants on floors 21 through 37 (17 floors) would have been in immediate danger had the building been fully occupied. At 170 people per floor, 17 floors could be occupied by up to 2890 people. The eight floors that were destroyed could have held 1360 people.

The One Meridian Plaza fire most certainly points out the potential for a large-loss-of-life fire in a nonsprinkler-protected high-rise building. The building was being retrofitted with sprinklers at the time of the fire. Floors 22 through 29 were not sprinkler protected and were lost to the fire. It is interesting to note that the fire was suppressed on the 30th floor by 10 sprinkler heads operating at the points of penetration. Can the value of sprinkler protection in high-rise buildings be underestimated? This fire serves as a testimonial to the value of automatic sprinkler systems in high-rise buildings.

The First Interstate Bank Building Fire

On May 4, 1988, at 10:25 PM, a fire similar to the One Meridian Plaza fire occurred in Los Angeles, California, at the 62-story First Interstate Bank Building (Figure 12-16).[14,15] At this fire, four floors were completely destroyed and a fifth floor was heavily damaged. The building was unoccupied except for security and maintenance personnel. The First Interstate fire burned floors 12 through 15 and was stopped at the 16th floor. Again, to consider the potential loss of life, we can look to the NFPA *Life Safety Code*. According to the *Life Safety Code*, the maximum number of occupants permitted per floor would have been 175. The fire was on the 12th floor of a 62-story building. Therefore, 51 floors were exposed to the fire and smoke conditions. This yields a potential occupant load of 8925 people above the fire. In this case, the actual number of building occupants was listed as 4000. If the actual occupant load is averaged equally among the floors, as many as 3290 people could have been above the fire during normal working hours.

Comparing the One Meridian Plaza and First Interstate Bank Building Fires

Examining similarities between these two fires enhances the learning experience that would be gained by studying a single fire. Table 12-2 compares the two incidents.

When studying case histories, the effect of delayed alarms should be noted. A 16-minute delay occurred at the First Interstate Bank Building fire.

Figure 12-16 First Interstate Bank Building, Los Angeles, California.

© CRAIG MATHEW/AP Images

Had the fire department been notified 16 minutes earlier, there is a high probability that the fire could have been extinguished without exposing fire fighters to prolonged and extreme danger and with much less damage. Some would speculate that this delay is unlikely when the building is fully occupied. Francis L. Brannigan, in one of his monthly *Fire Engineering* articles, took exception to this theory. Someone suggested that during normal working hours the employees at the First Interstate Bank Building would use the occupant-use hose line to extinguish the fire. Brannigan states the following in response:

Table 12-2 Similarities Between the One Meridian Plaza and First Interstate Bank Fires

	One Meridian Plaza	First Interstate Bank
Deaths	3 fire fighters	1 maintenance person
Occupancy	Office building	Office building
Time of day	8:30 PM, unoccupied	10:25 PM, 40 occupants
Alarm	Delayed, > 4 minutes	Delayed, > 16 minutes
Sprinklers	Partial, being retrofitted	Partial, being retrofitted
Floor of origin	22	12
Area of involvement	8 floors, 17,000 ft²/floor (1579 m²/floor)	5 floors, 17,500 ft²/floor (1626 m²/floor)
Responding fire fighters	316	383

I think this answer is unrealistic. On a trading floor, operatives are dealing in multi-million-dollar split-second transactions. I think it would be more like the Bradford Stadium fire in England, where the fans kept alternately watching the fire and the game until they died. In addition, to the typical computer nerd, the thought of water on his computer is anathema.[16]

There are many documented large-loss fires in occupied buildings in which delayed alarms played a significant role. In hotel and office occupancies, security or maintenance personnel may first attempt to investigate the fire before calling the fire department. Sometimes, they become involved in firefighting operations with occupant-use hose lines or fire extinguishers and delay calling the fire department until the fire is out of control.

A fire that occurred on January 6, 1995, in North York, Ontario, Canada, verifies Brannigan's assertion that delayed alarms should be anticipated in occupied buildings.[17] It also demonstrates that multifatality high-rise fires are not limited to office and hotel occupancies. This fire, which occurred in a 29-story apartment building, killed six occupants. The occupant in the apartment of origin first attempted to extinguish a couch cushion fire with water. Other occupants smelled smoke and investigated, then assisted in trying to extinguish

the fire. Finally, one of the occupants called 911. The door to the apartment of origin was left open, allowing the fire to spread into the hallway and extend into one of the two stairways located at the central core.

This high-rise apartment building in Ontario is not a unique incident. According to Dr. John Hall, NFPA Assistant Vice President (retired), from 1989 through 1998, high-rise apartment buildings and hotels averaged 10,500 reported structure fires a year, or 86% of the combination of high-rise apartment buildings, hotels, office buildings, and hospitals. During this same reporting period, high-rise apartment buildings and hotels averaged 51 reported civilian fire deaths a year, or 96% of the combination of the four types of high-rise properties. The vast majority of these residential high-rise fires and deaths are specifically in apartment buildings. Also, the percentage of reported high-rise fires in buildings with sprinklers in 1998—the latest year in the reporting period (and the one in which the percentages were highest)—was 36% for apartment buildings (compared to only 5% for apartment buildings that were not high-rise), 77% for hotels, 80% for hospitals, and 63% for office buildings. It is pretty clear that most of the fire problem needing attention in high-rise buildings is specifically in apartment buildings, where—no coincidence—the usage of sprinklers lags far behind the usage in other high-rise buildings.

Experienced fire officers have speculated that a working fire in a large high-rise building during periods of full occupancy could require the services of 500 or more fire fighters. The fires in Philadelphia and Los Angeles make this estimate appear low, given that both occurred when the building was occupied by only a small staff of security and maintenance personnel.

Fire fighters in Los Angeles successfully mounted a manual attack on the 16th floor, whereas the One Meridian Plaza fire was not contained until it reached a sprinkler-protected floor. The difference was the standpipe system. At One Meridian Plaza, pressure-reducing valves reduced the water pressure below the operating pressures needed to properly supply the automatic nozzles being used by the fire department. The codes governing standpipe installations required reducing valves to avoid overpressurization. The flow-reduction valves that were used to meet code requirements were incorrectly set at a pressure below the required pressure and could not be changed at the time of the fire. Manufacturers now provide nozzles that will effectively operate at lower pressures. Smooth-bore nozzles are another option when faced with low pressures.

The Peachtree Plaza Fire

On June 30, 1989, a much smaller fire in a 10-story office building in Atlanta, Georgia, occurred at 10:30 AM on the sixth floor while the building was occupied Figure 12-17 .[18,19] Five of the 40 occupants on the sixth floor were killed, and at least six others were rescued using aerial ladders. Eleven of 40 occupants (28% of the occupants on the sixth floor) probably would have perished if the fire floor had been beyond the reach of aerials. Additional occupants were evacuated from other floors using the interior stairways. Many of the occupants were able to survive by breaking windows and waiting for exterior rescue. If this fire had occurred above the reach of aerials, a fatality rate of at least 25% for the involved floor is a reasonable assumption. Translate that to the 170 and 175 people per floor in One Meridian Plaza and the First Interstate Bank Building, and the result would have been a major catastrophe.

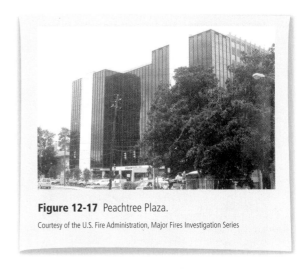

Figure 12-17 Peachtree Plaza.
Courtesy of the U.S. Fire Administration, Major Fires Investigation Series

Terrorist Attacks at the World Trade Center

The two largest high-rise incidents in the United States to date were the result of terrorist attacks on the World Trade Center in New York City Figure 12-18 .

World Trade Center Bombing: February 26, 1993

The first attack involved a truck bomb.[20,21] Although this incident was the result of a terrorist's bomb, fire and smoke were the major concerns. A quote from Chief Anthony L. Fusco after the first attack identifies the fire danger and reinforces the contention made throughout this book that the most important life safety measure is often extinguishment. In his remarks, Chief Fusco states, "The decision to attack the basement fires in the initial stages of the incident was the most important decision of the incident [commander] in that hundreds, maybe thousands of lives were saved owing to timely extinguishment of the fires."

The first World Trade Center attack should have removed any doubts about the potential for death and injury in high-rise building fires. The World Trade Center was a seven-building complex that included two 110-story towers. The area of each floor was 40,000 ft^2 (3716 m^2). It was estimated at the time of the truck bomb incident that 60,000 people

Figure 12-18 New York City skyline view including the World Trade Center.

© Jemini Joseph/ShutterStock, Inc.

were working in the buildings and there were thousands of visitors. Six deaths and 1042 injuries were reported at this incident, and 50,000 people were evacuated from the building.

This incident was the equivalent of simultaneous multiple-alarm incidents. Five alarms were sounded for the Vista Hotel operation, five alarms were sounded for the Tower 1 operation, and four alarms were sounded for the Tower 2 operation. Another alarm was sounded for additional resources. Forty-five percent of the department's on-duty resources were required at this incident. In addition, a large contingent of police officers responded, and 174 ambulances treated and transported the injured.

Few fire departments in the world have comparable resources. In fact, very few fire departments could provide the staffing needed to search the 99 elevators in each tower.

The World Trade Center was sprinkler protected, but the blast destroyed the sprinkler piping in the garage area where the fire occurred. With the sprinkler system out of service in the garage, the fire continued to burn and push smoke to the floors above. Once the fire was extinguished, the smoke production stopped. However, the smoke that had already entered the building lingered. As previously mentioned, venting a high-rise is always problematic and takes considerable time.

World Trade Center Attack: September 11, 2001

The second attack on the World Trade Center resulted in the largest number of civilian and fire fighter fatalities ever recorded in a building fire. Lessons are still being learned from this disaster. Although aircraft flying into buildings is not a typical high-rise fire scenario, it occurs frequently enough to deserve consideration. Small aircraft have flown into high-rise buildings on several occasions. A B-25 bomber flew into the Empire State Building in 1945. World Trade Center building designers had actually planned for an accidental aircraft accident involving a Boeing 707 aircraft making a landing approach. World Trade Center structures were designed to withstand the impact and fire from this slightly lighter aircraft assumed to be low on fuel and traveling at approximately 180 miles per hour (mph) (290 kmph). This "planned for" accident would produce much less impact and fire than the Boeing 767s that were nearly full of fuel and traveling at an estimated 470 to 590 mph (756 to 950 kmph) when they were purposely flown into the buildings on September 11, 2001.

The 767 airplanes caused considerable structural damage on impact, but the two towers withstood these initial impact forces and remained standing. However, the aircraft debris moving through the building removed sprayed-on fire-resistive coatings that protected steel structural components. The ensuing fire weakened the unprotected steel, resulting in the collapse of Towers 1 and 2, killing 2749 people including 340 fire fighters. NIST theorizes that the two buildings would not have collapsed had the thermal insulation remained in place.[22] It is estimated that Towers 1 and 2 at the World Trade Center were 33% to 50% occupied at the time of the attack. Had they been fully occupied, NIST estimates that it could have taken as long as 3 hours to evacuate the buildings and as many as 14,000 people could have lost their lives.

Prior to the collapse of the two World Trade Center buildings, high-rise buildings had not been known to collapse even when subjected to intense fires on multiple floors. The Meridian Plaza and Interstate Bank fires discussed earlier in this chapter both withstood severe fires on multiple floors and neither collapsed.

The MGM Grand Fire

During the MGM Grand (now Bally's) fire in Las Vegas, Nevada, occupants displayed what we now call convergence cluster behavior Figure 12-19 .[23,24] They gathered in certain rooms as groups, gaining a feeling of safety in the presence of others. What implications does this have for the fire department? Searching fire fighters might not find anyone in several rooms or in an entire floor area, while one room may contain far more victims than anticipated.

The occupancy type has a direct bearing on the behavior and awareness level of occupants as well as the population density. The MGM Grand had 3400 registered guests at the time of the fire. The death and injury toll was due to smoke infiltrating the 21-story tower due to a fire that originated in the first floor casino. The death toll of 85 people made this the second most deadly hotel fire in U.S. history (119 died in the Winecoff Hotel in Atlanta, Georgia, in 1946).

The fire in the MGM Grand started at the rear of the casino and eventually involved the entire 150- × 450-ft (46- × 137-m) casino floor. The ceiling height was not given in the reports, but the

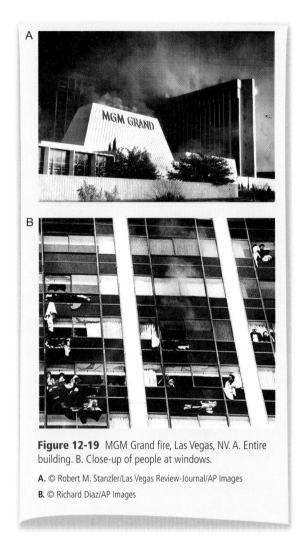

Figure 12-19 MGM Grand fire, Las Vegas, NV. A. Entire building. B. Close-up of people at windows.

A. © Robert M. Stanzler/Las Vegas Review-Journal/AP Images

B. © Richard Diaz/AP Images

ceilings were probably a minimum of 20 ft (6 m) high. Using the V/100 formula for 20-ft (6-m) ceilings yields a rate of flow of 13,500 GPM (gallons per minute) (51,103 L/min). The size of the fire area and the required rate of flow led to the conclusion that this fire was not going to be extinguished by manual means, and there were no sprinklers in the fire area. Therefore, the best tactic was to keep the fire out of the stairways and towers, and this was accomplished. However, smoke did spread throughout the structure, killing 85 people.

Chapter Summary

- Tactics and strategic objectives used in high-rise firefighting are the same as those that apply to any other structure fire, but with special considerations because of the height of the building.
- High-rise fires are some of the most challenging incidents a fire department encounters because of the high occupant load and building height/configuration.
- Preincident planning, code enforcement, use of NIMS, and development of high-rise SOPs improve the chance for successful operations in high-rise buildings.
- Logistics and access problems increase with building height.
- Avoid using elevators unless they will substantially improve operations, and even then, use with extreme caution.
- During a structure fire, never use an elevator to travel to or above the fire floor; exit the elevator at least two floors below the fire.
- All elevators should be under the control of fire fighters.
- When in doubt about the safety of elevators, a possible option is to use the elevator to send equipment (rather than personnel) to the floors above.
- Getting fire fighters and equipment to an interior staging area via stairs is essential when elevators are unsafe or unavailable and returning to ground level is impractical.
- Stairway transport of equipment is done via relay (ground support) to conserve personnel.
- When the fire is on an upper floor of a high-rise building, interior staging should be set up two or more floors below the fire.
- Helicopter rescues are extremely dangerous and usually unnecessary. Partial or sequential evacuation strategies can be determined in advance of a fire and incorporated into the alarm system.
- An emergency voice/alarm communications system (EVACS) is available in some buildings, but occupants do not always follow the instructions given.
- Most exterior defensive fire control tools are ineffective above the eighth floor. Fire forces must prevent fire extension into the stairways, which are the safest, most effective, and sometimes only available egress routes.
- Pressure-reducing valves and fire pumps both affect standpipe hose pressures.
- Due to large numbers of people and other distractions in the lobby, a high-rise lobby may be a poor choice of location for a command post.
- A command post's location should be transmitted to all responders regardless of whether it is inside or away from the building.
- Subtle differences in interior building configurations can have a major impact on firefighting and rescue operations.
- Weather conditions, fire intensity, stack effect, wind, and atmospheric pressure can have a significant effect on smoke movement within a high-rise structure.
- Dedicating a stairway with standpipe outlets for firefighting reduces the possibility of opposing lines on the fire floor.
- To avoid exposing people and equipment at ground level to falling hazards, a wide fire perimeter is essential.
- Communications is the most frequently cited problem during major emergency operations.
- Controlling the lobby, elevators, stairways, and HVAC facilitates both fire fighter access to and successful evacuation of the upper floors of a high-rise building.

Flames: © Dxc/Dreamstime.com; Steel texture: © SharpShot/Dreamstime.com

Key Terms

lobby control A high-rise assignment in which a crew is responsible for managing the stairways; elevators; heating, ventilation, and air-conditioning systems; and related duties.

smoke-proof tower A stairway that is designed to be separated from the building by a landing. This creates a separation that will limit the spread of smoke into the stairway and keep it clear for evacuation.

stack effect The vertical airflow within buildings caused by temperature differences between the building interior and the exterior; depending on conditions, stack effect could be positive or negative, causing smoke to move upward or downward.

Challenging Questions and Activities

1. Use the following RGB Management high-rise fire scenario to do the following:

   A. Size up the fire using information from this chapter and the size-up factors discussed in the *Procedures, Preincident Planning, and Size-Up* chapter.

   B. Establish a strategic mode: offensive, defensive, or non-attack.

   C. Outline an incident action plan.

   D. Assign/reassign companies to tactics and tasks necessary to carry out the incident action plan.

   E. Organize the operation using NIMS. Use high-rise positions described in this chapter as necessary, such as ground support, lobby control, staging, and base.

   Figure 12-20 is the preincident plan narrative for the RGB high-rise. **Figure 12-21**, **Figure 12-22**, and **Figure 12-23** are drawings showing floor layouts at various levels. Use the preplan information and scenario to guide your response.

   ### RGB High-Rise Fire Scenario

   Assume the role of IC and retain that role, even though command may change.

   Using the following information from the alarm card, call whatever assistance you need. Start by using on-duty resources. Then request

   mutual aid assistance as needed and call in off-duty personnel. Background information includes the following:

   - Fire report from dispatch: "Smoke detector alarms were received from floors 19, 20, 21, 22, and 23 and the 27th floor equipment room. In addition, calls were received from the 20th floor reporting a fire in the office area. A call from 21st floor reported heavy smoke conditions. We are getting multiple calls on this one."
   - Day/date: Tuesday, August 1
   - Time: 1600 hours (4:00 PM)
   - Weather conditions:
     - 90°F (32.2°C)
     - 90% humidity
     - Partly cloudy
     - Wind from the southwest at 1 mph (1.6 kmph)
   - The report from Engine 1 is as follows: "Engine 1 and Truck 1 are on the scene of a 26-story office building with a working fire. Fire is visible from several windows on an upper floor on the Central Parkway side. We are proceeding to the 18th floor using the elevator. Engine 1 is RGB command."

Flames: Dvv/Dreamstime.com; Steel texture: © Sharpshot/Dreamstime.com

RGB Building
500 Vine Street

Life Safety

Occupancy Type
- Business Occupancy (Office Building) with attached garage
- First floor Lobby, Restaurant, Bank, Gift Shop
- 26th floor storage area, 27th floor mechanical room

Estimated Number of Occupants
- Possibly 9000 weekdays 0700 to 1800 hrs
- 2 Security Personnel at all times
- 5 to 40 Maintenance Personnel outside working hours

Primary and Alternative Egress Routes
- 3 Stairways (see floor plan)
- Doors from stairway to hallways are locked on stairway side (no re-entry to hallways)

Access to Building
- Paved street access on sides A, B and C
- Parking garage on side D
- Building lobby open weekdays 0600 to 2000 hrs.
- Security maintains occupant log at other times

Construction Type
- Fire Resistive, central core high-rise
- 26 stories (no 13th floor), height 316 ft (96 m)
- Basement and Sub-basement
- Area at base (sub-basement to 3rd floor) 144 ft X 185 ft. (44 m X 56 m)
- Area on tower floors (4th to 27th) 60 ft. X 160 ft. (18 m X 49 m)
- Open stairs and escalators between 1st and 2nd floors

Condition
- Inspected every 6 months, minor violations only
- Building well maintained

Live and Dead Loads
- Typical Office Occupancy
- Large fire load on 26th floor storage area
- Large live load in mechanical room on 27th floor

Enclosures and Fire Separations
- First floor lobby open layout
- Floors 3 to 23 mostly open layout with cubicle separations between work stations
- Floors 24 and 25 executive offices with additional security. Subdivided into large offices around periphery, work stations near core.

Age
- Built in 1970

Extinguishment

Probability of Extinguishment
- Most building areas will require multiple 2 ½" lines for extinguishment

Figure 12-20 RGB high-rise pre-incident plan narrative. (*Continued*)

External Exposures
- Wide streets on Sides A, B, and C

Internal Exposures
- Attached garage
- Non-involved floors, some utility openings not sealed

Calculated Rate of Flow V/100
- Basement/sub basement = 2000 GPM
- 1st floor lobby = 4500 GPM
- 1st floor bank = 480 GPM
- 1st floor restaurant = 320 GPM
- 2nd floor lobby = 920 GPM
- Tower Floors w/open layout = 960 GPM
- Garage all levels = 760 GPM

Water Supply
- 20,000 GPM available from primary water supply
- Can be cross tied to suburban system adding 20,000 GPM
- Hydrants located on sides A, B, and C of building. Hydrant spacing 250'

Total Staffing/Apparatus Available

Alarm	Engine Companies	Truck Companies	Other
1st	1, 2, 3 and 4	1, 2, 3	Heavy Rescue 1, Bat. Chief 1
2nd	5, 6, 7, and 8	4 and 5	Bat. Chief 2, ALS 1
3rd	9, 10, 11, and 12	6	Fire Chief, 4 Asst. Chiefs, Heavy Rescue 2
4th	13, 14, 15, and 16	7	2 Bat. Chief Administrative Staff, ALS 2 and ALS 3
5th	17, 18, 19, and 20	8	Air Supply Truck, Command Van

Each fire company including Heavy Rescue staffed with an officer, driver and 2 fire fighters.
Department High-Rise SOP requires a full one alarm compliment be staged in Base when a working fire is declared.
Fire Department Has 40 Engine Companies, 18 Truck Companies, 3 heavy rescues, 12 ALS Units, 4 Assistant Chiefs, 5 on-duty Battalion Chiefs, 5 Administrative Battalion chiefs.
Approximately 700 off duty fire fighters can be recalled via a telephone relay system.

Fire Protection Systems
- Standpipe System with 1000 GPM pump
- FDC on Vine St. side of building
- 100 psi at each standpipe outlet, center stairs
- Field adjustable pressure-regulating valves, PRV tool in lock box.
- Dry Chemical hood system in kitchen area of restaurant on 1st floor

Property Conservation
- Computers and paper files - very susceptible to water damage

General Factors
Utilities (Water, Gas, Electric)
- Main Electric in Sub-basement (DO NOT ATTEMPT TO DE-ENERGIZE)
- Gas and Water Street Valves on Vine St.
- Gas and water equipment/meters in basement

Figure 12-20 RGB high-rise pre-incident plan narrative.

Figure 12-21 RGB high-rise typical floor plans—floors 1 to 3 of high-rise and floors 1 to 6 of garage.

Figure 12-22 RGB office high-rise plan view—first floor.

Figure 12-23 RGB office high-rise plan view—upper levels.

- Engine 1 is carrying a high-rise pack including 2½-in. (65-mm) hose to the fire area.

You arrive on the scene at the same time as Engine 2, Truck 2, and the heavy rescue unit. You notice that Engine 1 has a water supply and that the pump operator is at the apparatus hooking up to the hydrant. The security people approach you and verify that people leaving the 20th floor saw visible flames on the Central Parkway side.

Truck 1's officer is now on the radio: "Truck 1 reporting that Engine 1 and Truck 1 are hooking up to the standpipe on the 19th floor for a confirmed working fire on the 20th floor. The 20th floor appears to have an open layout and heavy fire. Elevator 2 is free of smoke and operational."

As the IC, you ask security, "Who occupies the 20th floor?" Security replies, "I will get back with you." You request that security also find out as much about the floor layout as possible.

You set up command in the first floor lobby alarm/command room, noticing that the annunciator panel is in the alarm mode for floors 19 to the penthouse due to smoke detector activation.

It is now 4:08 PM hours. You strike the second alarm, and Engines 3 and 4 as well as Truck 2 are approaching the command post. The heavy rescue unit is in the elevator lobby suggesting they assist the Engine 1, Truck 1 attack team on the 20th floor; you approve and they proceed toward the 20th floor with additional SCBA cylinders. All of the elevators have been recalled to the first floor. Truck 3 is assuming their normal role as the rapid intervention crew (RIC) and proceeding via an elevator to the 18th floor with extra SCBA cylinders and RIC equipment.

In accordance with SOPs, Truck 2 is setting up PPV in all three stairwells beginning with the evacuation stairway on the Vine Street side. They will continue as the vent team to improve conditions in the stairways and assist in evacuation in the stairways as needed.

Engine 1 is now reporting back to the command post via radio: "Engine 1 and Truck 1 are now on the 20th floor. We estimate the fire to cover the entire Central Parkway side of an open-layout floor plan. We have heavy smoke throughout the floor and we are not making progress on the fire. We need additional hose streams or a master stream up here"

Truck 1 is now reporting back to the command post via radio: "On the way up to the 20th floor we ran into at least 10 people evacuating via the stairway. The occupants looked okay but a few were coughing. The stairway has moderate smoke from the 19th floor to at least the 21st floor."

2. Consider the RGB scenario, but with the following changes to the situation and occupancy. Explain how these changes will affect the operations you described in Question 1:

- Place the fire in a residential occupancy for the elderly.
- Place the fire in a hotel.
- Change the situation as follows:
  – Elevators are out of service.
  – There is a 45-mph (72-kmph) wind from the north (Central Parkway side).
  – The fire area is subdivided into small office areas.
  – The fire occurs on Sunday afternoon in an office building.
  – The building is protected by a working sprinkler system.
  – There are heavy smoke conditions on the fire floor and all floors above.

Flames: Drx/Dreamstime.com; Steel texture: © Sharpshot/Dreamstime.com

Endnotes

1. Hall, J. R. Jr. *High-Rise Building Fires*. Quincy, MA: NFPA; 2011.

2. National Institute of Standards and Technology. *NIST, NCSTAR 1 Federal Building and Fire Safety Investigation of the World Trade Center Disaster, Final Report*. Gaithersburg, MD: 2005.

3. National Fire Protection Association. *NFPA 1710: Standard for the Organization and Deployment of Fire Suppression Operations, Emergency Medical Operations, and Special Operations to the Public by Career Fire Departments*. Quincy, MA: NFPA; 2010.

4. National Institute of Standards and Technology. *Technical Note 1661, Report on Residential Fireground Field Experiments*. Gaithersburg, MD: 2010.

5. Averill, J., Moore-Merrill, L. *Report on High-Rise Fireground Field Experiments, NIST Technical Study Note 1797*. Gaithersburg, MD: National Institute of Standards and Technology; 2013.

6. Demers, D. P. *Investigative Fire Report on the Las Vegas Hilton Hotel Fire*. Quincy, MA: NFPA; 1982.

7. Independent Commission appointed by Cook County Board President John H. Stronger Jr, *Report of the Cook County Commission, Investigating the 69 West Washington Building Fire*; October 17, 2003.

8. National Fire Protection Association. *NFPA 101: Life Safety Code*. Quincy, MA: 2012.

9. National Fire Protection Association. *NFPA Fire Protection Handbook* (19th ed.). Quincy, MA: 2003: 12–119.

10. Kerber, S. *NISTIR 7412, Evaluating Positive Pressure Ventilation in Large Structures: High-Rise Pressure Experiments*. Gaithersburg, MD: National Institute of Standards and Technology; March 2007.

11. Madrzykowski, D. *Technical Note 1618, Fire Fighting Tactics Under Wind Driven Conditions: Laboratory Experiments*. Gaithersburg, MD: National Institute of Standards and Technology; January 2009.

12. Klem, T. J. *One Meridian Plaza, Three Fire Fighter Fatalities, Philadelphia, PA*. Fire Investigation Report. Quincy, MA: NFPA; 1992.

13. Routley, G. J., Jennnings, C., Chubb, M. *High-Rise Office Building Fire, One Meridian Plaza, Philadelphia, Pennsylvania*. Emmitsburg, MD: U.S. Fire Administration; 1991.

14. Klem, T. J. *First Interstate Bank Building Fire, Los Angeles, CA*. Fire Investigation Report. Quincy, MA: NFPA; 1988.

15. Routley, G. J. *Interstate Bank Building Fire, Los Angeles, California*. Emmitsburg, MD: U.S. Fire Administration; 1988.

16. Brannigan, F. L. *Paranoia may save your life*. *Fire Engineering*. July 1998.

17. NFPA Fire Investigations Department. *Residential High-Rise, North York, Ontario, Canada, January 6, 1995*. Fire Investigation Report, Quincy, MA: 1995.

18. Isner, M. S. *Fatal Office Building Fire, Atlanta, GA*. NFPA Fire Investigation Report. Quincy, MA: NFPA.

19. Jennings, C. *Five-Fatality Office Building Fire—Atlanta, Georgia*. Emmitsburg, MD: U.S. Fire Administration; 1989.

20. Isner, M. S., Klemm, T. J. *World Trade Center Explosion and Fire*. Fire Investigation Report. Quincy, MA; NFPA.

21. Manning, W. A. *The World Trade Center Bombing: Report and Analysis*. Emmitsburg, MD: U.S. Fire Administration; 1993.

22. National Institute of Standards and Technology. *NIST, NCSTAR1 Federal Building and Fire Safety Investigation of the World Trade Center Disaster, Final Report*. Gaithersburg, MD: 2005.

23. Best, R., Demers, D. P. *MGM Grand Hotel Fire, Las Vegas, NV*. NFPA Fire Investigative Report. Quincy, MA: NFPA; 1980.

24. NFPA Fire Investigations Department. The MGM Hotel fire, Part 1. *Fire Service Today*. January 1982: 18–23.Figure CS8-1 Six-unit, three-story apartment building—side C.

Flames: Drx/Dreamstime.com; Steel texture: © Sharpshot/Dreamstime.com

This is an example of a Standard Operating Procedure (SOP) for structure fires. This SOP example would need to be modified to meet local conditions and resources. This structure fire SOP is provided to be evaluated by readers as part of the text's activities or as a text reference only.

Structure Fires Standard Operating Procedure

Scope

To establish standard operations to be used at all structure fires

When there is a conflict, pre-incident plans take precedence over this procedure.

A. Priorities

The incident priorities for structure fires are:

1. Life safety
2. Extinguishment
3. Property conservation

B. Safety

Full protective clothing, including SCBA, shall be worn at the scene of all structure fires until the incident commander (IC) directs otherwise.

However, members raising ladders, operating lines outside a building, or conducting other external activities are not mandated to use respiratory protective equipment as long they are in an area declared free of respiratory hazards by the IC and/or safety officer.

When in doubt, full protective clothing including SCBA shall be worn.

Members responding to working fires or multiple alarm fires must don full protective equipment, including SCBA backpack when reporting to the command post (CP) unless otherwise directed.

All units must work under the direction of "Command." No freelancing.

If an immediate evacuation of the structure by operating forces becomes necessary, an emergency message calling for retreat will be transmitted on all radio channels in use at the incident scene.

The radio message transmitted shall state: *"Emergency, all units at (location) retreat."*

This message shall be repeated five times.

In addition, an apparatus located near the building will signal the retreat by sounding ten (10) three (3)-second blast on the apparatus air horn.

Upon hearing this retreat message and/or signal, members shall immediately leave the structure and reassemble outside the fire perimeter.

Company officers shall account for all members and report evacuation status to their division/group commander if the fire ground is sectored. If divisions and/or groups have not been organized, company officers shall report their evacuation status to the IC.

Exclusion zones, fire zones and fire perimeters will be established as necessary. See Procedure 202.03 Fire Zones/Perimeters.

C. Water Supply

A source of water supply shall be secured by the first-arriving engine company unless the pre-plan preempts this requirement.

The water supply should be a 5" supply line connected to a fire hydrant.

A second source or water should be secured to augment the water supply and provide an emergency backup in case the first arriving engine company experiences difficulties in maintaining a reliable water supply.

Additional water supplies should be established that exceed the expected flow requirement.

The more pumped water, the higher the overall attack capability.

D. Rescue

Whenever it is safe to make entry, a primary search will be conducted at all structural fires. Primary search means that companies quickly go through all affected areas and verify the removal and/or safety of all occupants.

Time is the critical factor in the primary search process.

Secondary searches follow the primary search.

During the secondary search, companies thoroughly search the interior of the fire area.

Secondary search should be conducted by different companies from those involved in primary search activities in a specific area.

Thoroughness, rather than time, is the critical factor in secondary search.

Danger to occupants is most acute on the fire floor, the floor immediately above the fire, the top floor, and then intermediate floors, in that order. In high-rise fires, smoke may not rise to the top of the structure. See SOP 203.20 High-Rise Buildings.

Initial efforts must be directed toward supporting rescue efforts and hose lines must be placed in a manner to control interior access, confine the fire, and protect avenues of escape, until the primary and secondary searches are complete.

Hose line placement becomes a critical factor in rescue operations, and all operating companies must realize that the operation is in a rescue mode.

Extinguishment is the second priority; however, it is often the best life safety tactic.

Normal means of interior access should be used to remove victims whenever possible.

The most urgent reason for calling additional alarms is life safety.

E. Search Marking System

When it becomes necessary to evacuate a large structure or specific areas within a large structure, the chalk marking system will be used to identify areas searched.

Every member shall carry a piece of dry chalk in a plastic tube on their person.

When an entire apartment or other area has been searched, an "X" will be placed on the door or the wall next to the door at eye level.

Once an entire floor has been searched, an "X" will be placed on doors leading from stairways to that floor and on the wall opposite elevators.

F. Engine Company Operations

Establish a water supply as explained in section "C" of this procedure.

Attempt to determine fire location and extent and best position for fire attack.

The main observable factors that influence choice of tactics are:

- Life hazard
- Fire location
- Fire stage
- Ventilation status
- Extension probability
- Extension path

A risk versus benefit analysis is essential prior to commencing operations.

An aggressive interior attack (offensive strategy) will be initiated whenever possible.

When conducting an offensive attack the first hose line will be at least 1 ¾".

This first line will be backed up with a minimum size line of 1 ¾".

Do not operate fire streams into ventilation openings when firefighters are operating on the interior of the fire building.

An offensive attack is preferred, but it is not always possible to conduct a safe and effective interior attack.

When the risk versus benefit analysis indicates that a defensive attack is the best choice, only 2 ½" or master streams will be deployed.

If the defensive attack was initiated after an offensive attack, ALL members shall be evacuated from the interior of the fire building prior to commencing the defensive attack.

Do not operate exterior streams into a building after firefighters have made entry.

G. Ladder Company Operations

Forcible entry, search, rescue, ventilation, salvage, and overhaul are all ladder company tasks. Engine companies do their work with hose lines; ladder companies do their work with tools.

First arriving ladder company tasks include:

- Remove obviously endangered persons.
- Search fire floor for endangered persons.
- Check sides and rear of building for endangered persons and fire extension.
- Force entry if necessary.
- Ventilate as needed while searching.
  - Either isolate the area being searched or coordinate the search with fire stream protection.
- Ventilate at roof when safe and appropriate using a ladder or other platform to support firefighters assigned to the roof.
  - Vertical ventilation should occur only when hose lines are ready to attack the fire.
- Lower fire escape ladders.
- Raise ladders as needed.

Ladder company members must be observant and exercise initiative. They should be encouraged to use good judgment and allowed to take actions they think necessary; however, they must remain within the NIMS structure.

H. Apparatus Placement

Apparatus function should regulate placement.

Effective placement must begin with the arrival of the first units, based on initial size up, pre-incident plans, and general conditions upon arrival.

Avoid blocking access to the fire building and surrounding structures.

Engine companies should not block ladder company access or place the pumper in a position that hinders removal of ground ladders.

Additional units summoned as a tactical reserve should be staged.

Avoid placing apparatus in positions where they cannot be quickly repositioned.

Beware of overhead power lines that may fall on apparatus.

Medic units should be placed in a safe position that will provide the most effective treatment of fire victims and firefighting personnel, while not blocking movement of other apparatus or interfering with firefighting operations.

Imperial and Metric Conversions

Table B-1 Length

1 inch = 0.08333 foot, 1,000 mils, 25.40 millimeters

1 foot = 0.3333 yard, 12 inches, 0.3048 meter, 304.8 millimeters

1 yard = 3 feet, 36 inches, 0.9144 meter

1 rod = 16.5 feet, 5.5 yards, 5.029 meters

1 mile = (U.S. and British) 5,280 feet, 1.609 kilometers, 0.8684 nautical mile

1 millimeter = 0.03937 inch, 39.37 mils, 0.001 meter, 0.1 centimeter, 100 microns

1 meter = 0.094 yards, 3.281 feet, 39.37 inches, 1,000 millimeters

1 kilometer = 0.6214 mile, 1.094 yards, 3,281 feet, 1,000 meters

1 nautical mile = 1.152 miles (statute), 1.853 kilometers

1 micron = 0.03937 mil, 0.00003937 inch

1 mil = 0.001 inch, 0.0254 millimeters, 25.40 microns

1 degree = 1/360 circumference of a circle, 60 minutes, 3,600 seconds

1 minute = 1/60 degree, 60 seconds

1 second = 1/60 minute, 1/3600 degree

Table B-2 Area

1 square inch = 0.006944 square foot, 1,273,000 circular mils, 645.2 square millimeters

1 square foot = 0.1111 square yard, 144 square inches, 0.09290 square meter, 92,900 square millimeters

1 square yard = 9 square feet, 1,296 square inches, 0.8361 square meter

1 acre = 43,560 square feet, 4,840 square yards, 0.001563 square mile, 4,047 square meters, 160 square rods

1 square mile = 640 acres, 102,400 square rods, 3,097,600 square yards, 2.590 square kilometers

1 square millimeter = 0.001550 square inch, 1.974 circular mils

1 square meter = 1.196 square yards, 10.76 square feet, 1,550 square inches, 1,000,000 square millimeters

1 square kilometer = 0.3861 square mile, 247.1 acres, 1.196,000 square yards, 1,000,000 square meters

1 circular mil = 0.7854 square mil, 0.0005067 square millimeter, 0.0000007854 square inch

Table B-3 Volume (Capacity)

1 fluid ounce = 1.805 cubic inches, 29.57 milliliters, 0.03125 quarts (U.S.) liquid measure

1 cubic inch = 0.5541 fluid ounce, 16.39 milliliters

1 cubic foot = 7.481 gallons (U.S.), 6.229 gallons (British), 1,728 cubic inches, 0.02832 cubic meter, 28.32 liters

1 cubic yard = 27 cubic feet, 46,656 cubic inches, 0.7646 cubic meter, 746.6 liters, 202.2 gallons (U.S.), 168.4 gallons (British)

1 gill = 0.03125 gallon, 0.125 quart, 4 ounces, 7.219 cubic inches, 118.3 milliliters

1 pint = 0.01671 cubic foot, 28.88 cubic inches, 0.125 gallon, 4 gills, 16 fluid ounces, 473.2 milliliters

1 quart = 2 pints, 32 fluid ounces, 0.9464 liter, 946.4 milliliters, 8 gills, 57.75 cubic inches

1 U.S. gallon = 4 quarts, 128 fluid ounces, 231.0 cubic inches, 0.1337 cubic foot, 3.785 liters (cubic decimeters), 3,785 milliliters, 0.8327 Imperial gallon

1 Imperial (British and Canadian) gallon = 1.201 U.S. gallons, 0.1605 cubic foot, 277.3 cubic inches, 4.546 liters (cubic decimeters), 4,546 milliliters

1 U.S. bushel = 2,150 cubic inches, 0.9694 British bushel, 35.24 liters

1 barrel (U.S. liquid) = 31.5 gallons (various industries have special definitions of a barrel)

1 barrel (petroleum) = 42.0 gallons

1 millimeter = 0.03381 fluid ounce, 0.06102 cubic inch, 0.001 liter

1 liter (cubic decimeter) = 0.2642 gallon, 0.03532 cubic foot, 1.057 quarts, 33.81 fluid ounces, 61.03 cubic inches, 1,000 milliliters

1 cubic meter (kiloliter) = 1.308 cubic yards, 35.32 cubic feet, 264.2 gallons, 1,000 liters

1 cord = 128 cubic feet, 8 feet × 14 feet × 4 feet, 3.625 cubic meters

Table B-4 Weight

1 grain = 0.0001428 pound

1 ounce (avoirdupois) = 0.06250 pound (avoirdupois), 28.35 grams, 437.5 grains

1 pound (avoirdupois) = the mass of 27.69 cubic inches of water weighed in air at 4°C (39.2°F) and 760 millimeters of mercury (atmospheric pressure), 16 ounces (avoirdupois), 0.4536 kilogram, 453.6 grams, 7,000 grains

1 long ton (U.S. and British) = 1.120 short tons, 2,240 pounds, 1.016 metric tons, 1,016 kilograms

1 short ton (U.S. and British) = 0.8929 long ton, 2,000 pounds, 0.9072 metric ton, 907.2 kilograms

1 milligram = 0.001 gram, 0.000002205 pound (avoirdupois)

1 gram = 0.002205 pound (avoirdupois), 0.03527 ounce, 0.001 kilogram, 15.43 grains

1 kilogram = the mass of 1 liter of water in air at 4°C and 760 millimeters of mercury (atmospheric pressure), 2.205 pounds (avoirdupois), 35.27 ounces (avoirdupois), 1,000 grams

1 metric ton = 0.9842 long ton, 1.1023 short tons, 2,205 pounds, 1,000 kilograms

Steel: © Sharpshot/Dreamstime.com

Table B-5 Density

1 gram per millimeter = 0.03613 pound per cubic inch, 8,345 pounds per gallon, 62.43 pounds per cubic foot, 998.9 ounces per cubic foot

Mercury at 0°C = 0.1360 grams per millimeter basic value used in expressing pressures in terms of columns of mercury

1 pound per cubic foot = 16.02 kilograms per cubic meter

1 pound per gallon = 0.1198 gram per millimeter

Table B-6 Flow

1 cubic foot per minute = 0.1247 gallon per second, 0.4720 liter per second, 472.0 milliliters per second = 0.028 m³/min, lcfm/ft² 0.305 m³/min/m²

1 gallon per minute = 0.06308 liter per second, 1,440 gallons per day, 0.002228 cubic foot per second

1 gallon per minute per square foot = 40.746 mm/min, 40.746 l/min · m²

1 liter per second = 2.119 cubic feet per minute, 15.85 gallons (U.S.) per minute

1 liter per minute = 0.0005885 cubic foot per second, 0.004403 gallon per second

Table B-7 Pressure

1 atmosphere = pressure exerted by 760 millimeters of mercury of standard density at 0°C, 14.70 pounds per square inch, 29.92 inches of mercury at 32°F, 33.90 feet of water at 39.2°F, 101.3 kilopascal

1 millimeter of mercury (at 0°C) = 0.001316 atmosphere, 0.01934 pound per square inch, 0.04460 foot of water (4°C or 39.2°F), 0.0193 pound per square inch, 0.1333 kilopascal

1 inch of water (at 39.2°F) = 0.00246 atmosphere, 0.0361 pound per square inch, 0.0736 inch of mercury (at 32°F), 0.2491 kilopascal

1 foot of water (at 39.2°F) = 0.02950 atmosphere, 0.4335 pound per square inch, 0.8827 inch of mercury (at 32°F), 22.42 millimeters of mercury, 2.989 kilopascal

1 inch of mercury (at 32°F) = 0.03342 atmosphere, 0.4912 pound per square inch, 1.133 feet of water, 13.60 inches of water (at 39.2°F), 3.386 kilopascal

1 millibar (1/1,000 bar) = 0.02953 inch of mercury. A bar is the pressure exerted by a force of one million dynes on a square centimeter of surface

1 pound per square inch = 0.06805 atmosphere, 2.036 inches of mercury, 2.307 feet of water, 51.72 millimeters of mercury, 27.67 inches of water (at 39.2°F), 144 pounds per square foot, 2,304 ounces per square foot, 6.895 kilopascal

1 pound per square foot = 0.00047 atmosphere, 0.00694 pound per square inch, 0.0160 foot of water, 0.391 millimeter of mercury, 0.04788 kilopascal

Absolute pressure = the sum of the gage pressure and the barometric pressure

1 ton (short) per square foot = 0.9451 atmosphere, 13.89 pounds per square inch, 9,765 kilograms per square meter

Table B-8 Temperature

Temperature Celsius = 5/9 × (temperature Fahrenheit − 32°)

Temperature Fahrenheit = 9/5 × temperature Celsius + 32°

Rankine (Fahrenheit absolute) = temperature Fahrenheit + 459.67°

Kelvin (Celsius absolute) = temperature Celsius 273.15°

Freezing point of water: Celsius = 0°; Fahrenheit = 32°

Boiling point of water: Celsius = 100°; Fahrenheit = 212°

Absolute zero: Celsius = 273.15°; Fahrenheit = −459.67°

Table B-9 Sprinkler Discharge

1 gallon per minute per square foot (gpm/ft²) = 40.75 liters per minute per square meter (Lpm/m²) = 40.75 millimeters per minute (mm/min)

Strategy and Tactics FESHE Course Outcomes	Structural Firefighting: Strategy and Tactics, Third Edition Chapter Correlation
1. Discuss fire behavior as it relates to strategies and tactics.	2, 3, 5, 6, 8, 9, 12
2. Explain the main components of pre-fire planning and identify steps needed for a pre-fire plan review.	2, 7, 8, 11, 12
3. Identify the basics of building construction and how they interrelate to pre-fire planning and strategy and tactics.	2, 5, 6
4. Describe the steps taken during size-up.	2, 3, 12
5. Examine the significance of fireground communications.	1, 3, 5, 12
6. Identify the roles of the NIMS and ICS as it relates to strategy and tactics.	1, 8, 9, 12
7. Demonstrate the various roles and responsibilities in ICS/NIMS.	1, 8, 9, 12

Steel © Sharpshot/Dreamstime.com

GLOSSARY

10-codes Numeric codes used to communicate predefined situations or conditions. For example, "10-4" generally means "finished communicating." The use of 10-codes is discouraged.

20-minute rule A rule of thumb used for estimating the length of time until structural collapse occurs. It states that when a heavy volume of fire is burning out of control on two or more floors for 20 minutes or longer in a building of ordinary construction, structural collapse should be anticipated.

accountability system A system established on the fire ground to ensure that everyone entering the area has a specific assignment, to track all personnel at the scene, and to identify the location of any missing personnel if a catastrophic event should occur.

area of refuge A floor area with at least two rooms separated by smoke-resisting partitions in a building protected by a sprinkler system, or a space located in an egress path that is separated from other building spaces.

assembly occupancy An occupancy used for the gathering of 50 or more persons for deliberation, worship, entertainment, eating, drinking, amusement, awaiting transportation, or similar uses; or used as a special amusement building, regardless of occupant load.

attack pumper The first-arriving engine company that goes directly to the fire building without securing a water supply.

backdraft A fire condition that occurs when oxygen (air) is introduced into a superheated, oxygen-deficient compartment charged with smoke and pyrolytic emissions, resulting in an explosive ignition.

backpressure Also known as elevation pressure, the pressure required to overcome the weight of water in a piping or hose system. Each vertical foot of water in a pipe, hose, or tank exerts a pressure of 0.434 psi (3 kPa) at the base.

balloon-frame construction An older type of wood-frame construction in which the wall studs extend vertically from the basement of the structure to the roof.

branches The tactical-level management units immediately subordinate to sections in NIMS hierarchy. Units are subordinate to the logistics, finance/administration, and planning sections. Divisions and groups are subordinate to the operations section. Branches are used to reduce the span of control at very large operations or to manage a particular function/agency.

business occupancy An occupancy used for account and record keeping or the transaction of business other than mercantile.

chain of command An organizational structure establishing a line of authority and responsibility along which orders and instructions are passed—e.g., incident commander to operations section to branch director to division supervisor to company officer to fire fighter, or a reply to a higher level in reverse order

Class A foam Foam for use on fires in Class A fuels such as vegetation, wood, cloth, paper, rubber, and some plastics.

cold zone An area where personal protective clothing is not required; command post, rehabilitation, and medical treatment should be located in the cold zone.

collapse zone The area endangered by a potential building collapse; generally considered to be an area 1½ times the height of the involved building.

compartmentation Subdividing of a building into small areas (rooms) capable of limiting the spread of fire and the products of combustion.

complex preincident plan A plan used when a property has more than three buildings or when it is necessary to show the layout of the premises and relationship between buildings on the site.

Flames: © Dra/Dreamstime.com; Steel texture © Sharpshot/Dreamstime.com

convergence cluster phenomenon A reaction to fire conditions in which groups take shelter together to provide mutual support.

critical time The time available until the structure becomes untenable.

dead load The weight of a building; consists of the weight of all materials of construction incorporated into a building, including but not limited to floors, roofs, ceilings, stairways, built-in partitions, finishes, cladding, and other similarly incorporated architectural and structural items, as well as fixed service equipment.

decay phase The phase of fire development in which the fire has consumed either the available fuel or oxygen to a point that the fire begins to diminish in intensity.

defend-in-place A tactic utilized during a structure fire when it is very difficult to remove occupants from the building. Occupants are either protected at their present location or moved to a safe location within the building.

defensive A type of fire attack in which exterior fire suppression operations are directed at protecting exposures.

deluge system A sprinkler system in which all sprinklers or applicators are open. When an initiating device, such as a smoke detector or heat detector, is activated, the deluge valve opens and water discharges from all of the open sprinklers simultaneously.

detention and correctional occupancy An occupancy used to house four or more persons under varied degrees of restraint or security where such occupants are mostly incapable of self-preservation because of security measures not under the occupants' control.

direct attack Firefighting operations involving the application of extinguishing agents directly onto the burning fuel.

division-of-labor principle The principle that an incident or task should be broken down into smaller, more manageable tasks and personnel assigned to complete those tasks (developing job skills in a concentrated area to allow for more productivity).

division The tactical-level management unit in charge of a geographic area.

draft curtains Walls designed to limit horizontal spread of the fire that extend partially down (usually no more than 20% of the height of the compartment) from the underside of the roof.

drilling down Using a computer to navigate by pointing and clicking through a series of drop-down menus in a graphical user interface.

dry pipe sprinkler systems Systems in which the pipes are normally filled with compressed air or nitrogen. When a sprinkler is activated, it releases the air from the system, which opens a valve so the pipes can fill with water.

educational occupancy An occupancy used for educational purposes through the 12th grade by six or more persons for 4 or more hours per day or more than 12 hours per week.

extension Fire that moves into areas not originally involved, including walls, ceilings, and attic spaces; also, the movement of fire into uninvolved areas of a structure.

external exposures Buildings, vehicles, or other property threatened by fire that are external to the building, vehicle, or property where the fire originated.

finance/administration section The section that tracks and provides financial and administrative services required to compensate people or organizations providing goods and services at the incident scene.

fire perimeter A wide area beyond the hazard control zones, usually staffed by police to keep unauthorized people away from the scene.

flashover An oxygen-sufficient condition in which room temperatures reach the ignition temperature of the suspended pyrolytic emissions, causing all combustible contents to suddenly ignite.

flow path The movement of heat and smoke from the higher pressure fire area toward lower pressure areas on the interior and exterior of the structure.

foam house A fixed facility consisting of an enclosure housing a foam concentrate supply tank, a foam solution proportioning system, a pump, and sometimes an extra supply of foam concentrate that can be added to the proportioning system.

formal preincident plan A plan for a property with a substantial risk to life and/or property; includes a drawing of the property, specific floor layouts, and a narrative describing important features.

freelancing The act of performing tasks outside the incident's organizational structure.

frequency As related to fire fighter injuries, a measure of how often an injury occurs; for example, sprains and strains are the most frequent fire fighter injuries.

fuel-controlled Refers to a fire in which the heat release rate and growth rate are controlled by the characteristics of the fuel, such as quantity, chemistry, and geometry, and in which adequate air for combustion is available.

fuel load Fuels provided by a building's contents and combustible building materials; also called fire load.

fully developed phase The phase of fire development at which the fire is free-burning and consuming much of the fuel.

group The tactical-level management unit in charge of a function.

growth phase The phase of fire development at which the fire is spreading beyond the point of origin and beginning to involve other fuels in the immediate fire area.

hazard control zone The area in which emergency responders are working; it can be subdivided into no-entry, hot, warm, and cold zones.

health care occupancy An occupancy used for purposes of medical or other treatment or care of four or more persons, where such occupants are mostly incapable of self-preservation due to age, physical or mental disability, or security measures not under the occupants' control.

hot zone An operating area considered safe only for individuals wearing appropriate levels of personal protective clothing; established by the IC and safety officer.

hovering Accessing text or graphic items on a computer screen by placing the computer cursor over a specific area or by touching the screen in a touch-screen environment.

ignition phase The phase of fire development at which the fire is limited to the immediate point of origin.

immediately dangerous to life and health (IDLH) Exposure to airborne contaminants that are likely to cause death or immediate or delayed permanent adverse health effects or prevent escape from such an environment. Examples include smoke or other poisonous gases at sufficiently high concentrations.

incident action plan The objectives for the overall incident strategy, tactics, risk management, and member safety that are developed by the incident commander and updated throughout the incident.

incident commander (IC) The individual responsible for all incident activities, including the development of strategies and tactics and the ordering and the release of resources.

incident management team (IMT) The incident commander and appropriate command and general staff personnel assigned to an incident.

indirect attack Firefighting operations involving the application of extinguishing agents to reduce the buildup of heat released from a fire without applying the agent directly onto the burning fuel; ventilation is kept to a minimum while a fog stream is directed at the ceiling. This approach is most useful in unoccupied, tightly enclosed spaces.

industrial occupancy An occupancy in which products are manufactured or in which processing, assembling, mixing, packaging, finishing, decorating, or repair operations are conducted.

initial rapid intervention crew (IRIC) A stand-by team of at least two fire fighters located outside the hazard area available to provide assistance to fire fighters operating within the hazard area.

internal exposures Areas within the structure where the fire originates or within buildings directly connected to the building of origin that were not involved in the initial fire ignition.

liaison officer A member of the command staff, responsible for coordinating with representatives from cooperating and assisting agencies.

live loads The weight of the building's contents, people, or anything that is not part of or permanently attached to the structure.

lobby control A high-rise assignment in which a crew is responsible for managing the stairways; elevators; heating, ventilation, and air-conditioning systems; and related duties.

logistics section The section that obtains needed supplies, equipment, and facilities.

mercantile occupancy An occupancy used for the display and sale of merchandise.

mixed multiple occupancy A multiple occupancy where the occupancies are intermingled.

multiple occupancy A building or structure in which two or more classes of occupancy exist.

National Fire Academy formula A rate-of-flow calculation that calculates the rate of flow as the area in square feet divided by 3 (A/3).

National Incident Management System (NIMS) A system mandated by HSPD-5 that provides a consistent, nationwide approach for federal, state, local and tribal governments; the private sector and nongovernmental organizations (NGOs) to work effectively and ef ciently together to prepare for, respond to, and recover from domestic incidents, regardless of cause, size, or complexity.

no-entry zone An area that is unsafe regardless of the level of personal protective equipment and that must be cleared of all personnel, including emergency response personnel.

notation A piece of information about the premises, such as damage to the building from a previous fire. This information may accompany a preincident plan or may be available when the building does not have a preincident plan.

occupant density A measure of the number of people in a given area; usually stated as the number of square feet per person.

offensive A type of fire attack in which fire fighters advance into the fire building with hose lines or other extinguishing agents to overpower the fire.

operations section The section that manages all tactical units deployed at an incident scene.

overhaul Examination of all areas of the building and contents involved in a fire to ensure that the fire is completely extinguished.

oxygen- (ventilation-) controlled fire A fire in which the heat release rate and/or growth of the fire are controlled by the amount of air available to the fire.

planning section The section that gathers and evaluates information, assists the incident commander in developing the incident action plan, and tracks progress; also tracks resource status.

platform-frame construction A construction technique using separate components to build the frame of a structure (one floor at a time). Each floor has a top and bottom plate that act as fire-stops.

preaction sprinkler A dry sprinkler system that uses a deluge valve instead of a dry pipe valve and requires activation of a secondary device before the pipes fill with water.

prefire condition Factors that can contribute to a collapse in a building that is heavily involved in fire; these factors include construction types, weight, fuel loads, damage, renovations, deterioration, support systems, and related factors such as lightweight truss ceilings and floors.

preincident plans Written documents resulting from the gathering of general and detailed information to be used by public emergency response agencies and private industry for determining the response to reasonable anticipated emergency incidents at a specific facility.

primary damage Damage caused by the products of combustion.

primary factors The most important factors, assessed during size-up, which change from incident to incident and depend on specific incident conditions.

public information officer (PIO) A member of the command staff responsible for interfacing with the public and media or with other agencies with incident related information requirements.

pyrolysis The chemical decomposition of a compound into one or more other substances by heat alone; the process of heating solid materials until combustible vapors are emitted.

pyrolysis The chemical decomposition of a compound into one or more other substances by heat alone; the process of heating solid materials until combustible vapors are emitted.

quad A multifunction apparatus that is equipped to provide for the following four functions: water, pumps, hose, and ground ladders. In other words, this is a quint company minus the aerial ladder.

quint A multifunction apparatus that is equipped to perform both engine and truck company operations. It is equipped to provide for the following five functions: water, pumps, hose, ground ladders, and an aerial ladder.

rapid intervention crew (RIC) A minimum of two fully equipped personnel on-site, for immediate rescue of injured or trapped fire fighters.

rate of flow The minimum water application rate required for extinguishment

rational decision making (RDM) A form of decision making in which input is obtained from diverse sources and all options are carefully considered.

recognition-primed decision making (RPD) A form of decision making in which the incident commander must decide on the proper course of action with limited information available in a relatively short period of time.

rehabilitation The process of providing rest, rehydration, nourishment, and medical evaluation to members involved in extended incident scene operations and/or extreme weather conditions.

residential board and care occupancy A building or portion thereof that is used for lodging and boarding of four or more residents, not related by blood or marriage to the owners or operators, for the purpose of providing personal care services.

residential occupancy An occupancy that provides sleeping accommodations for purposes other than health care or detention and correctional.

risk-versus-benefit analysis The process of weighing predicted risks to fire fighters against potential benefits for owners/occupants and making decisions based on the outcome of that analysis.

Royer/Nelson formula A rate-of-flow calculation that calculates the rate of flow as the volume in the fire area in cubic feet divided by 100 (V/100). This formula is based on the assumption that structure fires are primarily oxygen controlled.

secondary damage Damage caused by fire-ground activities or the operation of a fire suppression system.

secondary factors Less important factors at an incident, which change from incident to incident and depend on specific incident conditions.

separated multiple occupancy A multiple occupancy where the occupancies are separated by fire resistance–rated assemblies.

severity The extent of an injury's consequence, usually categorized as death, permanent disability, temporary disability, and minor.

single command A command structure in which one person is designated as the incident commander (IC). This person is responsible for the development and implementation of the incident action plan. The IC can delegate staff and command positions as needed to assist in command and control functions.

smoke-proof tower A stairway that is designed to be separated from the building by a landing. This creates a separation that will limit the spread of smoke into the stairway and keep it clear for evacuation.

softening the target Cooling hot fire gases using rapid, short-duration application (usually 15 to 60 seconds) from a straight or smooth bore fire stream aimed at a steep angle toward the ceiling from the safest effective location. When using this tactic, water often is applied from the exterior to the interior of a building.

span of control The number of people reporting to a supervisor. The span of control should not exceed seven people reporting to a single supervisor under emergency conditions. As an example, a fire captain supervising an apparatus operator and three fire fighters would have a four-to-one span of control.

sprinkler system calculations Specific rate-of-flow calculations for sprinkler systems; based on the fuel load. These calculations can be found in various publications, including NFPA documents and Factory Mutual Data Sheets.

stack effect The vertical airflow within buildings caused by temperature differences between the building interior and the exterior; depending on conditions, stack effect could be positive or negative, causing smoke to move upward or downward.

standard operating procedures (SOPs) Written rules, policies, regulations, and procedures intended to organize operations in a predictable manner.

standpipe systems Piping systems with discharge outlets at various locations; in high-rise buildings an outlet will normally be located in the stairway on each floor level. Most are connected to a water source and the pressure is boosted by a fire pump.

storage occupancy An occupancy used primarily for the storage or sheltering of goods, merchandise, products, vehicles, or animals.

strike team: A set number of resources of the same kind and type that have an established minimum number of personnel—e.g., five engine companies. Strike teams always have a leader (usually in a separate vehicle) and have common communications among resource elements.

Superfund Amendments and Reauthorization Act (SARA) A federal law enacted in 1986 and also known as the Emergency Planning and Community Right to Know Act. Title III of this law requires businesses that handle or store hazardous chemicals in quantities above specific limits to report the location, quantity, and hazards of those chemicals to the State Emergency Response Commission (SERC), Local Emergency Planning Committee, and local fire department. Some of this information has been classified since the attack on the World Trade Center on September 11, 2001.

task forces Any combination of resources that can be temporarily assembled for a specific mission. All resource elements within a task force must have common communications and a leader. Task forces should be established to meet specific tactical needs and should be demobilized as single resources. A typical task force would be two engine companies and a truck company under the supervision of a chief officer.

tenders NIMS term for vehicles that transport water from a water source to the fire scene; also referred to as a tanker or mobile water supply apparatus.

trench cut An opening in the roof that extends from bearing wall to bearing wall to prevent horizontal fire spread in a building.

two-in/two-out rule A rule mandating that fire fighters working inside the hazard area work in crews of at least two people (two-in). The two people working in the hazard area must be backed up by at least two people outside the hazard area (two-out) who are properly equipped and immediately available to come to the aid of the inside crew. This rule applies to the first-arriving unit. As more units arrive, the two-out team is replaced by a rapid intervention crew.

type I construction Construction method in which the structural members, including walls, columns, beams, girders, trusses, arches, floors, and roofs are of approved noncombustible or limited-combustible materials and have the highest level of fire-resistance ratings.

type II construction Construction method in which the structural members, including walls, columns, beams, girders, trusses, arches, floors, and roofs are of approved noncombustible or limited-combustible materials but the fire resistance rating does not meet the requirements for Type I construction.

type III construction Construction method in which exterior walls and structural members that are portions of exterior walls are of approved noncombustible or limited- combustible materials, and interior structural members, including walls, columns, beams, girders, trusses, arches, floors, and roofs, are entirely or partially of wood or smaller dimensions than those required for Type IV construction or of approved noncombustible, limited-combustible, or other approved combustible materials.

type IV construction Construction method in which exterior walls and structural members that are portions of exterior walls are of approved noncombustible or limited-combustible materials. Other interior structural members, including walls, columns, beams, girders, trusses, arches, floors, and roofs are of solid or laminated wood without concealed spaces. Wood columns supporting floor loads are not less than 8-in. (203 mm) in any dimension; wood columns supporting roof loads only are not less than 6-in. (152 mm) in the smallest dimension and not less than 8-in. (203 mm) in depth. Wood beams and girders supporting floor loads are not less than 6-in.

(152 mm) in width and not less than 10-in. (254 mm) in depth. Wood beams and girders and other roof framing supporting roof loads only are not less than 4-in. (102 mm) in width and not less than 6-in. (152 mm) in depth. Specifics for other structural members are required to be large-dimension lumber as well.

type V construction Construction method in which exterior walls, bearing walls, columns, beams, girders, trusses, arches, floors, and roofs are entirely or partially of wood or other approved combustible material smaller than material required for Type IV construction.

unified command An application of the incident command system (ICS) that allows all agencies with jurisdictional responsibility for an incident or planned event, either geographical or functional, to manage an incident or planned event by establishing a common set of incident objectives and strategies.

unity of command A pyramidal command system ensuring that no one reports to more than one supervisor.

ventilation controlled A fire within a compartment or building that is limited due to a lack of air (oxygen) even though sufficient vapor fuel is available to support continued burning.

warm zone An intermediate area between the hot and cold zones where personal protective equipment is required, but at a lower level than in the hot zone.

wet pipe sprinkler system A sprinkler system in which the pipes are normally filled with water.

wood truss An assembly made up of small-dimension lumber joined in a triangular configuration that can be used to support either roofs or floors.

Z